AF830900

Genetic Diseases in Pregnancy

Contributors

Richard Apostol
Clyde H. Beck, Jr.
Sherman Elias
Elliot S. Gershon
Mitchell S. Golbus
William R. Griswold
Judith G. Hall
Margaret W. Hilgartner
Joel M. Lamon
Russell K. Laros, Jr.
Roger R. Lenke
Harvey L. Levy
Stanley A. Mendoza
Connie H. Miller
Menachem Nitzan
John T. Queenan
Robert W. Rebar
Jerome I. Rotter
James R. Schreiber
Joseph D. Schulman
Vivian E. Shih
Joe Leigh Simpson
Steven P. Targum
Ruth M. Yanagi

Genetic Diseases in Pregnancy

MATERNAL EFFECTS AND FETAL OUTCOME

Edited by

JOSEPH D. SCHULMAN

Section of Human Genetics
National Institute of Child Health
and Human Development
National Institutes of Health
Bethesda, Maryland
and
Department of Obstetrics and Gynecology
George Washington University
Washington, D.C.

JOE LEIGH SIMPSON

Department of Obstetrics and Gynecology
Section of Human Genetics
Northwestern University Medical School
Chicago, Illinois

ACADEMIC PRESS
A Subsidiary of Harcourt Brace Jovanovich, Publishers
NEW YORK LONDON TORONTO SYDNEY SAN FRANCISCO 1981

COPYRIGHT © 1981, BY ACADEMIC PRESS, INC.
ALL RIGHTS RESERVED.
NO PART OF THIS PUBLICATION MAY BE REPRODUCED OR TRANSMITTED IN ANY FORM OR BY ANY MEANS, ELECTRONIC OR MECHANICAL, INCLUDING PHOTOCOPY, RECORDING, OR ANY INFORMATION STORAGE AND RETRIEVAL SYSTEM, WITHOUT PERMISSION IN WRITING FROM THE PUBLISHER.

ACADEMIC PRESS, INC.
111 Fifth Avenue, New York, New York 10003

United Kingdom Edition published by
ACADEMIC PRESS, INC. (LONDON) LTD.
24/28 Oval Road, London NW1 7DX

Library of Congress Cataloging in Publication Data
Main entry under title:

Genetic diseases in pregnancy.

Includes bibliographies and index.
1. Pregnancy, Complications of--Genetic aspects.
2. Medical genetics. I. Schulman, Joseph D.
II. Simpson, Joe Leigh. [DNLM: 1. Hereditary Diseases--Complications. 2. Pregnancy complications. WQ 240 G328]
RG580.G45G46 618.3'042 80-68562
ISBN 0-12-630940-X AACR2

PRINTED IN THE UNITED STATES OF AMERICA

81 82 83 84 9 8 7 6 5 4 3 2 1

To Carol and Sandee

Contents

4. Genetic Disorders of Hemostasis

CONNIE H. MILLER AND MARGARET W. HILGARTNER

5. Rh and Other Blood Group Incompatibilities

JOHN T. QUEENAN

6. Cardiovascular Defects

SHERMAN ELIAS AND RUTH M. YANAGI

7. Respiratory System Disorders

SHERMAN ELIAS

12. Pregnancy, Genetic Counseling, and the Major Psychiatric Disorders

STEVEN P. TARGUM AND ELLIOT S. GERSHON

13. Pregnancies in Women with Chromosomal Abnormalities

JOE LEIGH SIMPSON

List of Contributors

Numbers in parentheses indicate the pages on which the authors' contributions begin.

Richard Apostol (269), Department of Pediatrics and Medicine, Harbor General Hospital, University of California at San Diego, La Jolla, California 92093

Clyde H. Beck, Jr. (247), Department of Medicine, University of California at San Diego, La Jolla, California 92097

Sherman Elias (197, 229), Department of Obstetrics and Gynecology, Northwestern University Medical School, Chicago, Illinois 60611

Elliot S. Gershon (413), Section on Psychogenetics, Biological Psychiatry Branch, National Institute of Mental Health, National Institutes of Health, Bethesda, Maryland 20205

Mitchell S. Golbus (89), Departments of Obstetrics, Gynecology, Reproductive Sciences, and Pediatrics, School of Medicine, University of California, San Francisco, California 94143

William R. Griswold (247), Department of Pediatrics, School of Medicine, University of California at San Diego, La Jolla, California 92093

Judith G. Hall (57), Department of Pediatrics, School of Medicine, University of Washington, Seattle, Washington 98195

Margaret W. Hilgartner (123), Department of Pediatrics, Cornell University Medical College, New York, New York 10021

Joel M. Lamon* (1), Human Genetics Section, National Institutes of Health, Bethesda, Maryland 20205

Russell K. Laros, Jr. (89), Department of Obstetrics, Gynecology, and Reproductive Sciences, University of California, San Francisco, California 94143

*Present address: Division of Hematology, and Department of Clinical Research, Scripps Clinic and Research Foundation, La Jolla, California 92037

Roger R. Lenke (1), Departments of Pediatrics and Neurology, Harvard Medical School, Boston, Massachusetts 02115

Harvey L. Levy* (1), Department of Neurology, Harvard Medical School, Boston, Massachusetts 02115

Stanley A. Mendoza (247), Department of Pediatrics, School of Medicine, University of California at San Diego, La Jolla, California 92093

Connie H. Miller (123), Department of Pediatrics, Cornell University Medical College, New York, New York 10021

Menachem Nitzan (339), Department of Pediatrics, Beilinson Medical Center, Petah-Tikva, Tel-Aviv, Israel

John T. Queenan† (169), Department of Obstetrics and Gynecology, University of Louisville, Louisville, Kentucky 40202

Robert W. Rebar (367), Department of Reproductive Medicine, School of Medicine, University of California at San Diego, La Jolla, California 92093

Jerome I. Rotter (269), Departments of Pediatrics and Medicine, Harbor General Hospital, University of California at Los Angeles, Torrance, California 90509

James R. Schreiber (367), Department of Reproductive Medicine, School of Medicine, University of California at San Diego, La Jolla, California 92093

Joseph D. Schulman (1), Section of Human Genetics, National Institute of Child Health and Human Development, National Institutes of Health, Bethesda, Maryland 20205 and Department of Obstetrics and Gynecology, George Washington University, Washington, D.C. 20037

Vivian E. Shih (1), Departments of Pediatrics and Neurobiology, Harvard University Medical School, Boston, Massachusetts 02115

Joe Leigh Simpson (439), Section of Human Genetics, Department of Obstetrics and Gynecology, Northwestern University Medical School, Chicago, Illinois 60611

Steven P. Targum‡ (413), National Institute of Mental Health, National Institutes of Health, Bethesda, Maryland 20205

Ruth M. Yanagi (197), Departments of Pediatrics, and Obstetrics and Gynecology, Northwestern University Medical School, Chicago, Illinois 60611

*Present address: Massachusetts General Hospital, Boston, Massachusetts 02114

†Present address: Georgetown University Hospital, Department of Obstetrics and Gynecology, 3800 Reservoir Road, N.W., Washington, D.C. 20007

‡Present address: The Psychiatric Institute of Washington, Washington, D.C. 20006

Preface

With improving care, an increasing number of women with genetic diseases survive until the reproductive age. Many genetic defects compatible with reduced or normal levels of female reproduction are modified in their expression by gestation. Furthermore, in some heritable disorders, gestation may be altered by direct or indirect effects of the disease on the mother or fetus. Women with genetic diseases may thus pose particularly complicated problems when reproduction is contemplated, both with respect to medical problems and to genetic counseling.

This volume represents the first attempt to review comprehensively the interactions of genetic disorders with pregnancy. We have attempted to summarize current knowledge on this subject after scanning thousands of articles selected by conventional and computer-assisted methods and by soliciting information regarding relevant cases from our colleagues. The editors and contributors believe the present work will meet a need long felt by those involved with caring for pregnant patients with heritable diseases.

In the first chapter, Joel M. Lamon, Roger R. Lenke, Harvey L. Levy, Joseph D. Schulman, and Vivian Shih discuss selected metabolic disorders affecting a variety of organ systems. A particularly detailed section is devoted to a discussion of the major problem of maternal phenylketonuria. Judith G. Hall next summarizes the complex disorders of connective tissue and skeletal dysplasias, which may give rise to problems during pregnancy. This review is followed by three chapters on the genetic hematological disorders of pregnancy, all written by outstanding authorities: Mitchell S. Golbus and Russell Laros discuss hemoglobinopathies and hereditary red cell defects; Connie Miller and Margaret Hilgartner systematically analyze genetic disorders of hemostasis; and John T. Queenan summarizes current knowledge about Rh and other blood group incompatibilities.

Sherman Elias and Ruth Yanagi next review genetic defects of the cardiovascular system, and in the subsequent chapter Dr. Elias examines disorders of the respiratory system. Stanley A. Mendoza, Clyde H. Beck, Jr., and William R. Griswold summarize important aspects of renal pathophysiology during pregnancy as a background to a lucid discussion of genetic defects that may affect the maternal kidney. Jerome I. Rotter, an expert on the genetics of gastrointestinal disorders, discusses their relationship to pregnancy.

Endocrine disorders are reviewed in two chapters: Menachem Nitzan discusses diabetes mellitus, and James R. Schreiber and Robert W. Rebar systematically review endocrine genetic disorders other than diabetes mellitus and provide a comprehensive discussion of this diverse subject. In the penultimate chapter, Steven P. Targum and Elliott S. Gershon examine the interaction of pregnancy and psychiatric disorders that are genetic in etiology. This chapter is of a special value because of its treatment of the genetics of the major psychiatric diseases, a subject on which the authors are acknowledged authorities.

Finally, Simpson exhaustively reviews current knowledge about pregnancies in women with chromosomal abnormalities. This chapter includes a clear discussion of the mechanisms by which such abnormalities may arise and be transmitted.

Despite efforts by the editors and contributors to be comprehensive, many genetic disorders occurring in adult women are nevertheless not discussed in this volume. In some cases this is because there appears to be no interaction of the disease with gestation, but in other instances failure to discuss a disorder merely indicates that no satisfactory data are currently available.

Especially because of the incompleteness of present information on this subject, we hope that the present volume will lead to increasing attention by physicians and investigators to interactions between genetic diseases and pregnancy. Inadequate and highly selected reporting of cases in the medical literature should be replaced by a growing awareness of the need to evaluate critically and to describe relevant observations. Scholarly reviews of genetic diseases should pay more attention to interactions with pregnancy, a subject often neglected even in otherwise outstanding contributions. We are convinced that understanding the interactions of pregnancy and genetic diseases is important both clinically and scientifically. The subject is worthy of future thought, synthesis, and investigation.

This book is not intended as a substitute for standard reference works on human genetics, genetic counseling, or prenatal diagnosis. Its perspective is different, and as such should satisfy many additional requirements of geneticists, obstetricians, pediatricians, internists, family practitioners, and others who must provide help to women with genetic diseases.

The editors and authors would be most grateful for information supplied to them providing new data and pointing out their omissions and errors.

Joseph D. Schulman
Joe Leigh Simpson

1

Selected Metabolic Diseases

JOEL M. LAMON, ROGER R. LENKE, HARVEY L. LEVY,
JOSEPH D. SCHULMAN, AND VIVIAN E. SHIH

Copyright © 1981 by Academic Press, Inc.
All rights of reproduction in any form reserved.
ISBN 0-12-630940-X

I. PHENYLKETONURIA

Phenylketonuria (PKU) is a metabolic disorder in which the conversion (hydroxylation) of phenylalanine (Phe) to tyrosine is impaired due to reduction in activity of the enzyme phenylalanine hydroxylase. Discovered by Fölling (1934), this disease may cause mental retardation and other manifestations of brain damage (e.g., seizures, hyperactivity, behavior disturbances, increased deep tendon reflexes). The major biochemical findings include increased concentrations of Phe and secondary phenolic metabolites (e.g., phenylpyruvate, phenylacetate) in blood and urine.

Newborn screening for PKU begun in the 1960s has led to the recognition that there is heterogeneity in this disorder. In addition to "classical" PKU in which the blood Phe concentration is usually greater than 20 mg/dl, patients with milder degrees of hyperphenylalaninemia have been discovered. Those with persistent hyperphenylalaninemia of 12 mg/dl or less are asymptomatic, whereas those whose blood Phe level is 12–20 mg/dl (sometime referred to as "atypical" PKU) may show intellectual deficit. In addition to the above primary defects in phenylalanine hydroxylase, occasional hyperphenylalaninemic patients have defects in the synthesis of the specific pteridine cofactor for this enzyme with secondary deficiencies in hydroxylation of Phe, tyrosine, and tryptophan. These latter two amino acids are precursors of brain neurotransmitters. Such patients, representing approximately 1% of hyperphenylalaninemic newborns, have a severe and progressive neurologic course sometimes culminating in death at 4–6 years of age. There is no effective therapy for patients with defects in pteridine synthesis.

If begun in early infancy, a low-phenylalanine diet that controls the biochemical abnormalities prevents mental retardation and most of the other neurologic sequelae of PKU. As a result of routine newborn screening, there are now thousands of children and adolescents with PKU who have received this therapy and who are of normal intelligence. Since PKU is an autosomal recessively inherited disorder, approximately one-half of these individuals are female and will potentially be subject to the maternal PKU syndrome.

The problem of maternal PKU first came to attention in 1957 when Dent recorded the occurrence of severe mental retardation in all offspring of a woman with PKU. Subsequently, Mabry and colleagues described several mothers with PKU whose non-PKU children were retarded (Mabry *et al.*, 1963, 1966; Denniston, 1963). Perhaps the most convincing evidence that PKU in the mother during pregnancy produces fetal damage is a family studied by Mabry *et al.* (1966) in which the mother with PKU had three mentally retarded non-PKU children; one of these children subsequently had three mentally normal offspring by a retarded non-PKU man. Biochemical evidence of this relationship between maternal PKU and fetal brain damage has been supplied by Menkes and Aeberhard (1969) who found the abnormalities of cerebral lipids in a non-PKU offspring of a PKU woman to be similar to the abnormalities in patients with PKU. These findings suggest that the fetal damage is related to prenatal exposure to the high concentrations of Phe and/or Phe metabolites in the mother.

The maternal PKU syndrome is now a defined complication of pregnancy. Several reviews on this subject have been published (Mabry *et al.*, 1966; MacCready and Levy, 1972; Hansen, 1978; Mabry, 1978), the most recent and most extensive by Lenke and Levy (1980). The latter information includes 524 pregnancies in 155 mothers with PKU or some degree of hyperphenylalaninemia. Readers are referred to the above reviews for detailed information.

The major clinical manifestations in non-PKU offspring of women with PKU are mental retardation, microcephaly, congenital heart disease, and low birth weight. Neurologic problems such as convulsions and spasticity may develop in childhood (Mabry, 1978). Women with PKU may have an increased frequency of spontaneous abortions.

When the mother has a blood Phe concentration of 20 mg/dl (1.2 μmole/ml) or greater and increased urinary excretion of abnormal phenolic metabolites (positive ferric chloride test in urine) brain damage in her offspring is the rule (Table I). When the maternal blood Phe concentration is in the range of 16 to 19 mg/dl (1.0–1.1 μmole/ml) the fate of the fetus is less predictable. Approximately ¾ of the offspring from such women are mentally retarded and about ⅔ have microcephaly (Table I). The risk of effects on the fetus is far less when the maternal blood phenylalanine concentration is between 11 and 15 mg/dl and may be no greater than normally expected with lower levels (10 mg/dl or less) of maternal hyperphenylalaninemia.

Congenital cardiac defects in the offspring follow a similar pattern though are less frequent (Table I). These defects are often major and result in neonatal death. No particular cardiac lesion seems to predominate. Anomalies of the skeletal system, gastrointestinal tract, eye, spleen, and lung have occasionally been reported (Mabry, 1978) but seem not to be particularly frequent. An exception to this may be esophageal atresia (Lenke and Levy, 1980). A surprisingly large number of offspring have themselves had PKU or a lesser degree of hyper-

TABLE I
Frequency of Spontaneous Abortions and Offspring Abnormalities among Untreated Pregnancies of Women with PKU or Hyperphenylalaninemia

	Maternal blood phenylalanine (mg/dl)			
	⩾ 20	16–19	11–15	3–10
Abortions (spontaneous)	24% (72/297)	30% (20/66)	0 (0/33)	8% (4/48)
Offspring[a]				
Mental Retardation[b]	92% (158/172)	73% (27/37)	22% (5/23)	18% (5/28)
Microcephaly[c]	73% (100/138)	68% (30/44)	54% (7/13)	24% (5/21)
Congenital Heart Disease	10% (22/225)	11% (5/46)	6% (2/33)	0 (0/44)
Birth Weight <2500 g	40% (36/89)	50% (14/28)	56% (5/9)	13% (2/16)

[a] Offspring with PKU or hyperphenylalaninemia are excluded.
[b] IQ < 75, or observed "mental retardation."
[c] Head circumference > 2 S.D. below the mean.

phenylalaninemia (Lenke and Levy, 1980) but whether this represents a truly increased frequency or is due to ascertainment bias remains to be seen.

It should be emphasized that most of this information is from retrospective studies. The reported values of blood Phe in these women were obtained at the time of investigation and may not reflect their metabolic status during pregnancy. These values were also obtained with different analytical methods, some perhaps more accurately quantitative than others. Consequently, correlation between these blood levels and effects on the fetus may be subject to error, particularly at maternal blood Phe concentrations of 10–20 mg/dl.

Treatment with a low-phenylalanine diet has been given during 30 pregnancies, the majority in women with pretreatment blood Phe levels 20 mg/dl or greater. The blood Phe concentration during treatment was generally maintained within the range of 4 to 12 mg/dl with some levels as low as <1 mg/dl or as high as 16 mg/dl. Data from these pregnancies are summarized in Table II, catagorized according to the time of gestation when therapy was initiated. In over 80% of the cases treatment was started in the first or second trimester. It appears that treatment begun in the first trimester results in better intellectual development of offspring than when the treatment is begun later in pregnancy. Even with first trimester treatment, however, there was no clearly beneficial effect in preventing microcephaly or congenital heart disease. That four of the 11 offspring from these mothers died of congenital heart disease is disconcerting (Bush *et al.*, 1979; Buist *et al.*, 1979; Smith *et al.*, 1979; Bovier-Lapierre *et al.*, 1974). Furthermore, the outcome of a treated pregnancy cannot be predicted solely on the basis of the degree of biochemical control, as reflected by periodic maternal blood Phe levels. Two equally well-treated pregnancies resulted in one case in a normal child (Farquhar, 1979) and in the other in an infant who died at 17 days of age with a severe cardiac defect (Smith *et al.*, 1979).

TABLE II
Summary of Offspring Results as Related to Dietary Treatment during Pregnancy

		Offspring[a]			
Treatment started	No. of pregnancies	IQ or DQ[b]	Microcephaly	Cardiac defect	Other
Prior to conception	3	90 (80–100) [N = 2]	33% (1/3)	0 (0/3)	0 (0/3)
First trimester	11	107 (85–127) [N = 6]	33% (3/9)	36% (4/11)[c]	18% (2/11)
Second trimester	16	86 (53–108) [N = 3]	60% (9/15)	13% (2/15)	31% (5/16)[d]

[a] See Lenke and Levy (1980) for individual references; offspring with PKU are excluded (two cases).
[b] Mean and range; number of cases in brackets.
[c] All four died in early infancy.
[d] Including one stillbirth.

Beginning treatment prior to conception might be a reasonable approach. The offspring from two pregnancies treated in this manner are seemingly normal (Nielsen *et al.*, 1979; Bush *et al.*, 1979) while the third pregnancy resulted in a child with borderline microcephaly, an IQ 80 at 5½ years of age, and no congenital heart disease (R. Warner, personal communication, 1979). However, no conclusion on the effectiveness of such treatment can be derived from this small series (Table II).

Some have suggested that women with PKU should be advised against having children (Perry *et al.*, 1973). In any case, it would seem prudent for women with PKU who do want children to plan their pregnancies. This would allow initiation of treatment before conception or at least very early in the first trimester. Complete information about the risks to the fetus from maternal PKU and the possible benefits (or lack of benefits) from treatment should be given to all teenage girls and other women with PKU and their families so that they may make informed decisions about pregnancies (Komrower *et al.*, 1979). This information must be carefully reviewed annually to reflect newer data, and should be provided by centers specializing in PKU and genetics.

It is unlikely that the presence of PKU will be known in all affected women of childbearing age. Some will have been born before newborn screening for PKU. Though most such women will be mentally retarded and perhaps diagnosed on the basis of specific testing, a few will be mentally normal and remain undetected (Perry *et al.*, 1973). Furthermore, most women with PKU who had been treated with a diet during infancy and childhood will have been terminated from this special diet by middle or late childhood and will enter the childbearing years on a normal diet. As a result of ignorance about the potential consequences of maternal PKU, reluctance to begin the difficult diet again, or for other reasons, these women may not call attention to their PKU. To meet the need for identification of PKU in women of childbearing age, a national register (Smith and Wolff, 1979) and routine screening have been proposed.

There is as yet no general agreement on the optimal time for this type of routine screening. Several prenatal blood screening programs are in progress (Buist *et al.*, 1979; J. S. Stevenson, personal communication, 1979) and are effective in identification. However, this type of ascertainment may not allow for treatment to begin before the second trimester. Similarly, routine prenatal testing of urine using ferric chloride reagent would also result in diagnosis relatively late in pregnancy. Whether premarital or other screening is feasible and practical remains to be determined.

II. HOMOCYSTINURIA

Homocystinuria was first discovered in 1962 and denoted the inborn error of sulfur amino acid metabolism cystathionine synthase deficiency. Subsequently, two other heriditary metabolic disorders, one in folate metabolism ($N^{5,10}$-methylenetetrahydrofolate reductase deficiency) and one in cobalamin metabolism [defect(s) leading to inadequate synthesis of the vitamin B_{12} coenzyme methylcobalamin] have also been found to cause homocystinuria (Mudd and Levy, 1978). Cystathionine synthase deficiency is far more common than the other two, and information regarding the outcome of pregnancy is available only in patients with this disorder. Thus, the term *homocystinuria* in this chapter will be interchangeable with cystathionine synthase deficiency.

Cystathionine synthase deficiency is inherited in an autosomal recessive manner and can be divided into two clinical types, pyridoxine responsive and nonresponsive. Generally, patients with pyridoxine-responsive homocystinuria have relatively milder clinical disease. The clinical manifestations of homocystinuria involve multiple organ systems and may resemble some features of the Marfan syndrome. Dislocation of the ocular lenses is the most consistent finding and has been the only abnormality noted in mildly affected patients. Major additional features include mental retardation, skeletal abnormalities, and vascular occulsions. Thromboembolic episodes involving intracranial, pulmonary, renal, and intestinal blood vessels are often the cause of death in these patients. Investigations of coagulation and fibrinolysis have not revealed the basic mechanism of the thromboembolic propensity.

There have been reports of 45 pregnancies in 17 homocystinuric women (McKusick *et al.*, 1971; Van Sprang, 1971; Ritchie and Carson, 1973; Hilden *et al.*, 1974; Brenton *et al.*, 1977; Kurcynski *et al.*, 1980; S. K. Wadman and M. Duran, personal communication, 1978; J. D. Schulman, unpublished data, 1979; D. P. Brenton, personal communication, 1979; Komrower, 1979; Drayer *et al.*,

1980). The majority (37 pregnancies) occurred in pyridoxine-responsive homocystinuric women. Eighteen pregnancies ended in fetal losses: 6 in the first trimester, 1 at 22 weeks of gestation, 4 in the third trimester due to fetal death, and 7 at an unspecified gestational age. One woman had a therapeutic abortion at 18 weeks of gestation. It is noteworthy that all but one (Drayer *et al.*, 1980) of the pregnancies ending in spontaneous fetal losses occurred before the diagnosis of homocystinuria was made and, hence, were untreated. Ten women with pyridoxine-responsive homocystinuria who received pyridoxine therapy during pregnancy bore 19 children. Of these, 17 were normal (including one premature), one had brain damage presumably secondary to a precipitous delivery and one had trisomy 21. One patient with pyridoxine-nonresponsive homocystinuria delivered a normal full term infant (J. D. Schulman, unpublished data, 1979). She received anticoagulant therapy (aspirin 900 mg/day and dipyridamole 100 mg/day) and folic acid (4 mg/day) for at least 6 months prior to conception, the drugs were continued until late in the third trimester, and reinstituted immediately after delivery. The pregnancy and delivery were uneventful; however, ileofemoral thrombophlebitis occurred 2 weeks postpartum. One other pyridoxine-nonresponsive patient, untreated during pregnancy, had 3 normal children (Drayer *et al.*, 1980). Additionally, 2 women with unclassified homocystinuria had 3 untreated pregnancies resulting in 2 normal children and 1 hydrocephalic stillbirth (McKusick *et al.*, 1971).

Placentas from three unsuccessful pregnancies were abnormal because of absence of chorionic vessels and the presence of infarcts (Van Sprang, 1971). On the other hand, no pathological abnormalities were found either in the fetal tissues recovered following a therapeutic abortion at 18 weeks gestation (Kurczynski *et al.*, 1978) or from two fetuses who died *in utero* in the seventh month of gestation (Hilden *et al.*, 1974). It appears possible that vascular complications of the placenta and collagen defects in the uterus may contribute to fetal losses.

Children of men with homocystinuria have had no abnormalities (McKusick *et al.*, 1971; Brenton *et al.*, 1977).

All these findings suggest a very high risk of fetal loss in women with the pyridoxine-responsive type of homocystinuria in its untreated state, and perhaps also in those with the pyridoxine-nonresponsive type. Administration during pregnancy of pyridoxine in combination with folic acid greatly improves the outcome of pregnancy in women with pyridoxine-responsive homocystinuria, and these women are generally expected to have normal offspring. The large doses of pyridoxine used in the treatment of these patients does not seem to have adverse effects on the fetus. The possible effectiveness and the safety of a low-methionine diet during pregnancy have not been determined.

There are more limited data on pregnancy in pyridoxine-nonresponsive homocystinuric women. The disorder probably carries a high risk for the fetus, if one extrapolates from the experience in untreated pyrixodine responders. There is presumably some increased maternal risk, as illustrated by the case with postpartum thrombophlebitis, yet in this instance the fetal outcome appears good and the parents feel the decision to have a child was the correct one.

As with the other autosomal recessive disorders discussed here, the risk of the fetus having the disease is very low (<1%) if the mother is affected and there is no parental consanguinity nor history of disease in the family of the unaffected parent.

Oral contraceptives are presumably contraindicated in homocystinuria as these drugs may increase the hazards of vascular complications.

III. HISTIDINEMIA

Histidinemia is due to a reduced capacity to metabolize histidine to urocanic acid as a result of heriditary histidase deficiency, an autosomal recessive disorder. Histidine and its imidazole metabolites are accumulated and excreted in the urine. The urine reacts with ferric chloride reagent to give a grayish green color.

Histidinemia is a relatively common amino acid metabolic disorder occurring once in approximately 24,000 newborns (Levy, 1973). The clinical significance of this disorder is still somewhat controversial. Recent studies on these patients indicate that histidinemia is not associated with any symptoms and treatment is not necessary (Levy *et al.*, 1974). Although increased histidine in the body seems innocuous to infants and adults, information on the effect of such a biochemical abnormality on the developing fetus is sparse. Several years ago Bulfield and Kacser (1973; Kacser *et al.*, 1977) studied a strain of histidase-deficient mice. Homozygous histidinemic mice show no obvious abnormalities; however, offspring of the histidinemic female, but not of the histidinemic male, have a balance defect which results in circling behavior. These observations suggest that the effects of maternal histidinemia in humans should be carefully examined.

There have been six recorded cases of maternal histidinemia (a total of nine pregnancies). The first case was reported by Neville *et al.* (1971). This woman developed seizures at 15 years of age. During her pregnancy she had intermittent hypertension, was treated for an overdose of barbiturate, and received psychiatric treatment. Labor and delivery were at term and uneventful. Her nonhistidinemic son was normal and had an IQ of 107 at 4 years of age. Armstrong (1975) described a normal woman whose histidinemia was an incidental finding. Her four sons, 12–21 years of age, were normal. Levy (1981) found two histidinemic women by routine screening of the cord blood for increased histidine. Each of the women had one child; both children are now under 2 years of age and have shown normal psychomotor development. Another woman found to have histidinemia by a maternal screening program had two nonhistidinemic children; their IQs were 89 and 94 at 4½ years and 19 months of age, respectively

(I. B. Sardharwalla, personal communication, 1979). The only evidence suggesting that maternal histidinemia may be harmful to the fetus comes from New Zealand. Lyon and his colleagues (1974) showed that the mean IQ of four of the five children of a histidinemic woman was 20 points less than the expected mid-parent value (89–97 vs. 107–117) and suggested that the lower IQ scores of the children is causally related to maternal histidinemia.

One additional family in which maternal histidinemia was reported probably represents a variant of histidinemia (Bruckman *et al.*, 1970). This family is atypical in that histidinemia was found in two successive generations and a speech defect and abnormal EEG were present in nonhistidinemic members. Four histidinemic sisters in this family had a total of 37 pregnancies (including seven miscarriages). Information about the 30 offspring was incomplete; five were histidinemic, and one of these was slightly retarded.

Additional information is needed before any firm conclusions can be reached on the effect of maternal histidinemia on the fetus. This question will undoubtedly be asked with increasing frequency as a generation of histidinemic girls detected by newborn screening enters the childbearing years and seeks counseling. In addition to routine screening for histidinemia using cord blood specimens, the discovery of new cases will be facilitated if metabolic screening is routinely performed at the first antenatal visit.

IV. GYRATE ATROPHY OF THE CHOROID AND RETINA ASSOCIATED WITH HYPERORNITHINEMIA

Gyrate atrophy of the choroid and retina is a choreoretinal degeneration with an autosomal recessive mode of inheritance (Takki and Simell, 1974). Patients report night blindness and loss of peripheral vision between the ages of 10 and 20 years. Progressive constriction of the visual field and cataract formation lead to eventual blindness between the ages of 40 and 50 years. Mentality does not appear to be affected. The electroencephalogram may be abnormal (Takki, 1974). Histopathological and electromyographical changes of the muscle have been found in some patients (Sipilä *et al.*, 1979). The biochemical defect in these patients is a marked hyperornithinemia secondary to a deficiency of ornithine ketoacid transaminase activity. There are two variants, pyridoxine-responsive and nonresponsive (Shih *et al.*, 1978). Lowering the plasma ornithine concentration by either a low-protein and low-arginine diet (Valle *et al.*, 1980; Kennaway *et al.*, 1980; V. E. Shih *et al.*, 1980) or pyridoxine therapy (Berson *et al.*, 1978; Kennaway *et al.*, 1980) may prove of clinical benefit.

Five women with gyrate atrophy of the choroid and retina have had a total of six children (Takki and Simell, 1974; Valle *et al.*, 1977; E. Berson and V. E. Shih, unpublished observation, 1979); all appeared normal and without visual impairment. No complications of pregnancy and delivery have been reported.

V. ARGININOSUCCINIC ACIDURIA

Argininosuccinic aciduria is due to a deficiency of argininosuccinate lyase (argininosuccinase) activity. This results in a markedly increased excretion of argininosuccinic acid, mild citrullinemia, and intermittent hyperammonemia. Affected patients are often retarded and may have episodic ataxia, lethargy, and seizures during acute hyperammonemic attacks (due to stress or increased protein intake). Early treatment with a low-protein diet may prevent the clinical symptoms. Argininosuccinic aciduria is one of the more common urea cycle disorders. The incidence of this disorder as obtained by routine newborn screening is approximately 1 in 169,000 (Levy, 1973). For some, yet to be explained reason, there is a female preponderance in reported cases, with a male to female ratio of 2 to 3.

There has been only one instance of maternal argininosuccinic aciduria (A. M. O. Veale and I. C. T. Lyon, personal communication, 1978). This woman was slightly retarded and had a short attention span. No diet therapy was instituted during her pregnancy. She had no symptoms of hyperammonemia and showed an improvement in her antisocial behavior. She gave birth to a biochemically normal girl whose developmental age at 16 to 18 months was consistent with her chronological age. Determining the effect of maternal argininosuccinic aciduria on the ultimate intellectual achievement of the progeny will require long-term follow-up.

VI. HYDROXYPROLINEMIA

Hydroxyprolinemia is a rare autosomal recessively inherited disorder of hydroxyproline metabolism in which the first step of hydroxyproline oxidation is impaired (Scriver, 1978). Two of the four patients described so far are mentally retarded. The other two patients are siblings of normal intelligence. One had a nodular goiter, but was otherwise well (Pelkonen and Kivirikko, 1970). She had three children, two girls and one boy, all apparently in good health, The hydroxyprolinemic brother of this woman had one son and one daughter, both clinically normal.

VII. FAMILIAL RENAL IMINOGLYCINURIA

Familial renal iminoglycinuria is a benign transport disorder of proline, hydroxyproline, and glycine in which urinary excretion of these three compounds is greatly increased. No clinical symptoms are associated with this condition. It is inherited in an autosomal recessive manner and is relatively common, occurring in approximately 1 of 14,000 newborn infants (Levy, 1973).

Three cases of maternal iminoglycinuria (Scriver, 1968; Whelan and Scriver, 1968; V. E. Shih and C. T. Coulombe, unpublished observation, 1979) are known. Of the five offspring, four were normal and one had cystinosis (Scriver, 1968). Two men with iminoglycinuria had a total of five normal children (Scriver, 1968; V. E. Shih and C. T. Coulombe, unpublished observation, 1979).

VIII. HARTNUP DISORDER

Hartnup disorder is a defect in the renal and intestinal transport of neutral amino acids. It is an autosomally inherited recessive trait with an incidence of about 1 in 24,000 live newborns (Levy, 1973). The finding of a characteristic neutral hyperaminoaciduria is the only reliable criterion for diagnosis. The clinical manifestations of this disorder are intermittent and quite variable even within the same family and range from no symptoms to a clinical triad of mental retardation, intermittent ataxia, and photosensitive rashes (Jepson, 1978). It is noteworthy that the majority of the patients discovered in the past 10 years were detected by routine amino acid screening of the urine, and have been asymptomatic.

The pathogenesis of the symptoms in some of the Hartnup patients is unknown. The ataxia and photosensitive rashes reminiscent of pellagra have been thought to be related to niacin deficiency secondary to increased loss of its precursor tryptophan (a neutral amino acid) in the urine and stool. This hypothesis is supported by the observation that these conditions improve in patients with Hartnup disorder following niacin administration.

There have been three known cases of maternal Hartnup disorder (Pomeroy *et al.*, 1968; S. K. Wadman and M. Duran, personal communication, 1979). Their obstetrical histories were unremarkable. Of eight offspring, seven were considered normal and healthy and one coincidentally had lethal methylmalonic acidemia. Additionally, a man with Hartnup disorder fathered two normal children (Pomeroy *et al.*, 1968).

IX. GALACTOSEMIAS

The term *galactosemia* describes two autosomal recessively inherited inborn errors of metabolism in which blocks exist in the normal galactose-to-glucose metabolic pathway and in which galactose levels become elevated in blood and urine. Classic galactosemia is due to deficient galactose-1-phosphate uridyltransferase activity, and is the most widely recognized of these two metabolic disorders. Conventionally, galactosemia is used to indicate this transferase deficiency. The incidence of galactosemia, as obtained from newborn screening data, is about 1 in 65,000 (Levy, 1973).

In addition to galactose, galactose 1-phosphate (gal-1-P) and other metabolites accumulate in the body. Affected infants are clinically normal at birth but vomiting, hepatomegaly, and jaundice frequently develop in early infancy, and there is no weight gain while on formulas containing milk. An as yet undetermined percentage of patients may die in the first weeks of life. Cirrhosis, cataracts, and mental retardation are usually noted within the first few years of life. Galactitol accumulation appears to contribute to cataract formation. With early diagnosis and early treatment with a galactose-free (lactose-free) diet, the symptoms and signs of the disease may be prevented or reversed. Infants so treated are expected to survive into adulthood in comparatively good health; however, perhaps because fetuses experience elevated gal-1-P levels *in utero* (Komrower, 1979), subtle neurological disturbances are often seen even with prompt postnatal therapy.

Galactokinase deficiency is the second of these inborn metabolic errors and is much less common than transferase deficiency. In this recently discovered disorder, early onset of cataracts is the only known manifestation. Even the carriers for this trait have been found to have a higher incidence of cataract than the general population (Beutler *et al.*, 1973).

A. Galactosemia (Galactose-1-Phosphate Uridyltransferase Deficiency)

We are aware of five reported cases of maternal galactosemia (Roe *et al.*, 1971; Tedesco *et al.*, 1972; Samuels *et al.*, 1976; Komrower, 1979). The pregnancy, delivery, and offspring appeared normal in all patients. During pregnancy these treated women had no appreciable accumulation of galactose in the blood; however, one had an unexplained increase of erythrocyte gal-1-P in the second trimester while Komrower's two patients had abnormal gal-1-P levels which rose further near term. Cord (fetal) gal-1-P content of red cells was normal in these latter cases. Two of the five patients were black women and may have a variant of galactosemia that has been reported in American blacks in which the liver has a limited residual capacity to metabolize galactose. Lactase deficiency, which is common in blacks and was established in one of these black mothers, may also protect the fetus against galactose toxicity. We are unaware of pregnancies in untreated or partially treated maternal galactosemia.

Of considerable interest are the recent reports of Kaufman *et al.* (1979, 1980) and of Hoefnagel *et al.* (1979) on ovarian failure in galactosemia. The former investigators reported that nine of 12 galactosemic females manifested hypergonadotrophic hypogonadism accompanied by primary or secondary amenorrhea or oligomenorrhea. The karyotype was studied in three patients and was 46,XX and no antiovarian antibodies were found. Hoefnagel and colleagues (1979) described two adult female galactosemics. One 26 year old with mental retardation had been treated by galactose restriction since early childhood and had normal menses since age 13; the other patient was evaluated for primary amenorrhea at age 21, and had hypergonadotrophic hypogonadism, a 46,XX peripheral blood karyotype, and bilateral streak ovaries with absent fallopian tubes ("XX-pure gonadal dysgenesis").

The possibility that aberrant galactose metabolism and/or galactose deprivation may cause premature ovarian failure is of great importance and will merit considerable investigative effort.

B. Galactokinase Deficiency

The only reported case of maternal galactokinase deficiency occurred in the first family with galactokinase deficiency studied by Gitzelmann (1967). A 64-year-old woman with both galactokinase deficiency and neurofibromatosis had six children, four of whom were alive and healthy at the time of the study.

Harley *et al.* (1974), in their study of erythrocyte galactokinase and galactose-1-phosphate uridyltransferase of infants and children with unexplained cataracts and their parents, showed that many of the mothers, but not the fathers, had noticeably low levels of activity of either or both enzymes. They suggested that maternal partial deficiency of these enzymes may contribute to the formation of cataracts in fetal life. These observations need to be confirmed, but may suggest desirability of measuring these enzyme levels in women who have had children with unexplained cataracts; if an abnormality were detected, milk restriction during subsequent pregnancies could be considered. However, Gitzelmann's galactokinase-deficient patient had not received diet therapy, and although she had cataracts these were apparently not present in her four living heterozygous offspring.

X. GLYCOGEN STORAGE DISEASE TYPE I (GLUCOSE-6-PHOSPHATASE DEFICIENCY)

Deficiency of glucose-6-phosphatase activity (EC 3.1.3.9) is the cause of the autosomal recessive disorder called von Gierke's disease. The defect has its clinical onset in infancy and is characterized by hepatic enlargement, adiposity, flabby musculature, renal enlargement, hyperlipidemia and xanthomas, hyperuricemia and gout, and a bleeding diathesis perhaps related to impaired platelet function. Hypoglycemia is characteristic, of variable severity, and may cause convulsions. Growth may be impaired. Renal compromise and uremic deaths may occur (Howell, 1978).

The disease is compatible with survival into adult life and with both male and female reproduction. We are unaware of reported cases studied carefully during pregnancy. There are several indications of apparently uncomplicated pregnancies in affected patients, but an enzymatic diagnosis is not always present, especially in older reports (van Crevald, 1961; Jeune *et al.*, 1959; J. B. Sidbury, unpublished observation, 1979). Alepa *et al.* (1967) have described a 35-year-old patient who had two term pregnancies with the birth of normal children; the gestations were marked by anemia and pyelonephritis and "extensive" blood loss accompanied both deliveries. Holling (1963) reports an affected woman who had a first trimester spontaneous abortion at age 26 years. At age 29 years, her second pregnancy resulted in toxemia, requiring cesarean section at 8 months gestation; no excessive bleeding was described. The child weighed 1560 g and was apparently healthy. Postpartum, menses remained infrequent and lactation did not occur. This patient later developed gout and uremia.

Pregnant patients with von Gierke's disease should theoretically be predisposed to a variety of serious complications. Renal disease should increase the risk of toxemia. Reduced stature would lead to a greater frequency of cesarean sections. Operative or vaginal deliveries might be accompanied by excessive blood loss. Hypoglycemic convulsions, though rare beyond childhood, could cause fetal damage. Pregnancy in the presence of substantial renal disease might lead to further deterioration of renal function. More clinical and metabolic details on pregnancy in affected patients are required before risks can be more reliably assessed.

XI. GLYCOGEN STORAGE DISEASE TYPE II (ACID MALTASE DEFICIENCY)

Deficient activity of the enzyme α-1,4-glucosidase (EC 3.2.1.20) leads to intralysosomal glycogen accumulation. In its most typical form (Pompe's disease), the autosomal recessive disorder which results is characterized by hypotonia and cardiac failure usually causing death before age 1 year. Widespread glycogen accumulation in many tissues is observed at autopsy. A milder form of the disease has its onset in childhood, with a more protracted course leading to death before age 20 years. A still milder variant is characterized by onset of muscle weakness in adult life and is often misdiagnosed initially as a progressive muscular dystrophy. A gross deficiency of acid maltase activity is present in all variants (Howell, 1978).

We are aware of two published reports of pregnancies in women with the adult variant. Hudgson *et al.* (1968) described a patient whose symptoms began at age 31 and who experienced two apparently uneventful pregnancies, one before and one 3 years after onset of clinical disease. Engel (1970) describes a 44-year-old woman whose disease had been manifest for 13 years and who had had five children and four spontaneous abortions over the past 23 years. There were no indications in these reports that labor or delivery were abnormal although this was not directly commented upon, nor did it seem that disease progression was modified by pregnancy. The offspring were apparently healthy. J. B. Sidbury (unpublished observations, 1979) notes two other patients with apparently uncomplicated gestations.

XII. GLYCOGEN STORAGE DISEASE TYPE V (MYOPHOSPHORYLASE DEFICIENCY)

McArdle's syndrome is a recessively inherited disorder characterized by deficient activity of muscle phosphorylase (EC 2.4.1.1) and manifests clinically by muscular fatigue, cramps during exercise, and occasional myoglobinuria. Apparently successful pregnancies were reported briefly by Schmid and Hammaker (1961) and by Dawson *et al.* (1968) but no details of these six pregnancies were given.

Cochrane and Alderman (1973) describe in detail a pregnancy in a single patient. The gestation was uncomplicated and spontaneous labor occurred at term. Contractions were normal during the first stage but after 18 hours the labor required stimulation with oxytocin despite artificial rupture of the membranes; an epidural anesthetic was used. The fetal head presented as an occiput-posterior and forceps delivery was required. No muscle cramps were noted and myoglobinuria was absent postpartum, in contrast to the case reported by Dawson *et al.* (1968). The child was normal.

This pattern of labor is probably compatible with normal uterine function, especially in view of the occiput posterior position of the fetal head and the generous size of the infant (3200 g). The uterine myophosphorylase level has not been reported in affected patients, but it may be normal since vascular smooth muscle phosphorylase activity appears unimpaired in McArdle's syndrome (Engel *et al.*, 1962).

XIII. GOUT

Primary gout is a form of acute and chronic arthritis associated with the presence of sodium urate crystals and frequently marked by an elevated serum urate level. The disease has been the subject of many reviews and two comprehensive discussions are especially recommended (Wyngaarden and Kelley, 1978; Seegmiller, 1980).

While gout is primarily a disease of adult males, it may occur in women. In various series, the frequency of disease in women averages approximately 5% of all cases. In the Framingham study (Wyngaarden and Kelley, 1978) in a population with a mean age of 58 about 15% of cases were in women.

When the disease occurs in women, the mean age of the first attack, as in men, is 45–50 years (Turner *et al.*, 1960). However, in various series a minority percentage of women ranging from approximately 10% to 40% will have the onset of disease while still actively menstruating (Turner *et al.*, 1960; Delbarre *et al.*, 1967). In a large French series, about 15% of gouty women were having completely normal periods at the time of first attack (Delbarre *et al.*, 1967).

There are suggestions that the disease is more severe when it occurs in women (Wyngaarden and Kelley, 1978) although the fundamental features are similar to those in men.

Nephrolithiasis has been reported in gouty women but may be less frequent than in males (Delbarre *et al.*, 1967; Schilling, 1968). However, there is a strong suggestion that renal compromise from urate nephropathy is more severe in gouty women than men (Delbarre *et al.*, 1967) and there may be an increased tendency to hypertension. In some cases, acute attacks of gouty arthritis may coincide with the menses (Talbott, 1967) or estrogen withdrawal (Friedman and Little, 1958).

There are a limited number of reports of gout associated with pregnancy. These represent highly selected patients, and it is difficult to determine from the literature the frequency with which unselected gouty females conceive and bear children. Suggestions that fecundity may be reduced (Delbarre *et al.*, 1967; Talbott, 1967; Batt *et al.*, 1963) require substantiation and better quantitation then can be reconstructed from the existing literature. This problem is discussed by Friedman and Little (1958).

Reported cases of pregnancy in gout have been reviewed by Talbott (1967) and Batt *et al.* (1963). We are aware of one additional report (Mertz *et al.*, 1972) in which mention is made of a 27 year old with a 5-year history of acute gout associated with type IV hyperlipoproteinemia, in whom pregnancy occurred at age 23 years with delivery 8 months later of a premature infant who died neonatally; no other details of this pregnancy are provided.

Certain gouty patients have had gestations unmarked by major complications (Mace, 1956; Francon *et al.*, 1956), sometimes with a partial amelioration of painful arthritis. Postpartum exacerbations (Friedman and Little, 1958; Lee and Loeffler, 1962) of acute gouty arthritis were noted in some patients. Mace (1956) described in his patient a postpartum rise in serum urate without increased symptoms. A few patients have had premature deliveries (Mertz *et al.*, 1972) or recurrent early abortions (McQueen, 1951) probably not related to their gout. In another patient described by McQueen (1951), acute arthritis was first observed after birth of a third child, at age 25 years, but "kidney trouble" not further described was recorded during her first and fourth pregnancies at ages 20 years and 31 years, respectively; at age 36 years this patient had hypertension and renal disease presumably secondary to urate nephropathy.

Of considerable interest are several documented cases of gouty nephropathy in which pregnancies were associated with azotemia.

Greenhut and associates (1953) describe a 28-year-old woman whose gout may have been primary or secondary to sarcoidosis (Batt *et al.*, 1963). Gouty arthritis had been present since age 26 years, and her two pregnancies at ages 22 years and 26 years were marked by preeclampsia. Severe anemia and uremia occurred during the first pregnancy and slowly remitted with health remaining "good" until the second pregnancy.

Friedman and Little (1958) reported a 29-year-old woman with a positive family history and 6-year duration of gout. She had had two uncomplicated pregnancies and two terminating in spontaneous abortion. At 34 weeks gestation, preeclampsia developed followed by overt uremia. The patient was in grave condition and labor was induced with oxytocin. A 1550-g infant was delivered and survived. There was rapid postpartum improvement with eventual apparent normalization of renal function. A subsequent pregnancy was marked by several gouty attacks in the first trimester; elective hysterotomy and tubal ligation were performed. At the time of surgery cardiovascular and renal function appeared unremarkable.

Weingold (1960) reported a gouty patient whose prior obstetrical history included three spontaneous abortions, a premature delivery, and a term delivery. In the two pregnancies described in more detail mild azotemia, and in the second a striking hyperuricemia, were observed. Preeclampsia did not occur and the gestations were otherwise uncomplicated. Both these pregnancies were followed by a postpartum gouty exacerbation.

Lee and Loeffler (1962) reported a patient with apparent gouty nephropathy who had anemia, azotemia, and mild hypertension. Pregnancy was terminated at 37 weeks by cesarean section. Like the patient reported by Weingold this woman was never critically ill during pregnancy; she had had several uncomplicated term pregnancies prior to onset of clinical gout.

Batt *et al.* (1963) report a patient who had gouty nephropathy with two pregnancies complicated by anemia and severe azotemia. In the second of these gestations, salt wasting renal disease was present, and progressive renal insufficiency was partially ameliorated by increasing salt intake; induction of labor was performed because of persistent hypertension.

Probenecid, high fluid intake, and colchicine as needed are recommended by several authors for gestational gout. Weingold (1960) observed hyaline membrane disease in a term infant whose mother was treated with probenecid and colchicine, but this does not seem to have occurred in other instances. We are unaware of reported experience with allopurinol in gestational gout.

The material summarized above makes it clear that gout and pregnancy may coexist, that they probably do so more frequently than generally recognized by physicians or reported in the literature, and that pregnancy may pose hazards in some instances in which gouty nephropathy exists. There is little good evidence for reduced fertility in gouty women, although some reduction would not be unexpected in individuals with a chronic illness. Pregnancy may, but does not always, reduce the frequency of arthritic attacks; the reverse has been reported in some patients postpartum.

Gout often has a familial nature, and inheritance of the disease and of asymptomatic hyperuricemia is well documented. However, with the exception of certain rare forms of gout inherited as X-linked recessive traits, the genetics has been poorly defined and dominant and multifactorial inheritance have been invoked. This subject is critically explored by Wyngaarden and Kelley (1978) who emphasize the heterogeneity of gout and the need to study mechanisms and patterns of transmission in individual families.

Genetic counseling is not usually of major importance in this disease, which is generally comparatively benign. As a composite figure, it appears (Smyth, 1957) that about 25% of the relatives of gouty patients have hyperuricemia and about 20% of these (approximately 5% of relatives) will have clinical gout. It is generally accepted (Seegmiller, 1980; Wyngaarden and Kelley, 1978) that gout is likely to be more severe than usual in affected sons of an affected mother.

XIV. XANTHINURIA

This is a rare disorder of purine metabolism associated with reduced tissue xanthine oxidase activity. Xanthine oxidase catalyzes the oxidation of hypoxanthine to xanthine and of xanthine to uric acid, and can also facilitate oxidation of certain other purines. The enzyme is normally present in abundance in human liver and small intestinal mucosa, and jejunal biopsy has been used to confirm the diagnosis of xanthinuria. Patients may be discovered on the basis of a routine serum urate screen and be otherwise asymptomatic, or may manifest urinary xanthine stones or a skeletal myopathy associated with hypoxanthine and xanthine crystal deposits in muscle (Seegmiller, 1980; Wyngaarden, 1978). The characteristic laboratory findings are low serum urate (almost invariably less than 1 mg/100 ml) and increased serum oxypurines, grossly reduced urinary urate, and increased urinary xanthine and hypoxanthine. Xanthine is considerably less soluble at the urinary pH of 5–7 than uric acid, explaining the prospensity to xanthine stones; in one case such stones caused hydronephrosis and eventual nephrectomy. There is some evidence that xanthinuria is inherited as an autosomal recessive trait (Seegmiller, 1980; Wyngaarden, 1978).

Pregnancy has occurred in at least three xanthinuric women. No data are available in the literature regarding gestation(s) of the grandmother described by Bradford *et al.* (1968) or the pregnancy mentioned briefly by Sperling *et al.* (1971). Recently, Curiel and Bandinelli (1979) documented an uncomplicated pregnancy in a 22-year-old asymptomatic xanthinuric female who had required cervical cerclage because of cervical incompetence causing two previous premature deliveries. None of this patient's gestations was marked by urolithiasis. The amniotic fluid urate concentration was about one-half the normal mean and well below the normal range at 34 weeks gestation but was normal at 40 weeks; presumably these values largely reflect fetal urinary excretion. The one living child appeared normal at birth.

Although urolithiasis has not characterized xanthinuric pregnancies so far, the risk must be borne in mind. A high fluid intake, reduction of dietary purines, alkalinization of the urine, and administration of allopurinol [which reversed the high xanthine/hypoxanthine ratio in one patient (Wyngaarden, 1978)] are reasonable proposals to reduce the risk of urinary stones, but their role in xanthinuric pregnancy apparently remains untested.

XV. REFSUM'S DISEASE (PHYTANIC ACID STORAGE DISEASE)

In 1946, Refsum identified a familial neurologic syndrome, which he considered a hereditary atonic polyneuropathy. It is now recognized that this disorder, inherited in an autosomal recessive manner, is due to accumulations of the 20 carbon branched chain phytanic acid (of exogenous origin) secondary to defective activity of the enzyme which normally converts phytanic acid to α-hydroxyphytanic acid. The syndrome may involve retinitis pigmentosa, polyneuropathy, ataxia, elevated cerebrospinal fluid protein, nerve deafness, anosmia, pupillary abnormalities, nonspecific EEG changes, ichthyosis, and epiphyseal dysplasia. Night blindness is usually the earliest symptom and may appear in childhood or adulthood. Death may occur suddenly, perhaps due to arrhythmias, respiratory paralysis, or pneumonia. Diets deficient in phytanate are under investigation therapeutically. The subject is exhaustively reviewed by Steinberg (1978).

Of particular interest here, dramatic worsening of symptomatology has been described in association with febrile illnesses, surgery, or pregnancy (Steinberg, 1978). While there is no evidence that maternal disease adversely effects fetal outcome, there are several instances in which there were pregnancy-associated exacerbations. Toussaint *et al.* (1959) describe a 26-year-old patient with a "brutal" worsening of visual acuity at the time of childbirth. Fleming (1957) notes a worsening of peripheral neuropathy 2 months after childbirth in a 24-year-old woman. The patient of Fryer *et al.* (1971) had an apparent reversible exacerbation of skin changes postpartum at age 25 years. Campbell and Williams (1967) describe an increase in weakness and ataxia during two pregnancies in one of their patients, and feel that after hysterectomy and left oophorectomy she "markedly and steadily" improved.

In contrast, some patients with Refsum's syndrome may have uneventful pregnancies (Edström *et al.*, 1959), while others may have an unpredictable course; Nevin *et al.* (1967) describe a patient who had had three prior pregnancies resulting in two cesarean sections and one first trimester miscarriage apparently without neurological worsening, yet who deteriorated in the last month of her fourth pregnancy and died several days postpartum.

Although the data are limited, it seems that the risk of gestational worsening is a real one in Refsum's disease, and this may occur during pregnancy or in the puerperium. Fetal damage is not described. The impact on mother and fetus of a low-phytanate diet during pregnancy is undetermined.

XVI. GAUCHER'S DISEASE

Of the three major forms of Gaucher's disease, all of which are characterized by autosomal recessive inheritance and sphingolipid storage secondary to glucocerebrosidase deficiency, only the chronic nonneuropathic adult variant (type I) is of interest with regard to pregnancy. The disease is unusually common in Askenazic Jews. The details of all Gaucher's disease variants are extensively reviewed by Brady (1978).

Patients with type I Gaucher's disease have characteristic lipid-laden cells in various organs and the bone marrow. The earliest manifestations are hepatosplenomegaly and bone pain; lung involvement may result in cor pulmonale and pneumonia. Fractures may occur. Hypersplenism may lead to hemorrhagic episodes and anemia. Although the liver and kidney are sites of glucocerebroside storage, hepatic or renal dysfunction is rare. In the adult variant, neurological disease appears to be absent.

There has been considerable attention in various reports to possible interactions of Gaucher's disease and pregnancy. There are some impressions that sterility, repeated abortions, neonatal deaths, and postpartum sepsis are increased in Gaucher's patients (Groen, 1948; Ross, 1969; Tenton and Treadwell, 1957) but other more critical studies make it clear that this is almost surely not correct (Hsia *et al.,* 1962; Fried, 1973; Greenwald and Fenton, 1959; Decker and McWhorter, 1956; Bromberg *et al.,* 1953; Mendel and McCullough, 1968). The interested reader should review these studies to form his own conclusions, but this reviewer (JDS) agrees with the view of Bromberg *et al.* (1953) that "childbirth does not exert any deleterious effect, either immediate or remote, upon the woman suffering from Gaucher's disease" nor does the maternal disease adversely affect the offspring, many of whom have been entirely normal. Therapeutic interruption of pregnancy is rarely, if ever, indicated but since patients may have marked hematologic or cardiopulmonary compromise close observation is certainly required. There appears to be no special risk of postpartum hemorrhage. The role of splenectomy, steroids, analgesics, orthopedic procedures, transfusions, and enzyme replacement in the management of Gaucher's patients is reviewed in several of the references, especially Brady (1978).

XVII. HYPERLIPOPROTEINEMIAS

The genetic causes of hyperlipoproteinemia comprise several familial disorders. Lipoproteins are categorized into five groups. These are chylomicrons, very low density lipoproteins (VLDL), low density lipoproteins (LDL), high density lipoproteins (HDL), and pre beta or floating beta VLDL. Increase of one or several of these lipoproteins is reflected in plasma lipid elevation. Hyperlipidemia can be defined as an elevation above normal of plasma cholesterol and/or triglyceride after an overnight fast (Fredrickson *et al.*, 1978). Plasma lipid measurement (cholesterol and triglyceride) and clinical information from history and physical examination provide an efficacious initial approach to establishing the diagnosis. More specific diagnostic comparisons and classifications using lipoprotein measurements and specific enzyme and receptor assays are described in a recent review (Fredrickson *et al.*, 1978). The most significant consequence of increased cholesterol and triglycerides is an augmented risk of atherosclerosis. Each of the genetic causes of hyperlipoproteinemia is manifested by a different combination and magnitude of plasma lipid increase. These lipid changes can be modified by endogenous factors including estrogen, alcohol, or fat intake.

As is the case for several metabolic disorders, sex hormones can modulate the expression of the basic metabolic abnormality. These effects are typically magnifications of the response of normal lipid metabolism to sex hormones. Exogenous estrogen given to postmenopausal women increases the plasma concentration of VLDL. This effect is presumably the result of an increase of hepatic VLDL production (Glueck *et al.*, 1975). Since VLDL is composed predominantly of triglycerides and to a lesser extent of cholesterol, increase proportionately. The major increase should be in the triglyceride fraction, as illustrated by Molitch *et al.*, (1974).

In addition to estrogen and estrogen-progestin combinations (oral contraceptives), numerous other exogenous factors elevate plasma triglycerides. These include glycogen storage disease, pancreatitis, insulinopenic diabetis mellitus, alcohol, dysglobulinemia and the nephrotic syndrome (Fallat and Glueck, 1976). Although cyclic changes of lipids in relation to the menstrual cycle have not been reported, normal pregnancy is associated with moderate elevation of cholesterol and triglycerides. These increases occur during the third trimester and persist through term. Free cholesterol may increase after delivery and this has persisted for over 1 month (de Alvarez *et al.*, 1959). Fabian *et al.* (1968) observed a significant decrease in the activity of lipoprotein lipase during the third trimester of pregnancy in 21 normal women. These changes reverted to normal following delivery. Thus, alterations in plasma lipids in pregnant women with one of the genetic hyperlipoproteinemias must be considered in the context of VLDL and chylomicron increases that can occur with uncomplicated pregnancy.

A. Maternal Effects

Clinical manifestations present in some of the familial hyperlipoproteinemias include atherosclerosis and its attendant cardiovascular effects, cutaneous signs of increased cholesterol and triglycerides, and lipemia with the possibility of pancreatitis. Inherited lipoprotein abnormalities are not, as a group, infrequent, but there is surprisingly limited information on the course of pregnancy in affected women.

Abdominal pain and acute pancreatitis have been observed in several patients with both lipoprotein lipase deficiency (type 1) and those with familial type 5 hyperlipoproteinemia (Aldersberg *et al.*, 1959; Bremmer and Third, 1978; Kaplan *et al.*, 1957; Statzer, 1961). One patient with lipoprotein lipase deficiency abandoned her low-fat diet when she became pregnant. Pancreatitis complicated by retroperitoneal hemorrhage occurred at 24 weeks gestation. The fetus died following premature labor but the mother recovered and resumed her previous diet. She has been asymptomatic since, but she has not become pregnant again (Bremmer and Third, 1978). In 1961, Statzer reported a pregnancy complicated by what he termed *essential hyperlipemia*. That patient had acute pancreatitis during the first trimester. She recovered and subsequently followed a low-fat, low-cholesterol diet. At the 36th week she experienced abdominal pain and fever despite this therapy. Nonetheless, a healthy child was born at term following a spontaneous labor and delivery. Kaplan *et al.* (1957) also reported late third trimester pancreatitis with continuation of pregnancy to term and delivery of a normal child.

Familial hypercholesterolemia (type II) in its heterozygous state may be undetected clinically, although cholesterol levels are generally elevated. The effect of pregnancy on cholesterol metabolism for the heterozygous female is not known. The homozygous individual experiences severe problems with atherosclerosis in the first decades of life. The significant morbidity and early death due to complications of atherosclerosis limit the opportunity for reproduction. However, apparently homozygous females have become pregnant. Tsang *et al.* (1978) reported normal gestation, parturition, and postpartum lactation in such a patient and carefully studied lipid patterns in the mother, neonate, breast milk, and placenta. They also cite one other report of a term gestation in familial hypercholesterolemia (Stein *et al.*, 1975).

Hyperlipoproteinemia may be recognized for the first time during pregnancy on the basis of cutaneous stigmata (Fredrickson *et al.*, 1978), lipemia, increased plasma triglycerides, and/or cholesterol on routine testing or in the course of evaluating abdominal pain. A positive family history may not be evident in these patients. Nonetheless an inherited lipoprotein elevation may be present. Goldstein *et al.* (1973) suggested that a new type of familial hyperlipoproteinemia may exist which they termed familial combined hyperlipidemia. It is an important diagnosis to consider because the family history may be unremarkable, and lipid and lipoprotein studies may reveal variable abnormalities among the family members. Therapeutic considerations for the patient and responsible preventive medical practice require complete lipoprotein evaluation in the patient and her family. Familial hyperlipoproteinemia should be considered whenever hyperlipidemia of pregnancy is diagnosed.

B. Fetal Outcome

Examples were cited above of the successful completion of pregnancy in the face of complicated hyperlipidemia. The children were considered to be normal. The amount of fetal cholesterol that can be attributed to maternal plasma levels at term is not known. The transplacental movement of cholesterol is probably dependent on gestational age (Tsang and Glueck, 1975). Prenatal diagnosis of a type II hyperlipoproteinemia is currently feasible, since cells in this disorder are characterized by gross reduction or other altered properties of LDL receptors. Brown *et al.* (1978) utilized LDL receptor measurements on amniotic fluid cells to evaluate a fetus at risk for homozygous familial hypercholesterolemia. The accuracy of their diagnosis was confirmed by documenting an elevation of LDL in the blood of the aborted fetus. The future wider application of this technique, as well perhaps as lipoprotein lipase measurements on fetal blood samples and other newer technologies for prenatal diagnosis of inherited forms of hyperlipoproteinemia, may be anticipated.

XVIII. PORPHYRIAS

The porphyric diseases of man include six genetically distinct disorders. Each type of porphyria represents a primary disorder of heme biosynthesis (Fig. 1). Although the division of the porphyrias into erythropoietic and hepatic types can be useful, it may be more clear in this clinical discussion to classify them as either acute attack or cutaneous types of disease.

Acute intermittent porphyria (AIP), variegate porphyria (VP), and hereditary coproporphyria (HCP) are acute attack forms of porphyria. The characteristic chemical abnormality is excretion of excess amounts of porphyrin precursors in the urine. Porphobilinogen (PBG) is the immediate precursor of uroporphyrinogen formation (Fig. 1), and increased urine PBG excretion is the characteristic finding during the acute attack phase of these diseases. Urine δ-aminolevulinic acid (ALA) excretion is typically increased in tandem with PBG, but to a lesser degree. Porphyria cutanea tarda (PCT), VP, HCP, erythropoietic protoporphyria (EPP), and congenital erythropoietic porphyria (CEP) are cutaneous forms of porphyria. The characteristic chemical finding in this group is the tissue accumulation and excretion of excess porphyrins (uroporphyrin, coproporphyrin, and/or protoporphyrin) in the red cells, plasma, urine, and feces. VP and HCP may have either acute attack or cutaneous manifestations or both. A detailed survey of the

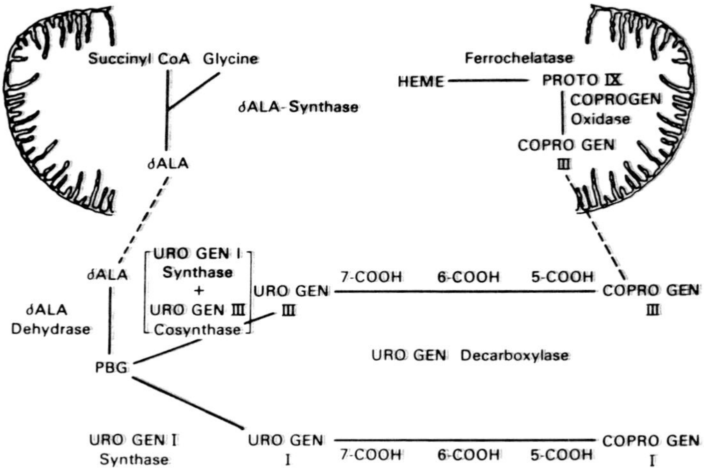

Acute Intermittent Porphyria	URO GEN I Synthase
Porphyria Cutanea Tarda	URO GEN Decarboxylase
Erythropoietic Protoporphyria	Ferrochelatase
Congenital Erythropoietic Porphyria	URO GEN III Cosynthase
Variegate Porphyria	Ferrochelatase
Hereditary Coproporphyria	Coproporphyrinogen III Oxidase

Fig. 1. Heme biosynthesis and the porphyrias. The cellular compartmentalization of the pathway is represented. δ-ALA-Synthase, coproporphyrinogen oxidase, and ferrochelatase are located within the mitochondrion. δ-ALA-Dehydrase, uroporphyrinogen I synthase, uroporphyrinogen III cosynthase, and uroporphyrinogen decarboxylase are located in the cytosol. The various forms of porphyria and possible enzymes that may be deficient in each disease are listed. Recently, Brenner and Bloomer (1979) reported the association of a partial deficiency of protoporphyrinogen oxidase with variegate porphyria. (This step catalyzes the oxidation of protoporphyrinogen IX to protoporphyrin IX, and it is not depicted here.)

porphyrias and specific diagnostic evaluations are the subject of two recent reviews (Meyer and Schmid, 1978; Tschudy and Lamon, 1980).

All of the porphyrias are inherited diseases, and in recent years the genetic bases for several of them have been resolved at the enzymatic level. Enzyme steps in the heme biosynthetic pathway at which deficiencies are presumed or suspected of being associated with specific porphyrias are illustrated in Fig. 1. CEP is inherited in a recessive manner, but the five other prophyrias are dominant traits. To this time enzyme deficiencies have been confirmed in four of the porphyrias and studies of the inheritance of the enzyme deficiency in affected families have confirmed autosomal dominant inheritance for AIP (Meyer *et al.*, 1972; Lamon *et al.*, 1979a), HCP (Elder *et al.*, 1976; Nordmann *et al.*, 1977),

PCT (Kushner *et al.*, 1976; Felsher *et al.*, 1978), and EPP (Bonkowsky *et al.*, 1975; Bottomly *et al.*, 1975).

A. Acute Attack Types of Porphyria

The three types of porphyria which comprise this group (AIP, VP, and HCP) are ascertained in two ways. First, the diagnosis is established during the acute attack phase of the disease. Exacerbations are typically intermittent. They occur over hours to days, each attack reaching maximum intensity in two to three days and persisting several days to several weeks. The clinical features are those of neurologic disease; however, the specific manifestations are protean in different patients. Symptoms may include abdominal pain, nausea, vomiting, constipation, urinary retention, palpitations, depression, and weakness. Signs of the acute attack involve most frequently hypertension and tachycardia and to a lesser extent hyponatremia, paresis, paralysis, and diaphoresis. The neurological manifestations have been recently reviewed (Becker and Kramer, 1977). Despite the protean nature of the clinical picture there is a characteristic abnormality detectable in the urine during an attack, an excess amount of PBG. Screening tests are available to detect elevated urinary PBG and are well suited to the acute attack situation (Lamon *et al.*, 1977). Precipitating factors in the disease include certain drugs, starvation, endogenous hormone variation in females, and infection. These factors are thoroughly discussed by Tschudy (Tschudy *et al.*, 1975; Tschudy and Lamon, 1980).

A diagnosis of porphyria may also be established via family studies prompted by the ascertainment of an index case. Although measurement of porphyrin precursor and porphyrin excretion may be useful for detecting latent disease in family members, false negative studies are possible during quiescent periods. If the index case has an enzyme deficiency involving either uroporphyrinogen I synthase or coproporphyrinogen oxidase, the diagnosis of AIP or HCP, respectively is confirmed. These specific enzymes are excellent tools for family evaluation. At the present time, quantitative porphyrin studies are the only available approach to confirm the diagnosis of VP.

Unfortunately, most acute porphyria cases are diagnosed during a symptomatic interval. This situation can occur during pregnancy.

1. Acute Intermittent Porphyria

AIP is inherited in an autosomal dominant manner. Studies in families using uroporphyrinogen I synthase measurement show that males and females are affected in equal proportion (Lamon *et al.*, 1979a). Despite this fact, the preponderance of females with clinical manifestations, the onset of clinical disease during adolescence or in the early twenties in females (approximately two decades earlier than males) and the known influence of exogenous and endogenous hormones on precipitating acute attacks (Tschudy and Lamon, 1980; Stein and

Tschudy, 1970; Mustajoki and Koskelo, 1976) clearly establish the postmenarchial premenopausal female with latent AIP to be at greatest risk of developing porphyric symptomatology. Although severe and sometimes chronic disease can interfere with normal menstrual function and thereby affect fertility, this problem is infrequent in AIP. In the author's experience (JML) the few occasions in which normal menses were interrupted occurred during acute attack periods and normal menstrual function resumed following resolution of the episode.

a. Maternal Effects. From the reported clinical experience and literature reviews concerning pregnancy in AIP (Mustajoki and Koskelo, 1976; Brodie *et al.*, 1977; Shapiro *et al.*, 1969; Huber, 1967; Tschudy and Lamon, 1980), the general impression is that the disease worsens during pregnancy. Clinical investigations by Welland *et al.* (1964) demonstrated the "porphyrinogenic" nature of exogenous female hormones in females with AIP. Serial measurements of urine porphyrin and porphyrin precursor excretion in parallel with estrogen, estradiol, and estriol levels during the course of pregnancy demonstrated a correlation of chemical exacerbation with hormonal changes. *In vitro* studies with chick embryo liver cells demonstrated the porphyrinogenic nature of progesterone (Kappas *et al.*, 1969). Although early reports of pregnancy in AIP were discouraging with respect to the disease in the pregnant female (Neilson and Neilson, 1958; Seckler and Rovinsky, 1965), more recent reports indicate that pregnancy can be well tolerated by females with AIP (Mustajoki and Koskelo, 1976; Brodie *et al.*, 1977; Tschudy and Lamon, 1980). A gestation with acute attack symptoms and a totally asymptomatic pregnancy have been noted successively (Hunter, 1971).

AIP is well known not only for variable clinical manifestations but also for variable sensitivity to precipitating factors. Therefore, several issues must be considered in attempting to evaluate the effect of pregnancy on the activity of AIP. These include differences between patients with previously symptomatic AIP and patients uncovered during pregnancy; the use of medication during pregnancy; and acute attacks during previous pregnancies. Brodie *et al.* (1977) observed no difference in the attack rate during pregnancy for those women ascertained as AIP before pregnancy as opposed to those women diagnosed while pregnant. Nonetheless, an awareness of the presence of AIP can have mitigating effects on the frequency of clinical exacerbation. A major responsibility in this regard belongs to the physician and concerns the use of medication during gestation and delivery. The types of medication with greatest potential for problems are the barbiturates, which may exacerbate porphyria but are used in many settings during pregnancy (Hunter, 1971; Shapiro *et al.*, 1969; Kerr, 1973; Huber, 1967). Hunter's case (1971) has pertinence since his patient's second pregnancy was observed with the knowledge that a barbiturate-associated exacerbation of AIP occurred in the first pregnancy. During the second pregnancy, porphyrin excretion increased from the 16th week but the only symptom was morning sickness. Contraindicated medications were avoided. A clinically normal female child was delivered spontaneously at term. Urinary excretions of ALA, PBG, uroporphyrin, and coproporphyrin were all increased at term. Brodie *et al.* (1977) demonstrated that pregnancy exacerbated AIP in 54% of his female patients, but only 24% of the pregnancies were affected by significant symptoms. There was only one maternal death in 1954. Mustajoki and Koskelo (1976) reported even better results with 100 full term pregnancies without clinical symptoms in 46 porphyric women; significant symptoms occurred in three women during pregnancy and in another three women in the puerperium.

b. Fetal Outcome. Zimmerman *et al.* (1966) noted that there was a "high incidence" of prematurity and spontaneous abortions coincident with acute exacerbations of porphyria. However, "most" premature infants survived the neonatal period. A recent British study (Brodie *et al.*, 1977) clarified the observations further. In that investigation, the total fetal wastage from spontaneous abortion and stillbirth in 95 pregnancies of AIP mothers was 9%. If perinatal deaths are included, the fetal loss rate was 13%. The authors cited a 12% spontaneous abortion rate in Great Britain. Nonetheless, the birth weights of infants born to AIP mothers who experienced an acute attack during their gestation were significantly lower than birth weights of infants born to AIP mothers who were asymptomatic for porphyria during gestation (2.7 ± 0.7 kg vs. 3.5 ± 0.6 kg) ($P < .001$). These data indicate that the acute attack process was associated with the fetal effects; however, the significance of elevated porphyrin precursors (ALA and PBG) with respect to these fetal effects is not clear. Asymptomatic AIP patients may have increased urine excretion of ALA and PBG (Tschudy and Lamon, 1980; Lamon *et al.*, 1979a,b) and porphyrin precursor excretion increases in pregnancy (Shapiro *et al.*, 1969; Huber, 1967; Hunter, 1971). No information is available concerning the activity of uroporphyrinogen I synthase in aborted, stillborn, premature, or low birth weight fetuses. The question of how the maternal chemical environment pertinent to AIP might affect the gestation of a fetus with a deficiency of uroporphyrinogen I synthase as compared to a fetus with normal enzyme activity has yet to be answered.

c. Genetic Considerations. Uroporphyrinogen I synthase deficiency is detectable in nonhepatic tissues such as red cells and skin fibroblasts. Cultured amniotic fluid cells should reflect the activity of that enzyme in other fetal tissues. This was confirmed by Sassa *et al.* (1975) who demonstrated uroporphyrinogen I synthase deficiency in the amniotic fluid cells of a fetus from an AIP mother; after a normal term delivery, the deficiency was confirmed using red cells and cultured skin fibroblasts from the newborn.

The presence of the biochemical marker of AIP in the fetus does not imply that AIP will become chemically manifest in later life. Rather, the deficiency creates a risk of disease that is modified by endocrine, metabolic, and other factors. Manifested disease can be of variable severity in the adult years.

2. *Variegate Porphyria*

Similar to AIP and HCP, which will be discussed below, VP is an autosomal dominant disease. This conclusion regarding its mechanism of inheritance is based upon family studies utilizing clinical symptoms and porphyrin analyses. The assignment of a specific site for primary dysfunction of heme biosynthesis is controversial. It was suggested to be at ferrochelatase by Becker *et al.* (1977), and Brenner and Bloomer (1979) recently demonstrated a partial deficiency of protoporphyrinogen oxidase in fibroblasts from patients with variegate prophyria. A deficiency of ferrochelatase had been assigned previously to erythropoietic protoporphyria. The clinical manifestations of the disease may include either an acute attack neurological disease that is indistinguishable from AIP, or a bullous dermatosis on sun exposed areas of the body, or both. The chemical evaluation for suspected VP is described elsewhere (Meyer and Schmid, 1978; Tschudy and Lamon, 1980).

a. Maternal Effects. The experience with the acute attack phenomena of VP in pregnancy is indistinguishable from the comments for AIP. The occurrence of symptomatic illness is not frequent (Mustajoki and Koskelo, 1976).

b. Fetal Outcome. The discussion for fetal outcome in AIP applies for VP.

c. Genetic Considerations. As described above, VP is inherited as an autosomal dominant trait. At this time, a specific enzyme lesion has not been confirmed, and there have been no reports of prenatal diagnosis of VP.

3. *Hereditary Coproporphyria*

HCP, like VP, may manifest either acute attack or cutaneous disease. The chemical features which are unique to HCP are described elsewhere (Meyer and Schmid, 1978; Tschudy and Lamon, 1980). Recent reports indicate a 50% deficiency of the enzyme coproporphyrinogen oxidase and family segregation of this enzyme deficiency in an autosomal dominant manner.

Compared to AIP, HCP is exceedingly rare. Because of the similarities to AIP and VP with respect to acute attack symptomatology and increased PBG excretion during an attack it is reasonable to assume that maternal and fetal effects would also be similar. Also, the cutaneous illness has been precipitated by the use of oral contraceptives. Application of recent enzyme technology to prenatal diagnosis has not been reported to this time.

4. *Treatment of Acute Attack Forms of Porphyria: Implications for Maternal and Fetal Outcome*

Numerous treatments have been proposed for the acute attack forms of porphyria. In any discussion of treatment, prevention must be mentioned first and avoidance of precipitating factors is the most important aspect of prevention. Brodie agrees that pregnancy has some deleterious effects in acute porphyria, yet he notes that the outcomes of porphyric pregnancies are much better than earlier reports suggested (Brodie *et al.*, 1977). Gajdos *et al.* (1969) feel that a good result of pregnancy is most likely when the patient and the physicians are informed and cautious. Thus, avoidance of pregnancy is not necessarily an important aspect of prevention. There is no evidence that therapeutic abortion during an acute attack is beneficial to the mother. Discontinuing medication and providing a high carbohydrate intake are currently the only widely accepted treatments. Recently, hematin (Lamon *et al.,* 1979b) and propranolol (Douer *et al.,* 1978) have been reported to have therapeutic efficacy. The possible effects of these newer treatments on the fetus are unknown.

B. Cutaneous Forms of Porphyria

The fact that porphyria does not necessarily connote severe neurological disease is not often appreciated. The cutaneous forms of porphyria are best characterized by the excretion in urine and/or feces and the accumulations in tissues of porphyrins. The three major porphyrin types are uroporphyrin, coproporphyrin, and protoporphyrin. The specific forms of porphyria and the appropriate samples for measurement are summarized elsewhere (Meyer and Schmid, 1978; Tschudy and Lamon, 1980). The characteristic symptom of these disorders is photosensitivity, especially to light in the near visible ultraviolet region (about 405 nm).

1. Variegate Porphyria and Hereditary Coproporphyria

These two forms of acute attack porphyria also can manifest cutaneous abnormalities in addition to the neurological disease or as an isolated problem. The skin disorder may include a bullous dermatosis with increased skin fragility and hypertrichosis. The chemical features of porphyrin excretion have been observed to increase in severity in association with pregnancy (Gajdos *et al.,* 1969), and flares of the cutaneous disease have been noted in associations with pregnancy (Zimmerman *et al.,* 1966; Handa *et al.,* 1975). Two children of VP mothers appeared to be affected by the disease. In one child, there was elevated fecal porphyrins without cutaneous disease (Gajdos *et al.,* 1969) and in the other a generalized bullous skin disease without photosensitivity was present since birth. No further chemical information was available (Handa *et al.,* 1975).

2. Porphyria Cutanea Tarda

This disease is classified as a hepatic porphyria, but unlike AIP, VP, and HCP it is a pure cutaneous disease. Clinically it resembles those lesions of VP and HCP. The inherited basis of the disease has recently been identified as a partial deficiency of the heme synthetic enzyme, uroporphyrinogen decarboxylase (Kushner *et al.,* 1976; Felsher *et al.,* 1978), and family studies are consistent with autosomal dominant inheritance. The expression of clinical disease is influenced by factors that include iron and estrogens.

Until recently, PCT would be excluded from any discussion involving pregnancy because the typical patient was an older male with siderosis or receiving estrogen treatment for prostate cancer. The use of oral contraceptives in recent years has exposed a group of young fertile women who are predisposed to PCT, presumably on the basis of a partial deficiency of uroporphyrinogen decarboxylase. There is no doubt that the frequency of pregnancy in PCT is much higher than the three reported cases uncovered in a recent literature search (Zimmerman *et al.*, 1966; Brugsch, 1973; Gilchrest *et al.*, 1975). This attests both to the rarity of symptomatic exacerbation and to the benign nature of PCT in pregnancy. This might seem puzzling in view of the increased levels of estrogen and progesterone in pregnancy. It is possible that the altered hormonal state in pregnancy could lead to increased uroporphyrin excretion and worsened symptoms. However, the maternal loss of iron to the fetus could act in a therapeutic manner. This situation has been observed recently (Lamon and Frykholm, 1979c).

Although uroporphyrin might be deposited in developing calcium containing structures, the rapid renal clearance, relatively small increase in uroporphyrin excess and falling iron stores during pregnancy would all favor absence of noticeable porphyrin deposition in the fetus. Erythrodontia was not seen in two offspring of a woman with PCT followed by J. M. L.

3. *Erythropoietic Protoporphyria*

This autosomal dominant disease is striking because of the nature and severity of the photosensitivity experienced by affected patients. Within minutes of sun exposure, patients will complain of itching, burning, and swelling of the sun-exposed areas of skin. This reaction is accompanied by erythema which may persist for several days. Five EPP females followed by Schmidt *et al.* (1974) reported marked decrease or disappearance of the cutaneous symptoms during pregnancy. No information has been reported on the outcome of gestation or on fetal effects. The lack of these comments may imply a generally unremarkable gestation, delivery, and fetal outcome.

4. *Congenital Erythropoietic Porphyria*

Unlike other forms of porphyria, CEP is inherited as an autosomal recessive trait. Severe photosensitivity is manifested as early as the neonatal period. The characteristic chemical abnormality is the excretion of massive amounts of uroporphyrin via the urine and feces. In addition to the severe photosensitivity with bullae, pigment changes in the skin, and facial deformity of the hands and face as a result of repeated photosensitivity and scarring, the exposure to excess uroporphyrin during development results in erythrodontia.

The small number of reports of offspring from CEP mothers reflects the rarity of the disease, shortened life span for unclear reasons, and the difficulties for affected people in finding mates because of the severe cutaneous disease (Townes, 1965; Tschudy and Lamon, 1980). Offspring of individuals with CEP would almost invariably be obligate heterozygotes, and therefore would not be expected to manifest the parents' disease. Townes (1965) reported a female with

CEP who experienced a normal uncomplicated pregnancy and delivery. The child was not affected at birth and experienced normal growth and development without major illnesses. At 13 months the four deciduous teeth that thad erupted were a deeply pigmented red-purple color and fluoresced in Wood's light (long wavelength ultraviolet) exposure. No information on eventual characteristics of the permanent dentition was reported. No other gestations of a CEP mother have been studied (Kench *et al.*, 1953). The first child was clinically normal following a term delivery. Increased levels of red cell, plasma, and meconium porphyrins were present initially in the infant, but by the sixth day the porphyrin abnormalities were absent. This confirms the impression from the previous case that plasma porphyrins can be transferred to the fetus across the placenta. No porphyrin elevation was detected in the breast milk. No description of subsequent development of the infant is provided. Although the mother tolerated her first pregnancy without difficulty, during her second pregnancy she experienced rapid hepatic deterioration and expired despite termination of the pregnancy. The infant died 4 days later, and was not autopsied. A coarse lobular cirrhosis was present in the maternal liver. There is insufficient experience and few well-documented reported to make an authorative statement regarding the likely cause of death in CEP. Besides the potential for liver disease, as noted above, patients often have had a splenectomy for hemolytic anemia and may be at increased risk for infection.

5. *Treatment for Cutaneous Porphyrias*

This subject is discussed at length in recent reviews (Meyer and Schmid, 1978; Tschudy and Lamon, 1980). In general, during pregnancy cutaneous porphyria is of far less concern than acute attack porphyria. The possible fetal effects from long term oral β-carotene treatment in EPP are not known.

XIX. WILSON'S DISEASE

Wilson's disease or hepatolenticular degeneration is an autosomal recessive disease affecting copper metabolism. The clinical manifestations may involve at least five organ systems. In addition to the well-known triad of cirrhosis, neurological deterioration and Kayser-Fleischer corneal rings, glomerular and tubular kidney function may be affected and hemolytic anema is often present. The laboratory feature common to all affected patients is a diminished level of serum ceruloplasmin, a copper-binding protein synthesized in the liver. Variability of the clinical disease has been observed with respect to the age of onset and the organ systems that are affected. In general, hepatic involvement is evident with early symptom onset, and neurologic disease is a prominent clinical feature with late onset. Unexpected psychiatric and hemolytic episodes are being associated with Wilson's disease with greater frequency in recent years, probably reflecting increased awareness of this disorder. Sass-Kortsak and Bearn (1978) recently prepared a thorough discussion of copper metabolism and Wilson's disease.

There would have been little to discuss in a chapter concerning pregnancy in Wilson's disease prior to the introduction of penicillamine therapy (Walshe, 1956) because of the poor reproductive fitness of untreated females. Their fertility problems may include primary or secondary amenorrhea, oligomenorrhea, and neuropsychiatric symptoms that can impair normal social behavior. Presumably the menstrual irregularities are the result of or may be exacerbated by superimposed chronic liver disease. Prior to effective therapy, pregnancy was achieved rarely and then it was typically in females who remained asymptomatic until their later adult years (Scheinberg and Sternlieb, 1975; Toaff *et al.*, 1977). It has been suggested that excess tissue copper in the uterus may prevent implantation, analogous to the mode of action of copper-containing intrauterine devices (Scheinberg and Sternliev, 1975; Walshe, 1977). At present, with early therapeutic intervention directed at depletion of tissue copper, more normal reproductive capacity is anticipated.

A. Maternal Effects

As noted above, pregnancy in females with Wilson's disease will be in large part restricted to asymptomatic individuals, either because of delayed manifestations of the disease or because of the institution of effective treatment and subsequent improvement. Since tissue accumulation of copper is a continuous process in this disorder, penicillamine must be administered throughout the patients life. The effect of pregnancy on the woman with Wilson's disease must be considered in the context of the possibility of using a material with a narrow therapeutic index for many years before conception and throughout gestation. The literature can be confusing regarding use of penicillamine in the fertile female. Many reports describe temporary discontinuation of the drug for part or all of some pregnancies (Deiss *et al.*, 1970; Walshe, 1977; Scheinberg and Sternlieb, 1975; Laver and Fairley, 1971; Arbisser *et al.*, 1976). The impetus for discontinuing penicillamine during pregnancy was the potential teratogenic effects of a chelating agent (Mjolnerod *et al.*, 1971; Marsh and Fraser, 1973), which will be discussed in more detail below.

Some reports claim that Wilson's disease actually improved during pregnancy (Dreifuss and McKinney, 1966; Albukerk, 1973). This improvement was characterized by a diminution in Kayser-Fleischer rings, diminished need for chelation therapy (Dreifuss and McKinney, 1966), and regression of tremor that returned a few weeks after delivery (Albukerk, 1976). Explanations for these observations have included mobilization of maternal copper and uptake by the fetus, and the increased serum ceruloplasmin concentration observed in pregnancy. Scheinberg and Sternlieb (1975) cautioned that the fetal uptake of copper during pregnancy was only about 12 mg (about one-tenth the chelating ability of penicillamine during gestation), and that clinical improvement in Wilson's disease is not correlated with serum ceruloplasmin concentration.

Continuation of penicillamine chelation during pregnancy has never been associated with maternal deterioration (Arbisser *et al.*, 1976; Scheinberg and Sternlieb, 1975; Walshe, 1977). However, some women who discontinued penicillamine prior to conception and/or during gestation have experienced recurrence of clinical manifestations prior to delivery (Walshe, 1977; Marecek and Graf, 1976; Deiss *et al.*, 1970). A brief withdrawal of chelation therapy, between the sixth and twelfth weeks of gestation, was not associated with maternal difficulties (Walshe, 1977). Because of concern for the fetus, Marecek and Graf (1976) recommended continuing penicillamine therapy during pregnancy, but at a reduced dose (250–500 mg/day). Scheinberg and Sternlieb (1975) recommended a reduction in dose to about 250 mg per day approximately 6 weeks before delivery if a cesarean section is planned. Their rationale is that this dose would be less likely to lead to poor wound healing that might be caused by the treatment (as a result of interference with collagen cross-linking), while at the same time maintaining cupriuresis. Walshe provides pyridoxine (vitamin B6) to combat the antipyridoxine effects of D-penicillamine (Walshe, 1977), and because of the possible interference of D-penicillamine with collagen cross-linking withholds the drug between the sixth and twelfth weeks of gestation (Walshe, 1977; Komrower, 1979). Although pregnancy in patients with Wilson's disease who have improved with D-penicillamine treatment has been successful, several investigators feel that pregnancy is not advisable for women with hepatic insufficiency or evidence of portal hypertension (Scheinberg and Sternlieb, 1975; Marecek and Graf, 1976; Walshe, 1977).

B. Fetal Outcome

The question of the effect on the fetus of untreated maternal Wilson's disease has not been addressed in the literature. This may be considered moot because of the fertility problems of symptomatic women but clearly gestations do occur and early or even presymptomatic chelation therapy is now increasingly common. In 1971, Mjolnerod *et al.* reported the birth of a child with a generalized connective tissue defect who was exposed during its entire gestation to maternal D-penicillamine at a dosage of 2 g per day. The mother was being treated for cystinuria; however, the possible relevance to Wilson's is obvious. We are aware of another similar fatal experience in a fetus at a cystinuric mother receiving penicillamine during gestation (K. Rosenbaum, unpublished observation, 1980). (Pregnancy in cystinuria is reviewed elsewhere in this volume.)

Prior to 1979, connective tissue defects were not reported in offspring of mothers with Wilson's disease who received D-penicillamine treatment throughout their pregnancies at dosages comparable to and sometimes higher than that in the case of Mjolnerod *et al.* (Walshe, 1977; Scheinberg and Sternlieb, 1975;

Laver and Fairley, 1971; Arbisser *et al.*, 1976). D-Penicillamine has also been discontinued in anticipation of becoming pregnant (Deiss *et al.*, 1970; Marecek and Graf, 1976) and during pregnancy (Fukuda *et al.*, 1977; Marecek and Graf, 1976), and children of these pregnancies were all normal. Prolonged withdrawal of D-penicillamine resulted in symptomatic exacerbations in two mothers (Deiss *et al.*, 1970; Marecek and Graf, 1976), but the fetuses exposed to these circumstances were normal at birth. Both were delivered by cesarean section, and one was delivered early (32 weeks).

More recently, an infant with a generalized connective tissue disorder characterized by intrauterine growth retardation, flat face, broad nasal bridge, lax and wrinkled skin, and bilateral inguinal hernias was reported after D-penicillamine had been administered to the mother (900 mg daily) throughout gestation (Solomon *et al.*, 1977). This mother, similar to that reported by Mjolnerod *et al.*, was not affected by Wilson's disease. She suffered from severe rheumatoid arthritis. Solomon *et al.* (1977) noted that their case, and the first report of infant abnormalities by Mjolnerod *et al.* (1971) were from non-Wilson's disease mothers. To explain this possible dichotomy, they proposed that penicillamine, after chelating the unbound copper in Wilson's disease, was excreted in the urine, preventing free D-penicillamine accumulation to the levels possible in cystinuria and rheumatoid arthritis. However, this hypothesis is not consistent with an earlier report by Wenzl and Kludas (1970) of five cystinuria cases treated with D-penicillamine during pregnancy without untoward effects on the fetus. In addition, the report of reversible generalized cutis laxa and a right inguinal hernia in the child of a woman with Wilson's disease treated throughout gestation with 1.5 g per day of penicillamine (Linares *et al.*, 1979) now makes it apparent that maternal Wilson's disease does not preclude fetal toxicity from penicillamine. Whether or not there is, on average, a lower fetal risk from penicillamine in Wilson's disease than in other entities not characterized by increased tissue copper remains to be determined. No data are yet available on the effects, if any, of copper chelators other than penicillamine on human gestation.

C. Genetic Considerations

Since the individual with Wilson's disease is the mother in this discussion, all of her offspring would be heterozygous for the recessive allele; that is, they would be carriers. An exception to this prediction for all the children would occur if in addition the children were fathered by a carrier for Wilson's disease. Such a mating of two individuals who were, respectively, homozygous and heterozygous for the same recessive trait would yield a 50% probability for each child to be affected (homozygous) and a 50% probability for each child to be a carrier (heterozygous). The likelihood for this circumstance would be enhanced in a consanguinous union, but is otherwise remote. Genetic counseling, must, nonetheless, include a careful search for the possibility of Wilson's disease in the father's family.

Parents of an affected person are obligate heterozygotes for the allele of Wilson's disease. As noted above, the patient's siblings are at risk of being affected. Because the clinical disease is preventable and often reversible with treatment with D-penicillamine, the physician must ensure that appropriate family evaluations are undertaken.

XX. FAMILIAL MEDITERRANEAN FEVER

This rare recessively inherited disease is most common in races living in the Mediterranean area, especially Arabs, Armenians, and Sephardic Jews. The onset of symptoms may begin in childhood or in the early adult years. There are recurrent episodes, sometimes predictable, of acute painful sterile peritoneal inflammation. The signs of an acute abdomen with tenderness, diminished bowel sounds, and fever may suggest problems such as acute appendicitis, cholecystitis, neoplasia, tuberculosis, collagen disease, porphyria, or malaria. The white count is usually normal. The correct diagnosis often requires the elimination of these other possibilities, a positive family history and/or racial background, and a therapeutic trial of colchicine, which may help to confirm the diagnosis (Reimann, 1975). Attacks may be precipitated by activity, stress or cold exposure, and periods of remission have been noted that may last greater than 11 years. Other clinical features of the disease can be attributed to a secondary amyloidosis. This feature of the disease may be unique to certain ethnic groups, such as Sephardic Jews. There is a virtual absence of amyloidosis in affected people of Armenian ancestory (Schwabe and Peters, 1974).

Females with familial Mediterranean fever (FMF) frequently experience exacerbations in association with their menstrual periods. Approximately 70% of women with FMF from an Israeli study experienced menses-associated exacerbations (Ehrenfeld and Polishuk, 1970), while only about 10% of Armenian women with FMF were so affected (Schwabe and Peters, 1974).

The disease may impair the fertility of both sexes. This was not clearly the experience of Ehrenfeld and Polishuk (1970), who observed only one infertile female among seven they studied. Five of those women had a total of 15 pregnancies, and four of these 15 terminated in spontaneous abortion. However, a larger study of female fertility by Ismajovich *et al.* (1973) demonstrated that infertility may be more prominent. Fifteen of the 45 women (33%) were married and had failed to conceive. Extensive evaluation of those women showed adnexal adhesions in only two and evidence for anovulation in 12. Clomiphene citrate treatment assisted three of these 12 women to conceive. Two of the three pregnancies went to term and the children were normal, and one pregnancy spontaneously aborted in the first trimester. Human menopausal gonadotropin treatment in four of the infertile women led to pregnancy in one woman, but this pregnancy aborted in the first trimester. Male infertility has been documented in one case, and it was related to febrile episodes. Semen volume remained relatively constant, but the sperm count diminished to less than 20,000,000 per ml approximately 15 days after a febrile episode. Thirty-five days later the sperm count returned to the previous level of 60,000,000–80,000,000 per ml (French *et al.*, 1973).

Maternal Effects and Fetal Outcome

The reports from the literature are variable, but in general the recurrent attacks seem to be less frequent and often absent during gestation and lactation (Ehrenfeld and Polishuk, 1970; Graff and Sharf, 1971; Schwabe and Peters, 1974; Reimann, 1975). No relationship has been established between maternal morbidity or mortality and pre-existing FMF. Eclampsia developed in a young woman with documented amyloidosis during her second pregnancy. A 2-kg child was delivered by cesarean section at 32 weeks gestation. The placenta weighed only 300 g and there was no histologic evidence of amyloid in that organ (Graff and Sharf, 1971). Sohar *et al.* (1967) reported three pregnancies with normal children born at term. There was no amyloid in those placentas. There are no reports of colchicine administration during pregnancy, perhaps because of the salutory effect of pregnancy on the maternal disease.

XXI. HEREDITARY ANGIOEDEMA

Hereditary angioedema (HAE) is an autosomal dominant disease that affects the complement, kinin, and fibrinolytic systems. The biochemical defect has been identified as a deficiency of the serum protein inhibitor of the first component of complement, the C1 esterase inhibitor (Donaldson and Evans, 1963). The clinical illness is characterized by intermittent episodes of edema that may affect the limbs, face, or airway, and may cause abdominal pain. Often antecedent minor trauma or dental manipulation is identified as the precipitating factor. The biochemical and clinical complexities of the disorder will not be discussed further, but this has been reviewed in detail (Frank *et al.*, 1976).

Endogenous hormones play an as yet undefined role in the precipitation of attacks. Gelfand noted attacks were more prevalent during menstrual periods in 5 of 12 female patients (Frank *et al.*, 1976). Androgen administration has been associated with prompt and prolonged remission (Spaulding, 1960). Recently, a weak synthetic androgen called danazol (Danocrine), which has minimal virilizing effects, has been used to attempt to prevent attacks. The results for both males and females have been impressive (Gelfand *et al.*, 1976). Danazol suppressed ovulation in females, and increased serum levels of C1 esterase inhibitor were observed in both men and women with the disorder. This response was confined to those patients whose serum C1 esterase activity was low as a result of deficient levels of the esterase protein. A phenocopy with normal levels of an abnormal protein is not responsive to danazol. This latter variant is a much less frequent cause of HAE.

In view of the effects of endogenous hormonal changes in females, it is not surprising that pregnancy would have a definite effect on the clinical activity of HAE. Gelfand and associates (1976) noted that among 25 pregnancies in 10 women there were many fewer or no attacks during the last two trimesters. There were no episodes of angioedema at delivery, despite the significant tissue trauma associated with childbirth (Frank *et al.*, 1976). Although pregnancy may allow a symptomatic hiatus, frequent attacks that occur in association with normal menstrual cycles may present a difficult therapeutic problem. The interruption of ovulation with oral contraceptives does not assure that recurrent menstrual attacks will subside, since symptomatic illness has occurred during oral contraceptive (estrogen and progestin) use. Hysterectomy and oophorectomy were performed in one patient who subsequently remained in remission, but this is insufficient evidence to presume that hysterectomy and oophorectomy has therapeutic efficacy. As already noted, danazol (Danocrine) is a promising agent, which has minimal maternal virilizing effects and has been reasonably well tolerated in females (Gelfand *et al.*, 1976).

The fetal outcome of pregnancies in women with HAE would be expected to be normal. Approximately one-half of the offspring should be affected by the disorder, although the clinical disease is rarely expressed before puberty. However, the theoretical possibility that female fetuses could become virilized as a result of maternal danazol therapy must be considered.

XXII. MUSCULAR DYSTROPHY

Muscular dystrophy is a diagnostic term that encompasses several distinct inherited diseases of progressive muscle degeneration. No specific metabolic defects have as yet been associated with these diseases. At the present time they are categorized on the basis of affected muscle groups, age of onset, mode of inheritance, and clinical course. Until specific defects are determined which can be used for classifying the muscular dystrophies, a useful approach is via the mode of inheritance (Table III).

As with any severely debilitating disease, the likelihood that affected individuals will have children is diminished by social as well as organic factors. However, X-linked inheritance does allow for a group of heterozygous females who are clinically normal and can transmit the gene to one-half of their offspring. The male recipients of the trait would be expected to have complete expression of the

TABLE III
Genetic Classification of the Muscular Dystrophies

X-linked
Duchenne
Becker's
Autosomal dominant
Myotonic
Facioscapulohumeral
Distal
Oculopharyngeal
Ocular
Autosomal recessive
Limb girdle (also autosomal dominant and sporadic)

disease. Duchenne muscular dystrophy is inherited as an X-linked recessive trait, and it is the most common of the muscular dystrophies. Its frequency is estimated at 15–30 per million live births (Gardner-Medwin, 1970).

Diseases transmitted as autosomal dominant traits are well known for their variable phenotypic expression. Myotonic dystrophy is an interesting example of an autosomal dominant dystrophy because of its nonmuscle findings and because of increasing experience with pregnancy in affected women. Clinical and genetic features of these and the other muscular dystrophies have been recently reviewed (Appel and Rose, 1978). Only Duchenne and myotonic dystrophies will be discussed here.

A. Duchenne Muscular Dystrophy

Pregnancy only concerns the female carrier for Duchenne muscular dystrophy. The phenotype of pseudohypertrophic proximal muscle weakness that begins as early as age 3 years and progresses rapidly to create severe motor disability and early sudden death, often from cardiac dysrhythmia, is essentially restricted to males. However, the ostensibly clinically normal yet heterozygous female may demonstrate a more subtle expression of the phenotype. Although primary medical complications resulting from the interaction of pregnancy with the heterozygous genotype must be considered, there is no basis in published experiences to indicate that pregnancy is deleterious to these females with respect to expression of muscle disease. Regardless of the genotype of the fetus, a full term pregnancy with normal *in utero* growth and development is the rule.

The female at risk of carrying but not confirmed as a carrier of the X-linked Duchenne trait may also present some difficult genetic questions to her physician, often after she has become pregnant. No absolutely reliable method for carrier detection is available.

Carrier ascertainment utilizes quantitative genetics for risk assessment and attempts to identify biological abnormalities associated with the genotype. The pedigree of any X-linked recessive trait, which is completely expressed and is lethal, should show only affected males with no male to male transmission. Bayes theorem affords a quantitative approach to assessing a probability (relative risk) that a female consultant from a family with a confirmed case of Duchenne dystrophy is a carrier for the Duchenne gene. This approach is described in detail by Murphy and Chase (1975). Recently, a "relative probability method" has been discussed and compared with the Bayesian method (Bundey, 1978).

In part to assess the utility of possible markers for the carrier state, mothers who are obligate carriers have been evaluated by a variety of methods, seeking changes resembling the more profound abnormalities in affected males. These have included muscle testing (Appel and Rose, 1978; Morgan, 1974), electromyography with nerve conduction and related studies (Morgan, 1974; Oepkes and van Weerden, 1979), muscle biopsy (Morgan, 1974), serum creatine phosphokinase (CPK) (Morgan, 1974; King *et al.*, 1972; Appel and Rose, 1978), steroid-stimulated serum creatine phosphokinase (Sen and Das, 1978), serum myoglobin (Adornato *et al.*, 1978a and b), and leukocyte capping abnormalities (Pickard *et al.*, 1979).

To this time, only serum CPK has been extensively evaluated in the pregnant Duchenne carrier. Serum CPK varies over the course of a normal woman's life. CPK is highest in premenarcheal teenagers and diminishes successively in postmenarcheal teenagers, mature women, and pregnant women. It rises after menopause. Approximately 70% of healthy female carriers of Duchenne dystrophy have an elevated CPK (Morgan, 1974; King *et al.*, 1972). However, when women at risk for being carriers seek counseling during pregnancy, the utility of the serum CPK determination may be reduced (Blyth and Hughes, 1971; Emery and Pasccasio, 1965). In normal women, serum CPK declines during the first 20 weeks of pregnancy, and during the remainder of gestation and postpartum is not significantly different from values in nonpregnant women (King *et al.*, 1972). The mean CPK of pregnant women (16 week or less gestation) is less than one-half the level of teenagers (Bundey, 1978). Despite these findings, which diminish the efficacy of CPK determination for carrier detection during pregnancy, CPK measurements are apparently not affected by endogenous (menstrual cycle) or exogenous (oral contraceptive) hormone fluctuations (Simpson *et al.*, 1974).

In women definitely considered to be carriers or at very high risk of being carriers, the geneticist can assist in providing more information regarding the genotype of the fetus. A fetal karyotype can provide important, albeit not diagnostic, information. A female fetus, detected by chromosome studies, would have negligible risk of being affected, while a male fetus would have an approximately 50% risk of being affected with Duchenne dystrophy. Termination of pregnancy for male fetuses is a possible option. Laboratory studies such as measurement of amniotic fluid or fetal blood CPK (Mahoney *et al.*, 1977; Thompson, 1978) and myoglobin (Hansen *et al.*, 1978) might permit theoretically the identification of affected male fetuses. Preliminary data on plasma CPK determination in fetal blood removed from the placenta at 18 and 20 weeks gestation in two at-risk fetuses suggested efficacy (Mahoney *et al.*, 1977). However, reports of false positive amniotic fluid CPK (Thompson, 1978) and false negative fetal serum CPK determinations (Emery and Pasccasio, 1965) have appeared. Fetal blood sampling for CPK measurement has subsequently not proven to be reliable in prenatal diagnosis of Duchenne dystrophy (J. Hobbins and M. J. Mahoney, unpublished).

B. Myotonic Dystrophy

The autosomal dominant forms of muscular dystrophy present a challenge for maternal care as well as for counseling concerning the genotype of the fetus. Unlike X-linked inheritance, both males and females can be affected. Although sterility is frequently commented upon in males, successful pregnancies have been reported for females with myotonic dystrophy. For affected patients there is a 50% probability that each offspring will be affected. Although each dominant type of musuclar dystrophy (Table III) shares similar genetic considerations, each type is unique and is characterized by its specific clinical features. The experience with these disorders to date warrants only a detailed discussion of myotonic dystrophy.

The nonmyopathic features of myotonic dystrophy include mental retardation, frontal balding, cataracts (independent of diabetes), diabetes mellitus, gonadal atrophy, dysphagia, abdominal pain, and cranial dysostosis. The myopathy affects skeletal, cardiac muscle and smooth muscle (Kohn *et al.*, 1964). The classical finding is myotonia, or the sustained contraction of a muscle or group of muscles. This is illustrated as "percussion myotonia," especially of the thenar eminence, and "action myotonia," as witnessed after a handshake. These phenomena are exacerbated by cold and diminished by repetition (Appel and Rose, 1978). Although it is an autosomal dominant trait, the sex ratio for affected individuals favors males by five to one.

1. Maternal Effects

The problem of infertility is well-recognized in myotonic dystrophy. Severely affected women rarely become pregnant. The testicular atrophy with decreased libido and impotence seen in males by age 25–30 can be paralleled in females, and it is signaled by amenorrhea (Sciarra and Steer, 1961). Conception is possible, but typically in younger females (Gardy, 1963).

The women who become pregnant experience three types of problems. First, a worsening of their disease (Davis, 1958; Hopkins and Wray, 1967; Sciarra and Steer, 1961; Hilliard *et al.*, 1977; Gardy, 1963); second, an increased frequency of spontaneous abortions and premature labor (Hilliard *et al.*, 1977; Shore and MacLachlan, 1971; Watters and Williams, 1967); and third, problems during labor (Shore and MacLachlan, 1971; Webb *et al.*, 1978; Davis, 1958; Gardy, 1963; Hopkins and Wray, 1967; Sciarra and Steer, 1961; Hilliard *et all*, 1977. The accelerated course of the disease associated with pregnancy does not occur regularly, and it may not occur with every pregnancy in any one patient. The most frequently encountered feature of worsening disease is weakness with difficulty walking (Davis, 1958; Gardy, 1963). Despite this problem, it is recommended that moderate activity during pregnancy should be encouraged (Shore and MacLachlan, 1971). Premature labor occurred in 5 of 23 pregnancies reported by Webb *et al* (1978). Polyhydramnios has been observed in 6 of 23 pregnancies (Webb *et al.*, 1978), and another report noted polyhydramnios in 18 of 35 pregnancies (Hilliard *et al.*, 1977). Both Hilliard and Webb observed that polyhydramnios indicated that the fetus was probably affected with myotonic dystrophy. In fact, approximately 46% (25 of 54) of offspring with myotonia had gestations complicated by polyhydramnios (Hilliard *et al.*, 1977). As mentioned with respect to fetal outcome, children with myotonic dystrophy born to mothers with the disease are generally more ill at birth than nondystrophic siblings, This is manifested by nonspecific findings of a ''floppy'' baby with poor feeding and respiratory distress.

All stages of labor have been noted to be abnormal in women with myotonic dystrophy. The first stage is often prolonged; there may be poor voluntary assistance during the second stage; and uterine hemorrhage associated with poor contractions may occur in the third. The uterine contractions are myotonic in that they are short lived and have a prolonged interval from the peak contractions to relaxation (Hopkins and Wray, 1967; Sciarra and Steer, 1961). Oxytocin administration may improve the character of the contractions by shortening the relaxation interval in some women (Hilliard *et al.*, 1977; Sciarra and Steer, 1961). However, Hopkins and Wray (1967) observed only short-lived contractions after 24 hours of oxytocin. The addition of 200 mg of quinine sulfate along with the oxytocin resulted in strong labor (Hopkins and Wray, 1967; Webb *et al.*, 1978). Further, the postpartum complications of uterine atonia and hemorrhage may be reduced by administration of oral quinine sulfate (Hopkins and Wray, 1967; Webb *et al.*, 1978). The hazard of sedation and anesthesia was illustrated by the experience of Gardy (1963), using mild sedation with meperidine and promethazine. Following the third stage of labor, his patient became flaccid, areflexic, nonresponsive to pain, lost her pupillary responses, had bilateral Babinski reflexes, and nuchal rigidity. Her respirations remained 8–10 per minute. After 4 hours she was again alert and fully responsive. The risks of anesthesia increase in proportion to the degree of muscle wasting. Loss of accessory muscles of respiration can create difficulty with recovering from anesthesia. Hypoventilation may occur (Gardy, 1963; Kaufman, 1960).

2. *Fetal Outcome*

The rate of fetal problems is high. Webb *et al.* (1978) noted that 5 of 23 births were breech presentation; there was one stillbirth, two hydrocephalic fetuses, and six neonatal deaths. Hilliard *et al.* (1977) reported fetal or neonatal death in 9 of 35 pregnancies. The neonatal morbidity of affected children, characterized by respiratory distress and feeding problems, may correlate with the severity of the maternal disease. For reasons that are not clear, affected children are more likely to have these problems when born to affected mothers than to affected fathers. As mentioned, polyhydramnios suggests that the fetus has inherited the gene for myotonic dystrophy.

3. *Prenatal Diagnosis*

It is not yet possible to determine if a fetus is affected with myotonic dystrophy. The recognition that the gene for myotonic dystrophy is closely linked to the Lutheran blood group and ABH-secretion loci may permit some information pertinent to prenatal diagnosis (Harper *et al.*, 1972; Harper, 1974). However, the approach is limited by crossing over, and by the percentage of parents at risk with genotypes permitting useful deductions regarding the genotype of the fetus.

XXIII. PERIODIC PARALYSIS

The periodic paralyses are a group of disorders characterized by episodic attacks of weakness with paralysis of skeletal musculature. Attacks may cause death from respiratory failure and less often from cardiac or other causes.

Periodic paralysis may be sporadic or associated with hyperthyroidism, but three hereditary forms are known: hypokalemic, hyperkalemic, and normokalemic. These are classified with regard to the serum potassium levels during attacks. The fundamental molecular lesions in all forms are unknown.

The inherited periodic paralyses all have an autosomal dominant pattern. In the hypokalemic form, there is evidence of some reduced expression in females; there are fewer females than males affected in many families, and the disease is generally milder in affected females (Pearson and Kalyanaraman, 1972). Such sex differences are not evident in the other two variants.

The diseases appear not to be expressed at birth. The periodic paralyses are clinically manifest prior to the reproductive years in most cases, and gestations have occurred in affected females. We are unaware of any reports of difficulties in fetuses of patients, nor of postnatal episodes in genetically affected or unaffected newborns.

There is limited information on the effects of gestation on expression of periodic paralysis. Talbott (1941) cites very old reports by Janota and Weber (1928) and Cousot (1896) suggesting that gestation may exert a favorable influence on the frequency of attacks. Talbott regarded periodic paralysis as always associated with hypokalemia, but it is not clear if the reports definitely apply to this variant. This same review also cites a postpartum attack occurring 2 days after a ''normal'' childbirth.

In contrast, in hyperkalemic periodic paralysis Egan and Klein (1959) describe one female in each of two unrelated families who reported increasing severity of attacks during the last two trimesters of gestation. The attacks continued with decreasing severity postpartum, but did not suddenly abate after birth. In another familial case of periodic paralysis, attacks began after infancy, ceased at age 16–17 years and reappeared during the fourth month of the first gestation with subsequent intermittent episodes. This individual had no alteration in serum potassium during attacks and oral potassium administration had not modified disease expression. She had 3 affected children and a possibly affected father (Ecker and Carson, 1953). The metabolic relationship of this patient to the family described by Poskanzer and Kerr (1961), who showed normokalemic paralysis made worse by potassium administration is unclear; no mention is made of the characteristics of gestation in this large kindred.

While it is tempting to suggest that gestation may worsen hyperkalemic and ameliorate hypokalemic periodic paralysis, any conclusion in this direction must be regarded as extremely tentative and further data derived from study of a larger number of unselected patients is required.

Since rest after exertion, cold, and stress have been implicated in the etiology of periodic attacks in all forms of the disease it might be expected that attacks would tend to occur postpartum, but no such consistent association has been noted in the literature.

However, attacks associated with anesthesia, including the anesthesia used at childbirth, have been reported. Egan and Klein (1959) noted that each of their three hyperkalemic families had members with anesthesia-induced paralysis lasting 2–5 hours. In one instance an unidentified general anesthetic had been used at childbirth; in a dental case the offending agent was pentothal. One patient delivered a child by spinal anesthesia without experiencing an attack.

In the review of Miller and Katz (1973) there were no reports of anesthesia administration to patients with hypokalemic paralysis. The subsequent experiences of several authors (Siler and Discavage, 1975; Horton, 1977; Bashford, 1977) make it clear that anesthesia, perhaps especially when accompanied by the use of muscle relaxants, may induce mild or severe attacks in the hypokalamic variant.

If complications of pregnancy requiring surgery arise in patients with periodic paralysis it would seem desirable to use regional rather than general anesthesia whenever possible. Miller and Katz (1973) and Horton (1977) review the management of patients and prevention of attacks if general anesthesia is required; careful regulation of serum potassium and avoidance of chilling may be especially important.

Based on present data there seems to be no reason that pregnancy need modify the medical measures customarily used to prevent or treat attacks in any form of periodic paralysis, including the use of acetazolamide, but more information is required. Approaches to medical therapy have been reviewed by Pearson and Kalyanaramen (1972).

XXIV. MALIGNANT HYPERTHERMIA

This is a dominantly inherited defect whose familial nature was originally established by Denborough and Lovell (1960) and Denborough *et al.* (1962). Administration of succinyl choline and/or general anesthetic agents leads to fulminant and often fatal hyperthermia, frequently associated with and possibly caused by muscular rigidity. We have been unable to locate reports describing hyperthermic attacks during gestation, although Britt *et al.* (1969) mention three instances of hyperthermia associated with ''gynaecological and obstetrical'' operations. Affected females apparently reproduce normally. Individuals with this disease require regional anesthesia if surgery is mandatory.

In affected pedigrees, it is important to determine if a specific individual who has never received anesthesia in the past manifests the disorder. Elevated circulating creatinine phosphokinase levels and muscle wasting can be used to detect some cases; abnormalities of skeletal muscle obtained by biopsy or of platelet ATP levels may also establish the diagnosis (McKusick, 1978), although absolute exclusion is not currently possible.

Since pregnancy complications may rapidly require surgical intervention, an anesthesiologist with extensive experience with regional anesthesia should be available during gestation in affected females. Affected mothers or fathers would be expected to transmit the disorder to approximately 50% of their offspring. There is no recorded experience in which a patient received anesthesia *in utero* (e.g., during surgery on an unaffected mother). If the father has the disease it would seem prudent to avoid fetal exposure to anesthesia, even though there is some evidence that the abnormal response to anesthesia is not manifested during very early childhood in affected subjects (Britt *et al.,* 1969).

Very recently the United States Food and Drug Administration has approved the use of intravenous dantrolene sodium in the treatment of malignant hyperthermia. We await data on use of this drug during pregnancy (Food and Drug Administration, 1979).

XXV. PSEUDOCHOLINESTERASE DEFICIENCY

Serum pseudocholinesterase catalyzes degradation of succinylcholine, which is used as a muscle relaxant in obstetrical procedures. Certain procaine derivatives, such as chloroprocaine and tetracaine, used in paracervical or other regional obstetrical anesthesia appear to be hydrolyzed by the same enzyme. In normal gestations, mean plasma pseudocholinesterase decreases to about 80% of the nonpregnant level at term but this is insufficient reduction to lead to prolonged neuromuscular blockade from succinylcholine (Blitt *et al.*, 1977). Similar mild reductions are present in a variety of disease states (Oropollo, 1978).

Genetic pseudocholinesterase deficiency is an autosomal recessive trait associated with extreme sensitivity to succinylcholine and prolonged apnea after its administration. Several genotypic variants of pseudocholinesterase deficiency are known, associated with reduced inhibition by dibucaine, reduced inhibition by flouride, or complete absence of cholinesterase activity (McKusick, 1978). Aagesen and Ronquist (1977) reported prolonged apnea after cesarian section in a woman who had been given succinylcholine during the operation; she was subsequently demonstrated to have an inherited gross pseudocholinesterase deficiency with altered dibucaine inhibition. The infant, who had a normal serum enzyme activity, cried within seconds after birth and had a normal Apgar score.

XXVI. FRIEDREICH'S ATAXIA

This disorder is a form of spinocerebellar degeneration of unknown etiology. Inheritance is generally autosomal recessive, although autosomal dominant and X-linked recessive inheritance has been suggested in some pedigrees. Onset occurs before the end of puberty in virtually all cases. Clinical and other features are extensively reviewed by Tyrer (1975), who references many older reviews, and in a series of articles in the *Canadian Journal of Neurological Sciences* (Quebec Cooperative Study, 1976). A cardiomyopathy and diabetes mellitus frequently accompany neurological symptoms.

The relatively early onset of the disease and its severe features make it clear that reproductive fitness should be impaired; however, fitness is not "negligible" as suggested by Skre (1975). Pregnancies have been reported, and there are comments (Steinberg, 1978; Kinnier Wilson, 1954) that pregnancy may be associated with a worsening of the disease. We are unaware of any systematic review of the reproductive history of Friedreich's patients or of convincing cases to justify this conclusion. Most reviews do not seem to comment on this feature. Hewer (1968) in his study of 82 fetal cases (most of whom died of the associated myopathy, pneumonia, or diabetes) makes no comments regarding precipitation of a fatal outcome during gestation. A computer search of the literature from 1965 to 1979 revealed only three reports of gestation in Friedreich's ataxia. The patient of Ziegler *et al.* (1974) had disease onset prepubertally, and a seizure disorder since age 9 years. During pregnancy her seizures worsened. A "normal" infant was eventually delivered, and this infant developed seizures and "poor coordination" at age 4½ years. The patient died about 4 months after delivery and neuropathological examination showed dentatorubral atrophy of a type previously associated with some forms of cerebellar disturbance. The authors do not focus on the possible association of worsening with pregnancy, which may or may not have been coincidental. Kolb and Nietmann (1965) report a 21-year-old patient who had an uncomplicated spontaneous term delivery of an apparently normal infant associated with a normalization of her own electrocardiogram during pregnancy. No exacerbation of the disease was noted. Moneta *et al.* (1976) report a pregnancy in a 32-year-old patient with severe neurological impairment and diabetes associated with Friedreich's ataxia. Urinary estriols were normal. A cesarean section was performed at 38 weeks because of diabetes and weakness of abdominal musculature. A vigorous infant of 3690 g was born, but died shortly thereafter because of cardiac anomalies (subaortic stenosis and aortic hypoplasia).

These cases and opinions in review articles seem inadequate to permit satisfactory conclusions on possible interactions of pregnancy and Friedreich's ataxia. Should pregnancy occur, it would seem reasonable that cardiac, neurological, respiratory, and endocrine status should be closely observed. An unaffected fetus should generally be anticipated in the recessive form, in the absence of parental consanquinity.

XXVII. HUNTINGTON'S CHOREA

There is an extensive literature, referenced in all standard works on human genetics, on this progressive neurological disorder which is inherited in an autosomal dominant manner. Disease manifestations may appear before or after reproduction occurs. Genetic counseling in Huntington's chorea families is of great significance because of the approximately 50% risk to offspring of affected individuals. No effort will be made to review here a subject of this magnitude, on which many other sources of information exist. Experienced geneticists should always be involved in the complex counseling required for individuals who have or may be at risk for this disease.

We are unaware of specific pregnancy complications or altered disease expression during gestation in Huntington's chorea. There is evidence, however, that reproductive performance is modified by the mutant gene and its expression in the social framework. This subject is effectively discussed by Reed and Neel (1959) and Wallace (1971, 1976). The fundamental conclusions appear to be that choreic individuals have significantly *more* offspring than their normal sibs, but fewer than the general population. Affected females are responsible for most of the comparatively increased reproductive capacity of affected compared to unaffected sibs. This appears to reflect a decreased likelihood of marriage by affected males and voluntary restriction of family size by normal siblings. Wallace (1971, 1976) convincingly argues that these reproductive phenomena can be understood by a consideration of the profound social consequences of the mutant gene.

REFERENCES

Aagesen, G., and Ronquist, G. (1977). Prolonged succinylcholine-induced paralysis in connection with Caesarian section. *Acta Anaesth. Scand.* **21,** 379.

Adornato, B. T., Engel, W. K., and Fordart, M. (1978a). *Arch. Neurol. (Chicago)* **35,** 577–580.

Adornato, B. T., Kagen, L. J., and Engel, W. K. (1978b). *Lancet* **2,** 499–501.

Albukerk, J. N. (1973). *Fertil. Steril.* **24,** 494–497.

Albukerk, J. N. (1976). *N. Engl. J. Med.* **294,** 670–671.

Aldersberg, D., Feldman, E. B., Kaplan, S., and Wang, C.-I. (1959). *Am. J. Obstet. Gynecol.* **78,** 851–859.

Alepa, F. P., Howell, R. R., Klinenberg, J. R., and Seegmiller, J. E. (1967). Relationships between glycogen storage disease and tophaceous gout. *Am. J. Med.* **42,** 58.

Appel, S. H., and Rose, A. D. (1978): *In* "The Metabolic Basis of Inherited Diseases" (J. B. Stanbury, J. B. Wyngaarden, and D. S. Fredrickson, eds.), pp. 1260–1281. McGraw-Hill, New York.

Arbisser, A. I., Scott, C. I., Jr., and Howell, B. R. (1976). *Lancet* **1,** 312–313.

Armstrong, M. D. (1975). Maternal histidinaemia. *Arch. Dis. Child.* **50,** 830.

Bashford, A. C. (1977). Case report: Anesthesia in familial hypohalemic periodic paralysis. *Anaesth. Intensive Care Med.* **5,** 74.

Batt, R. E., Cirksena, W. J., and Lebherz, T. B. (1963). Gout and salt wasting renal disease in pregnancy. *J. Am. Med. Assoc.* **186,** 835.

Becker, D. M., and Kramer, S. (1977). *Medicine* (*Baltimore*) **56,** 411–423.

Becker, D. M., Vilgoen, J. D., Katz, J., and Kramer, S. (1977). *Br. J. Haematol.* **36,** 171–179.

Berson, E. and Shih, V. E. Unpublished observation.

Beutler, E., Matsumoto, F., Kuhl, W., Krill, A., Levy, N., Sparkes, R., and Degnan, M. (1973). Galactokinase deficiency as a cause of cataracts. *N. Engl. J. Med.* **288,** 1203–1206.

Blitt, C. D., Petty, W. C., Allerternst, E. E., and Wright, B. J. (1977). Correlation of plasma cholinesterase activity and duration of action of succinylcholine during pregnancy. *Anesth. Analg.* (*Cleveland*) **56,** 78.

Blyth, H., and Hughes, G. P. (1971). *Lancet* **1,** 855–856.

Bonkowsky, H. L., Bloomer, J. R., Ebert, P. S., and Mahoney, M. J. (1975). *J. Clin. Invest.* **56,** 1139–1148.

Bottomly, S. S., Tanake, M., and Everett, M. A. (1975). *J. Lab. Clin. Med.* **86,** 126–131.

Bovier-Lapierre, M., Sain-Dizier, C., Freycon, F., David, M., Dorche, C., and Jeune, M. (1974). Two infants born to a mother with phenylketonuria. Failure of a low phenylalanine diet during the second pregnancy. *Pediatrie* **29,** 51–72.

Bradford, M. J., Krakoff, I. M., Leeper, R., and Balis, M. F. (1968). Study of purine metabolism in a xanthinuric female. *J. Clin. Invest.* **47,** 1325.

Brady, R. O. (1978). Glucosylceramide lipidosis: Gaucher's disease. *In* "The Metabolic Basis of Inherited Disease" (J. Stanbury, J. Wyngaarden, and D. S. Fredrickson, eds.), 4th ed., Chapter 33, p. 688. McGraw-Hill, New York.

Bremmer, W. F., and Third, J. L. H. C. (1978). *Am. J. Med.* **64,** 912–913.

Brenner, D. A., and Bloomer, J. R. (1979). *Clin. Res.* **27,** 274A.

Brenton, D. P. Personal communication.

Brenton, D. P., Cusworth, D. C., Biddle, S. D., Garrod, P. J., and Lasley, L. (1977). Pregnancy and homocystinuria. *Ann. Clin. Biochem.* **14,** 161–162.

Britt, B. A., Locher, W. G., and Kalow, W. (1969). Hereditary aspects of malignant hyperthermia. *Can. Anaesth. Soc. J.* **16,** 89.

Brodie, M. J., Moore, M. R., Thompson, G. G., Goldberg, A., and Low, R. A. (1977). *J. Obstet. Gynaecol. Br. Commonw.* **84,** 726–731.

Bromberg, Y. M., Toaff, R., and Diengott, D. (1953). Pregnancy and Gaucher's disease. *Br. Med. J.* **2,** 761.

Brown, M. S., Goldstein, J. L., Van den Berghe, K., Fryns, J. P., Kovanen, P. T., Eckels, R., Van den Berghe, H., and Cassisman, J. J. (1978). *Lancet* **1,** 526–529.

Bruckman, C., Berry, H. K., and Dasenbrock, R. J. (1970). Histidinemia in two successive generations. *Am. J. Dis. Child.* **119,** 221–227.

Brugsch, J. (1973). *Muench. Med. Wochenschr.* **115,** 1547–1548.

Buist, N. R. M., Lis, E. W., Tuerck, J. M., and Murphey, W. H. (1979). Maternal phenylketonuria. *Lancet* **2,** 589.

Bulfield, G., and Kacser, H. (1974). Histidinaemia in mouse and man. *Arch. Dis. Child.* **49,** 545–552.

Bundey, S. (1978). *J. Med. Genet.* **15,** 249–253.

Bundey, S., Crawley, J. M., Edwards, J. H., and Westhead, R. A. (1979). *J. Med. Genet.* **16,** 117–121.

Bush, R. T., Hoby, A. M., and Dukes, P. L. (1979). Successful planned pregnancy and dietetic management in maternal phenylketonuria. *In* "Proceedings of the Fifth Congress of International Association for the Scientific Study of Mental Deficiency" (P. Mittler, ed.). In press.

Campbell, A. G. G., and Williams, E. R. (1967). Natural history of Refsum's syndrome in a Gloucestershire family. *Br. Med. J.* **3,** 777.

Carson, N. A. J. Personal communication.

Cochrane, P., and Alderman, B. (1973). Normal pregnancy and successful delivery in myophosphorylase deficiency (McArdle's syndrome). *J. Neurol. Neurosurg. Psychiatry* **36,** 225.

Cousot, G. (1896). Cas de paralysie périodique. *Bull. Acad. R. Med. Belg.* **20,** 791.

Curiel, P., and Bandinelli, R. (1979). Pregnancy in a woman with xanthinuria: Study of amniotic fluid uric acid. *Am. J. Obstet. Gynecol.* **134,** 721.

Davis, H. A. (1958). *J. Obstet. Gynaecol. Br. Emp.* **65,** 479–480.

Dawson, D. M., Spong, F. L., and Harrington, J. F. (1968). McArdle's disease: Lack of muscle phosphorylase. *Ann. Intern. Med.* **69,** 229.

de Alvarez, R. R., Gaiser, D. F., Simkins, D. M., Smith, E. K., and Bratvold, G. E. (1959). *Am. J. Obstet. Gynecol.* **77,** 743–759.

Decker, B., and McWhorter, C. A. (1956). Gaucher's disease and pregnancy. *Ann. Intern. Med.* **44,** 1219.

Deiss, A., Lee, G. R., and Cartwright, G. E. (1970). *Ann. Intern. Med.* **73,** 312–313.

Delbarre, F., Braun, S., and Saint-Georges-Chammet, F. (1967). La goutte féminine. Analyse de quarante observation. *Sem. Hop.* **43,** 623.

Denborough, M. A., and Lovell, R. R. H. (1960). Anaesthetic deaths in a family. *Lancet* **2,** 45.

Denborough, M. A., Forster, J. F. A., Lovell, R. R. H., Maplestone, P. A., and Villiers, J. D. (1962). Anesthetic deaths in a family. *Br. J. Anaesth.* **34,** 395.

Denniston, J. D. (1963). Children of mothers with phenylketonuria. *J. Pediatr.* **63,** 461–462.

Dent, C. E. (1957). Discussion of M. D. Armstrong. Relation of biochemical abnormality to development of mental defect in phenylketonuria. *In* "Etiologic Factors in Mental Retardation," pp. 32–33. Ross Laboratories, Columbus, Ohio.

Donaldson, V. H., and Evans, R. R. (1963). *Am. J. Med.* **35,** 37–44.

Douer, D., Weinberger, A., Pinkhas, J., and Atsmon, A. (1978). *J. Am. Med. Assoc.* **204,** 766–768.

Drayer, J. I. M., Cleophas, A. J. M., Trijbels, J. M. F., Smals, A. G. H., and Kloppenborg, P. W. C. (1980). Symptoms, diagnostic pitfalls, and treatment of homocystinuria in seven adult patients. *Neth. J. Med.* **23,** 89–94.

Dreifuss, F. E., and McKinney, W. M. (1966). *J. Am. Med. Assoc.* **195,** 960–962.

Ecker, I., and Carson, M. J. (1953). Familial periodic paralysis. *J. Pediatr.* **42,** 751.

Edström, R., Grontoft, O., and Sandring, H. (1959). *Acta Psychiatr. Scand.* **34,** 40.

Egan, T. J., and Klein, R. (1959). *Pediatrics* Hyperkalemic familial periodic paralysis. **24,** 761.

Ehrenfeld, E. N., and Polishuk, W. Z. (1970). *Isr. J. Med. Sci.* **6,** 9–13.

Elder, G. H., Thomas, N., Evans, J. O., Cox, R., Brodie, M. J., and Goldberg, A. (1976). *Lancet* **2,** 1217–1219.

Emery, A. E. H., and Pasccasio, F. M. (1965). *Am. J. Obstet. Gynecol.* **91,** 18–22.

Engel, A. G. (1970). Acid maltase deficiency in adults. *Brain* **93,** 599.

Engel, W. K., Eyerman, E. L., and Williams, H. E. (1962). Late-onset type of skeletal muscle phosphorylase deficiency. *New Engl. J. Med.* **268,** 135.

Fabian, E., Stork, A., Kucerova, L., and Sponarova, J. (1978). *Am. J. Obstet. Gynecol.* **100,** 904–907.

Fallat, R. W., and Glueck, C. J. (1976). *Atherosclerosis* **23,** 41–62.

Farquhar, J. W. (1979). Conception in a phenylketonuric woman. *Lancet* **1,** 322.

Felsher, B. F., Norris, M. E., and Shih, J. C. (1978). *N. Engl. J. Med.* **299,** 1095–1098.

Fleming, R. (1957). Refsum's syndrome. *Neurology* **7,** 476.

Fölling, A. (1934). Phenylpyruvic acid as a metabolic anomaly in connection with imbecility. *Hoppe-Seyler's Z. Psysiol. Chem.* **227,** 169–176.

Food and Drug Administration (1979). *FDA Drug Bull.* November, p. 27.

Francon, F., Durand, M. *et al.* (1956). Trois nouvelles observation de goutte feminine tophacée. *Rev. Rhum. Mol. Osteo-Articulaires* **23,** 322.

Frank, M. M., Gelfand, J. A., and Atkinson, J. P. (1976). *Ann. Intern. Med.* **84,** 580–593.

Fredrickson, D. L., Goldstein, J. L., and Brown, M. S. (1978). *In* "The Metabolic Basis of Inherited Disease" (J. B. Stanbury, J. B. Wyngaarden, and D. L. Fredrickson, eds.), pp. 604–655. McGraw-Hill, New York.

French, D. J., Leeb, C. S., and Jecht, E. W. (1973). *Fertil. Steril.* **24,** 490–493.

Fried, K. (1973). Population study of chronic Gaucher's disease. *Isr. J. Med. Sci.* **9,** 1396.

Friedman, E. A., and Little, W. A. (1958). Pregnancy and gout: A case report. *Am. J. Obstet. Gynecol.* **76,** 913.

Fryer, D. G., Winckleman, A. C., Ways, P. O., and Swanson, A. G. (1971). Refsum's disease: a clinical and pathological report. *Neurology* **21,** 162.

Fukuda, K., Ishii, A., Matsue, Y., Funaki, K., Hoshiai, H., and Maeda, S. (1977). *Tohoku J. Exp. Med.* **123,** 279–285.

Gajdos, A., Gajdos-Torok, M., Guillard, A., and Nouailhat, F. (1969). *Presse Med.* **77,** 589–590.

Gardner-Medwin, O. (1970). *J. Med. Genet.* **7,** 334–337.

Gardy, H. H. (1963). *Obstet. Gynecol.* **21,** 441–445.

Gelfand, J. A., Sherins, R. J., Alling, D. W., and Frank, M. M. (1976). *N. Engl. J. Med.* **295,** 1444–1448.

Gilchrest, B., Pathak, M. A., and Parrish, J. A. (1975). *Arch. Dermatol.* **3,** 263–264.

Gitzelmann, R. (1967). Hereditary galactokinase deficiency, a newly recognized cause of juvenile cataracts. *Pediatr. Res.* **1,** 14–23.

Glueck, C. J., Fallat, R. W., and Scheel, D. (1975). *Metab., Clin. Exp.* **24,** 537–545.

Goldstein, J. L., Schrott, H. G., Hazzard, W. R., Bierman, E. L., and Motulsky, A. G. (1973). *J. Clin. Invest.* **52,** 1544–1568.

Graff, G., and Sharf, M. (1971). *J. Obstet. Gynaecol. Br. Commonw.* **78,** 284–286.

Greenhut, I. J., Silver, R. A., and Campbell, J. A. (1953). Occurrence of gout in a female. *Radiology* **60,** 257.

Greenwald, J. C., and Fenton, A. N. (1959). Gaucher's disease and pregnancy. *Obstet. Gynecol.* **14,** 79.

Groen, J. (1948). The hereditary mechanism of Gaucher's disease. *Blood* **3,** 1238.

Handa, F., Kumar, K., and Kumar, R. (1975). *Br. J. Dermatol.* **92,** 347–350.

Hansen, H. (1978). Variability of reproductive casualty in maternal phenylalaninemia. *Early Hum. Dev.* **2,** 51–71.

Hansen, K. N., Lademan, A., and Norgaard-Pedersen, B. (1978). *Lancet* **2,** 1250–1251.

Harley, J. D., Mutton, P., Irvine, S., and Gupta, J. D. (1974). Maternal enzymes of galactose metabolism and the "inexplicable" infantile cataract. *Lancet* **2,** 259–262.

Harper, P. S. (1974). *Birth Defects, Orig. Artic. Ser.* **10**(10), 120–125.

Harper, P. S., Rivas, M. L., Bias, W. B., Hutchinson, J. R., Dyken, P. R., and McKusick, V. A. (1972). *Am. J. Hum. Genet.* **24,** 310–316.

Hewer, R. L. (1968). Study of fatal cases of Friedreich's ataxia. *Br. Med. J.* **3,** 649.

Hilden, M., Brandt, N. J., Nilsson, I. M., and Schonheyder, F. (1974). Investigations of coagulation and fibrinolysis in homocystinuria. *Acta Med. Scand.* **195,** 533–535.

Hilliard, G. D., Harris, R. E., Gilstrap, L. C., and Shoumarker, R. D. (1977). *South. J. Med.* **70,** 446–447.

Hoefnagel, D., Wurster-Hili, D., and Child, E. L. (1979). Ovarian failure in galactosaemia. *Lancet* **2,** 1197.

Holling, H. E. (1963). Gout and glycogen storage disease. *Ann. Intern. Med.* **58,** 654.

Hopkins, A., and Wray, S. (1967). *Neurology* **17,** 166–168.

Horton, B. (1977). Anesthetic experiences in a family with hypokalemc familial periodic paralysis. *Anesthesiology* **47,** 308.

Howell, R. R. (1978). The glycogen storage diseases. *In* "The Metabolic Basis of Inherited Disease" (J. B. Stanbury, J. B. Wyngaarden, and D. S. Fredrickson, eds.), 4th ed., p. 137. McGraw-Hill, New York.

Hsia, D. Y., Naylor, J., and Bigler, A. (1962). The genetic mechanism of Gaucher's disease. *In* "Cerebral Sphingolipidoses" (S. M. Aronson and B. Volk, eds.), p. 327. Academic Press, New York.

Huber, F. B. (1967). *Schweiz. Med. Wochenschr.* **97,** 1498–1501.

Hudgson, P., Gardner-Medwin, D. G., Worsfold, M., Pennington, M. J. T., and Walton, J. A. (1968). Adult myopathy from glycogen storage disease due to acid maltase deficiency. *Brain* **91,** 435.

Hunter, D. J. (1971). *J. Obstet. Gynaecol. Br. Commonw.* **78,** 746–750.

Ismajovich, B., Zemer, D., Revach, M., Serr, D. M., and Sohar, E. (1973). *Fertil. Steril.* **24,** 844–847.

Janota, O., and Weber, K. (1928). Die paroxysmale Lahmung. *Abh. Neurol. Psychiatr. Psychol. Ihren Grenzgeb.* **46,** 1.

Jepson, J. B. (1978). Hartnup disease. *In* "The Metabolic Basis of Inherited Disease" (J. B. Stanbury, J. B. Wyngaarden, and D. S. Fredrickson, eds.), 4th ed., pp. 1563–1577. McGraw-Hill, New York.

Jeune, M., François, R., and Jarlot, B. (1959). Etude des polycories glycogéniques du fois. *Rev. Int. Hepatol.* **9,** 1.

Kacser, H., Mya, K. M., Duncker, M., Wright, A. F., and Bulfield, G. (1977). Maternal histidine metabolism and its effect on foetal development in the mouse. *Nature* (*London*) **265,** 262–266.

Kaplan, S., Bossak, E. T., Wang, C.-I., and Aldersberg, D. (1957). *J. Mt. Sinai Hosp., N.Y.* **2,** 39–44.

Kappas, A., Song, C. S., Levere, R. D., Sachson, R. A., and Granick, S. (1969). *Proc. Natl. Acad. Sci. U.S.A.* **61,** 509–513.

Kaufman, L. (1960). *Proc. R. Soc. Med.* **53,** 183–188.

Kaufman, F., Kogut, M. D., Donnell, G. N., Koch, R., and Goebelsmann, U. (1979). Ovarian failure in galactosaemia. *Lancet* **2,** 737–738.

Kaufman, F., Kogut, M. D., Donnell, G. N., Koch, R., Goebelsmann, U., and March, C. M. (1980). Ovarian failure in galactosemia. *Pediatr. Res.* **14,** (4, Part 2), 523 (Abstr. No. 587).

Kench, J. D., Langley, F. A., and Wilkinson, J. F. (1953). *Quart. J. Med.* **22,** 285–294.

Kennaway, N. G., Weleber, R. G., and Buist, N. R. M. (1980). Gyrate atrophy of the choroid and retina with hyperornithinemia: biochemical and histologic studies and response to vitamin B_6. *Am. J. Hum. Genet.* **32,** 529–541.

Kerr, G. D. (1973). *Proc. R. Soc. Med.* **66,** 763-764.
King, B., Spikesman, A., and Emery, A. E. (1972). *Clin. Chim. Acta* **36,** 267-269.
Kinnier Wilson, S. A. (1954). "Neurology," p. 1078. (Revised by A. N. Bruce.) Williams and Wilkins, Baltimore, Maryland.
Kohn, N. N., Faires, J. S., and Rodman, T. (1964). *N. Engl. J. Med.* **271,** 1179-1183.
Kolb, P., and Nietmann, E. (1965). Elektrokardiographische Verlaufs beobachtung eines Falles von Friedreichscher Ataxie wahrend einer Graviditat. *Z. Kreislaufforsch.* **54,** 611.
Komrower, G. M. (1979). Pregnancy and inherited metabolic disease: An emerging medical problem. Presented at the Annual Meeting of the Canadian Paediatric Society, Edmonton, Alta. *June 23-27.*
Komrower, G. M., Sardharwalla, I. B., Coutts, J. M. J., and Ingram, D. (1979). Management of maternal phenylketonuria: An emerging clinical problem. *Br. Med. J.* **1,** 1383-1387.
Kurczynski, T. W., Muir, W. A., Fleisher, L. D., Palomaki, J. F., Gaull, G. E., Rassin, D. K., and Abramowsky, C. (1980). Maternal homocystinuria: studies of an untreated mother and fetus. *Arch. Dis. Child.* **55,** 721-722.
Kushner, J. P., Barbuto, A. J., and Lee, G. R. (1976). *J. Clin. Invest.* **58,** 1089-1097.
Lamon, J. M., Frykholm, B. C., and Tschudy, D. P. (1977). *Arch Neurol. (Chicago)* **34,** 709-712.
Lamon, J. M., Frykholm, B. C., and Tschudy, D. P. (1979a). *J. Med. Genet.* **16,** 134-139.
Lamon, J. M., Frykholm, B. C., Hess, R. A., and Tschudy, D. P. (1979b). *Medicine (Baltimore)* **58,** 252-269.
Laver, M., and Fairley, K. F. (1971). *Lancet* **1,** 1019-1020.
Lee, F. I., and Loeffler, F. E. (1962). Gout and pregnancy. *J. Obstet. Gynaecol. Br. Commonw.* **69,** 299.
Lenke, R. R., and Levy, H. L. (1980). Maternal phenylketonuria and hyperphenylalaninemia. *N. Engl. J. Med.* **303,** 1202-1208.
Levy, H. L. (1973). Genetic screening. *Adv. Hum. Genet.* **4,** 1-104.
Levy, H. L. (1981). Maternal histidinaemia in the human. *J. Inherited Metab. Dis.* in press.
Levy, H. L., Shih, V. E., and Madigan, P. M. (1974). Routine newborn screening for histidinemia. Clinical and biochemical results. *N. Engl. J. Med.* **291,** 1214-1219.
Linares, A., Zarranz, J., Rodrigues-Alarcon, J., and Diaz-Perez, J. (1979). Reversible cutis laxa due to maternal D-penicillamine treatment. *Lancet* **2,** 43.
Lyon, I. C. T., Gardner, R. J. M., and Veale, A. M. O. (1974). Maternal histidinaemia. *Arch. Dis. Child.* **49,** 581-583.
Mabry, C. C. (1978). Maternal phenylketonuria. Presented at the 13th Conference of Collaborative Study of Children Treated for Phenylketonuria. Tahoe, Nevada.
Mabry, C. C., Denniston, J. C., Nelson, T. L., and Son, C. C. (1963). Maternal phenylketonuria: Cause of mental retardation in children without metabolic defect. *N. Engl. J. Med.* **269,** 1404-1408.
Mabry, C. C., Denniston, J. C., and Coldwell, J. G. (1966). Mental retardation in children of phenylketonuric mothers. *N. Engl. J. Med.* **275,** 1331-1336.
MacCready, R. A., and Levy, H. L. (1972). The problem of maternal phenylketonuria. *Am. J. Obstet. Gynecol.* **113,** 121-128.
Mace, B. E. W. (1956). Goutte et grossesse. *Rev. Rhum. Mal. Osteo-Articulaires* **23,** 156.
McKusick, V. A. (1978). "Mendelian Inheritance in Man," 5th ed. Johns Hopkins Press, Baltimore, Maryland.
McKusick, V. A., Hall, J. G., and Char, F. (1971). The clinical and genetic characteristics of homocystinuria. *In* "Inherited Disorders of Sulphur Metabolism" (N. A. J. Carson and D. N. Raine, eds.), pp. 179-203. Churchill-Livingstone, London and New York.
McQueen, E. G. (1951). The syndrome of gout. *Med. J. Aust.* **1,** 644.
Mahoney, M. J., Haseltine, F. P., Hobbins, J. C., Banker, B. Q., Caskey, C. T., and Golbus, M. S. (1977). *N. Engl. J. Med.* **297,** 968-973.

Marecek, Z., and Graf, M. (1976). *N. Engl. J. Med.* **295,** 841–842.
Marsh, L., and Fraser, F. C. (1973). *Lancet* **2,** 846.
Mendel, E. B., and McCullough, J. A. (1968). Gaucher's disease and pregnancy. *Obstet. Gynecol.* **32,** 607.
Menkes, J. H., and Aeberhard, E. (1969). Maternal phenylketonuria; the composition of cerebral lipids in an affected offspring. *J. Pediatr.* **74,** 924–931.
Mertz, D. P., Henninges, D., and Schwoerer, P. (1972). Gicht und Schwangerschaft. *Muench. Med. Wochenschr.* **114,** 658.
Meyer, U. A., and Schmid, R. (1978). *In* "The Metabolic Basis of Inherited Disease" (J. B. Stanbury, J. B. Wyngaarden, and D. S. Fredrickson, eds.), 4th ed., pp. 1166–1220. McGraw-Hill, New York.
Meyer, U. A., Strand, L. J., Doss, M., Rees, C. A., and Marver, H. S. (1972). *N. Engl. J. Med.* **286,** 1277–1282.
Miller, J., and Katz, R. L. (1973). Muscle diseases. *In* "Anesthesia and Uncommon Diseases" (J. Katz and L. B. Kadis, eds.), Chapter 13, p. 425. Saunders, Philadelphia, Pennsylvania.
Mjolnerod, O. K., Dommerud, S. A., Rasmussen, K., and Gjeruldsen, S. T. (1971). *Lancet* **1,** 673–675.
Molitch, M. E., Oill, P., and Odell, W. D. (1974). *J. Am. Med. Assoc.* **227,** 522–525.
Moneta, E., Cocchi, A., and Scoppetta, C. (1976). Morbo di Friedreich, diabete e gravidanza. *Minerva Med.* **67,** 459.
Morgan, G. (1974). *Aust. N.Z. Med.* **4,** 196–197.
Mudd, S. H., and Levy, H. L. (1978). Disorders of transsulfuration. *In* "The Metabolic Basis of Inherited Disease" (J. B. Stanbury, J. B. Wyngaarden, and D. S. Fredrickson, eds.), 4th ed., pp. 458–503. McGraw-Hill, New York.
Murphy, E. A., and Chase, G. A. (1975). "Principles of Genetic Counseling." Yearbook Publ., Chicago, Illinois.
Mustajoki, P., and Koskelo, P. (1976). *Acta Med. Scand.* **200,** 171–178.
Neilson, D. R., and Neilson, R. P. (1958). *West. J. Surg., Obstet. Gynecol.* **66,** 133–149.
Neville, B. G. R., Harris, R. F., Stern, D. J., and Stern, J. (1971). Maternal histidinaemia. *Arch. Dis. Child.* **46,** 119–121.
Nevin, N. C., Cumings, J. N., and McKeown, F. (1967). Refsum's syndrome. *Brain* **90,** 419.
Nielsen, K. B., Wamberg, E., and Weber, J. (1979). Successful outcome of pregnancy in a phenylketonuric woman after low-phenylalanine diet introduced before conception. *Lancet* **1,** 1245.
Nordmann, Y., Grandchamp, B., Phung, N., de Verneuil, H., Grelier, M., and Noire, J. (1977). *Lancet* **1,** 140.
Oepkes, C. T., and van Weerden, T. W. (1979). *Neurology* **29,** 523–524.
Oropollo, A. T. (1978). Abnormal pseudocholinesterase levels in a surgical population. *Anesthesiology* **48,** 284.
Pearson, C. M., and Kalyanaraman, K. (1972). The periodic paralyses. *In* "The Metabolic Basis of Inherited Disease" (J. B. Stanbury, J. B. Wyngaarden, and D. S. Fredrickson, eds.), 3rd ed., p. 1181. McGraw-Hill, New York.
Pelkonen, R., and Kivirikko, K. I. (1970). Hydroxyprolinemia. An apparently harmless familial metabolic disorder. *N. Engl. J. Med.* **282,** 451–456.
Perry, T. L., Hansen, S., Tischler, B., Richards, F. M., and Sokol, M. (1973). Unrecognized adult phenylketonuria. Implications for obstetrics and psychiatry. *N. Engl. J. Med.* **289,** 395–398.
Pickard, N. A., Gruemer, H. D., Verrill, H. L., Issacs, E. R., Robinow, M., Nance, W. E., Myers, E. C., and Goldsmith, B. (1979). *N. Engl. J. Med.* **299,** 841–846.
Pomeroy, J., Efron, M. L., Dayman, J., and Hoefngel, D. (1968). Hartnup disorder in a New England family. *N. Engl. J. Med.* **278,** 1214–1216.

Poskanzer, D. C., and Kerr, D. N. S. (1961). A third type of periodic paralysis, with normokalemia and favourable response to sodium chloride. *Am. J. Med.* **31**, 328.

Quebec Cooperative Study (1976). *Canad. J. Neurol. Sci.* **3**, 271–396.

Reed, T. E., and Neel, J. V. (1959). Huntington's chorea in Michigan. 2. Selection and mutation. *Am. J. Hum. Genet.* **11**, 107.

Reimann, H. A. (1975). *J. Am. Med. Assoc.* **231**, 64–66.

Ritchie, J. W. K., and Carson, N. A. J. (1973). Pregnancy and homocystinuria. *J. Obstet. Gynecol.* **80**, 664–669.

Roe, T. F., Hallatt, J. G., Donnell, G. N., and Ng, W. G. (1971). Childbearing by a galactosemic woman. *J. Pediatr.* **78**, 1026–1030.

Rosenbaum, K. Unpublished data.

Ross, L. (1969). Gaucher's cells in kidney glomeruli. *Arch. Pathol.* **87**, 164.

Samuels, S., Sun, S. C., and Verasestakul, S. (1976). Normal infant birth in white galactosemic women. *J. Med. Soc. N.J.* **73**, 309–311.

Sardharwalla (1979). Personal communication.

Sassa, S., Solish, G., Levere, R. D., and Kappas, A. (1975). *J. Exp. Med.* **142**, 722–731.

Sass-Kortsak, A., and Bearn, A. G. (1978). Hereditary disorders of copper metabolism. *In* "The Metabolic Basis of Inherited Disease" (J. B. Stanbury, J. B. Wyngaarden, and D. S. Frederickson, eds.), 4th ed., pp. 1098–1126. McGraw-Hill, New York.

Scheinberg, I. H., and Sternlieb, I. (1975). *N. Engl. J. Med.* **293**, 1300–1302.

Schilling, A. (1968). Gicht bei der Frau. *Z. Rheumaforsch.* **27**, 192.

Schmid, R., and Hammaker, L. (1961). Hereditary absence of muscle phosphorylase. *N. Engl. J. Med.* **264**, 223.

Schmidt, H., Smitker, G., Thomsen, K., and Lintrup, J. (1974). *Arch. Dermatol.* **110**, 58–64.

Schulman, J. D. Unpublished data.

Schwabe, A. D., and Peters, R. S. (1974). *Medicine (Baltimore)* **53**, 453–462.

Sciarra, J. J., and Steer, C. M. (1961). *Am. J. Obstet. Gynecol.* **82**, 612–615.

Scriver, C. R. (1968). Renal tubular transport of proline, hydroxyproline and glycine. III. Genetic basis for more than one mode of transport in human kidney. *J. Clin. Invest.* **47**, 823–835.

Scriver, C. R. (1978). Disorders of proline and hydroxyproline metabolism. *In* "The Metabolic Basis of Inherited Disease" (J. B. Stanbury, J. B. Wyngaarden, D. S. Fredrickson, eds.), 4th ed., pp. 336–361. McGraw-Hill, New York.

Seekler, S. G., and Rovinsky, J. J. (1965). *In* "Medical, Surgical and Gynecological Complications of Pregnancy" (A. F. Guttmacher and J. J. Rovinsky, eds.), pp. 782–790. Williams & Wilkins, Baltimore, Maryland.

Seegmiller, J. E. (1980). Diseases of purine and pyrimidine metabolism. *In* "Duncan's Diseases of Metabolism" (P. K. Bondy and L. E. Rosenberg, eds.), 8th ed., pp. 777–937. Saunders, Philadelphia, Pennsylvania.

Sen, S., and Das, P. K. (1978). *Eur. Neurol.* **17**, 258–264.

Shapiro, H. I., Mazur, J. R., and Emich, J. P., Jr. (1969). *Obstet. Gynecol.* **34**, 189–193.

Shih, V. E., and Coulombe, C. T. (1979). Unpublished observation.

Shih, V. E., Berson, E., Mandell, R., and Schmidt, S. Y. (1978). Ornithine ketoacid transaminase deficiency in gyrate atrophy of the choroid and retina. *Am. J. Hum. Genet.* **30**, 174–179.

Shih, V. E., Berson, E. L., and Gargiulo, M. Reduction of hyperornithinemia with a low protein, low arginine diet and pyridoxine in patients with a deficiency of ornithine-ketoacid transaminase (OKT) activity and gyrate atrophy of the choroid and retina. In press.

Shore, R. N., and MacLachlan, T. B. (1971). *Obstet. Gynecol.* **38**, 448–454.

Sidbury, J. B. (1979). Unpublished observations.

Siler, J. N., and Discavage, W. J. (1975). Anesthetic management of hypokalemic periodic paralysis. *Anesthesiology* **43**, 489.

Simpson, J., Zellweger, H., Burmeister, L. F., Christee, R., and Nielsen, M. R. (1974). *Clin. Chim. Acta* **52,** 219–223.

Sipila, I., Simell, O., Rapola, J., and Tuuteri, L. (1979). Gyrate atrophy of the choroid and retina with hyperornithinemia: Tubular aggregates and type 2 fiber atrophy in muscle. *Neurology* **29,** 996–1005.

Skre, H. (1975). Friedreich's atoxia in Western Norway. *Clin. Genet.* **7,** 287.

Smith, I., and Wolff, O. H. (1979). Maternal phenylketonuria. *Br. Med. J.* **2,** 127.

Smith, I., Erdohazi, M., Macartney, F. J., Pincott, J. R., Wolff, O. H., Brenton, D. P., Biddle, S. A., Fairweather, D. V. I., and Dobbing, J. (1979). Fetal damage despite low-phenylalanine diet after conception in a phenylketonuria woman. *Lancet* **1,** 17–19.

Smyth, C. J. (1957). Hereditary factors in gout. A review of the recent literature. *Metab. Clin. Exp.* **6,** 218.

Sohar, E., Gafine, J., Pras, M., and Heller, H. (1967). *Am. J. Med.* **43,** 227–253.

Solomon, L., Abrams, G., Dinner, M., and Berman, L. (1977). *N. Engl. J. Med.* **296,** 54–55.

Spaulding, W. B. (1960). *Ann. Intern. Med.* **53,** 739–745.

Sperling, O., Liberman, U. A., Frank, M., and DeUries, A. (1971). Xanthinuria. An additional case with demonstration of xanthine oxidase deficiency. *Am. J. Clin. Pathol.* **55,** 351.

Statzer, D. E. (1961). *Obstet. Gynecol.* **17,** 587–592.

Stein, E. A., Meany, C., Spitz, L., *et al.* (1975). Portacaval shunt in 4 patients with homozygous hypercholesterolemia. *Lancet* **1,** 832.

Stein, J. A., and Tschudy, D. P. (1970). *Medicine (Baltimore)* **49,** 1–16.

Steinberg, D. (1978). Phytanic acid storage disease. *In* "The Metabolic Basis of Inherited Disease" (J. B. Stanbury, J. B. Wyngaarden, and D. S. Fredrickson, eds.), 4th ed. Chapter 33, p. 688. McGraw-Hill, New York.

Stevenson, J. S. (1979): Personal communication.

Takki, K. (1974). Gyrate atrophy of the choroid and retina associated with hyperornithinaemia. *Br. J. Ophthalmol.* **58,** 3–23.

Takki, K., and Simell, O. (1974). Genetic aspects in gyrate atrophy of the choroid and retina with hyperornithinaemia. *Br. J. Ophthalmol.* **58,** 907–916.

Talbott, J. H. (1941). Periodic paralysis. *Medicine (Baltimore)* **20,** 85.

Talbott, J. H. (1967). "Gout," 3rd ed. Grune & Stratton, New York.

Tedesco, T. A., Morrow, G., and Mellman, W. J. (1972). Normal pregnancy and childbirth in a galactosemic woman. *J. Pediatr.* **81,** 1159–1162.

Tenton, J. B., and Treadwell, N. C. (1957). Gaucher's disease in pregnancy. *Am. J. Obstet. Gynaecol. Br. Commonw.* **74,** 1363.

Thompson, C. E. (1978). *N. Engl. J. Med.* **298,** 1479.

Toaff, R., Toaff, M. E., Peyser, M. R., and Streifler, M. (1977). *Obstet. & Gynecol. Surv.* **32,** 497–507.

Toussaint, D., Coers, C., and Toppet, N. (1959). Heredopathia atactica polyneuritiformis. *Bull. Soc. Belge Ophtalmol.* **121,** 383.

Townes, P. L. (1965). *J. Pediatr.* **67,** 600–602.

Tsang, R. C., and Glueck, C. J. (1975). *Clin. Perinatol.* **2,** 275–294.

Tsang, R. C., Glueck, C. J., McLain, C., Russell, P., Joyce, T., Bové, K., Mellies, M., and Steiner, P. M. (1978). *Metab., Clin. Exp.* **27,** 823–829.

Tschudy, D. P., and Lamon, J. M. (1980). *In* "Metabolic Control and Disease" (P. K. Bondy and L. E. Rosenberg, eds.), 8th ed., pp. 993–1008. Saunders, Philadelphia, Pennsylvania.

Tschudy, D. P., Valsamis, M., and Magnussen, C. R. (1975). *Ann. Intern. Med.* **83,** 851–864.

Turner, R. E., Frank, M. J., van Ausdal, D., and Bollet, A. J. (1960). Some aspects of the epidemiology of gout. *Arch. Intern. Med.* **106,** 400.

Tyrer, J. M. (1975). Friedreich's ataxia. *Handb. Clin. Neurol.* **21,** 319.

Valle, D., Kaiser-Kupfer, M. I., and Del Valle, L. A. (1977). Gyrate atrophy of the choroid and retina: Deficiency of ornithine aminotransferase in transformed lymphocytes. *Proc. Natl. Acad. Sci. U.S.A.* **74,** 5159-5161.

Valle, D., Walser, M., and Brusilow, S. W. (1980). Gyrate atrophy of the choroid and retina: Amino acid metabolism and correction of hyperornithinemia with an arginine deficient diet. *J. Clin. Invest.* **65,** 371-378.

van Crevald, S. (1961). Clinical course of glycogen storage disease. *Chem. Weekbl.* **57,** 445.

Van Sprang, F. J. (1971). General discussion on the treatment of homocystinura. *In* "Inherited Disorders of Sulphur Metabolism" (N. A. J. Carson and D. N. Raine, eds.), p. 305. Churchill-Livingstone, London and New York.

Veale, A. M. O., and Lyon, I. C. T. (1978). Personal communication.

Wadman, S. K., and Duran, M. (1979). Personal communication.

Wallace, D. C. (1971). Huntington's chorea: A partial model of the aging process. *Med. J. Aust.* **2,** 1275.

Wallace, D. C. (1976). The social effect of Huntington's chorea on reproductive effectiveness. *Ann. Hum. Genet.* **39,** 375.

Walshe, J. M. (1956). *Am. J. Med.* **21,** 487-495.

Walshe, J. M. (1977). *Quart. J. Med.* **46,** 73-83.

Warner, R. (1979). Personal communication.

Watters, G. V., and Williams, T. W. (1967). *Arch. Neurol. (Chicago)* **17,** 137-152.

Webb, D., Muir, I., Faulkner, J., and Johnson, G. (1978). *Am. J. Obstet. Gynecol.* **132,** 265-270.

Weingold, A. B. (1960). Gout and pregnancy. *Obstet. Gynecol.* **16,** 309.

Welland, F. H., Hellman, E. S., Collins, A., Hunter, G. W., and Tschudy, D. P. (1964). *Metab., Clin. Exp.* **13,** 251-258.

Wenzl, H., and Kludas, M. (1970). *Muench. Med. Wochenschr.* **112,** 2187-2190.

Whelan, D. T., and Scriver, C. R. (1968). Cystathioninuria and renal iminoglycinuria in a pedigree: A perspective on counseling. *N. Engl. J. Med.* **278,** 924-927.

Wyngaarden, J. B. (1978). Hereditary xanthinuria. *In* "The Metabolic Basis of Inherited Disease" (J. B. Stanbury, J. B. Wyngaarden, and D. S. Fredrickson, eds.), 4th ed., p. 1037. McGraw-Hill, New York.

Wyngaarden, J. B., and Kelley, W. N. (1978). Gout. *In* "The Metabolic Basis of Inherited Disease" (J. B. Stanbury, J. B. Wyngaarden, and D. S. Fredrickson, eds.), 4th ed., p. 916. McGraw-Hill, New York.

Ziegler, D. K., van Speybroech, N. W., and Seitz, E. F. (1974). Myoclonic epilepsy portialis continua and Friedreich's ataxia. *Arch. Neurol. (Chicago)* **31,** 308.

Zimmerman, T. S., McMillin, J. M., and Watson, C. J. (1966). *Arch. Intern. Med.* **118,** 229-240.

2

Disorders of Connective Tissue and Skeletal Dysplasia

JUDITH G. HALL

Copyright © 1981 by Academic Press, Inc.
All rights of reproduction in any form reserved.
ISBN 0-12-630940-X

During the last decade significant advances have been made in the recognition and understanding of heritable connective tissue disorders and skeletal dysplasias (McKusick, 1972; Spranger *et al.*, 1974). The establishment of clinical criteria for defining specific entities is leading to the definition of natural history, inheritance, and biochemical basis of each disorder. Much confusion is still generated by the older medical literature, in which diagnostic designations were used loosely and ''lumping'' of patients with different conditions results in misleading statements. Because most genetic disorders are relatively rare, they have, also, been considered curiosities. In reality, however, these disorders are experiments of nature from which we can learn much about ''normal'' processes.

Very little has been reported about the complications of pregnancy in specific skeletal dysplasias and disorders of connective tissue. Moreover, in the few reports concerning obstetric or gynecologic problems, one often cannot be certain of the specific diagnosis in the mother. Thus, personal experience and the communications of associates must frequently be employed in the present chapter. Undoubtedly, both this anecdotal experience and the literature are skewed toward unusual events and dramatic complications because routine and uncomplicated cases are seldom recognized. Hopefully, with improving ability to recognize distinct entities, documentation and recognition of characteristic complications in pregnancy in each of the heritable skeletal dysplasias and connective tissue disorders will be possible.

I. HERITABLE DISORDERS OF CONNECTIVE TISSUE

A. Ehlers-Danlos Syndrome

Over the last 15 years, what had been called the Ehlers-Danlos syndrome has been recognized to include at least eight different specific entities (Bornstein and Byers, 1978; Beighton, 1970). It is important to determine into which subtype of Ehlers-Danlos syndrome a given patient fits.

''Classic''' Ehlers-Danlos type I (gravis) is characterized by hyperextensible skin, hypermobile joints, fragile tissues, and a bleeding diatheses. Other specific types may have ocular fragility (type VI), vascular rupture (type IV), or severe orthopedic problems secondary to recurrent joint dislocations (type VII). ''Classic Ehlers-Danlos syndrome'' or type I is the only type in which any extensive experience is available concerning complications of pregnancy. Type I is an autosomal dominant trait; thus, an affected mother will have a 50% chance of having an affected child. Ehlers–Danlos syndrome type II (mitis) and type III (benign hypermobility) are also autosomal dominant, but neither show an increased risk of complications during pregnancy. To our knowledge, pregnancies have not been reported in women with any of the other types of Ehlers–Danlos syndromes, nor have the pregnancies of normal women delivered of affected fetuses been complicated. Ehlers–Danlos type IV, characterized by increased bruisability, rupture of the great vessels, and rupture of the bowel and menorrhagia in afflicted women, may be heterogeneous. Usually considered autosomal recessive, it may be composed of several different disorders, including an autosomal dominant type. Although no information is available about pregnancies in this condition, one would anticipate rupture of bowel and major vessels would make pregnancy hazardous. Ehlers–Danlos syndrome type V, an X-linked recessive disorder, does not affect women, and there is no suggestion that the pregnancies carrying an affected male are complicated for the fetus.

Pregnancy in Ehlers–Danlos syndrome type I involves a risk for both the mother and the fetus (Beighton, 1970; McKusick, 1972). During pregnancy affected mothers may (1) manifest increased bruisability (Samuel *et al.*, 1953), (2) develop abdominal wall, inguinal, umbilical, and femoral hernias, and (3) develop varicosities of both leg veins and the vulva (Beighton, 1970). Rupture of the great vessels has occurred during pregnancy (McKusick, 1972), as well as severe postpartum hemorrhage, bleeding from the episiotomy site, and perineal hematomas. Persistent bleeding postpartum has even required hysterectomy (Smith *et al.*, 1968; Stoddard and Myers, 1968). Other complications in Ehlers–Danlos syndrome type I include separation of the symphysis pubis during delivery with prolonged postpartum discomfort, uterine prolapse and slow wound healing (Beighton, 1970), vaginal rupture, and avulsion of the bladder (McKusick, 1972). Episiotomy sites and cesarean section incision suture may heal slowly, and may spontaneously extend. All sutures should be left in for at least twice the usual length of time to avoid wound dehiscence. Perineal hematomas may occur even without episiotomy. If an affected woman had a previous cesarean section, she may be more likely to have a rupture in the scar site of the uterine wall during labor (McKusick, 1972); therefore, elective repeat cesarean section is suggested prior to the onset of labor. Spontaneous rupture of the intact uterus has apparently not been observed in Ehlers–Danlos syndrome. Because of abnormal connective tissue, forceps should be used with caution (Beighton, 1970).

Interestingly, women with Ehlers–Danlos syndrome apparently do not develop striae gravidarum, nor do they usually have any problems with cardiovascular compromise unless they have scoliosis. There is no contraindication to any particular type of anesthesia.

In addition to the above complications, the pregnancies frequently terminate prematurely, either because of premature rupture of fetal membranes (if the child is affected) or because of lax cervical tissues (incompetent cervix). Rapid effacement and dilation of the cervix may lead not only to premature, but also to precipitous delivery.

Kanof (1952), Mories (1954), and Barabas (1966) found that premature rupture of fetal membranes usually occurs between 32 and 35 weeks gestation, thus predisposing infants to all the complications of prematurity. There is some suggestion that birth weight is decreased in affected infants (Barabas, 1966). Prematurity is apparently specifically related to premature rupture of fetal membranes; thus, if the fetus is not affected, this complication should not occur with increased frequency. There seems to be a correlation between prematurity and the development of several symptoms of Ehlers-Danlos syndrome in adulthood (Barabas, 1966). Affected infants also tend to be hyperextensible and prone to develop congenitally dislocated hips. For example, Thomas *et al.* (1952) reported an infant with dislocation of the shoulder and a breech presentation; McFarland and Fuller (1964) described an Erb's palsy, believed related to excessive joint mobility. In the newborn period floppiness and bleeding tendencies have led affected infants to be misdiagnosed as having neurologic conditions or hemophilia (McKusick, 1972).

Beighton (1970) observed pregnancies in 29 affected women, and noted 27 affected and 34 unaffected infants. Although this is not a statistically significant difference, one would have anticipated equal numbers of affected and unaffected. There is no other suggestion of fetal wastage.

Prenatal diagnosis is currently not available for type I, II, or III Ehlers-Danlos syndrome, although in the types of Ehlers-Danlos in which a specific enzyme deficiency or collagen abnormality is known, prenatal diagnosis might be possible. Prenatal diagnosis of the classic Ehlers-Danlos (type I) awaits recognition of the underlying specific biochemical abnormality.

Other gynecologic problems noted in women with Ehlers-Danlos syndrome include prolapse of the uterus and bladder and abdominal hernia. The bleeding diathesis leads to metrorrhagia and menorrhagia. Complications of sexual intercourse have not been reported, nor do affected women undergo premature menopause, or manifest an increased susceptibility to malignancies. No complications with birth control have been described, although one might anticipate that intrauterine devices might be more likely to perforate the uterus.

Prevention of complications requires anticipation. Bedrest may be indicated for a mother who has previously had premature deliveries, inasmuch as prematurity seems to run in families (Barabas, 1966). Cesarean section should be avoided unless it is strictly indicated. Genetic counseling for each type of Ehlers-Danlos syndrome should be provided. If the father is affected, artificial insemination should be discussed.

B. Marfan Syndrome

The Marfan syndrome is characterized by abnormalities of the (1) skeleton (hyperextensibility, scoliosis, abnormalities of the chest wall and excessively long, thin limbs); (2) eye (dislocated lenses, myopia, and retinal detachments); and (3) cardiovascular system (primarily involving dilation and dissection of the ascending aorta, aortic valve dilation, and floppy mitral valve with mitral regurgitation (McKusick, 1972)). The condition is an autosomal dominant trait; however, there are many other conditions with which the Marfan syndrome has been confused in the past (e.g., homocystinuria, contractural arachnodactyly, Stickler syndrome). A few families suggest autosomal recessive inheritance, but in these gonadal mosaicism seems a more likely explanation; thus, for all practical purposes, Marfan syndrome should be considered an autosomal dominant disorder. Most cases are familial if parents are carefully examined; however, some cases of the Marfan syndrome represent new dominant mutations, and these are more likely to be associated with advanced paternal age (Murdoch *et al.,* 1972b).

The life-threatening cardiovascular complication seen in the Marfan syndrome is cystic medial necrosis of the wall of the aorta, which predisposes to aneurysm formation. The aortic ring tends to dilate, leading to aortic regurgitation and left ventricular failure. In addition, the mitral valve is often redundant, leading to mitral valve regurgitation (McKusick, 1972). Mandel *et al.* (1954) reported that 50% of the aortic aneurysms in women with Marfan syndrome under 40 years of age occur in women who are pregnant. Tricomi (1965), Sutinen and Piiroinen (1971), and Pedowitz and Perell (1957), have reported large series of pregnant Marfan patients. Dissection and rupture of the aortic aneurysms occurs with increased frequency during pregnancy (Hirst *et al.,* 1958; Mandel *et al.,* 1954; Sutinen and Piiroinen, 1971; Elias and Berkowitz, 1976; Cava and Dreier, 1970). The most stressful period on the aorta seems to be in the third trimester; rupture of aortic aneurysms is not particularly increased during labor or the postpartum period (Pedowitz and Perell, 1957). Rupture of the splenic artery also apparently occurs with increased frequency during pregnancy in the Marfan syndrome (Schnitker and Bayer, 1944; Sheehan and Falkiner, 1948), and coronary artery dissection has also been reported (McKusick, 1972). Aortic replacement has been performed successfully during pregnancy (Donaldson and de Alvarez, 1965).

The pregnant women with the Marfan syndrome at greatest risk appear to be those who already have aortic valve changes or dilation of the aorta prior to pregnancy because this type of dilation may progress to aortic aneurysm during the pregnancy. Those women without any valve changes are probably less prone to develop aneurysms. The exact risk is unknown. Some authors have suggested mortality rates as high as 50% in pregnant women with the Marfan syndrome, but this figure seems too high. Perhaps 10–20% of women with aortic valve involvement prior to pregnancy will die during the pregnancy because of aortic catastrophy. Women without aortic involvement probably have less than a 5% risk of dying during pregnancy. Thus, pregnancy should probably be avoided when the aortic involvement is already present. Donaldson and de Alvarez (1965) have suggested that since the risk of developing an aortic aneurysm seems to increase with age, women with Marfan syndrome should be encouraged to plan their pregnancies at a young age if they plan to have children (i.e., prior to the average age of dissection, which has been determined to be 30 years of age by one study (Halpern *et al.,* 1971)). Unfortunately, this seems too simplistic an approach. More data is needed about the relationship of vascular status prior to pregnancy and morbidity and mortality during pregnancy in the Marfan syndrome.

No explanation for the increased susceptibility to vascular aneurysm during pregnancy is available, although there seems to be some relationship between vascular changes and estrogens (Danforth *et al.,* 1964; Manalo-Estrella *et al.,* 1965; Manalo-Estrella and Barker, 1967; Irey *et al.,* 1970; Masi and Dugdale, 1970; Simpson and Neilson, 1973). Pregnancy increases both the circulating levels of estrogen as well as increasing cardiac output. The cystic medial necrosis seen in Marfan syndrome can be produced with estrogen in animal models. McKusick (1972), reported three cases without evidence of the Marfan syndrome in which changes in the aorta occurred in association with pregnancy, assumed to be due to the increased estrogens of pregnancy, resulting in aortic aneurysms and dissection of the aneurysms. Increased pulse pressure as a result of increased vascular output during pregnancy may place an unacceptably increased stress on the heart and aorta. Whatever the reason, there is no question about the increased association of aortic dilation, aneurysm, and dissection in some pregnant women with the Marfan syndrome. There is also an apparent correlation between severity of the clinical manifestations of the Marfan syndrome in the mother and likelihood of occurrence of aortic problems during pregnancy (McKusick, 1972).

In addition to life-threatening dissection of the aorta, respiratory compromise may occur during pregnancy if the mother has scoliosis; therefore, pulmonary function should be monitored both before and during pregnancy when scoliosis is present. Women with the Marfan syndrome do not seem more prone to problems related to cesarean section and have apparently normal wound healing. If they have mitral valve involvement prophylactic antibiotics should be used appropriately. An increase in hernias and stretch marks occur during pregnancy in the Marfan syndrome (McKusick, 1972).

Hypertension is not a common complication of pregnancy in the Marfan syndrome; however, blood pressure should be monitored and hypertension treated vigorously since this would increase the risk of rupture of the aortic aneurysms. Treatment with propranolol has been proposed to decrease the pulse pressure and thereby prevent aortic dilatation and subsequent dissection; however, the results of such treatment are inconclusive, and effects on placental blood flow and fetal status uncertain (Ose and McKusick, 1977).

There does not seem to be any particular concern about the anesthesia used for delivery. The hyperextensibility noted in Marfan syndrome has not been a problem for intubation or surgical procedures.

No specific complications for either the affected or unaffected newborn are expected. Affected children occur in the expected ratio of 50% of infants with equal numbers of males and females and can often be recognized in the newborn period (McKusick, 1972). There does not appear to be fetal wastage or infertility in Marfan syndrome. Infants born to women with Marfan syndrome do not appear to be more severely affected than those born to affected fathers. There is some evidence that new mutations for the Marfan syndrome may be more severely affected than the familial cases.

With regard to other gynecologic disturbances, women with Marfan syndrome enter puberty normally. Skovby and McKusick (1977) recently recommended use of estrogens at age 7–8 years to fuse epiphyses prematurely and decrease ultimate height in girls with Marfan syndrome. This has successfully decreased height, but girls must be monitored for scoliosis because that problem is more likely to occur in the Marfan syndrome during the growth spurt of early adolescence. The rationale for decreasing height in girls with the Marfan syndrome is both cosmetic and therapeutic. Decreasing the ultimate height might decrease required cardiovascular output and thus reduce the risk of aortic aneurysm. As mentioned earlier, there is some suggestion that estrogens may increase vascular complications in women and this must be considered when exogenous estrogens (including oral contraception) are used in women with the Marfan syndrome. Thus birth control pills should be used with caution in women with the Marfan syndrome. No specific complications of menses are known.

Prevention of the Marfan syndrome is not possible by prenatal diagnosis at this time. Indeed, until the basic defect is recognized, it will probably not be available. Artificial insemination can be considered if the father is affected.

C. Pseudoxanthoma Elasticum (PXE)

Pseudoxanthoma elasticum is a heritable disorder of connective tissue primarily involving skin, eye, gastrointestinal tract, and peripheral vessels. Most reported families are consistent with autosomal recessive inheritance (McKusick, 1972; Pope, 1974b); however, genetic heterogeneity exists, and several families consistent with autosomal dominant inheritance have been reported (Pope, 1974a; Altman *et al.*, 1974). Onset of clinical features is usually not noted until the second or third decade, and is progressive thereafter. The lax redundant and relatively inelastic skin has characteristic yellow xanthoma-like (pseudoxanthoma) papular skin lesions which give the appearance of a "plucked chicken." There is exaggeration of folds in the neck, nasolabial area, axillae, and antecubital fossa. Hyperkeratotic nodules and intradermal calcifications may occur. Angioid streaks of the retina are usually first noted in the second or third decade.

Breaks in Bruch's membrane may later cause hemorrhage and chorioretinal scarring, which often lead to decreasing vision and blindness. Vascular changes are characterized by arterial occlusion and calcification of peripheral vessels, resulting in loss of peripheral pulses, hypertension, claudication, and angina pectoris. Gastrointestinal hemorrhage is common, and hematuria occasionally occurs.

No specific complications of pregnancy have been reported. Indeed, many affected patients have had pregnancies with no apparent complications related to their disease or present in their offspring. Occasionally, postpartum uterine bleeding has been severe (Altman *et al.*, 1974), for reasons that are unclear. Bleeding may be related to the underlying arterial disease of PXE because affected women often have menorrhagia and may require hysterectomy (Altman *et al.*, 1974).

There appears to be no reduced fertility, and no particular precautions are necessary for the management of pregnancy except recognition that these patients are more susceptible to having gastrointestinal bleeding and postpartum hemorrhage. The leg cramps typical of pregnancy do not seem to be more severe in women with previous claudication as a result of PXE.

Although pseudoxanthoma elasticum involves elastic fibers, the biochemical nature of the basic defect is unknown; thus, prenatal diagnosis is not available.

D. Cutis Laxa

In cutis laxa, extreme laxity results in the skin hanging in loose folds over the entire body, particularly around the face and eyes, causing baggy jowls and "hangdog" facies. These skin changes give the face a prematurely aged appearance. The skin is hyperextensible but inelastic, i.e., it does not slip back into place. There is no increased bruisability. Involvement of lungs (emphysema and cysts), gastrointestinal tract (diverticula), and arteries may occur. Cutis laxa appears to be heterogeneous in that there are different ages of onset and different modes of inheritance—including autosomal dominant, autosomal recessive, and a recently described X-linked form (McKusick, 1972; Goltz *et al.*, 1965; Byers *et al.*, 1976). In the recessive form, severe pulmonary complications preclude affected children surviving until reproductive age. Emphysema and urologic diverticula are not usually observed in the dominant type of cutis laxa, but should be sought during pregnancy, preferably by comparison to base-line studies conducted prior to pregnancy.

In the X-linked form, a specific enzyme, lysyl oxidase, appears deficient (Byers *et al.*, 1976). In this form of cutis laxa, prenatal diagnosis is potentially possible both by sex determination (i.e., only males will be affected) and by enzyme studies since the enzyme is present in amniotic cells (P. H. Byers, personal communication). Prenatal diagnosis is probably not possible in the other two forms of cutis laxa. With fetoscopy, however, it eventually may be possible to determine on the basis of skin biopsy whether normal elastin is present in fetal skin in the autosomal recessive type (Dallaire *et al.*, 1976; Kaye *et al.*, 1974). Heterozygate detection is not now possible in the recessive type of cutis laxa.

Fertility does not seem to be decreased in the dominant type of cutis laxa, and there are apparently no particular problems related to pregnancy or delivery. Mother to daughter transmission has been described by Lewis (1948), without maternal complications during pregnancy. In the recessive and X-linked forms there have also been no complications of pregnancies reported related to the disease in the fetus. Prematurity does not occur, although some infants have been reported to be small for gestational age in the dominant and recessive forms.

E. Osteogenesis Imperfecta (OI)

Osteogenesis imperfecta is characterized by osteoporosis and fractures of long bones with minimal trauma. Many patients also have progressive deformities of the long bones, scoliosis, blue sclerae, short stature, dentinogenesis imperfecta, and deafness (McKusick, 1972; Schoenfeld *et al.*, 1975; Sillence *et al.*, 1978). Sillence *et al.* (1978) have described at least four types of osteogenesis imperfecta. Probably there is further heterogeneity among the four types, and perhaps other types are yet to be defined. In the past, there has been much confusion about osteogenesis imperfecta because of the designations *congenita* and *tarda* (McKusick, 1972). *Osteogenesis imperfecta congenita* was used to designate a condition characterized by fractures at birth, whereas *osteogenesis imperfecta tarda* designated a condition in which fractures occurred after birth. We now know that fractures can be present at birth in some individuals in a given family, whereas other individuals in the same kindred may not develop fractures until late childhood or even adulthood. Sillence's new categorization (Sillence *et al.*, 1978) has helped separate different types of osteogenesis imperfecta on a clinical and genetic basis. Sillence's type I is dominantly inherited with blue sclerae; type II is lethal neonatal osteogenesis imperfecta with "crumpled" bones and autosomal recessive inheritance; type III has normal sclerae and progressive deformities and probably autosomal recessive inheritance (although heterogeneity within this category may exist); type IV has white or normal sclerae with variable severity of fractures and autosomal dominant inheritance. The basic defect is unknown in all types of osteogenesis imperfecta, although there is a suggestion that abnormal collagen may play some role (Penttinen *et al.*, 1975; Müller *et al.*, 1975, 1977). Defective bone formation is seen in all forms.

Women with osteogenesis imperfecta apparently menstruate at a normal age and have no menstrual irregularity, although they may have menorrhagia.

Reproduction is probably hampered in all types of osteogenesis imperfecta. Patients sometimes may not marry because of their deformities, and may not have normal sexual relationships because of susceptibility to fractures or limited hip movement. However, mildly affected individuals may have no alteration in reproductive potential, and fertility does not appear to be diminished if the person survives through childhood. Spontaneous abortion does not seem to be increased, but perinatal loss is high if the child is severely affected and if there is difficulty with delivery. Much variability is seen in a given family, such that a very severely affected child may be born to a mildly affected parent and vice versa.

Severely affected individuals with osteogenesis imperfecta (particularly types I and III) who have short stature and kyphoscoliosis may have increasing respiratory compromise during pregnancy (Sengupta *et al.*, 1977; Willard, 1903), and would benefit from monitoring of pulmonary functions.

Many types of treatment have been tried for severe osteogeneis imperfecta—large doses of fluoride, magnesium (Solomons and Styner, 1969), vitamin C (Kurz and Eyring, 1974). However, none has had any definitive effect on the disease (McKusick, 1972). Certainly these types of therapy ought to be avoided during pregnancy since they might adversely affect a normal fetus. On the other hand, occasional OI patients develop hypocalcemia during pregnancy (Freda *et al.*, 1961); thus supplemental calcium should be part of the management of the pregnancy. Severely affected individuals may develop fractures with minimal trauma; thus, during pregnancy the change of body shape and imbalance may increase the likelihood of fractures from falls or trauma. There seems to be an increased risk of spontaneous fractures during pregnancy as well. Fractures occurring during pregnancy appear to heal normally. Women who have had previous fractures of their pelvis may have a contracted pelvis that will not permit vaginal delivery. In addition, the hyperextensibility of severe OI may make affected individuals very awkward during pregnancy. Some patients apparently have an increase in their hyperextensibility during pregnancy. An increased incidence of hernias also occurs during pregnancy. Surgical scars tend to be thin and stretch to give "cigarette paper" scars (Scott and Stiris, 1953). Women with severe OI may be more prone to uterine rupture and separation of the pubis (Sengupta *et al.*, 1977; Young and Gorstein, 1968). Stapels and Riva (1954) suggest routine cesarean section in affected women because of a possible increased risk of fractures during labor. Care must be taken in positioning a woman with OI for delivery in order to avoid fractures. General anesthesia is associated with an increased risk of malignant hyperthermia in osteogenesis imperfecta (Solomons and Myers, 1971). However, spinal anesthesia should be performed with great care because an anesthetic needle may fracture vertebrae with severe osteoporosis. A bleeding diathesis, apparently secondary to abnormal platelet function and increased capillary fragility (Siegel *et al.*, 1957), may occur. Placenta previa, hematomas, and postpartum hemorrhage have been reported (Young and Gorstein, 1968).

Since most affected women who become pregnant have dominantly inherited conditions (types I and IV), these women have a 50% chance of having an affected child. Prenatal diagnosis may be possible for the dominantly inherited variety of OI (Lachman and Hall, 1979). Radiographic diagnosis is probably not possible during the second trimester, but may be possible during the third trimester because of the presence of fractures *in utero* (Lachman and Hall, 1979). Radiographic prenatal diagnosis should be considered prior to birth because detection of an obviously affected infant might alter the mode of delivery, that is, cesarean section might be employed to prevent further fractures and head trauma, and the mother might be transferred to a large medical center where an intensive care nursery is available. Prenatal diagnosis by X-ray during the second trimester (early enough for elective termination of pregnancy) is feasible in the lethal perinatal autosomal recessive crumpled bone type of OI (type II) (Heller *et al.*, 1975). Pregnancies producing the autosomal recessive type II patients may also be complicated by polyhydramnios and stillbirth (Heller *et al.*, 1975; Monks, 1968). Prenatal diagnosis using amniotic fluid pyrophosphate (Roberts and Solomons, 1975) is unreliable.

Affected individuals with type I born to affected mothers neither seem to have earlier onset of fractures nor seem to be more severely affected than those born to affected fathers. There may be a slight increase in fetal wastage in the dominant types of OI. The most important neonatal problems are related to fractures, which may be present at birth in an affected infant; thus, the delivery should be in the least traumatic manner possible. Cesarean section should be considered if fractures are already known to be present *in utero*. Forceps should be avoided when delivering the infant with OI because the osteoporotic skull consists of wormian bones that do not provide the normal protection to the brain; therefore the skull does not have normal strength. The use of forceps may thus increase the chance of damaging the head and causing intracranial bleeding.

F. Arthrogryposis (AMC)

Arthrogryposis multiplex congenita (AMC) is a heterogeneous group of disorders in which individuals are born with multiple areas of joint limitation (Hall *et al.*, 1979). Some types of AMC have other associated anomalies. In addition, congenital joint contractures are seen in many specific syndromes.

The mother is rarely affected with contractures herself, but pregnancies carrying an affected individual can be associated with discomfort to the pregnant woman, often nausea and vomiting. Usually there is decreased fetal movement, or kicking only in one area. The birth of the affected child may be complicated because he does not assume the normal attitude. Breech and transverse lies are common. Many deliveries require cesarean section because of inadequate labor progression or abnormal fetal position. Affected infants often develop fractures in the process of delivery because their joints will not bend. In familial types of AMC, prenatal diagnosis may be possible by either "real time" ultrasound assessment of joint movement or by fetoscopy.

Women with arthrogryposis report normal onset of menarche. Sexual relationships may be difficult because of joint limitation and difficulty in finding comfortable positions, but fertility appears to be normal. A pregnancy in an affected woman may be complicated by respiratory compromise if scoliosis is present; however, pregnancies are usually normal. Delivery may be complicated by limitation of hip movement, and cesarean section may be necessary in an affected woman. Most known cases of affected women are delivered of normal infants.

II. CONDITIONS WITH GENERALIZED CONNECTIVE TISSUE INVOLVEMENT THAT ARE NOT USUALLY CONSIDERED HERITABLE DISORDERS OF CONNECTIVE TISSUE

A. Neurofibromatosis

Neurofibromatosis is a dominantly inherited condition characterized by tumors of peripheral nerves and abnormal cutaneous pigmentation. The basic defect is not known. One-half of the cases of neurofibromatosis represent new mutations (Crowe *et al.,* 1956). In "classic" neurofibromatosis, five or more cafe-au-lait spots, greater than 1 cm, are found associated with neurofibromas along peripheral nerves (Crowe *et al.,* 1956; Whitehouse, 1966). The cafe-au-lait spots are usually detectable shortly after birth, and almost always by 5 years of age. They increase in number throughout life. Neurofibromas usually are not detected until adolescence. Other serious complications include pseudoarthroses, scoliosis (20%), pheochromocytomas (5%), optic gliomas, malignant degeneration of neural tumors (5%), central nervous system tumors (10%), seizures (10%), and mental retardation (10%). There is a great deal of variation in severity.

No particular gynecologic problems have been observed, although both precocious and delayed puberty have been reported (Saxena, 1970). No complications of specific types of contraception have been reported. However, estrogens may increase the number of neurofibromas; thus, birth control pills should be used with careful observation. Leiomyomata of the uterus are not increased, nor are other genito-urinary malignancies.

Women with neurofibromatosis apparently have normal fertility, but their pregnancies may be complicated by increase in neurofibromas and pigmentation (Sharp and Young, 1937; Swapp and Main, 1973). Brickner (1906) called the lesions which first appear during pregnancy "fibroma molluscum gravidarum." The neurofibromas usually do not occur until after puberty and increase with increasing parity. In some case reports, large neurofibromas have dramatically increased in size during pregnancy (Ansari and Nagamani, 1976), often regressing after delivery. The hormonal basis for these changes is not known, but exaggeration of pigmentation during pregnancy can occur also (Swapp and Main, 1973) and may also regress after delivery. Burger and Kreutzinger (1966) also reported an increase in size of intracranial (central) neurofibromas during pregnancy, with resulting deafness and papilledema. Haynes (1969) reported the development and increase in size of both intraspinal and intracranial lesions during pregnancy. In such circumstances, termination of the pregnancy may markedly alleviate the symptoms. In addition, hypertension (Swapp and Main, 1973) has occurred with increased frequency, apparently related to catecholamine output. Pheochromocytomas have been reported in association with neurofibromatosis, but not specifically during pregnancy. Vascular changes (Reubi, 1944; Fye *et al.*, 1975) may lead to hemorrhage into large lesions, into the GI tract (Devereux *et al.*, 1975), or into the intracranial neurofibromas. Decreased fetal oxygenation might occur if a significant drop in maternal hematocrit occurred as a result of such a bleed. Haynes (1969) and Ansari and Nagamani (1976) suggested that there is an increased incidence of abortion and stillbirth in neurofibromatosis; however, this has not been well-documented in other studies. No special complications of labor and delivery have been reported.

There is recent evidence that affected infants born to affected mothers are more likely to have more severe complications with earlier onset of complications than those who either are born to affected fathers or are new mutations (Miller and Hall, 1979). Because of this observation, it may be possible in the future to predict from previous pregnancies whether offspring at risk will be affected with neurofibromatosis. From several studies, it appears if a mother has had one severely affected child, subsequent affected pregnancies are more likely to have a similar degree of severity and similar type of involvement. The child born to an affected mother needs careful observation in the newborn period and throughout his childhood to watch for serious or life-threatening complications of neurofibromatosis. At this time, prenatal diagnosis is not available; however, nerve growth factor assays may be helpful in the future.

B. Myotonic Dystrophy

Myotonic dystrophy is a neuromuscular disorder characterized by slowly progressive neurologic degeneration with wasting of muscle (88%), cataracts (75%), myotonia, early balding, hypogonadism, and mental retardation (60%). Obstetrical and gynecologic aspects are covered in Chapter 1. However, it should be noted that in this autosomal dominant condition linkage with secretor locus has been established (Mohr, 1954; Renwick *et al.*, 1971). Offspring of affected individuals have a 50% chance of inheriting the mutant gene. Affected children born to affected mothers may show more severe involvement than those born to affected fathers (Harper, 1975; Swift *et al.*, 1975). These severely affected children are born with joint limitation, mental retardation, hypotonia, and frequently die in the newborn period. Unaffected infants born to affected mothers are not abnormal. Prenatal diagnosis is possible in about 12% of families through linkage studies (Schrott *et all,* 1973; Omenn and Schrott, 1974). In severe congenital myotonic dystrophy, "real time" ultrasound may also be helpful in observing decreased fetal movement.

C. Systemic Lupus Erythematosis (SLE)

This condition is not considered a heritable, but rather an immunologic disorder of connective tissue. Many affected women show improvement of the inflammatory component of their illness during pregnancy, only to worsen again after delivery. Women with SLE may have hypertension and toxemia related to renal vascular compromise and may become very sick during pregnancy. In addition, antibodies to platelets, white cells, or red cells may cross the placenta and such antibodies may result in a neonate with thrombocytopenia or anemia (Vonderheid *et al.*, 1976). Infants of mothers with SLE may be born with skin lesions that resemble discoid lupus, implying some transplacental process during pregnancy (Hammar and Rönnerfält, 1977; Vonderheid *et al.*, 1976). More recently, it has been recognized that children born to mothers with lupus are at increased risk for having congenital heart block due to progressive cardiac fibrosis (Chameides *et al.*, 1977; McCue *et al.*, 1977). This congenital heart block apparently occurs because of placental transmission of antibodies that affect the fetal cardiac conduction system. Congenital structural cardiac anomalies also occur in infants born to mothers with SLE (Chameides *et al.*, 1977). Prenatal diagnosis of affected infants might be possible in maternal SLE by fetal EKG and Doppler studies of the fetal heart rate; however, neither method has reliably been applied.

III. SKELETAL DISORDERS

Until the last 15 years, the literature on skeletal dysplasias was extremely confusing because almost all types of disproportionate short stature were called achondroplasia. Distinctions between specific types of chondrodystrophies were not made. References to the mode of inheritance, the specific distinguishing characteristics of the type of dwarfism, and the complications of pregnancy were not documented. Unusual cases were reported out of curiosity. During the last 15 years, several classifications and reviews have helped to distinguish specific conditions (Lamy and Maroteaux, 1961; Rubin, 1964; Maroteaux *et al.*, 1968; Spranger *et al.*, 1974; Rimoin, 1975). These now serve as references for making the diagnoses of specific chondrodystrophies. Every effort should be made to arrive at a specific diagnosis in a given patient, since the inheritance pattern, natural history of the condition, and potential obstetrical and gynecologic problems depend on that diagnosis. In addition, if a chondrodystrophic woman is married to a chondrodystrophic man, the chances of having affected or even "doubly" affected children depend on the specific diagnoses of the parents. In 1941 Morch suggested there were specific complications associated with pregnancies in chondrodystrophic women, but not until the report by Tyson *et al.* (1970) was a study and comparison of the gynecologic and obstetrical problems of different chondrodystrophies available. Since then little has been published; thus, personal experience and personal communications are heavily relied upon for the conclusions of this chapter (L. E. Karp, private communication, 1978; D. L. Rimoin, private communication, 1978; C. I. Scott, private communication, 1978; J. E. Tyson, private communication, 1978).*

Disorders of short stature are most easily categorized into those conditions in which there is disproportionate short stature (chondrodystrophies or skeletal dysplasias) (Table I), and those conditions in which there is proportionate short stature (familial short stature, intrauterine growth retardation such as Russell-Silver dwarfism, and endocrine deficiencies). The lay public refers to the former as dwarfs and the latter as midgets. Among the types of disproportionate short stature (chondrodystrophies) are conditions which are recognizable in the newborn period and conditions which are not diagnosable until later in childhood. Of those recognized in the newborn period, there are many which are uniformly lethal, designated neonatal lethal chondrodystrophies, and there are some which may or may not be lethal, depending upon the severity in a given child. In addition, many conditions not recognized at birth may lead to severe or mild short stature. Table I is not intended to be all inclusive (nor could it be), but

*Lay information about specific types of short stature is available through the Little People of America, Box 126, Owatonna, Minnesota 55060 and the Human Growth Foundation, Maryland Academy of Science Building, 601 Light Street, Baltimore, Maryland 21230. Two booklets prepared for handicapped individuals by Planned Parenthood, *Toward Intimacy* (Shaul *et al.*, 1977) and *Within Reach* (Hale *et al.*, 1977), have useful information concerning chondrodystrophies. (Copies can be obtained by writing to Task Force on Concerns of Disabled Women, Planned Parenthood, Snohomish County, 2730 Hoyt, Everett, Washington 98201.) Special consideration during sexual counseling must be given to the need for individuals with chondrodystrophies to find positions of comfort for sexual relations.

TABLE I
Chondrodystrophies

Disorders of Short Stature	Condition	Characteristics	Genetics
I. Neonatal lethal			
A. Uniformly lethal	Achondrogenesis types I and II	Short trunk and extremities, absence of vertebral body ossification, very short underossified long bones, hydropic appearance	Autosomal recessive
	Thanatophoric dysplasia	Short limbs, relatively normal trunk with large head, flattened vertebral bodies, short broad long bones	Sporadic
	Noonan-Saldino	Short limbs, narrow chest, postaxial polydactyly, internal malformations	Autosomal recessive
	Majewski	Narrow chest, polysyndactyly, cleft lip or palate, short limbs, internal anomalies, short tibia	Unknown
B. Usually lethal	Chondrodysplasia punctata—rhizomelic type (rhizomelic Conradi syndrome)	Stippled epiphyses, disproportionate rhizomelic shortening of limbs, joint contractures, flat face, cataracts, skin changes	Autosomal recessive
	Infantile osteopetrosis	Dense bones, thick dense skull, metaphyseal irregularity, hepatosplenomegaly, macrocephaly, failure to thrive	Autosomal recessive
	Infantile hypophosphatasia	Short underossified long bones, absence of ossification of skull, low serum alkaline phosphatase, ↑ urinary phosphoethanolamine	Autosomal recessive
	Neonatal lethal osteogenesis imperfecta (crumpled bone)	Short deformed limbs, soft large skull, prominent eyes, blue sclerae, broad bones (crumpled) with multiple fractures	Autosomal recessive
	Homozygous achondroplasia	Short limbs, relatively normal trunk, large head, short broad long bones, both parents with achondroplasia	Autosomal recessive
II. Recognizable at Birth			
A. Variable survival	Asphyxiating thoracic dysplasia (Jeune)	Long narrow thorax, postaxial polydactyly, respiratory distress, spur-like projection of the acetabular roof	Autosomal recessive

	Diastrophic dysplasia	Multiple joint contractures, club feet, cleft palate, cauliflower ears, scoliosis, dislocated hips, club hands	Autosomal recessive
	Chondroectodermal dysplasia (Ellis–van Creveld)	Distal shortening of limbs, postaxial polydactyly, multiple frenula, dental anomalies, cardiac defects	Autosomal recessive
	Kneist syndrome	Short trunk, short limbs with prominent joints, lumbar lordosis, cleft palate, myopia, deafness, flat face	Autosomal dominant
	Spondyloepiphyseal dysplasia congenita	Short trunk, flat face, normal hands and feet, myopia, genu valgum, broad chest, delayed ossification, coxa vara	Autosomal dominant
	Chondrodysplasia punctata—asymmetric type (Conradi–Hunermann)	Stippled epiphyses, asymmetric shortening of limbs, scoliosis, ichthyosiform skin, flat face, cataracts	Autosomal dominant
B. Usually with normal life span	Achondroplasia	Short limbs with proximal shortening, large head, prominent forehead, trident hand, short pedicles	Autosomal dominant
	Hypochondroplasia	Short limbs, relatively normal skull, short broad long bones, occasional mental retardation	Autosomal dominant
	Cartilage hair hypoplasia (metaphyseal chondrodysplasia—McKusick type)	Short limbs, sparse fine hair, ligamentous laxity, metaphyseal irregularities, long fibulae	Autosomal recessive
III. Usually Not Recognized at Birth	Spondyloepiphyseal dysplasia (many types)	Vertebral and epiphyseal abnormalities, all combinations and severities	Variable
	Pseudoachondroplasia	Normal newborn, normal skull, short limbs, arthritis, hyperextensibility	Autosomal dominant
	Vitamin D-resistant rickets (hypophosphatemic familial rickets)	Metaphyseal irregularities, more severe in lower limbs, bowing of legs, dental changes, hypophosphatemia, elevated alkaline phosphatasia	X-Linked dominant

rather to give a point of reference for the following discussion. Specific diagnostic criteria should be obtained by consulting the appropriate references.

IV. NEONATAL LETHAL CHONDRODYSTROPHIES

Among those conditions which are uniformly lethal in the newborn period are achondrogenesis types I and II, thanatophoric dysplasia, and the short-ribbed polydactyly syndromes of Noonan-Saldino and Majewski (see Spranger *et al.*, 1974, for details of specific conditions). There are several conditions in which infants may live past the newborn period but never past infancy. These include chondrodysplasia punctata-rhizomelic type (rhizomelic type of Conradi syndrome), infantile osteopetrosis, infantile hypophosphatasia, the lethal perinatal osteogenesis imperfecta (crumpled bone type II, Sillence *et al.*, 1978), and homozygous achondroplasia. The inheritance of these conditions depends upon the specific condition, but in none is the mother similarly affected.

In all these conditions complications may be associated with pregnancies; however, the exact frequency of complications is unknown. Polyhydramnios is a common complication in the late second and third trimesters. Frequently, the affected infants are prenatally diagnosed because of polyhydramnios during the early third trimester. If a maternal abdominal X-ray is taken, or if ultrasound studies are performed because of polyhydramnios, abnormalities of the limbs, and head size or bones are frequently recognized (Cremin and Shaff, 1977; Anteby *et al.*, 1977). Many fetuses with neonatal lethal dwarfing conditions are spontaneously aborted during the second trimester or are stillborn during the early third trimester. In the past, these aborted fetuses were not recognized as having specific conditions, but rather were labeled with the term *achondroplasia,* leading to the misconception that neonates with achondroplasia show decreased survival.

Prenatal diagnosis is available for many of these conditions by radiography (Omenn *et al.*, 1977; Lachman and Hall, 1979), probably early enough in the second trimester for elective pregnancy termination. In the rhizomelic Conradi syndrome and infantile osteopetrosis (Golbus *et al.*, 1976), prenatal diagnosis might be difficult because the X-ray changes may not be severe enough to be seen during the second trimester. Infantile hypophosphatasia can probably be diagnosed by a combination of chemical and ultrasound studies (Rudd *et al.*, 1976).

The number of liveborn offspring of unaffected mothers may be decreased because of increased frequencies of stillborns and spontaneously aborted affected fetuses. However, unaffected mothers apparently do not have any increased trouble becoming pregnant. Management of pregnancies depends upon complications such as polyhydramnios. Early induction of labor and termination of the pregnancy may thus be appropriate. Since affected children are small, the delivery itself is usually relatively easy except for possible megalocephaly in thanatophoric dwarfism and homozygous achondroplasia.

No supportive efforts are recommended for these affected children since the conditions are inevitably lethal. However, autopsy should be done, both to establish the diagnosis and to learn as much as possible about the basic defects.

V. CHONDRODYSTROPHIES RECOGNIZED AT BIRTH WITH VARIABLE SURVIVAL

Many dwarfing conditions are sometimes lethal in the newborn period, but compatible with survival (see Table I). These include the autosomal recessive conditions asphyxiating thoracic dystrophy (Jeune syndrome), diastrophic dysplasia and Ellis-van Creveld (chondroectodermal dysplasia), and the autosomal dominant conditions Kniest syndrome, SED congenita and Conradi-Hunermann syndrome. (See Spranger *et al.*, 1974, for details of specific conditions.) Since these conditions are compatible with both life and normal intelligence, aggressive care should be given to the affected individuals in the newborn period. No specific complications occur in the pregnancies of normal women carrying affected fetuses. Polyhydramnios and toxemia have not been reported. In conditions with polydactyly, cleft lip or cleft palate, prenatal diagnosis may be possible by fetoscopy or ultrasonography (Mahoney and Hobbins, 1977). Conditions with significant shortening of long bones at birth may be diagnosable prenatally by radiography (Omenn *et al.*, 1977; Lachman and Hall, 1979) or ultrasound.

A. Asphyxiating Thoracic Dysplasia

The small chest usually causes severe respiratory compromise in the newborn period. If the infant survives, the chest expands as the child grows older and affected individuals usually show normal respiratory function by early childhood. Although smaller than their sibs, affected individuals may reach a height within normal adult limits. If an individual survives the respiratory compromise of the newborn period, cystic renal tubular dysplasia or progressive glomerulosclerosis may occur during childhood. There are no known complications of menarche or menses in affected women. The pelvis may be somewhat small and cesarean section would need to be considered. Asphyxiating thoracic dysplasia shows autosomal recessive inheritance; thus, children of an affected individual would not be expected to be affected unless the mother married a heterozygote or another homozygote. If the mother does not have renal or pancreatic complications there is no reason to expect adverse effects of pregnancy carrying unaffected fetuses; however, we are unaware of reported cases of pregnancies in affected women.

B. Diastrophic Dysplasia

Respiratory compromise due to small chest, cleft palate, or tracheomalacia may lead to death in the newborn period. If an affected individual survives the newborn period (as most do), diastrophic dysplasia is complicated by scoliosis, equinovarus club foot, club hands, limitation of joint movement, small pelvis, and dislocated hips; therefore, sexual intercourse may be difficult for such individuals. There are no complications of menarche, and fertility appears to be quite normal. Several affected women have had children; deliveries have all required cesarean section because of a small contracted pelvis; there have been no other complications. Occasional patients with diastrophic dwarfism have spinal stenosis and cervical vertebral dysplasia; therefore, spinal (either epidural or caudal) anesthesia should be attempted with care. Many individuals with diastrophic dysplasia have surgically repaired cleft palates and an occasional patient has tracheomalacia or incomplete tracheal cartilage; therefore, intubation may be difficult. Because of hip contractures and lumbar lordosis, the uterus may tip anteriorly and make ambulation difficult during pregnancy, and cesarean section technically more difficult. Anterior uterine displacement also predisposes to deflexion of the fetal head (face or brow position). Since diastrophic dysplasia is an autosomal recessive condition, it would be unlikely for a child of an affected woman to be affected unless she married a heterozygote or another affected individual. All children of affected individuals have been normal and apparently unaffected by the pregnancy. In cases of diastrophic dysplasia characterized by severe scoliosis, cardiac or respiratory compromise may occur.

C. Chondroectodermal Dysplasia (Ellis–van Creveld Syndrome)

This autosomal recessive trait is characterized by postaxial polydactyly, distal shortening of limbs, hypoplastic nails, small thorax, multiple frenula, and dental anomalies. Of affected individuals, 50% have congenital heart disease, primarily atrial septal defects of single atrium. Infants may thus die from respiratory or cardiovascular compromise. Menarche occurs at a normal age with no particular gynecologic problems (hypospadius and cryptorchidism occur in males). Adult height is often within normal limits, although small for the family. Few unusual problems are associated with pregnancy, unless the affected woman has congenital heart disease. However, the pelvis is contracted; thus, most affected women would be delivered by cesarean section. If congenital heart disease is present, prophylactic antibiotics are indicated. Respiratory function may be slightly decreased because of a small chest. Because the condition is autosomal recessive, affected children would not be expected. No adverse problems have been observed in normal fetuses born to affected women. Prenatal diagnosis is possible by fetoscopy or sonography (Hobbins, 1977).

D. Kniest Dysplasia

This disease is characterized by flat facies, short trunk, short limbs with limitation of joint movement and prominent joints, and cleft palate (50%). Chronic otitis media, deafness, myopia and retinal detachments, and umbilical and inguinal hernias have also been described. Kniest dysplasia may be a genetically heterogeneous condition, but in several families autosomal dominant inheritance has been observed. Severely affected infants die in the newborn period. However, many cases have survived to adulthood without evidence of infertility. If kyphosis and scoliosis are present, cardiovascular compromise may occur during pregnancy. Delivery usually requires cesarean section because of contracted pelvis, but no other specific complications have been reported. However, coexisting cleft palate and tracheomalacia may require special care during intubation. Failure of fusion of the odontoid to C2 has also been reported; thus, cervical stability must be evaluated prior to general anesthesia in order to avoid spinal cord compression. Affected children born to affected women are no more severely affected than those born to unaffected women.

E. Spondyloepiphyseal Dysplasia (SED) Congenita

This autosomal dominant chondrodystrophy shows significant variability in expression. It is characterized by a short trunk and neck, delayed ossification of pubis, normal hands and feet, flat facies, and myopia with retinal detachment (50%). Most individuals represent new mutations; however, dominant transmission has been observed. Severely affected individuals with significant platyspondyly or scoliosis have cardiac and respiratory compromise. Some infants die, but with growth the relative size of chest and heart improves. Individuals with severe dysplasia of femoral heads and limited hip movement may have difficulty positioning for sexual intercourse. The pelvis is small and cesarean section is necessary for delivery. Affected individuals are prone to retinal detachments; therefore, careful anesthetic management avoiding trauma to the eye is important during any surgical procedure. Many individuals with SED congenita have cleft palates or submucous clefts, and thus are more prone to chronic otitis. No other respiratory or anesthetic problems have been reported. Odontoid hypoplasia has been reported, so that care must be used in manipulation of the neck. Because of the narrow interspaces between the vertebrae, epidural and caudal anesthesia is probably contraindicated. Affected children born to affected women do not appear to be more severely affected than those born to unaffected women.

F. Conradi–Hunermann Syndrome (Chondrodysplasia Punctata—Dominant Asymmetric Type)

This is an autosomal dominant condition with extreme intrafamilial variability. It involves asymmetric shortening of the long bones associated with epiphyseal stippling, ichthyosiform skin changes, mid-face hypoplasia, scoliosis, cataracts, small depressed nose and alopecia. Menarche occurs at a normal age and there is no decrease in fertility.

Many affected women have had affected children. Affected children usually appear no more severely affected than their mothers, or than cases born to normal women. Bergström *et al.* (1972) reported a severely affected male born to a more mildly affected female. The diagnosis was by radiography at 32 weeks gestation; however, it is not clear whether the epiphyseal stippling is present by 24 weeks to facilitate prenatal diagnosis in time to consider termination of pregnancy. Severely affected infants may be stillborn or die during infancy. Polyhydramnios and hydrops fetalis are seen in such cases. After the newborn period, life expectancy is normal. Other maternal complications of pregnancy include cardiorespiratory compromise related to scoliosis, limitation of hip movement due to femoral head dysplasia, and small dysplastic pelvis that may necessitate cesarean section.

VI. CHONDRODYSTROPHIC CONDITIONS WITH APPARENTLY NORMAL LIFE SPAN THAT ARE RECOGNIZED IN THE NEWBORN PERIOD

A few general considerations are applicable to all chondrodystrophy patients discussed in this section. Most will need cesarean sections because of small or distorted pelvises. The vagina is usually short and narrow, and because of pelvic tilt and contracted pubis the sacral-vertebral prominence is often at a depth of only 7–9 cm. Many babies present in breech. In addition, early delivery may be necessary because of maternal respiratory compromise.

A. Achondroplasia

"Classic" achondroplasia has been clearly defined in recent years (Langer *et al.*, 1967; Spranger *et al.*, 1974). The term *achondroplasia;* however, has been used in the past as a "catch-all" term applied to many disparate forms of short-limbed dwarfism, resulting in such misleading statements as "the majority of individuals affected with achondroplasia died in the newborn period." We now recognize neonatal deaths represented other neonatal lethal chondrodystrophies, e.g., achondrogenesis and thanatophoric dwarfism. A child with classic achondroplasia should not have a complicated newborn period unless trauma to the head occurs during delivery. Indeed, life expectancy is probably normal in achondroplasia unless spinal cord compression occurs. Such compression occurs in about 40–50% of cases and is incapacitating in 5–10%.

Achondroplasia is autosomal dominantly inherited; however, as many as 75–80% of cases represent new dominant mutations (Murdoch *et al.*, 1970). New mutations are often associated with advanced paternal age. There is a 50% risk that an infant born to a mother or a father with achondroplasia who is married to an average sized individual will also be affected with achondroplasia. Because achondroplasia is the most frequent type of disproportionate short stature, a number of achondroplastic individuals are married to each other; thus, homozygous achondroplasia (Hall *et al.*, 1969) and other genetic compounds (McKusick *et al.*, 1973) may occur. Individuals with two achondroplasia alleles (homozygous achondroplasia) are so severely affected that they almost always die in the newborn period. However, we are aware of two homozygous cases that have survived past the newborn period; both were severely retarded and severely dwarfed.

Individuals with heterozygous classic achondroplasia undergo puberty at a normal to slightly early age, but occasionally have premature menarche. There may be an increased sensitivity to estrogen in achondroplasia. In addition to early menarche, perhaps 60% of women with achondroplasia have an increased number of uterine leiomyomata, may be particularly susceptible to migraine headaches associated with the menstrual cycle, and frequently request reduction mamoplasties because of the relatively large size of their breasts. 10–20% of all achondroplastic women undergo menopause in their mid-thirties.

During pregnancy achondroplastic women find it difficult to move around because of their awkward shape and size. They tend to be overweight before pregnancy and to gain weight easily during pregnancy. There may be an increased incidence of diabetes in achondroplastic individuals. There seems to be an increased incidence of glucosuria during pregnancy, but not at other times (Tyson *et al.*, 1970). Symptoms of spinal compression may also increase during pregnancy. The pregnancy itself may be complicated by cardiorespiratory compromise during the second trimester. Achondroplastic women may be unable to sleep supine during the last months of gestation. Cesarean section may be required at 36 weeks because of this cardiorespiratory compromise. Indeed, all women with achondroplasia must have cesarean sections because of their contracted pelves. Labor is inappropriate; however, fetal lung maturity must naturally be maximized before intervention. General anesthesia should be used for the cesarean section because spinal stenosis secondary to short pedicles makes it very difficult to give caudal, epidural, or spinal anesthesia. Care must also be exercised during general anesthesia because it is sometimes difficult to intubate an individual with achondroplasia. Careless manipulation of the neck during surgery or anesthesia may result in neural compromise because of spinal stenosis and a small foramen magnum.

An affected infant born to an affected mother may be somewhat larger in weight (J. E. Tyson, personal communication, 1978) than a normal or an achondroplastic infant born to an average sized woman. Since the achondroplasia fetus has a relatively large head with increased biparietal and occipitofrontal diameters (Horton *et al.,* 1978), its birth to an unaffected mother may nessitate cesarean section. An affected child in prolonged labor because of cephalopelvic disproportion may suffer neural damage, but otherwise would encounter no particular problems in the newborn period. Classic achondroplasia cannot be diagnosed prenatally (Golbus and Hall, 1974; Hall *et al.,* 1979); however, homozygous achondroplasia probably can be diagnosed by X-ray by 24 weeks of pregnancy (Lachman and Hall, 1979). Ultrasound may prove to be more useful than X-ray.

A comment concerning birth control in women with achondroplasia is in order because oral contraceptives increase the size of leiomyomata. Barrier contraception or possibly intrauterine devices may be more appropriate; however because of their short arms achondroplastic women may not be able to manage these methods of contraception.

B. Hypochondroplasia

This condition is very similar to achondroplasia, but milder in expression. The skull is normal and occasionally mental retardation is present. Hypochondroplasia is also an autosomal dominant condition, but there is no overlap between families with hypochondroplasia and achondroplasia. Most cases represent new dominant mutations. Gynecologic and obstetric complications are similar in the two disorders. Specifically all pregnant women require cesarean section, although the pelvis is generally somewhat larger in hypochondroplasia than in achondroplasia. No cases of severe spinal stenosis have been reported in hypochondroplasia.

C. Cartilage Hair Hypoplasia

This autosomal recessive metaphyseal dysplasia results in extremely short stature, joint hyperextensibility, and thin, sparse, slow-growing hair. The condition may be heterogeneous. Immune defects have been seen in some affected individuals. Fertility and age of menarche appear normal, and normal children born to affected women do not seem to have any particular problems. Women with cartilage hair hypoplasia may have cardiovascular and respiratory compromise during pregnancy because of their short trunks. All term deliveries require cesarean section. No particular anesthetic concerns are known. Because the condition is autosomal recessive, one would not expect affected women to have affected children unless an individual married a heterozygote or another affected individual (McKusick, 1964).

VII. CHONDRODYSTROPHIES NOT GENERALLY RECOGNIZED IN THE NEWBORN PERIOD

A. Spondyloepiphyseal Dysplasias

Spondyloepiphyseal dysplasias (SED) are a group of chondrodystrophies primarily involving the vertebrae and epiphyses. Many spondyloepiphyseal dysplasias have been described (Maroteaux *et al.*, 1968; Spranger *et al.*, 1974) and the genetic basis is frequently known; however, little is recorded about the natural histories or obstetrical complications. None of these dysplasias appear to have problems with fertility. Menarche seems to be at a normal time and apparently there are no other particular gynecologic problems. Hypothyroidism can masquerade as a spondyloepiphyseal dysplasia and must always be ruled out in the growing child. Problems with pregnancy reflect whether or not the individual has scoliosis or a small chest and, hence, respiratory or cardiovascular compromise. Women usually feel well during pregnancy, in contrast to the usual presence of joint pain from degenerative arthritis related to their epiphyseal dysplasia at other times. Particular care should be taken during general anesthesia in any woman with spondyloepiphyseal dysplasia because dysplasia of the vertebrae may be associated with odontoid hypoplasia and secondary instability of C1 and C2. Subluxation of C1 or C2 can also occur during general anesthesia.

B. Pseudoachondroplasia

This autosomal dominant condition is usually not diagnosed until the second or third year of life, and many cases represent new mutations. Affected individuals have trunks of relatively normal length, short limbs, normal craniofacial appearance, hyperextensibility, and significant epiphyseal dysplasia leading to degenerative arthritis. Some have scoliosis, "knock-knees," or bowed legs. The condition is relatively variable even within a family. Pregnancy is relatively uncomplicated. No special considerations of anesthesia seem to be necessary, although birth is always by cesarean section due to pelvic contraction. As in other individuals with short stature, there may be some respiratory or cardiovascular compromise during late stages of pregnancy, particularly if there is scoliosis. In those cases in which an affected woman has an affected child, the child does not seem to be more severely affected than those born to unaffected mothers.

C. Hypophosphatemic Familial Rickets (Vitamin D-Resistant Rickets, VDRR)

In this X-linked dominant condition, affected women, who show normal fertility, may be on therapy with phosphorus supplements and high doses of vitamin D. Because vitamin D could cause hypercalcemia in newborns, affected women should probably be withdrawn from such therapy during pregnancy, with phosphorus and calcium levels carefully monitored. However, we have observed at least one case in which the mother continued on therapy throughout pregnancy and was delivered of a normal child. Affected infants do not appear to be more severely affected when born to an affected mother than when born to an unaffected average sized mother. The degenerative arthritis associated with VDRR seems to improve during pregnancy. Cesarean section is not routinely necessary, the pelvic dimensions often being adequate for vaginal delivery. No particular gynecologic problems have been recognized, although one patient complained that when her phosphorus was low that her libido was decreased.

VIII. OTHER CHONDRODYSTROPHIES

It is impractical to offer an all-inclusive list of the skeletal dysplasias and their gynecologic and obstetric problems. The reader is referred to the comprehensive reviews for specific entities. Pregnancies have been reported in metatropic dwarfism, pyknodysostosis, osteopetrosis (dominant type, Delahaye *et al.*, 1976), Jansen's metaphyseal dysplasia, several spondylometaphyseal dysplasias, multiple epiphyseal dysplasia, osteoplastica, and cranial metaphyseal dysplasia. Unexpected complications were not observed during these pregnancies. However, many women with chondrodystrophies may require cesarean sections because of small pelvic size. Special considerations for anesthesia have been emphasized above, i.e., special care not to sublux C1 or C2, care during intubation, care to avoid trauma to the eye, and care with any type of spinal anesthesia to avoid trauma to the spinal cord. Pregnancies may also be complicated by cardiorespiratory compromise, but symptoms of degenerative arthritis usually improve with pregnancy. Menarche in women with other chondrodystrophies appears to be normal with no specific gynecologic problems reported.

IX. CONDITIONS LEADING TO MILD SHORT STATURE

There are many other genetic conditions in which mild generalized chondrodystrophic changes lead to mild short stature. These include nail-patella syndrome (osteo-onycho-dysostosis), trichorhinophalangeal syndrome, cleidocranial dysostosis, multiple exostosis, dyschondrosteosis, fibrodysplasia ossificans progressiva, the Weill-Marchesani syndrome, Larsen syndrome, Albright osteodysplasia syndrome, and the idiopathic osteolyses (see Spranger *et al.*, 1974). As might be expected, complications again are related to shape and size of the pelvis, severe changes requiring cesarean section. In hereditary multiple exostoses, exostosis may occur in the pelvis and produce dystocia during fetal descent. Women with the above conditions may also have limitation of movement that leads to cardiopulmonary compromise, difficulty during delivery, and difficulty during sexual intercourse. In conditions characterized by hyperextensibility or recurrent dislocations, there may be difficulty in ambulation during pregnancy. In other conditions the major concern may be genetic. For example, if two individuals affected with dyschondrosteosis marry, a homozygous child may be severely dwarfed but prenatally detectable by X-ray. Pregnancy by affected women does not seem to produce more severely affected children in any of these conditions.

A final genetic condition worth mentioning is the X-linked dominant Albright osteodystrophy (pseudohypoparathyroidism). In this trait severely affected children have seizures and mental retardation, but can be born to mildly affected parents. Albright osteodystrophy is not diagnosable prenatally, but severely affected and mildly affected individuals frequently occur in the same family. Calcium should be monitored during the pregnancy.

In summary, there are now recognized many specific skeletal dysplasias and heritable disorders of connective tissues. Many appear to have predictable complications of pregnancy, but there is little reported data. Any information about specific complications in these conditions should be reported, so that unusual complications can be monitored and prevented. In managing such patients, the first and most important step is establishing a specific diagnosis.

REFERENCES

Altman, L. K., Fialkow, P. J., Parker, F., and Sagebiel, R. W. (1974). Pseudoxanthoma elasticum. *Arch. Intern. Med.* **134,** 1048–1054.

Ansari, A. H., and Nagamani, M. (1976). Pregnancy and neurofibromatosis (von Recklinghausen's disease). *Obstet. Gynec.* **47,** Suppl., 25s–29s.

Anteby, S. O., Aviad, I., and Weinstein, D. (1977). Prenatal diagnosis of achondrogenesis. *Radiol. Clin. Biol.* **46,** 109–114.

Barabas, A. P. (1966). Ehlers-Danlos syndrome: Associated with prematurity and premature rupture of foetal membranes; possible increase in incidence. *Br. Med. J.* **2,** 682–684.

Beighton, P. (1970). "The Ehlers-Danlos Syndrome." Heinemann, London.

Bergström, K., Gustavson, K. H., and Jorulf, H. (1972). Chondrodystrophia calcificans congenita (Conradi's disease) in a mother and her child. *Clin. Genet.* **3,** 158–161.

Bornstein, P., and Byers, P. H. (1980). Disorders of collagen metabolism. *In* "Metabolic Control and Disease" (P. Bondy and L. Rosenberg, eds.), 8th ed., pp. 1089–1153. Saunders, Philadelphia, Pennsylvania.

Brickner, S. M. (1906). Fibroma molloscum gravidarum: A new clinical entity. *Am. J. Obstet. Gynecol.* **53,** 191.

Bundey, S. E. (1974). Detection of heterozygotes for myotonic dystrophy. *Clin. Genet.* **5,** 107–109.

Bundey, S. E., and Carter, C. O. (1970). Early recognition of heterozygotes for the gene for dystrophia myotonica. *J. Neurol., Neurosurg. Psychiatry* **33,** 279–293.

Burger, G., and Kreutzinger, E. (1966). Beitrag Zu morbus Recklinghausen und Schwangerschaft. *Aschr. Aerztl. Fortbild.* **60,** 384.

Byers, P. H., Narayanan, A. S., Bornstein, P., and Hall, J. G. (1976). An X-linked form of cutis laxa due to deficiency of lysyl oxidase: The collagen and elastin crosslinking enzyme. *Birth Defects, Orig. Artic. Ser.* **12,** 293-298.

Cava, E. F., and Dreier, R. L. (1970). The Marfan syndrome in pregnancy: A case report. *Trans. Pac. Coast Obstet. Gynecol. Soc.* **38,** 129-133.

Chameides, L., Truex, R. C., Vetter, V., Rashkind, W. J., Galioto, F. M., and Noonan, J. A. (1977). Association of maternal systemic lupus erythematosus with congenital complete heart block. *N. Engl. J. Med.* **297,** 1204-1207.

Cremin, B. J., and Shaff, M. I. (1977). Ultrasonic diagnosis of thanatophoric dwarfism in utero. *Radiology* **124,** 479-480.

Crowe, F. W., Schull, W. J., and Neel, J. F. (1956). "A Clinical, Pathological and Genetic Study of Multiple Neurofibromatosis." Thomas, Springfield, Illinois.

Dallaire, L., Cantin, M., Nelancon, S. B., Perreault, G., and Potier, M. (1976) A syndrome of generalized elastic fiber deficiency with leprechaunoid features: A distinct genetic disease with an autosomal recessive mode of inheritance. *Clin. Genet.* **10,** 1-10.

Danforth, D. N., Manalo-Estrella, P., and Buckingham, J. C. (1964). The effect of pregnancy and of Enovid on the rabbit vasculature. *Am. J. Obstet. Gynecol.* **88,** 952-962.

Delahaye, R. P., Metges, P. J., Anglade, J. P., Malmezat, X., and Pascal-Suisse, P. (1976). Découverte simultanée d'une ostéopétrose chez la mère et le foetus à l'occasion d'une radiopelvimétrie. *J. Radiol. Electrol., Med. Nucl.* **57,** 359-362.

Devereux, R. B., Koblenz, L. W., Cipriano, P., and Gray, G. F. (1975). Gastrointestinal hemorrhage—an unusual manifestation of neurofibromatosis. *Am. J. Med.* **58,** 135-138.

Donaldson, L. B., and de Alvarez, R. R. (1965). The Marfan syndrome and pregnancy. *Am. J. Obstet. Gynecol.* **92,** 629-641.

Elias, S., and Berkowitz, R. L. (1976). The Marfan syndrome and pregnancy. *Obstet. Gynecol.* **47,** 358-361.

Foukas, M., and Skouteris, A. (1966). Ober das Zusammentreffen von Morbus Recklinghausen: Hyperparathyroidismus und graviditat. *Zentralbl. Gynaekol.* **30,** 999.

Freda, V. J., Vosburgh, G. J., and DiLiberti, C. (1961). Osteogenesis imperfecta congenita. *Obstet. Gynecol.* **18,** 535-547.

Fye, K. H., Jacobs, R. P., and Roe, R. L. (1975). Vascular manifestations of von Recklinghausen's disease. *West. J. Med.* **122,** 110-116.

Golbus, M. S., and Hall, B. D., (1974). Failure to diagnose achondroplasia in utero. *Lancet* **1,** 629.

Golbus, M. S., Koerper, M. A., and Hall, B. D. (1976). Failure to diagnose osteopetrosis in utero. *Lancet* **2**(7997), 1246.

Goltz, R. W., Hult, A. M., Goldfarb, M., and Gorlin, R. J. (1965). Cutis laxa, a manifestation of generalized elastolysis. *Arch. Dermatol.* **92,** 373-387.

Hale, J., Norman, A. D., Bogle, J., and Shaul, S. (1977). Within reach: Providing family planning services to physically disabled women. Planned Parenthood Task Force on Concerns of Physically Disabled Women, 1977.

Hall, J. G., Dorst, J. P., Taybi, H., Scott, C. I., Langer, L. O., and McKusick, V. A. (1969). Two probable cases of homozygosity for the achondroplasia gene. *Brith Defects, Orig. Artic. Ser.* **5,** 24-34.

Hall, J. G., Graham, C. B., Golbus, M. S., Pagon, R. A., Luthy, D. A., and Filly, R. A (1979a). Failure of early prenatal diagnosis in classical achondroplasia. *Am. J. Med. Gen.* **3,** 371-375.

Hall, J. G., Arthrogrypopis. *In* "Klinische Genetik in der Pädiatrie, Symposium in Mainz," Spranger, J., Gehler, J., Tolksdorf, M., eds. New York: Georg Thieme Verlag Stuttgart, 1980.

Halpern, B. L., Char, F., Murdoch, J. L., Horton, W. B., and McKusick, V. A. (1971). A prospectus on the prevention of aortic rupture in the Marfan syndrome with data on survivorship without treatment. *Johns Hopkins Med. J.* **129,** 123–129.

Hammar, H., and Rönnerfält, L. (1977). Annular erythemas in infants associated with autoimmune disorders in their mothers, report on three cases. *Dermatologica* **154,** 115–127.

Harper, P. S. (1975). Congenital myotonic dystrophy in Britain. *Arch. Dis. Child.* **50,** 514–521.

Haynes, D. M. (1969). "Medical Complications During Pregnancy." McGraw-Hill, New York.

Heller, R. H., Winn, K. J., and Heller, R. M. (1975). The prenatal diagnosis of osteogenesis imperfecta congenita. *Am. J. Obstet. Gynecol.* **121,** 572–573.

Hirst, A. E., Johns, V. J., and Kime, S. W. (1958). Dissecting aneurysm of the aorta: A review of 505 cases. *Medicine (Baltimore)* **37,** 217–279.

Horton, W. A., Rotter, J. I., Rimoin, D. L., Scott, C. I., and Hall, J. G. (1978). Standard growth curves for achondroplasia. *J. Pediatr.* **93,** 435–438.

Irey, N. S., Manion, W. C., and Taylor, H. B. (1970). Vascular lesions in women taking oral contraceptives. *Arch. Pathol.* **89,** 1–8.

Kanof, A. (1952). Ehlers-Danlos syndrome; report of a case with suggestion of a possible causal mechanism. *Am. J. Dis. Child.* **83,** 197.

Kaye, C. I., Fisher, D. E., and Esterly, N. B. (1974). Cutis laxa, skeletal anomalies, and ambiguous genitalia. *Am. J. Dis. Child.* **127,** 115–117.

Kurz, D., and Eyring, E. J. (1974). Effects of vitamin C on osteogenesis imperfecta. *Pediatrics* **54,** 56–61.

Kushner, I. (1931). Pregnancy as a complication of neurofibromatosis (von Ricklinghausen's disease). *Am. J. Obstet. Gynecol.* **21,** 116.

Lachman, R., and Hall, J. G. (1979). The radiographic prenatal diagnosis of generalized bone dysplasias and other skeletal abnormalities. *Birth Defects, Orig. Artic. Ser.* XV(5A):3–24, 1979.

Lamy, M., and Maroteaux, P. (1961). "Les Chondrodystrophies Génotypiques." L'Expansion Scientifique Francaise, Paris.

Langer, L. O., Baumann, P. A., and Gorlin, R. J. (1967). Achondroplasia. *Am. J. Roentgenol., Radium Ther. Nucl. Med.* [N.S.] **100,** 12.

Lewis, E. (1948). Cutis laxa. *Proc. R. Soc. Med.* **41,** 864.

McCue, C. M., Mantakas, M. E., Tingelstad, J. B., and Ruddy, S. (1977). Congenital heart block in newborns of mothers with connective tissue disease. *Circulation* **56,** 82–90.

McFarland, W., and Fuller, D. E. (1964). Mortality in Ehlers-Danlos syndrome due to spontaneous rupture of large arteries. *N. Engl. J. Med.* **271,** 1309.

McKusick, V. A. (1964). Dwarfism in the Amish. *Trans. Assoc. Am. Physicians* **77,** 151–168.

McKusick, V. A. (1972). "Heritable Disorders of Connective Tissue," 4th ed. Mosby, St. Louis, Missouri.

McKusick, V. A., Kelley, T. E., and Dorst, J. P. (1973). Observations suggesting allelion of the achondroplasia and hypochondroplasia genes. *J. Med. Genet.* **10,** 11–16.

Mahoney, M. J., and Hobbins, J. C. (1977). Prenatal diagnosis of chondroectodermal dysplasia (Ellis-van Creveld syndrome) with fetoscopy and ultrasound. *N. Engl. J. Med.* **297,** 258–260.

Manalo-Estrella, P., and Barker, A. E. (1967). Histopathologic findings in human aortic media associated with pregnancy. A study of 16 cases. *Arch. Pathol.* **83,** 336.

Manalo-Estrella, P., Danforth, D. N., and Buckingham, J. C. (1965). Regression rate of vascular effects induced by pregnancy and by norethynodrelmestranol. *Fertil. Steril.* **16,** 81–84.

Mandel, W., Evans, E. W., and Walford, R. L. (1954). Dissecting aortic aneurysm during pregnancy. *N. Engl. J. Med.* **251,** 1059–1061.

Maroteaux, P., Wiedemann, R., Spranger, J., Koslowski, K., and Lenzi, L. (1968). "Essai de Classification des Dysplasies Spondylo-Epiphysaires." Simep Editions, Lyon.

Masi, A. T., and Dugdale, M. (1970). Cerebro-vascular diseases associated with the use of oral contraceptives. *Ann. Intern. Med.* **72,** 111–121.

Miller, M. E., and Hall, J. G. (1978). Possible maternal effect in neurofibromatosis. *Lancet* **ii,** 1071–1073.

Mohr, J. (1954). "A Study of Linkage in Man." Munksgaard, Copenhagen.

Monks, P. L. (1968). Intrauterine osteogenesis imperfecta: Report of a case associated with hydramnios and dystocia. *Aust. N. Z. J. Obstet. Gynecol.* **8,** 157–159.

Morch, E. T. (1941). "Chondrodystrophic Dwarfs in Denmark." Munksgaard, Copenhagen.

Mories, A. (1954). An investigation into the Ehlers-Danlos syndrome. Thesis, Edinburgh.

Müller, P. K., Lemmen, C., Gay, S., and Meigel, W. (1975). Disturbance in the regulation of the type of collagen synthesized in the form of osteogenesis imperfecta. *Eur. J. Biochem.* **59,** 97–104.

Müller, P. K., Raisch, K., Matzen, K., and Gay, S. (1977). Presence of type III collagen in bone from a patient with osteogenesis imperfecta. *Eur. J. Pediatr.* **125,** 29–37.

Murdoch, J. L., Walker, B. A., Hall, J. G. Abbey, H., Smith, K. K., and McKusick, V. A. (1970). Achondroplasia a genetic and statistical survey. *Am. J. Hum. Genet.* **33,** 227.

Murdoch, J. L., Walker, B. A., Halpern, B. L., Kuzma, J. W., and McKusick, V. A. (1972a). Life expectancy and causes of death in the Marfan syndrome. *N. Engl. J. Med.* **286,** 804–808.

Murdoch, J. L., Walker, B. A., and McKusick, V. A. (1972b). Parental age effects on the occurrence of new mutations for the Marfan syndrome. *Ann. Hum. Genet.* **35,** 331.

Omenn, G. S., and Schrott, H. G. (1974). Prenatal prediction of autosomal dominant diseases by linkage studies: myotonic dystrophy. *Acta Genet. Med. Gemellol.* **23,** 221–228.

Omenn, G. S., Hall, J. G., Graham, C. B., and Karp, L. E. The use of radiographic visualization for prenatal diagnosis. *Birth Defects, Orig. Artic. Ser.* **13,** 217–229.

Ose, L., and McKusick, V. A. (1977). Prophylactic use of propranolol in the Marfan syndrome to prevent aortic dissection. *Birth Defects, Orig. Artic. Ser.* **13,** 163–169.

Pedowitz, P., and Perell, A. (1957). Aneurysms complicated by pregnancy. *Am. J. Obstet. Gynecol.* **73,** 720–735.

Penttinen, R. P., Lichtenstein, J. R., Martin, G. R., and McKusick, V. A. (1975). *Proc. Natl. Acad. Sci. U.S.A.* **72,** 586.

Pope, F. M. (1974a). Autosomal dominant pseudoxanthoma elasticum. *J. Med. Genet.* **11,** 152–157.

Pope, F. M. (1974b). Two types of autosomal recessive pseudoxanthoma elasticum. *Arch. Dermatol.* **110,** 209–212.

Renwick, J. H., Bundey, S. E., Ferguson-Smith, M. A., and Izatt, M. M. (1971). Confirmation of linkage of the loci for myotonic dystrophy and ABH secretion. *J. Med. Genet.* **8,** 407.

Reubi, F. (1944). Les vaisseaux et les glandes endocrine dans la neurofibromatose: Le syndrome sympathicotonique dans la maladie de Recklinghausen. *Schweiz. Z. Pathol. Bakteriol.* **7,** 168.

Rimoin, D. L. (1975). the Chondrodystrophies. *Adv. Hum. Genet.* **5,** 1–118.

Roberts, J. M., and Solomons, C. C. (1975). Management of pregnancy in osteogenesis imperfecta: new perspectives. *Obstet. Gynecol.* **45,** 168–170.

Rubin, P. (1964). "Dynamic Classification of Bone Dysplasias." Yearbook Publ., Chicago, Illonois.

Rudd, N. L., Miskin, M., Hoar, D. I., Benzie, R., and Doran, T. A. (1976). Prenatal diagnosis of hypophosphatasia. *N. Engl. J. Med.* **295,** 146–148.

Samuel, M. A., Schwartz, M. L., and Meister, M. M. (1953). The Ehlers-Danlos syndrome. *U.S. Armed Forces Med. J.* **4,** 737.

Saxena, K. M. (1970). Endocrine manifestations of neurofibromatosis in children. *Am. J. Dis. Child.* **120,** 265–271.

Schnitker, M. A., and Bayer, C. A. (1944). Dissecting aneurysm of the aorta in young individuals, particularly in association with pregnancy, with report of case. *Ann. Intern. Med.* **20,** 486–511.

Schrott, H. G., Karp, L. E., and Omenn, G. S. (1973). Prenatal prediction in myotonic dystrophy: Guidelines for genetic counseling. *Clin. Genet.* **4,** 38–45.

Scott, D., and Stiris, G. (1953). Osteogenesis imperfecta tarda. A study of 3 families with special reference to scar formation. *Acta Med. Scand.* **145,** 237.

Sengupta, B. S., Sivapragasam, S., Wynter, H. H., Moule, N., and Gajraj, M. K. (1977). Osteogenesis imperfecta: Its physiopathology in pregnancy. *J. R. Coll. Surg. Edinburgh* **22,** 358–364.

Sharpe, J. C., and Young, R. H. (1937). Recklinghausen's neurofibromatosis. *Arch. Intern. Med.* **59,** 299–327.

Shaul, S., Bogle, J., Hale, J., and Norman, A. D. (1977). Toward intimacy: Family planning and sexuality concerns of physically disabled women. Planned Parenthood Task Force on Concerns of Physically Disabled Women, 1977.

Sheehan, H. L., and Falkiner, N. M. (1948). Splenic aneurysm and splenic enlargement in pregnancy. *Br. Med. J.* **2,** 1105–1106.

Shoenfeld, Y., Fried, A., and Ehrenfeld, N. (1975). Osteogenesis imperfecta: Review of the literature with presentation of 29 cases. *Am. J. Dis. Child.* **129,** 679–687.

Siegel, B., Friedman, I., and Schwartz, A. (1957). Hemorrhagic disease in osteogenesis imperfecta. Study of platelet function defect. *Am. J. Med.* **22,** 315.

Sillence, D. O., Rimoin, D. L., and Danks, D. M. (1978). Clinical variability in osteogenesis imperfecta—variable expressivity or genetic heterogeneity. In press.

Simpson, C. F. (1979).

Simpson, C. F., and Neilson, J. T. M. (1973). Aortic atherosclerosis of turkeys induced by a single treatment with diethylstilbestrol. *Atherosclerosis* **17,** 245–255.

Skovby, F., and McKusick, V. A. (1977). Estrogen treatment of tall stature in girls with the Marfan syndrome. *Birth Defects, Orig. Artic. Ser.* **13,** 155–161.

Smith, S. A., Powell, C., and Essin, E. M. (1968). Ehlers-Danlos syndrome and pregnancy, report of a case. *Obstet. Gynec.* **32,** 331–335.

Solomons, C. C., and Myers, D. N. (1971). Hyperthermia of osteogenesis imperfecta and its relationship to malignant hyperthermia. *In* "International Symposium on Malignant Hyperthermia" (R. A. Gordon, B. A. Gritt, and W. Kalow, eds.), pp. 319–330. Thomas, Springfield, Illinois.

Solomons, C. C., and Styner, J. (1969). Osteogenesis imperfecta: Effect of magnesium administration on pyrophosphate metabolism. *Calcif. Tissue Res.* **3,** 318–326.

Spranger, J. W., Langer, L. O., and Wiedemann, H. R. (1974). "Bone Dysplasias. An Atlas of Constitutional Disorders of Skeletal Development." Saunders, Philadelphia, Pennsylvania.

Staples, P. O., Jr., and Riva, H. L. (1954). Maternal osteogenesis imperfecta: Report of two cases in sisters. *Obstet. Gynecol.* **4,** 557–561.

Stoddard, F. J., and Myers, R. E. (1968). Connective tissue disorders in obstetrics and gynecology. *Am. J. Obstet. Gynecol.* **102,** 240.

Sutinen, S., and Piiroinen, O. (1971). Marfan syndrome, pregnancy, and fatal dissection of aorta. *Acta Obstet. Gynecol. Scand.* **50,** 295–300.

Swapp, G. H., and Main, R. A. (1973). Neurofibromatosis in pregnancy. *Br. J. Dermatol.* **80,** 431.

Swift, T. R., Ignacio, O. J., and Dyken, P. R. (1975). Neonatal dystrophia myotonica. *Am. J. Dis. Child.* **129,** 734–737.

Thomas, C., Cordier, J., and Algan, B. (1952). Une étiologie nouvelle du syndrome de luxation spontanée des cristallins: La maladie d'Ehlers-Danlos. *Bull. Soc. Belg. Ophtalmol.* **100,** 375.

Tricomi, V., ed. (1965). "Clinical Obstetrics and Gynecology," pp. 334–349. Harper, New York.

Tyson, J. E., Barnes, A. C., McKusick, V. A., Scott, C. I., and Jones, G. S. (1970). Obstetric and gynecologic considerations of dwarfism. *Am. J. Obstet. Gynecol.* **108,** 688–704.

Vonderheid, E. C., Koblenzer, P. J., Pen Ming, L. M., and Burgoon, C. F., Jr. (1976). Neonatal lupus erythematosus. *Arch. Dermatol.* **112,** 698–705.

Whitehouse, D. (1966). Diagnostic value of cafe-au-lait spots in children. *Arch. Dis. Child.* **41,** 316.

Willard, W. G. (1903). A case of pregnancy in a dwarf weighing thirty-nine pounds, abdominal hysterectomy. *Am. J. Obstet.* **48,** 289–295.

Young, B. K., and Gorstein, F. (1968). Maternal osteogenesis imperfecta. *Obstet. Gynecol.* **31,** 461–470.

3

Hemoglobinopathies and Hemolytic Anemias

MITCHELL S. GOLBUS AND RUSSELL K. LAROS, JR.

The modern concept of molecular genetics has unfolded rapidly over the past three decades, due in large part to the conceptual contributions resulting from studies of the abnormal human hemoglobins and the thalassemia syndromes. In this chapter, we will review the genetic and molecular biochemistry of the hemoglobinopathies and the thalassemia syndromes. In addition, we will consider the influence of hemoglobinopathies and of hemolytic anemias on human reproduction. Additionally, we will consider selected aspects of genetic counseling related to these disease states.

Copyright © 1981 by Academic Press, Inc.
All rights of reproduction in any form reserved.
ISBN 0-12-630940-X

I. BIOCHEMICAL GENETICS OF HEMOGLOBIN

Hemoglobin is a tetrameric protein consisting of four globin chains. In humans there are six structurally different types of globin chains, designated alpha (α), beta (β), gamma (γ), delta (δ), epsilon (ϵ), and zeta (ζ) (Bunn *et al.*, 1977). The early embryonic globins ϵ and ζ, are rapidly replaced as the fetus develops (Huehns *et al.*, 1964a,b). Two α chains and two non-α chains combine to form each of the different hemoglobins found in the human. Figure 1 illustrates the changes in distribution of globin chains during pregnancy and in the neonate.

The messenger RNA for each globin chain is transcribed from a varying number of genes, which may differ among populations. It appears that there are four genes controlling α-chain synthesis in Orientals and Caucasians (Rucknagel and Winter, 1974) and an inconstant number, from two to four, in Black populations (Orkin and Nathan, 1976).

Because β, γ, and δ chains differ in amino acid composition, they each possess an unique oxygen affinity, surface charge, structural configuration, and distinguishable electrophoretic mobility when placed in a charged field. This last characteristic allows electrophoresis on cellulose acetate or citrate agar to identify types of hemoglobin (Schmidt *et al.*, 1975). Table I summarizes the composition of the normal hemoglobins and their relative distribution in the adult. Fetal hemoglobin (Hb F) differs from adult hemoglobin (Hb A) in its ability to release oxygen at a lower pH and oxygen tension, its lesser solubility in acid buffer, and its resistance to denaturation by alkali. Thus, a blood smear with mixed fetal and

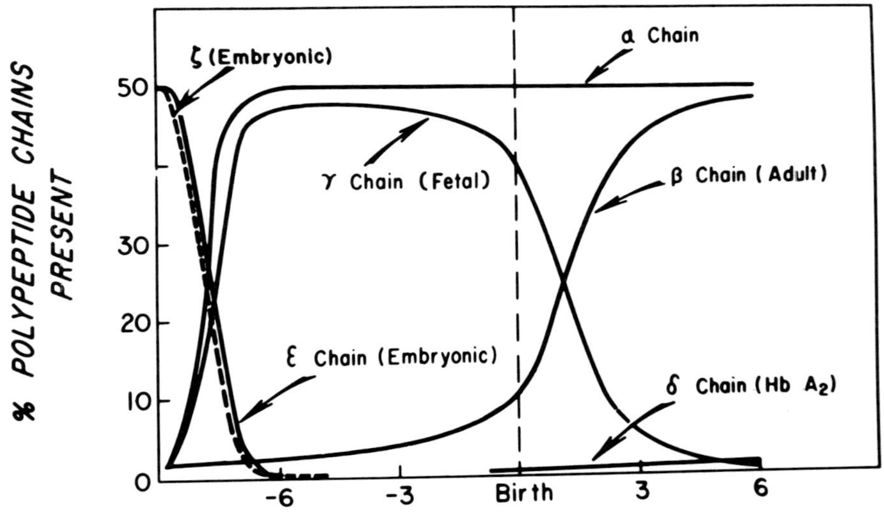

Fig. 1. Distributions of globin chains during fetal and early neonatal life.

TABLE I
Normal Hemoglobin Types in the Adult

Abbreviated name	Structure	Normal values (%)
Hb A	$\alpha_2\ \beta_2$	95
Hb A_2	$\alpha_2\ \delta_2$	2-3.5
Hb F	$\alpha_2\ \gamma_2$	<2

maternal cells can be washed with dilute sodium hydroxide with subsequent lysis of only maternal cells (Apt test). While at birth Hb F accounts for 60–80% of all hemoglobin present, by 12 months to 2 years of age the level has decreased to under 2%. Hb A_2 is normally at a low concentration in the adult, and levels above 3.5% may be a subtle manifestation of thalassemia.

A. Abnormal Hemoglobin: Polypeptide Chain Alterations

The concept of structural globin abnormalities was clarified in 1956 when Ingram, using the peptide mapping or ''fingerprinting'' technique, showed that in sickle hemoglobin's β-polypeptide chain a glutamic acid residue, now known to be the sixth from the free amino terminal end of the molecule, was replaced by a valine residue. Since the side chain of glutamic acid is negatively charged and that of valine is neutral, it follows that the charge of the Hb S molecule is positive with respect to Hb A, thus explaining the difference in electrophoretic mobility. To date, several hundred structural variants of α, β, γ, and δ chains have been identified. Most differ from normal by only one amino acid. The molecular location of the substitution has a marked influence on the functional effect. While most substitutions are without effect, those at the sixth amino acid residue from the free amino terminal end of the molecule lead to aggregation and reduced solubility. Examples include Hb S and Hb C. Substitutions in other areas of the molecule lead to instability and reduced solubility, methemoglobinemia, and increased or decreased oxygen affinity.

The nomenclature and frequency of the commonest hemoglobinopathies are outlined in Table II. An individual may be homozygous for the S or C hemoglobin gene. If so, all her adult hemoglobin will contain S (β^S) or C (β^C) globin chains instead of normal β chains. If heterozygous, one-half of the hemoglobin tetrameres will contain β^S or β^C globin instead of normal β^A chains. Because of altered structure and physiology, these hemoglobins are less stable in reducing environments such as are present in conditions of hypoxia, acidosis, or infection. In these situations, the hemoglobin molecule loses flexibility and the red blood cells (RBC) are unable to pass freely through the microcirculation with con-

TABLE II
Nomenclature and Frequency of the Commonest Hemoglobinopathies in Adult U.S. Blacks[a]

Hemoglobinopathy	Abbreviated name	Frequency
Sickle cell trait	Hb SA	1:13
Sickle cell anemia	Hb SS	1:1875
Sickle cell-hemoglobin C disease	Hb SC	1:1250
Hemoglobin C disease	Hb CC	1:4444
Hemoglobin C trait	Hb CA	1:33
Hemoglobin S-β-thalassemia	Hb S-β-Thal	1:3333
β-Thalassemia trait	Hb A-β-Thal	1:66

[a]Adapted from Motulsky (1973).

sequences to be discussed later. For the heterozygous individual, the normal hemoglobin protects the RBC except in unusual circumstances such as extreme hypoxia or acidosis.

B. Thalassemia

The thalassemia syndromes are named and classified by the type of chain that is represented inadequately. The two major groups are the α- and β- thalassemias, both of which affect the synthesis of Hb A. Reduced synthesis of γ or δ chains, and of combinations in which two or more globin chains are affected [e.g., $\delta\beta$, $\gamma\beta$, hereditary persistence of fetal hemoglobin (HPFH)], have been reported, but these are rare (Stamatoyannopoulos *et al.*, 1969; Kan *et al.*, 1972a; Conley *et al.*, 1963). In each instance, the thalassemia is a quantitative disorder of globin synthesis; structurally the chains are entirely normal. There is no single etiology of these diseases. On the molecular level, a variety of defects have been demonstrated, each involving a different step along the pathway of transcription of gene DNA into messenger RNA (mRNA) and translation of mRNA into globin chains.

In the most common form of α-thalassemia, one or more structural genes are physically absent, permanently deleted from the genome (Ottolenghi *et al.*, 1974). In the homozygous condition all four genes are deleted, no α chains are produced, and the fetus is unable to synthesize Hb F or any of the adult hemoglobins (Weatherall *et al.*, 1970). The predominant hemoglobin is Hb Bart (γ_4), which, owing to its high oxygen affinity, releases inadequate amounts of oxygen at the tissue level. The result is high-output cardiac failure, hydrops fetalis, and a stillborn fetus. The most severe form of α-thalassemia compatible with extrauterine life is Hb H disease, which results from the deletion of three α genes (Kan *et al.*, 1975a). In these patients abnormal quantities of both Hb H (β_4) and Hb Bart accumulate. Beta tetramers precipitate during the life of the red blood cell to form inclusion bodies and the "pitting out" of these bodies by the reticuloendothelial system causes a moderately severe hemolytic anemia. In α-thalassemia minor (α-thalassemia-1), two genes are deleted causing a mild hypochromic microcytic anemia, which must be differentiated from iron deficiency. A single gene deletion (α-thalassemia-2) is clinically undetectable or the "silent carrier state" (Orkin and Nathan, 1976). Alpha-thalassemia also may result from a nondeletion defect (Kan *et al.,* 1977a). The genes are present, yet the phenotype is that of α-thalassemia-1. Apparently, only a small amount of globin can be synthesized from these genes. Still another group of defects is exemplified by a mutant hemoglobin such as Hb Constant Spring wherein an elongated α-gene product is incorporated into the hemoglobin molecule (Milner *et al.,* 1971). A greatly reduced amount of this mutant product is synthesized.

The etiology of α-thalassemia in Blacks is less clear than for Orientals and Caucasians. The former do not demonstrate as marked or consistent an imbalance in globin synthesis. Hemoglobin H disease is very rarely seen, and thalassemia minor is clinically very mild, apparently because the α-thalassemia in this population involves a deletion on different chromosomes (trans) rather than on the same chromosome (cis).

In β-thalassemia, no gene deletions have been demonstrated. There are two forms of β-thalassemia, designated β^+ and β^0 depending on whether β-chain production is reduced or entirely absent (Conconi *et al.,* 1972; Forget *et al.,* 1974; Kan *et al.,* 1975b). In β^+ thalassemia, mRNA for β-globin chains is present but is quantitatively reduced. This might be due to a particularly unstable mRNA or to a decreased rate of transcription of mRNA (Housman *et al.,* 1973). Reduced β-chain synthesis may also be seen with Hb Lepore (a product of δ- and β-gene crossover), with HPFH, and in $\delta\beta$-thalassemia. Hereditary persistence of fetal hemoglobin and $\delta\beta$-thalassemia differ from β-thalassemia in that they both involve β-gene deletions (Ramirez *et al.,* 1976).

In the homozygous β-thalassemia condition, α-chain production is unimpeded and these highly unstable chains accumulate, and eventually precipitate. Marked ineffective erythropoiesis and severe hemolysis result in a condition known as thalassemia major or Cooley's anemia (Nathan and Gunn, 1966), the clinical manifestations of which will be discussed below.

II. CLINICAL IMPLICATIONS IN HUMAN REPRODUCTION

A. Structural Gene Abnormalities

1. Sickle Cell Trait

Approximately 10% of the black population in the United States is heterozygous for the hemoglobin S gene; this state is referred to as sickle cell trait (Motulsky, 1973). Although most practitioners consider this condition only in black patients, it is important to remember that the β^S gene pool stems from the Mediterranean area. Thus, the trait also may be present in Italians, Turks, Greeks, and Caribbean natives. Individuals with sickle cell trait appear quite normal and enjoy good health. They are rarely anemic, and their peripheral smear does not show sickling in most situations. Screening for the presence of Hb S may be performed inexpensively by a variety of techniques (Schmidt, 1973).

Whalley *et al.* (1963, 1964), Pritchard *et al.* (1973), and Blattner *et al.* (1977) have contrasted the pregnancy outcome of women with normal hemoglobin and those with sickle cell trait. They noted comparable incidences of spontaneous abortion, low birth weight infants, toxemia, low Apgar scores, and perinatal mortality. However, gravidas with sickle cell trait did have a twofold increase in the rate of urinary tract infections. Also, they may be unable to concentrate their urine (hyposthenuria), but gross hematuria is rare (Hatch *et al.,* 1967). These women do well during pregnancy and labor, but caution must be observed when using anesthesia to assure good oxygenation. We have managed these patients in our regular prenatal clinic without untoward effect. The only special care required is close observation for urinary tract infections (Whalley *et al.*, 1964). Sickle cell trait is not associated with the rapid marrow turnover characteristic of hemolysis; therefore, iron does not accumulate. Unlike sickle cell anemia, these patients may become iron deficient and supplementation is necessary during pregnancy.

2. *Sickle Cell Anemia*

The incidence of sickle cell anemia (SCA) is 1:625 (0.0189%) black births. The observed and theoretic incidence (based on calculations using the Hardy-Weinberg law) of this and the other structural gene abnormalities is identical. This finding suggests that the presence of such a hemoglobinopathy exerts no lasting deleterious effect on the fetus. The incidence falls to 1:1875 by reproductive age (Motulsky, 1973). The patient with SCA has no Hb A and may have greater than 90% Hb S. Hb F is elevated in variable degrees but is usually below 14%. Females with SCA have a somewhat higher concentration of Hb F than males, and Jackson and co-workers (1961) suggest a milder course in individuals with elevated Hb F levels. Patients with SCA suffer from life-long complications, in part due to the markedly short life span of their red blood cells. Improved blood banking techniques, better understanding of the underlying pathophysiology of the microcirculation, and prompt intervention in states of severe anemia, infection, and acidosis have led to improved survival (Henderson, 1950; Flakpui and Moran, 1973). Therefore, the coexistence of pregnancy and SCA is becoming more common.

The influence of pregnancy on this illness and, reciprocally, the influence of SCA on pregnancy and its outcome have been discussed at length. The recent

TABLE III
Reproductive Consequences of Sickle Cell Anemia (SCA)

Number of pregnancies	Reference	Abortion (%)	Maternal mortality (%)	Fetal salvage (%)	Infants <2500 g (%)
45	32	11	0	76	49
50	86	32	0	75	38
97	21	16	6	61	—
50	79	20	0	68	—
41	22	27	0	64	42
43	63,64	—	9	—	—
46	83	—	0	80	8

literature is summarized in Table III. While most observers describe no maternal mortality on modern obstetrical services, the perinatal mortality (approximately 28%) and the incidence of birth weights less than 2500 g (approximately 45%) are high. Most observers feel that the prepregnancy course of an individual is a good index of how she will do during pregnancy. Pregnancy termination because of the risk to the patient from gestation is debatable and usually unwarranted. However, each situation must be individualized. The spontaneous abortion rate does not appear to be increased.

Virtually all the signs and symptoms of SCA are secondary to either hemolysis, vascular occlusive disease, or an increased susceptibility to infection. Each of the pathologic events is produced by sickling of the red blood cell. The clinical manifestations of SCA are summarized in Table IV and will not be discussed in detail, however, we do wish to stress some of these features.

Sickling of the red cell can occur without significant hypoxia whenever the concentration of Hb S is greater than 40%. Thus, SCA becomes evident in early childhood as ischemic pain episodes involving the distal extremities. Dactylitis (swollen, painful dorsal surface of hands and feet) is the presenting symptom of SCA in one-third of patients. Affected children evidence significant growth retardation, and most develop classical skeletal changes which include prominence of the frontal bones and long bones. Unfortunately, there is not yet adequate data concerning mortality rates for individuals with SCA. Diggs' (1973) data indicate that the median survival is 14.3 years with the distribution of mortality being 20% during the first 2 years of life, 10% between ages 2 to 5 years, 50% between ages 5 to 30 years, and 20% after the age of 30 years. It has recently been demonstrated that immunization with polyvalent pneumococcal-polysaccharide significantly reduces the incidence of *Streptococcus pneumoniae* infections in children and young adults (Ammann *et al.*, 1977). It remains to be seen what impact this advance will have on mortality statistics.

TABLE IV
Clinical Manifestations of Sickle Cell Anemia

- I. Growth and development
  - A. Retarded growth
  - B. Skeletal changes
  - C. Decreased life span
- II. Sickle cell crisis
  - A. Painful vaso-occlusive episode: bones, abdomen, chest, back
- III. Cardiovascular—manifestations of hyperdynamic circulation
  - A. Cardiomegaly
  - B. Systolic murmurs
  - C. Failure
- IV. Pulmonary signs
  - A. Infection—pneumococcal, mycoplasma, hemophilus, salmonella
  - B. Vascular occlusion
- V. Abdominal involvement
  - A. Painful vaso-occlusive episodes
  - B. Hepatomegaly
  - C. Hepatitis
  - D. Cholecystitis
  - E. Splenic infarction
- VI. Bone and joint changes
  - A. Bone marrow infarction
  - B. Osteomyelitis—salmonella
  - C. Arthritis
- VII. Genitourinary signs
  - A. Hyposthenuria
  - B. Hematuria
  - C. Pyelonephritis
- VIII. Neurologic manifestations
  - A. Vascular occlusion
  - B. Convulsions
  - C. Hemorrhage
  - D. Visual disturbances
- IX. Ocular manifestations
  - A. Conjunctival vessel changes
  - B. Vitreous hemorrhage

A sickle cell crisis is an episode of uncomplicated painful vaso-occlusion. The pain is described as gnawing and increases in intensity over several hours. It usually involves the extremities, expecially the tibial and periarticular areas, but may also involve the abdomen, chest, or back. The crisis usually lasts several days with gradual abatement, but may last only a few hours. It is important not to let the diagnosis of sickle cell crisis blunt one's acuity in the search for other intercurrent disease in patients with SCA (Diggs, 1965).

Severe anemia subjects the cardiovascular system to repeated episodes of acute and chronic stress. Cardiomegaly, multiple murmurs, and high output failure are frequent findings. Horger (1972) described a hemoglobin concentration less than 9.5 g/dl in 87% of his patients, and 75% required transfusion at some point during pregnancy.

The major pulmonary disorders in patients with SCA are infections and vaso-occlusive episodes. Thus, the clinician must frequently differentiate between pneumonia and thrombosis and/or embolization. The organisms most commonly responsible for pneumonia are *Pneumococcus, Mycoplasma, Haemophilus, Salmonella,* and *E. coli.* Clinical features favoring the diagnosis of thrombosis are the absence of significant temperature elevation, a lack of chills, the presence of associated pain in the extremities or abdomen, and the presence of "blister cells" (Karayalcin *et al.*, 1972).

Chronic hemolysis leads to hepatomegaly with mildly elevated serum bilirubin levels. Barrett-Conner (1968) states that 30–40% of patients with SCA have gallstones, of which only 10% are symptomatic. The spleen and its function are always altered. In childhood the spleen is enlarged and splenic sequestration may occur. During sequestration, the spleen enlarges rapidly, entrapping RBCs with resulting systemic hypovolemia. Over time, there are multiple episodes of splenic infarction and the spleen becomes progressively smaller and is eventually nonfunctional. Thus, the spleen is not palpable in most adults with SCA.

The individual with SCA is very susceptible to chronic leg ulcers with subsequent scarring. Bone marrow infarctions, osteomyelitis, and joint involvement with avascular necrosis and septic arthritis are common problems. The usual offending organisms are *Salmonella* and *Staphylococcus*.

The kidneys are prone to embolic infarction with renal papillary necrosis, which leads to interference with the countercurrent mechanism and subsequent hyposthenuria and impaired hydrogen ion excretion. Hematuria is common and may be prolonged and lead to significant blood loss. Recurrent pyelonephritis is also very common. Although hypertension is not itself increased among nonpregnant individuals with SCA, toxemia occurs in about 33% of pregnancies associated with SCA (Pritchard *et al.*, 1973).

Because of chronic hemolysis and the body's ability to reutilize iron, the serum iron concentration is increased and these patients rarely need iron supplementation. In fact, individuals with SCA are prone to hemosiderosis because of hemolysis and multiple transfusions. On the other hand, folic acid is consumed in the process of red cell synthesis; thus, folate supplementation is essential. Anemia may be worsened by folic acid deficiency (megaloblastic crisis), and this must not be overlooked as a cause for a falling hemoglobin level (Lindenbaum and Klipstein, 1963). A true hemolytic crisis is rare in SCA, but painful crisis is very common. Treatment for painful crisis consists of bedrest, hydration, oxygen therapy, analgesia, and appropriate antibiotics when indicated. Use of urea or cyanate during pregnancy is discouraged because of questionable efficacy and possible teratogenicity.

Evidence in the literature does not support the practice of early elective induction of labor and, in fact, some studies have shown the use of oxytocin to be associated with decreased bone blood flow (Singer *et al.*, 1951; Fraser and Watt, 1964). Anesthesia-related hypovolemia and/or hypoxia is contraindicated; therefore, regional anesthetics should be administered with great caution. Patients with SCA must be watched closely for endometritis and anemia during the puerperium. Postpartum endometritis occurs in approximately 20% of patients (Horger, 1972). If present, both should be treated aggressively. Pyrexia may lead to dehydration and is to be avoided. Patients with SCA also may have pregnancy-related complications not associated with their hemoglobinopathy, and sickle cell crisis may be misdiagnosed in situations where the real problem is abruptio placenta, ectopic gestation, pyelonephritis, or pulmonary embolization. Exchange transfusion has generated significant excitement as a method by which to decrease maternal and fetal morbidity and mortality. This therapeutic modality will be discussed later.

3. *Sickle Cell Hemoglobin C Disease*

Women who are doubly heterozygous for both the β^S and β^C genes are referred to as having hemoglobin SC disease (Hb SCD). The incidence of Hb SCD is approximately 1:1250 adult Blacks in the United States (Motulsky, 1973). Hemoglobin electrophoresis reveals approximately 60% Hb C and 40% Hb S. Hb C has unique properties which, unfortunately, potentiate tendency of Hb S to sickle in reduced O_2 environments.

Patients with Hb SCD generally have a normal habitus, a healthy childhood, and a normal life span. Many are diagnosed only during the latter part of pregnancy, when complications are most likely to occur. At the beginning of pregnancy, the woman is mildly anemic (hematocrit approximately 30%), and splenomegaly (present in about 50% of patients) and target cells in the peripheral blood smear may be overlooked. Hemoglobin electrophoresis will insure the correct diagnosis.

Hb SCD during pregnancy is more worrisome than SCA because of the high mortality rate outlined in Table V. Although some investigators have reported markedly elevated maternal mortality (Horger, 1972; Fort *et al.*, 1971; Curtis, 1959), their results have not been confirmed by others (Laros, 1967; Pritchard *et al.*, 1973). Perhaps the high and often unanticipated mortality is related to less careful antenatal observation of previously undiagnosed patients, or to pathophysiologic problems unique to patients with Hb SCD. In contrast to patients with SCA, they frequently experience rapid and severe anemic crisis due to splenic sequestration. In addition, they have a greater tendency to experience bone marrow necrosis with the release of fat or marrow emboli. Fullerton and Watson-Williams (1962), in fact, recommend heparinization for patients with Hb SCD having bone pain crises. Because the efficacy of heparinization has not been documented by other studies, this practice has not been employed widely. The clinical manifestations of Hb SCD are otherwise similar to SCA but milder. The renal problems include papillary necrosis with hyposthenuria, urinary tract infections (70%), gross hematuria (30%), and pyelonephritis (21%) (McCurdy, 1964).

TABLE V
Reproductive Consequences of Hemoglobin SC Disease (Hb SCD)

Number of pregnancies	Reference	Abortions (%)	Maternal mortality (%)	Fetal salvage (%)	Infants <2500 g (%)
9	33	—	33	38	—
78	86	6	3	91	22
34	21	24	9	73	—
17	79	29	0	58	—
—	13	16	25	65	—
14	58	—	0	85	—
—	107	12	0	88	—
25	90	20	0	78	—
67	51	7	2	88	—
18	63,64	—	6	—	—

Pulmonary infection, embolization, or infarction occur in about 50% of these patients, with aseptic necrosis of the femoral head, vitreous hemorrhage, and postpartum endometritis also being common. Preeclampsia occurs in about 20% of parturients with Hb SCD. Anemia usually is not severe until the third trimester. Considerations for the management of labor are the same as with SCA. Additionally, these patients experience a high incidence of postpartum hemorrhage (12–22%). Thrombophlebitis occurs with increased frequency and should be watched for closely during the puerperium. Patients with Hb SCD require increased folate intake as in SCA. Unlike gravidas with SCA, those with Hb SCD do require iron supplementation and, if not given, they may develop a confounding iron deficiency anemia.

4. Hemoglobin C Trait and Disease

The estimated frequency of hemoglobin C trait is 1:33 in U.S. black adults and that of hemoglobin C disease is 1:4444. Hb C trait (Hb AC) is an asymptomatic state without reproductive consequences. Target cells are found in the peripheral blood smear, but anemia is not present. Hemoglobin C disease (Hb CD) is a mild disorder usually discovered during medical evaluation for other problems. There is a mild hemolytic anemia with the hematocrit in the range of 25 to 35%, and a constant, mild reticulocytosis. The red blood cells are characterized by targeting and many microspherocytes are present. Unlike SCA, there is no increased morbidity or mortality associated with pregnancy.

5. Hemoglobin S-β-Thalassemia

The gene controlling the rate of β-chain synthesis is intimately linked to those controlling globin chain structure. Hb S-β-thalassemia ($\alpha_2\beta^S\beta^{thal}$) is a condition in which the patient is heterozygous for both the sickle cell and the β-thalassemia genes. This is an uncommon disorder believed to occur with the frequency of 1:3333 in black adults in the United States (Motulsky, 1973). In addition to decreased β-chain production, there is a variably increased production of Hb F and Hb A_2. Because of this variable production rate, hemoglobin electrophoresis reveals a spectrum of hemoglobin concentrations. Hb S may account for 70–95% of the hemoglobin present, with Hb F rarely exceeding 20%. Because of the thalassemia influence, Hb S exceeds Hb A concentration. This is in sharp contrast to patients with sickle cell trait in which Hb A exceeds the concentration of Hb S. The diagnosis is made in an anemic patient by demonstrating an increased Hb A_2 and Hb F levels in association with levels of Hb S exceeding Hb A. The peripheral smear reveals hypochromia and microcytosis with anisocytosis, poikilocytosis, basophilic stippling, and targeting. The hemoglobin electrophoresis pattern is characteristic as noted above, and reticulocytosis is prominent.

The clinical manifestations of this disorder parallel SCA but are usually milder. The patient frequently is mildly anemic (hemoglobin 9-11 g/dl), painful crises may occur, and chronic hemolysis is present. The hemolysis leads to hepatosplenomegaly, an elevated serum bilirubin, and episodic jaundice. These patients have a normal body habitus and frequently enjoy an uncompromised life span. Renal complications may include hematuria, hyposthenuria, papillary necrosis, and pyelonephritis. Aseptic necrosis of the femoral head and thrombophlebitis also commonly occur.

Maternal and perinatal mortality may be elevated in Hb S-β-thalassemia patients. The small numbers of case reports make overall conclusions difficult. Pritchard *et al.* (1973) reported no maternal deaths and 91% fetal salvage in his study of 37 pregnancies. Laros (1971), on the other hand, described one maternal death in 38 pregnancies (2.6%) and an 81% rate of fetal salvage. These patients need close supervision in the third trimester and during the intrapartum and postpartum periods for problems similar to those seen with Hb SCD. They require folic acid but not iron supplementation. The clinical severity is usually inversely proportionate to the hemoglobin S concentration.

6. Partial Exchange Transfusion (PET)

The major risks to a pregnant woman with SCA, Hb SCD, and Hb S-β-thalassemia occur during the third trimester, the intrapartum period, and postpartum. These risks have been detailed above and are basically those related to severe anemia and vaso-occlusion. If one were able to remove a significant amount of the patient's whole blood and replace it with fresh carefully cross-matched packed red blood cells at a critical time in gestation, perhaps these sequelae might be avoided. Ricks (1968), and later Perkins (1976a), tested this hypothesis in a limited number of patients. However, the largest experience is

TABLE VI
Protocol for Partial Exchange Transfusion (PET)

I. Admit: at 26–28 weeks gestation
II. Baseline laboratory studies:
    A. Hb, Hct, WBC, reticulocyte count
    B. Hb electrophoresis
III. Type and crossmatch blood: 4 units of fresh, buffy coat poor, washed packed RBCs
IV. Exchange
    A. In morning
        1. Infuse 500 ml crystalloid over 1–2 hours
        2. Remove 500 ml by phlebotomy over same time period
        3. Infuse 2 units of packed RBCs over 1–2 hours
    B. In afternoon: repeat morning procedure
V. Repeat laboratory evaluation: following morning
    A. Hb, Hct
    B. Hb electrophoresis
VI. Additional exchange (2 to 4 units) if:
    A. Hct < 35%
        or
    B. Hb A level < 40%

that described by Morrison and Wiser (1976a,b). They utilized the protocol outlined in Table VI in 36 patients who were admitted to the hospital during the 28th week of gestation. If the initial four-unit transfusion achieved a hematocrit (Hct) of greater than 35% and an Hb A level of greater than 40%, the patient was discharged. If not, an additional two- or four-unit exchange was performed. The patient was followed on a weekly basis and the PET repeated at 36–39 weeks gestation. An early repeat PET was indicated by the occurrence of an Hct less than 25%, an Hb A level less than 20%, crisis, or labor.

They reported two cases of maternal hepatitis in 36 patients; in one of these the infant died. Only one patient experienced a transfusion reaction, and no premature labor was noted. One maternal death occurred in a patient who received a single PET at 28 weeks gestation, but refused further exchange despite the occurrence of crisis. Neonatal outcome was described for the 36 pregnancies, and compared to 160 gestations matched for age, race, gravidity, socioeconomic status, and length of gestation. The controls had SCA but were not treated by PET. They noted a significant decrease in the perinatal death rate, reduced prematurity, and fewer low birth weight infants in the treated group. The investigators postulated that sickling in the placenta leads to vaso-occlusion and intrauterine hypoxia. If partial, intrauterine growth retardation may occur; if massive, spontaneous abortion or stillbirth may result.

While Morrison and associates advocate PET for patients with SCA, Hb SCD, and Hb S-β-thalassemia, we have not used PET routinely in the latter disorders. Although Hb SCD and Hb S-β-thalassemia can be very severe and lead to both fetal and maternal morbidity and mortality, the outcome for both mother and fetus is usually substantially better than in SCA. Because of the inherent risks of PET and the lack of demonstrated benefit in these subgroups, we have reserved PET for those cases of Hb SCD and Hb S-β-thalassemia in which the mother has experienced severe crises or hemolysis leading to a Hct of less than 25%.

B. The Thalassemia Syndromes

Thalassemia is found throughout the world but is concentrated in the Mediterranean coastal areas, central Africa, and parts of Asia. In these regions it is believed that the high incidence represents a balanced polymorphism due to heterozygote advantage (Weatherall and Clegg, 1972). Heterozygotes are protected against malaria when compared to the homozygous normal individual. All forms of thalassemia are inherited as autosomal recessive traits. The lethal homozygous state results only when all of an individual's genes for the α or β chain are affected. Characteristics of the various thalassemias are summarized in Table VII.

1. β-Thalassemia Major (Cooley Anemia)

β-Thalassemia major is the homozygous state in which there is very little or no production of β chains. The fetus is protected from severe disease by γ-chain production. However, this protection disappears rapidly after birth, and the affected infant is anemic by 3 months of age. He or she has hepatomegaly, requires blood transfusions every 3–4 weeks, and dies by the third decade usually of

TABLE VII
The Thalassemia Syndromes

	Hb Level	Hb A_2	Hb F	Abnormal Hb	Clinical severity
Homozygotes					
α-Thalassemia	↓↓↓↓	0	Variable	Hb Barts	Hydrops—fatal
β-Thalassemia	↓↓↓	↑	↑	0	Severe
Heterozygotes					
α-Thalassemia trait	↓	N	N	Hb Barts as newborn	Mild
Hemoglobin H disease	↓↓	N	N	Hb H, Hb Barts as newborn	Moderate
β-Thalassemia	↓-↓↓	↑	N-↑	0	Moderate
Double heterozygote					
Hb S-β-Thalassemia	↓-↓↓	↑	↑	Hb S	Moderate

myocardial hemosiderosis. Female patients surviving until puberty are usually amenorrheic with severely impaired fertility (Necheles, 1973).

As in sickle cell anemia, consideration of basic pathophysiology allows prediction of clinical events. The marrow is in a high-turnover state in an effort to produce new red blood cells. This requires folate, and, therefore, these patients need adequate supplementation of folic acid. Failure to provide folate enhances the possibility of an aplastic crisis. Because of demands for erythropoiesis, the marrow is expanded and may result in a habitus with a tower skull and long extremities. Abnormal red cells with Heinz bodies are apparent in the peripheral smear as are very small cells with hypochromasia. Nucleated red cells reflect the ineffective erythropoiesis. Hyperbilirubinemia is frequently present and may promote gallstone formation.

The clinical severity of β-thalassemia is influenced by the extent to which γ-chain production can compensate for deficient β-chain synthesis. The rare instances of pregnancy in patients reported to have thalassemia major most likely occurred in individuals actually heterozygous for two different β-thalassemia genes, at least one of which is a "high F" form (Perkins, 1971a; Necheles, 1973).

2. *β-Thalassemia Minor*

β-Thalassemia minor results in a variable degree of illness depending upon the rate of β-chain production. It is subclassified into β-thalassemia intermedius and β-thalassemia minimus. While this distinction may be of more academic than clinical importance, it is important to realize that thalassemia trait may be quite subtle in clinical appearance, and is frequently diagnosed by appropriate lab tests only after the patient fails to respond to iron therapy or delivers a child found to have the homozygous disease state.

Characteristic findings include a relatively high red blood cell count, moderate to marked microcytosis, and a peripheral smear resembling iron deficiency with small cells and hypochromasia. While Hb electrophoresis characteristically shows Hb A_2 to be elevated above 3.5%, Hb F may or may not be elevated. Serum iron or ferritin concentration is high and the iron saturation is greater than 20%.

β-Thalassemia trait does not impair fertility. The incidence of prematurity, low birth weight infants, and infants of abnormal size for gestational age are identical to those delivered of normal women. Unlike previous studies (Beischer *et al.*, 1968; Fleming, 1973), Alger and associates (1979) did not find an increase in placental size or an abnormal fetal/placental weight ratio. The expected physiologic hemodilution of pregnancy is observed as a tendency for the Hct, Hb, and red cell count to fall after the first trimester. In the series of Alger *et al.* (1979), this decrease was no greater than that reported for normal pregnancy. These findings are in contrast to those of Fleming and Lynch (1969), who postulated that increasingly ineffective erythropoiesis and/or hypersplenism might contribute to the decline in hemoglobin levels observed between the first and second trimesters in thalassemic patients.

3. α-Thalassemia and Hemoglobin H Disease

In the most common form of α-thalassemia, the structural genes for α globin are absent. In the homozygous condition all four genes are deleted. Thus, no α chains are produced and the fetus is unable to synthesize Hb F or any adult hemoglobin. The result is severe anemia producing hydrops fetalis and stillbirth. The most noteworthy pregnancy complication is that approximately 50% of women carrying a fetus with α-thalassemia will develop severe toxemia (Rowley, 1976).

Hemoglobin H disease results from the deletion of three α-globin genes and is recognizable as a moderate hemolytic anemia with increased reticulocytosis and an abnormal quantity of Hb H seen on electrophoresis. α-Thalassemia-1, the heterozygous state with two deleted genes, is a diagnosis by exclusion in cases of a mild hypochromic microcytic anemia in which iron deficiency and β-thalassemia trait have been excluded as etiologies. Individuals with α-thalassemia trait and hemoglobin H disease appear to have normal pregnancy outcomes (Alger *et al.*, 1979).

III. PRENATAL DIAGNOSIS

In all of the above hemoglobinopathies the involved genes may be considered recessive in that an heterozygous individual has one-half the genes (i.e. either one or two) for a particular globin chain functioning to make the normal protein molecule and one-half functioning to make an abnormal one. The individual with a hemoglobinopathy is homozygous or doubly heterozygous for the abnormal gene(s). Until very recently, the obstetrician has had little to offer patients who were α- or β-thalassemic carriers or carriers of a gene for a structurally abnormal hemoglobin. They usually had uncomplicated pregnancies and the obstetrician's role was limited to genetic counseling and performing abortions for those couples fearful of delivering an affected child. The development of methods to diagnose hemogloginopathies *in utero* now allows the obstetrician to give parents definite answers rather than statistical probabilities. Couples at risk can avoid having affected children without remaining childless, and are spared months of anxiety awaiting the outcome of a pregnancy.

Since the homozygous state is invariably fatal, α-thalassemia does not produce the major medical and social problems associated with β-thalassemia. Even so, early diagnosis is desirable to minimize the emotional trauma of an unsuccessful pregnancy and to prevent the development of third trimester preeclampsia, which frequently accompanies this disease (Eng *et al.*, 1968).

The method for prenatal diagnosis of α-thalassemia involves measuring the number of intact α-globin structural genes present in the cellular genome using the technique of molecular hybridization (Kan *et al.*, 1976). Since any nucleated cell, including amniotic fluid fibroblasts, contains the full complement of globin genes, an amniocentesis provides fetal cells which can be cultured and then analyzed. Using reverse transcriptase, a viral enzyme which transcribes DNA from RNA, complementary α-DNA (c-DNA) can be transcribed from α-globin mRNA. This complementary DNA also is complementary to and capable of binding to α-globin structural genes in cells. The rapidity with which the c-DNA binds to DNA from amniotic fibroblasts is proportional to the number of copies of α-globin structural genes present in the cells. This technique clearly distinguishes the homozygous state from carrier or normal states. A new technique utilizing restriction endonucleases to diagnose α-thalassemia is now available, and this methodology will be discussed in considering sickle cell anemia.

To prenatally diagnose β-thalassemia or a structurally abnormal adult hemoglobin (e.g., sickle cell anemia) requires that Hb A normally be produced by the late first–early second trimester fetus. During the 1960s a number of investigators demonstrated hemoglobins other than Hb F in early second trimester fetuses (Walker and Turnbull, 1955; Thomas *et al.*, 1960). In 1972 Pataryas and Stamatoyannopoulos used peptide mapping to confirm that Hb A first appeared at the 11th week of gestation. They also detected abnormal adult hemoglobins, Hb C and Hb S, early in the second trimester. Several laboratories have examined the onset and type of hemoglobin synthesis in early gestations by measuring the uptake of radioactive amino acids into hemoglobin or globin chains, and this is now the most widely used methodology for such studies. The determination of the hemoglobin type of a fetus is more readily accomplished by examination of synthesis rather than content, and these techniques require only 10 μl (0.01 cc) of fetal blood (Kan *et al.*, 1972b).

The prenatal diagnosis of hemoglobinopathies other than α-thalassemia has required the ability to obtain fetal erythrocytes usable for synthetic studies. The fetal blood may be obtained under direct visualization with a modified cannula measuring 2.2 mm by 2.7 mm through which both the fetoscope and 27-gauge blood drawing needle can be inserted (Hobbins and Mahoney, 1975). Using this technique it has been possible to obtain sufficient fetal blood cells for hemoglobin synthesis studies from approximately 90% of the patients. A second technique for obtaining fetal blood is by sonographically directed placental aspiration through a 20-gauge needle (Golbus *et al.*, 1976). This technique also has been successful in obtaining samples from 90% of the patients. There is insufficient data at this time for either method of obtaining fetal blood to make accurate estimates of safety. Both methods require an experienced operator, and no attempt at prenatal diagnosis should be made until sufficient experience is gained by studying pregnancies undergoing elective second trimester termination. Regardless of technique, all investigators agree that fetal blood sampling should be done after 18 weeks gestation because, with increasing gestation, a given blood loss represents a smaller proportion of the fetal blood volume.

Both of the above methods for obtaining fetal cells produce a mixture of fetal and maternal blood. Since fetal red cell volume is greater than that of adult cells, the presence of fetal erythrocytes is determined by analysis of cell volume by a Coulter particle size analyzer (Kazazian *et al.*, 1972). The presence of maternal cells is not an insurmountable problem since the overall globin synthetic rate of first trimester fetal cells is more than 500 times greater than that of maternal cells (Nathan *et al.*, 1975). Thus, even though β-chain synthesis is approximately 50% of the total in maternal cells and only 5% of the total in fetal cells, the maternal cells do not make a great contribution to the sample β-chain radioactivity. Aspiration samples containing a low proportion of fetal cells (we use 50% as our criteria) can have the fetal cells separated from the mixture by differential agglutination using the antibody, anti-i. As the i antigen is found only on fetal cells, after concentration by differential agglutination, samples will contain 75–100% fetal cells with no selective loss of fetal cells containing Hb A (Kan *et al.*, 1974). The moderate loss of fetal cells which occurs during this procedure is well compensated by the enrichment.

A. Current Status

To date, prenatal diagnosis has been attempted on over 500 fetuses at risk for a hemoglobinopathy. The worldwide experience has demonstrated two significant difficulties. In one center, four out of nine pregnancies tested with placental aspiration and three out of 13 sampled with fetoscopy ended in spontaneous abortions. In experienced hands the spontaneous abortion rate after either procedure is now approximately 5% (Alter *et al.*, 1976; Kan *et al.*, 1977b). We attribute the poor outcome cited above to the inexperience of the operator.

Another difficulty has been errors in the prenatal prediction. In one center a fetus predicted to have homozygous β-thalassemia was a heterozygote when re-studied after pregnancy termination (Alter *et al.*, 1976). A second fetus predicted to have sickle trait proved to have sickle cell disease after delivery. These errors were early in their experience and, with compulsive attention to detail and more experience, such errors should be avoidable.

B. The Future

The prenatal diagnosis of the hemoglobinopathies is an accomplished fact. However, the difficulty in obtaining fetal erythrocytes indicates that a technique requiring only fetal fibroblasts would have much broader clinical application. The entire genome is present in the DNA of every fetal cell. The various differentiated and undifferentiated cell types in the body differ in which portion of that DNA is selectively derepressed in an otherwise fully repressed genome (Frenster and Herstein, 1973). The problem is one of activating unexpressed genes in the fetal fibroblasts obtained by amniocentesis.

The method envisioned is to fuse the fetal fibroblasts with "activator" erythrocyte cell lines. The hybridization of mammalian somatic cells has been accomplished. However, when a differentiated cell line expressing a specialized function such as hemoglobin synthesis is fused with an undifferentiated cell line, the specialized function is usually extinguished. A technique will have to be developed whereby hemoglobin synthesis is expressed in the undifferentiated amniotic fluid cell if prenatal diagnosis is to be accomplished using this methodology. That this is possible has been demonstrated by obtaining mouse albumin from mouse fibroblasts fused with rat hepatoma cells (Peterson and Weiss, 1972). The gene for albumin synthesis is normally "on" only in liver cells but is derepressed in the mouse fibroblasts by the hepatoma cells.

An alternate method of utilizing fetal fibroblasts for prenatal diagnosis of the hemoglobinopathies involves use of restriction endonucleases, enzymes that cleave DNA within specific base-sequence recognition sites. There is an ever increasing repertoire of restriction endonucleases, each specific for a particular short DNA sequence. The DNA fragments can be separated by agarose-gel electrophoresis, and the number and type of fragments allows genome mapping. Orkin *et al.* (1978) used these techniques to verify deletion of α-globin genes in homozygous α-thalassemia and the deletion of β-globin sequences in homozygous $\delta\beta$-thalassemia and HPFH. In theory, any point mutation also could be recognized, if there was a restriction endonuclease with the appropriate specificity. For example, the globin β^S represents a change in DNA sequence from GAGGAG to GTGGAG. Since there is an endonuclease specific for the sequence GAGG, the ability of this enzyme to cleave DNA may be useful to detect homozygosity for this β^S globin gene. Another possibility results from the fact that 64% of heterozygotes for the β^S gene have a second mutant approximately 5000 nucleotides further along the DNA (Y. W. Kan, personal communication). This second mutant is recognizable utilizing the endonuclease technique and signals the presence of the β^S gene. A fetus homozygous for β^S would demonstrate a specific pattern of DNA fragments after endonuclease treatment of its DNA.

IV. ANTENATAL SCREENING

The availability of prenatal diagnosis provides strong impetus for the creation of mass maternal screening programs. Since the primary aim of screening is the detection of individuals who already carry the gene and the prevention of homozygous individuals in the future, the ideal population for evaluation is reproductive-age women. A good screening method should have no false negatives and few false positives. It should be inexpensive, simple to perform, and acceptable to patients. Protocols which center upon the determination of red blood cell indices are particularly appealing, since they presently afford the greatest potential for meeting the above criteria.

Several proposals may be found in the literature. Healy (1965) suggests that all individuals in whom the function $(0.23 \times \text{MCV}) - (0.22 \times \text{RBC}) - (0.93 \times \text{Hb}) - 3$ is less than zero have thalassemia trait. England and Fraser (1973) simplified the formula to $\text{MCV} - \text{RBC} - 5\ \text{Hb} - 3.4 < 0$. Mentzer (1973) advocates using the ratio of MCV/RBC, values less than 13 corresponding to patients with thalassemia trait. Recently, Shine and Lal (1977) have developed a heuristic which they claim yields the lowest false negative rate and an acceptable number of false positives: $\text{MCV}^2 \times \text{MCH} < 1530$.

At our institution we have implemented a thalassemia screening program specifically designed to evaluate obstetrical patients. We believe this protocol provides the information most pertinent to the pregnant patient while retaining the virtues of simplicity and low cost. Each patient registering for prenatal care at our clinic has a complete blood count with red blood cell indices performed at the time of her first prenatal visit. This includes measurement of the MCV by a Coulter electronic counter. If the MCV is less than 80 fl [similar to Pearson *et al.*'s (1973) threshold of 79 fl], the patient is further evaluated following the scheme outlined in Fig. 2. In addition to thalassemia, a microcytic hypochromic anemia may be due to iron deficiency, chronic illness, or a sideroblastic anemia. To differentiate between these disorders serum iron studies are performed (serum iron concentration, iron binding capacity, percent saturation). If iron studies are normal (serum iron greater than 50 μg/dl, saturation greater than 15%) the Hb A_2 level is determined. A level greater than 3.5% is considered diagnostic of β-thalassemia trait. If the Hb A_2 is less than 3.5% all studies are repeated and the patient's relatives evaluated if possible. A diagnosis of α-thalassemia is made only when microcytosis, normal serum iron studies, and a normal level of Hb A_2 are confirmed. A diagnosis of Hb H disease is made if, in addition, hemoglobin electrophoresis demonstrates β_4.

Serum iron studies were chosen as the second diagnostic step for several reasons. By far the largest number of false positives using Pearson's threshold will be due to iron deficiency. Indeed, pregnant patients, including those with thalassemia trait, are at particular risk for this condition. Moreover, patients with thalassemia trait are often misdiagnosed and treated with iron for years without response. This may lead to iron overloading, particularly if parenteral iron is used (Fleming and Lynch, 1969). Since it is desirable to treat all iron-deficient patients but inappropriate to do so without first verifying the diagnosis, determination of iron status would properly follow even if the obstetrician was uninterested in thalassemia. By eliminating microcytosis secondary to iron deficiency fewer hemoglobin electrophoreses are required. Finally, in cases of coexistant

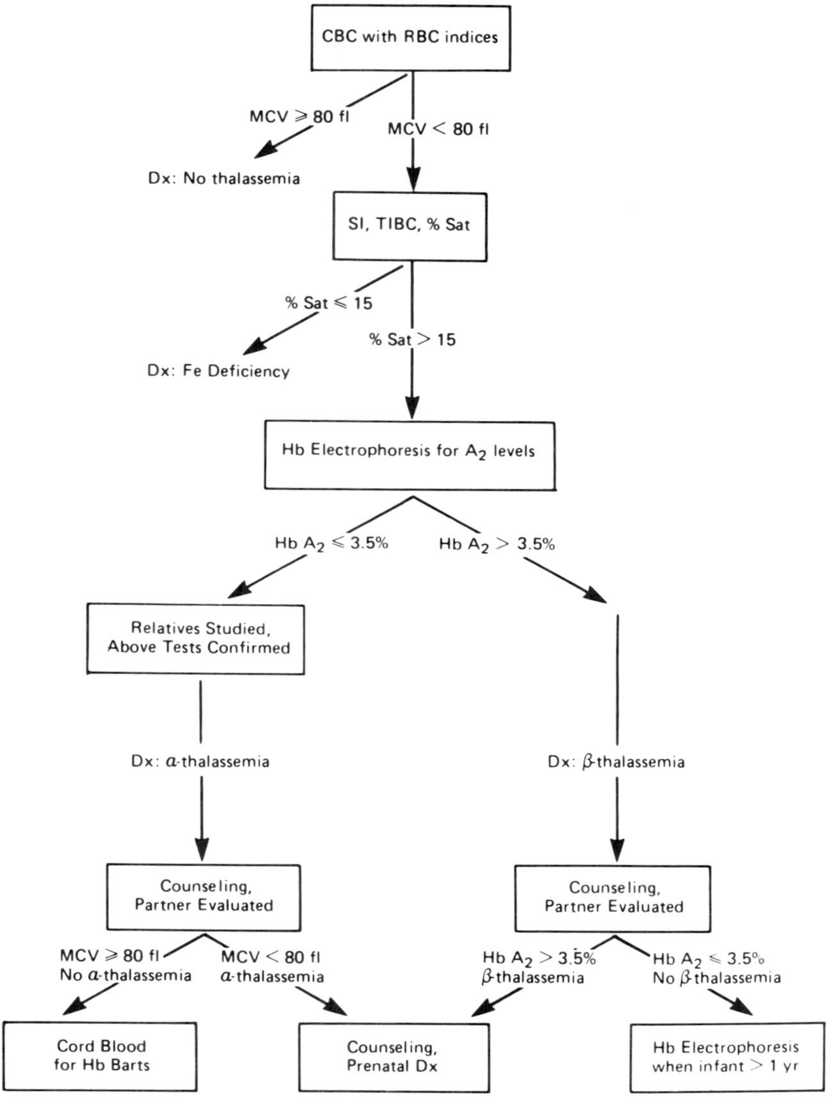

Fig. 2. Flow chart for antenatal thalassemia screening program.

β-thalassemia trait and iron deficiency, Hb A_2 levels may be depressed below 3.5% and the diagnosis missed. In this event Hb A_2 levels may become elevated only after iron replacement (Fleming and Lynch, 1969). If after appropriate iron therapy microcytosis persists, a repeat Hb A_2 level is performed.

Once a diagnosis of thalassemia trait has been made the patient is informed as to its possible significance, and her partner is evaluated by means of a complete blood count and hemoglobin electrophoresis. If he is heterozygous for a like thalassemia trait, or is heterozygous for a hemoglobinopathy that, in the double heterozygous state with thalassemia trait, causes clinical disability, the couple receives genetic counseling and is offered prenatal diagnosis as described above.

V. HEMOLYTIC ANEMIAS

A. Introduction

In normal individuals erythrocytes survive approximately 120 days. Thus, about 1% of the red cell mass is destroyed and replaced each day. This produces a normal reticulocyte count of 0.2–2.0%. In a hemolytic state the *in vivo* life span of the erythrocyte is shortened. Since the bone marrow is capable of a six- to eight fold increase in red cell production, anemia does not occur until the erythrocyte life span is markedly foreshortened. Hemolytic anemia may be classified as inherited or acquired. The inherited forms may be on the basis of either a structural or enzymatic defect in the red cell (Table VIII).

B. Structural Erythrocyte Defects

The most common form of inherited hemolytic anemia is *spherocytosis*. This familial anemia is characterized by intermittent jaundice, splenomegaly, cholelithiasis, and spherocytes in the peripheral blood smear. Classically, there is an increased red cell osmotic fragility. The inheritance is as an autosomal dominant trait, and both incomplete penetrance and phenocopies have been demonstrated (Morton *et al.*, 1962). Approximately one-fourth of the cases are sporadic, and the prevalence rate is 2.2×10^{-4}. This syndrome is etiologically heterogeneous, with the most commonly suggested defect being in the cell membrane.

On the basis of the known prevalence, over 650 pregnancies occur annually in the United States in women with hereditary spherocytosis. A hemolytic crisis can be precipitated by many conditions including infection, trauma, or pregnancy per se. The mechanism by which pregnancy induces hemolysis is not known, but may be related to the increased maternal blood volume and splenic blood flow (Moore *et al.*, 1976). Prenatal care of women with spherocytosis who have not had a splenectomy should include folic acid supplementation for the increased erythrocyte production and careful monitoring of the hemolytic state. A hemolytic crisis can be treated conservatively with replacement transfusion (Johnson and McAllister, 1951), or with splenectomy which will correct the shortened erythrocyte survival time. Splenectomy as therapy for hemolysis in spherocytosis has

TABLE VIII
Classification of Hereditary Hemolytic Anemias

I. Structural
  A. Spherocytosis
  B. Elliptocytosis
II. Enzymatic
  A. Glycolytic pathway defects
    1. Hexokinase
    2. Glucosephosphate isomerase
    3. Phosphofructokinase
    4. Triosephosphate isomerase
    5. Phosphoglycerate kinase
    6. Pyruvate kinase
    7. Others
  B. Hexose monophosphate shunt defects
    1. Glucose-6-phosphate dehydrogenase
    2. 6-phosphogluconic dehydrogenase
    3. Gluthathione synthetase, reductase, and peroxidase

been successfully performed during pregnancy in at least seven instances (McElin *et al.*, 1950; Moore *et al.*, 1976). The dangers of splenectomy are the perioperative morbidity and thrombosis secondary to the usual postsplenectomy thrombocytosis. In the absence of a severe untreated anemia, heredity spherocytosis does not contribute to prenatal or neonatal morbidity.

Genetic counseling should include an explanation of autosomal dominant inheritance and the 50% chance that each offspring will be affected. Prenatal diagnosis might be possible by examining a smear of fetal blood and determining fetal erythrocyte osmotic fragility. This would require fetal blood sampling by either fetoscopy or sonographic directed placental aspiration. Prenatal diagnosis has not been attempted yet, and, in view of the moderate incapacitation caused by this disorder and the readily available therapy by splenectomy, the appropriateness of such a diagnosis may be questioned.

Hereditary *elliptocytosis* is a somewhat milder hemolytic anemia state caused by an erythrocyte structural defect. The symptoms and signs are similar to those of spherocytosis, albeit not as severe. There are at least two genetically distinct varieties of elliptocytosis, one that is linked with the Rhesus locus and one that is not. Hemolysis is greater in the "unlinked" form than in the "linked" form. Hemolysis may be precipitated by pregnancy, with the patient manifesting no symptoms or signs between pregnancies. Although splenectomy will be almost universally successful at correcting the anemia, transfusion therapy has been all that was required in the few reported cases involving a pregnancy (Weiss, 1963; Breckenridge and Riggs, 1968). Hereditary elliptocytosis is inherited as an autosomal dominant trait, and the comments above on the genetics and prenatal diagnosis of spherocytosis also apply to this disorder.

C. Enzymatic Erythrocyte Defects

The reticulocyte can utilize either glucose or nonglycolytic substrates for energy metabolism, whereas the mature erythrocyte relies upon anerobic glycolysis for its energy needs. These pathways are summarized in Figs. 3 and 4. This attenuation of metabolic capacity coincides with the loss of mitochondria and microsomes from the reticulocyte. Glycolysis produces only 2 mol ATP/mol of glucose metabolized (steps 7a and 10 in Fig. 3), whereas mitochondrial oxidative phosphorylation produces 36 mol/ATP/mol of glucose. The mature red cell generates 93% of its ATP via the glycolytic pathway and the remainder from the hexose monophosphate shunt (Oski and Naiman, 1972). The most prevalent red cell enzyme deficiencies are those involving the hexose monophosphate shunt. Because this pathway supplies only a minor component of the cells' energy, these deficiencies usually are not associated with constant hemolytic anemias but rather episodic hemolysis secondary to oxidative damage. However, there are some chronic hemolyzing glucose-6-phosphate dehydrogenase variants. Enzyme deficiencies in the glycolytic pathway, albiet rarer, are usually associated with more severe hemolysis.

Prenatal diagnosis of enzymatic defects of erythrocytes is theoretically possible; however, it has not been attempted. It would require obtaining fetal blood by fetoscopy or sonographic directed placental aspiration and determining the enzyme activities present in the fetal erythrocytes. For each disease entity the normal range of fetal erythrocyte enzyme level would have to be determined. Each disorder and each family should be considered individually, so as to provide the genetic counseling appropriate to the situation.

D. Glycolytic Pathway Defects (Fig. 3)

The glycolytic pathway defects associated with hemolytic anemias will be discussed in the order in which they act in metabolizing glucose to lactic acid. The first enzyme is *hexokinase* (1) which catalyzes the conversion of glucose to glucose 6-phosphate. The theoretical relative lethality of a deficiency of this key enzyme may be reflected in the paucity of reported cases. Two clinical pictures have been reported. Four families had affected members with mild hemolysis, reticulocytosis of 5–21%, and decreased exercise tolerance disproportionate to the degree of anemia (Piomelli and Corash, 1976). Three other subjects had a Fanconi's anemia-like syndrome with hemolysis, pancytopenia, and chromosomal aberrations (Löhr *et al.*, 1965). Only one of the 10 reported cases of erythrocyte hexokinase deficiency is a woman of childbearing age, and no mention was made of the gestational history of this patient. Three of the affected neonates had hyperbilirubinemia requiring exchange transfusions (Gilman, 1974). This enzyme deficiency is inherited as an autosomal recessive trait.

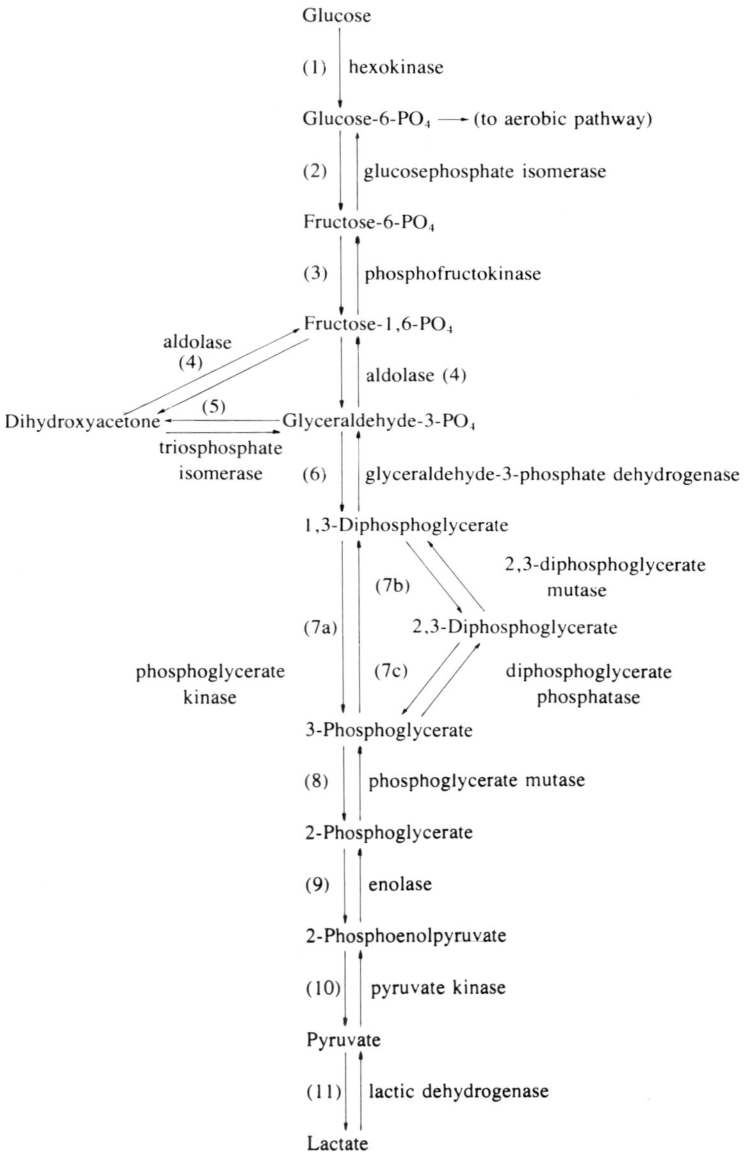

Fig. 3. Anaerobic glycolysis in the red blood cell.

The next enzyme in the glycolytic pathway is *glucosephosphate isomerase* (2) which catalyzes the conversion of glucose 6-phosphate to fructose 6-phosphate. This is the second most common glycolytic enzyme deficiency associated with a hemolytic anemia. Affected individuals have a moderately severe hemolysis and splenomegaly. Splenectomy results in marked clinical improvements. Only two of the 25 reported cases were women in their reproductive years and no gestational histories are available (Piomelli and Corash, 1976). At least four affected neonates had a hyperbilirubinemia requiring exchange transfusion. This enzyme deficiency also is inherited as an autosomal recessive trait.

The next enzyme in the glycolytic cascade associated with a hemolytic anemia is *phosphofructokinase* (3) which catalyzes the conversion of fructose 6-phosphate to fructose 1,6-diphosphate. This enzyme deficiency is associated with mild to moderate hemolysis. Some patients also have absent muscle phosphofructokinase activity and a resultant myopathy (Layser *et al.*, 1967). There was no mention in the case reports as to whether the two affected females had been pregnant. This enzyme deficiency is inherited as an autosomal recessive trait.

Triosephosphate isomerase (5) is associated with a hemolytic anemia. These patients have a severe chronic hemolysis and neurologic abnormalities. Only one patient survived beyond puberty, at which time her transfusion requirement ceased, but her neurologic symptoms persisted (Harris *et al.*, 1970). This autosomal recessive trait probably could be diagnosed *in utero,* and such a possibility should be discussed with parents of an affected child.

Phosphoglycerate kinase (7a) deficiency is associated with a severe chronic hemolysis amenable to therapy by splenectomy, and with neurologic symptoms. The structural gene for phosphoglycerate kinase is on the long arm of the X chromosome (Pearson *et al.*, 1975). Thus, only male offspring of a female carrier are at risk. Prenatal diagnosis would be by amniotic fluid cell karyotyping to establish the sex of the fetus, and then enzyme activity studies on erythrocytes of male fetuses.

Pyruvate kinase (10) deficiency is by far the most common form of red cell glycolytic defect associated with chronic hemolysis. Approximately 1:20,000–1:40,000 newborns have this disorder. Hemolysis is moderate to severe and preventable by splenectomy. Additional features may include cholelithiasis, jaundice, and widened bone marrow spaces with facial and roentgenographic abnormalities (Piomelli and Corash, 1976). As with many of the inherited hemolytic anemias, a hemolytic crisis may be precipitated by biological stress states including pregnancy (Collier *et al.*, 1966; Kendall and Charlow, 1977). There are two reports of oral contraceptives intensifying the hemolytic process (Nixon and Buchanan, 1967; Kendall and Charlow, 1977). The fetal effect is evident at birth, when accentuation of hemolysis results in severe jaundice, often requiring exchange transfusion. There are at least two classes of defective red cell pyruvate kinase mutants, one with defective activity and another with kinetic abnormalities. The disorder is inherited as an autosomal recessive trait, and affected individuals may be either homozygotes for the same mutant or double heterozygotes for two different mutant genes (Tanaka and Paglia, 1971).

Deficiencies of other enzymes in the glycolytic pathway (e.g., *glyceraldehyde-3-phosphate dehydrogenase* (6), *2,3-diphosphoglycerate mutase* (7b), *diphosphoglycerate phosphatase* (7c), and *phosphopyruvate enolase* (9) also have been associated with hemolytic anemias. For each of these only a few affected individuals are known, and no information related to pregnancy or fetal effects is available. Each is probably inherited as an autosomal recessive trait.

E. Hexose Monophosphate Shunt Defects

Glucose-6-phosphate dehydrogenase (G-6-PD) (Fig. 4,1) deficiency is the most common inherited form of hemolytic anemia, and is found in more than 100 million persons of all races throughout the world. More than 70 variants of the enzyme have been described, with at least 20 variants associated with chronic hemolysis. This enzyme is the first step of the hexose monophosphate shunt, which is important in maintaining the level of reduced glutathione in the red cell and thus protecting the cell against oxidants. Glutathione metabolism is depicted

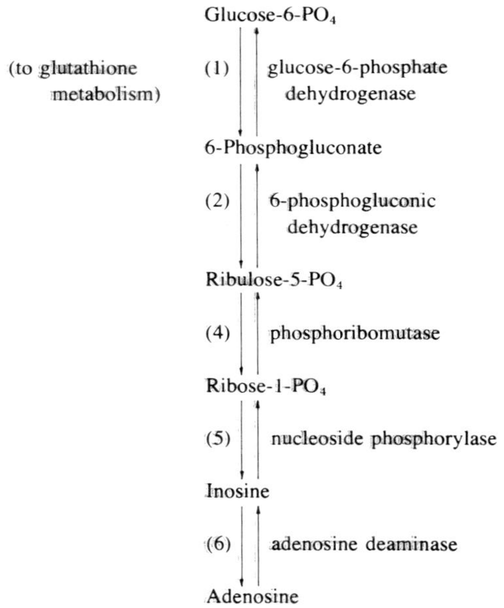

Fig. 4. Aerobic glycolysis in the red blood cell.

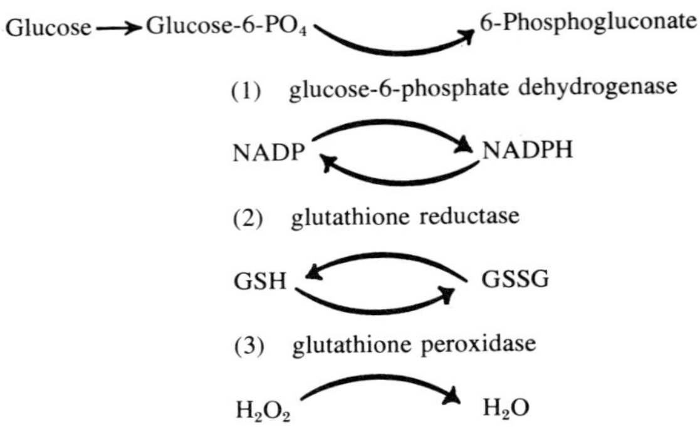

Fig. 5. Glutathione metabolism. NADP and NADPH, oxidized and reduced nicotinamide adenine dinucleotide; GSH, reduced glutathione; GSSG, oxidized glutathione.

in Fig. 5 and is critical in protecting hemoglobin, enzymes, and the RBC membrane from oxidative damage. The most important oxidant with which the RBC must contend is hydrogen peroxide (H_2O_2). When the RBC is exposed to unknown oxidant drugs, glutathione (GSH) serves to protect against damage. A list of medications that may trigger hemolysis in these patients is given in Table IX. The severity of hemolysis is related to the specific enzyme variant, the severity of deficiency, and exposure to biological stress or certain oxidizing agents. For example, the major G-6-PD variant, A+, is found in the Black population and deficiency of this variant occurs in 10% of American Black males and 3% of American Black females (Oski and Naiman, 1972). This specific deficiency is usually associated with episodic hemolysis but not a chronic hemolytic state. Alternatively, hemolysis induced by *Fava* beans (Favism) is seen in Caucasian and Oriental individuals with certain G-6-PD mutants, but is unusual in the Black population (Carson and Frischer, 1966).

Pregnancy in patients with heterozygous G-6-PD deficiency is associated with a number of specific complications. There is decreased G-6-PD activity in one-third of the patients during the last trimester (Vergnes and Clerc, 1968). This may predispose these women to hemolytic episodes at this time. An association with anemia is not unexpected, and a study of 180 G-6-PD-deficient pregnant women reported 62% had a hematocrit less than 30% (Silverstein *et al.*, 1974). Urinary tract infection also was suggested as being more common among enzyme-deficient women. This produces an additional risk because the drugs often employed to treat urinary tract infections are oxidants. A reasonable course would be to screen Black pregnant women with urinary tract infections for G-6-PD deficiency prior to initiation of therapy. One report has indicated an increased incidence of low birth weight infants born to G-6-PD-deficient mothers, but no correction for the effects of anemia or urinary tract infections was made (Perkins, 1976b).

TABLE IX
Agents to Be Avoided in Red Cell Enzyme Deficiency States

Analgesics
Acetanilid
Acetylsalicylic acid
Acetophenetidin (phenacetin)
Sulfonamides and sulfones
Sulfanilamide
Sulfapyridine
Diaphenylsulfone
N-Acetylsulfanilamide
Sulfisoxazole (Gantrisin)
Thiazolsulfone
Salicylazosulfapyridine (Azulfidine)
Sulfoxone
Sulfamethoxypyridazine (Kynex)
Sulfacetamide
Antimalarials
Primaquine
Pamaquine
Pentaquine
Quinocide
Quinacrine (Atabrine)
Nonsulfonamide antibacterial agents
Furazolidine
Nitrofurantoin (Furadantin)
Chloramphenicol
Para-aminosalicylic acid
Furmethonol
Nitrofurazone
Miscellaneous
Naphthalene (moth balls)
Methylene blue
Nalidixic acid
Phenylhydrazine
Ascorbic acid (massive doses)
Trinitrotoluene
Quinine
Quinidine
Niridazole

The glucose 6-phosphate deficient fetus is at risk both before and after delivery. Exposure to maternally ingested oxidants such as sulfisoxazole, ascorbic acid, or Fava beans may produce hemolysis, hydrops fetalis, and death (Perkins, 1971b; Mentzer and Collier, 1975). The neonate is at risk for anemia, hyperbilirubinemia, and kernicterus (Gilman, 1974). The incidence of severe jaundice in randomly selected G6PD-deficient newborn males is approximately 5%.

However, this incidence rises to 50% if there was a prior icteric sib (Fessas *et al.*, 1962).

Glucose 6-phosphate deficiency is inherited as an X-linked trait. Males are either hemizygous or unaffected. Females are homozygous, heterozygous, or unaffected. Clinical involvement may occur in heterozygous females as well as homozygous females and hemizygous males. This may be a reflection of variable X-chromosome inactivation, as there is a tremendous variation in enzyme levels among heterozygous females. Although genetic counseling and providing affected individuals with a list of oxidants to avoid is extremely important, prenatal diagnosis probably is not necessary in view of the usually benign nature of this enzyme deficiency.

A number of other enzymes in the hexose monophosphate shunt and related to glutathione metabolism have been reported to be absent in patients with hemolytic anemias. These include *6-phosphogluconic dehydrogenase* (Fig. 4,2), *glutathione reductase* (Fig. 5,2), *glutathione peroxidase* (Fig. 5,3), and *glutathione synthetase*. Glutathione synthetase catalyzes the production of GSH from γ-glutamyl-cysteine plus glycine. Each of these deficiencies is relatively rare and there is no available information about maternal effects during pregnancy. Most of these may cause sufficient hemolysis in the neonate to produce hyperbilirubinemia requiring exchange transfusion.

REFERENCES

Alger, L. S., Golbus, M. S., and Laros, R. K. (1979). Thalassemia and pregnancy. *Am. J. Obstet. Gynecol.* **134,** 662–673.

Alter, B. P., Modell, C. B., Fairweather, D., Hobbins, J. C., Mahoney, M. J., Frigoletto, F. D., Sharman, A. S., and Nathan, D. G. (1976). Prenatal diagnosis of hemoglobinopathies. *N. Engl. J. Med.* **295,** 1437–1443.

Ammann, A. J., Addiego, J., Wara, D., Luber, B., Smith, W. B., and Mentzer, W. C. (1977). Polyvalent pneumococcalpolysaccharide immunization of patients with sickle-cell anemia and patients with splenectomy. *N. Engl. J. Med.* **297,** 897–900.

Barrett-Connor, E. (1968). Cholelithiasis in sickle cell anemia. *Am. J. Med.* **45,** 889–898.

Beischer, N. A., Holsman, M., and Kitchen, W. H. (1968). Relation of various forms of anemia to placental weight. *Am. J. Obstet. Gynecol.* **101,** 801–809.

Blattner, P., Dar, H., and Nitowsky, H. M. (1977). Pregnancy outcome in women with sickle cell trait. *J. Am. Med. Assoc.* **238,** 1392–1394.

Breckenridge, R. L., and Riggs, J. A. (1968). Hereditary elliptocytosis with hemolytic anemia complicating pregnancy. *Am. J. Obstet. Gynecol.* **101,** 861–862.

Bunn, H. F., Forget, B. J., and Ranney, H. M. (1977). "Human Hemoglobins." Saunders, Philadelphia, Pennsylvania.

Carson, P. E., and Frischer, H. (1966). Glucose-6-phosphate dehydrogenase deficiency and related disorders of the pentose phosphate pathway. *Am. J. Med.* **41,** 744–761.

Collier, H. B., Ashford, D. R., and Bell, R. E. (1966). Three cases of hemolytic anemia with erythrocyte pyruvate kinase deficiency in Alberta. *Can. Med. Assoc. J.* **95,** 1188–1192.

Conconi, F., Rowley, P. T., Del Senno, L., Pontremoli, S., and Volpato, S. (1972). Induction of β globin synthesis in the β thalassemia of Ferrara. *Nature* (London), *New Biol.* **238,** 83-87.

Conley, C. L., Weatherall, D. J., Richardson, S. N., Shepard, M. K., and Charache, S. (1963). Hereditary persistence of fetal hemoglobin: A study of 79 affected persons in 15 Negro families in Baltimore. *Blood* **21,** 261-281.

Curtis, E. M. (1959). Pregnancy in sickle cell anemia, sickle-hb C disease, and variants thereof. *Am. J. Obstet. Gynecol.* **77,** 1312-1323.

Diggs, L. W. (1965). Sickle cell crises. *Am. J. Clin. Pathol.* **44,** 1-19.

Diggs, L. W. (1973). *In* "Sickle Cell Disease." Abramson, Bertus, and Wethers (eds.). C. V. Mosby, St. Louis.

Eng, L. I. L., Lopez, C. G., and Dutt, A. K. (1968). Pathological findings in hydrops foetalis due to alpha-thalassemia: A review of 32 cases. *Trans. R. Soc. Trop. Med. Hyg.* **62,** 874-879.

England, J. M., and Fraser, P. M. (1973). Differentiation of iron deficiency from thalassemia trait by routine blood count. *Lancet* **1,** 449-452.

Fessas, P. H., Doxiadis, S. A., and Valaes, T. (1962). Neonatal jaundice in glucose-6-phosphate dehydrogenase deficient infants. *Br. Med. J.* **2,** 1359-1366.

Flakpui, E. Z., and Moran, E. M. (1973). Pregnancy in the sickle hemoglobinopathies. *J. Reprod. Med.* **11,** 28-34.

Fleming, A. F. (1973). Maternal anemia and fetal outcome in pregnancies complicated by thalassemia minor and "stomatocytosis." *Am. J. Obstet. Gynecol.* **116,** 309-319.

Fleming, A. F., and Lynch, W. (1969). Beta-thalassemia minor during pregnancy with particular reference to iron status. *J. Obstet. Gynaecol. Br. Commonw.* **76,** 451-457.

Forget, B. G., Benz, E. J., Skoultchi, A., Baglioni, C., and Housman, D. (1974). Absence of messenger RNA for β globin chain in β^0-thalassemia. *Nature (London)* **247,** 379-381.

Fort, A. T., Morrison, J. C., Berreras, L., Diggs, L. W., and Fish, S. A. (1971). Counseling the patient with sickle cell disease about reproductions. *Am. J. Obstet. Gynecol.* **111,** 324-327.

Fraser, J. L., and Watt, H. J. (1964). Megaloblastic anemia in pregnancy and the puerperium. *Am. J. Obstet. Gynecol.* **89,** 532-534.

Freeman, M. G., and Ruth, G. J. (1969). SS disease, Sc disease, and CC disease—obstetric considerations of treatment. *Clin. Obstet. Gynecol.* **12,** 134-156.

Frenster, J. H., and Herstein, P. R. (1973). Gene de-repression. *N. Engl. J. Med.* **288,** 1224-1229.

Fullerton, W. T., and Watson-Williams, E. J. (1962). Haemoglobin sickle cell disease and megaloblastic anemia of pregnancy. *J. Obstet. Gynaecol. Br. Emp.* **69,** 729-735.

Gilman, P. A. (1974). Hemolysis in the newborn infant resulting from deficiencies of red blood cell enzymes: Diagnosis and management. *J. Pediatr.* **84,** 625-634.

Golbus, M. S., Kan, Y. W., and Naglich-Craig, M. (1976). Fetal blood sampling in mid-trimester pregnancies. *Am. J. Obstet. Gynecol.* **124,** 653-655.

Harris, S. R., Paglia, D. E., Jaffe, E. R., Valentine, W. N., and Klein, R. L. (1970). Triosephosphate isomerase deficiency in an adult. *Clin. Res.* **18,** 529.

Hatch, F. E., Culbertson, J. W., and Diggs, L. W. (1967). Nature of renal concentration defect in sickle cell disease. *J. Clin. Invest.* **46,** 336-345.

Healy, M. J. R. (1965). *In* "Mathematics and Computer Science in Biology and Medicine," pp. 93-102. HM Stationery Office, London.

Henderson, A. B. (1950). Sickle cell anemia. *Am. J. Med.* **9,** 757-765.

Hobbins, J. C., and Mahoney, M. J. (1975). Fetal blood drawing. *Lancet* **2,** 107-109.

Horger, E. (1972). Sickle cell and sickle cell-hb C disease during pregnancy. *Obstet. Gynecol.* **39,** 873-879.

Housman, D., Forget, B. G., Skoultchi, A., and Benz, E. J. (1973). Quantitative deficiency of chain specific globin mRNA in the thalassemia syndromes. *Proc. Natl. Acad. Sci. U.S.A.* **70,** 1809-1813.

Huehns, E. R., Dance, D., Beaven, G. H., Hecht, F., and Motulsky, A. G. (1964a). Human embryonic hemoglobins. *Cold Spring Harbor Symp. Quant. Biol.* **29,** 327-331.

Huehns, E. R., Dance, D., Beaven, G. H., Keil, J. V., Hecht, F., and Motulsky, A. G. (1964b). Human embryonic hemoglobins. *Nature* (*London*) **201,** 1095-1097.

Ingram, V. M. (1956). A specific chemical difference between globins of normal human and sickle-cell anemic hemoglobin. *Nature* **178,** 792-794.

Jackson, J. F., Odom, J. L., and Bell, W. N. (1961). Amelioration of sickle cell disease by persistent fetal hemoglobin. *Am. Med. Assoc.* **177,** 867-869.

Johnson, H. H. C., and McAllister, J. E. (1951). Familial hemolytic icterus occurring during pregnancy. *Can. Med. Assoc. J.* **65,** 377.

Kan, Y. W., Forget, B. G., and Nathan, D. G. (1972a). Gamma-beta thalassemia: A cause of hemolytic disease of the newborn. *N. Engl. J. Med.* **286,** 129-134.

Kan, Y. W., Dozy, A. M., Alter, B. P., Frigoletto, F. D., and Nathan, D. G. (1972b). Detection of the sickle cell gene in the human fetus. *N. Engl. J. Med.* **287,** 1-5.

Kan, Y. W., Nathan, D. G., Cividalli, G., and Crookston, M. C. (1974). Concentration of fetal red blood cells from a mixture of maternal and fetal blood by anti-i serum—an aid to prenatal diagnosis of hemoglobinopathies. *Blood* **43,** 411-415.

Kan, Y. W., Dozy, A. M., Varmus, H. E., Taylor, J. M., Holland, J. P., La-Injo, L. E., Ganeson, J., and Todd, D. (1975a). Deletion of α globin genes in haemoglobin-H disease demonstrates multiple α globin structural loci. *Nature* (*London*) **255,** 255-256.

Kan, Y. W., Holland, J. P., Dozy, A. M., and Varmus, H. E. (1975b). Demonstration of nonfunctional β globin m-RNA in homozygous β^0-thalassemia. *Proc. Natl. Acad. Sci. U.S.A.* **72,** 5140-5144.

Kan, Y. W., Golbus, M. S., and Dozy, A. M. (1976). Prenatal diagnosis of α-thalassemia. *N. Engl. J. Med.* **295,** 1165-1167.

Kan, Y. W., Dozy, A. M., Trecartin, M. S., and Todd, D. (1977a). Identification of a nondeletion defect in α-thalassemia. *N. Engl. J. Med.* **297,** 1081-1084.

Kan, Y. W., Golbus, M. S., Trecartin, R. F., Filly, R. A., Valenti, C., Furbetta, M., and Lao, A. (1977b). Prenatal diagnosis of β-thalassemia and sickle cell anemia. *Lancet* **1,** 269-271.

Karayalcin, G., Imran, M., and Rosner, F. (1972). Blister cells. *J. Am. Med. Assoc.* **219,** 1727-1729.

Kazazian, H. H., Kaback, M. M., Woodhead, A. P., Leonard, C. O., and Nersesian, W. S. (1972). Further studies in the antenatal detection of sickle cell anemia and other hemoglobinopathies. *Adv. Exp. Med. Biol.* **28,** 337-346.

Kendall, A. G., and Charlow, G. F. (1977). Red cell pyruvate kinase deficiency: Adverse effect of oral contraceptives. *Acta Haematol.* **57,** 116-120.

Laros, R. K. (1967). Sickle cell hemoglobin-C disease of pregnancy. *Pa. Med.* **70,** 73-77.

Laros, R. K., and Kalstone, C. (1971). Sickle cell-β-thalassemia and pregnancy. *Obstet. Gynecol.* **37,** 67-71.

Layzer, R. B., Rowland, L. P., and Ranney, H. M. (1967). Muscle phosphofructokinase deficiency. *Arch. Neurol.* (*Chicago*) **17,** 512-523.

Lindenbaum, J., and Klipstein, F. A. (1963). Folic acid deficiency in sickle cell anemia. *N. Engl. J. Med.* **269,** 875-882.

Löhr, G. W., Waller, H. D., Anschutz, F., and Knoop, A. (1965). Biochemische defecte in den Blutzellen bei familiarer Ponmyelopathie. *Humangenetik* **1,** 383-387.

McCurdy, P. R. (1964). Abnormal hemoglobins and pregnancy. *Am. J. Obstet. Gynecol.* **90,** 891-896.

McElin, T. W., Mussey, R. D., and Watkins, C. H. (1950). Splenectomy during pregnancy, with a report of 5 cases and review of the literature. *Am. J. Obstet. Gynecol.* **59,** 1036-1044.

Mentzer, W. C. (1973). Differentiation of iron deficiency from thalassemia trait. (Letter.) *Lancet* **1,** 882.

Mentzer, W. C., and Collier, E. (1975). Hydrops fetalis associated with erythrocyte G-6-PD deficiency and maternal ingestion of fava beans and ascorbic acid. *J. Pediatr.* **86,** 565-567.

Milner, P. F., Clegg, J. B., and Weatherall, D. J. (1971). Haemoglobin H disease due to a unique haemoglobin variant with an elongated α chain. *Lancet* **1,** 729-732.

Moore, A., Sherman, M. M., and Strongin, M. J. (1976). Hereditary spherocytosis with hemolytic crisis during pregnancy. *Obstet. Gynecol.* **47,** Suppl., 19-21.

Morrison, J. C., and Wiser, W. L. (1976a). The effect of maternal partial exchange transfusion on the infants of parents with sickle cell anemia. *J. Pediatr.* **89,** 286-289.

Morrison, J. C., and Wiser, W. L. (1976b). The use of prophylactic partial exchange transfusion in pregnancies associated with sickle cell hemoglobinopathies. *Obstet. Gynecol.* **48,** 516-520.

Morton, N. E., Mackinney, A. A., Kosower, N., Schilling, R. F., and Gray, M. P. (1962). Genetics of spherocytosis. *Am. J. Hum. Genet.* **14,** 170-184.

Motulsky, A. G. (1973). Frequency of sickling disorders in U.S. blacks. *N. Engl. J. Med.* **288,** 31-33.

Nathan, D. G., and Gunn, R. B. (1966). Thalassemia: The consequences of unbalanced hemolog synthesis. *Am. J. Med.* **41,** 815-830.

Nathan, D. G., Alter, B. P., and Frigoletto, F. (1975). Antenatal diagnosis of hemoglobinopathies: Social and technical considerations. *Semin. Hematol.* **12,** 305-321.

Nechelen, T. (1973). Obstetric complications associated with haemoglobinopathies. *Clin. Hematol.* **2,** 497-514.

Nixon, A. D., and Buchanan, J. G. (1967). Haemolytic anemia due to pyruvate kinase deficiency. *N. Z. Med. J.* **66,** 859-864.

Orkin, S. H., and Nathan, D. G. (1976). The thalassemias. *N. Engl. J. Med.* **295,** 710-714.

Orkin, S. H., Alter, B. P., Altay, C., Mahoney, M. T., Lazarus, H., Hobbins, J. C., and Nathan, D. G. (1978). Application of endonuclease mapping to the analysis and prenatal diagnosis of thalassemia caused by globin-gene deletion. *N. Engl. J. Med.* **299,** 166-172.

Oski, F. A., and Naiman, J. L. (1972). "Hematologic Problems in the Newborn," 2nd ed. Saunders, Philadelphia, Pennsylvania.

Ottolenghi, S., Lanyon, W. G., Paul, J., Williamson, R., Weatherall, D. J., Clegg, J. B., Pritchard, J., Pootrakul, S., and Boon, W. H. (1974). The severe form of α-thalassemia is caused by a haemoglobin gene deletion. *Nature (London)* **251,** 389-392.

Pataryas, H. A., and Stamatoyannopoulos, G. (1972). Hemoglobins in human fetuses: Evidence for adult hemoglobin production after the 11th gestational week. *Blood* **39,** 688-696.

Pearson, H. A., O'Brien, R. T., and McIntosh, S. (1973). Screening for thalassemia trait by electronic measurement of mean corpuscular volume. *N. Engl. J. Med.* **288,** 351-353.

Pearson, P. L., Sanger, R., and Brown, J. A. (1975). Report of the commission on the genetic constitution of the X-chromosome. *Cytogenet. Cell Genet.* **14,** 190-195.

Perkins, R. P. (1971a). Inherited disorders of hemoglobin synthesis and pregnancy. *Am. J. Obstet. Gynecol.* **111,** 120-159.

Perkins, R. P. (1971b). Hydrops fetalis and stillbirth in a male glucose-6-phosphate dehydrogenase-deficient possibly due to maternal ingestion of sulfisoxazole. *Am. J. Obstet. Gynecol.* **111,** 379-381.

Perkins, R. P. (1976a). Partial exchange transfusion in a pregnant patient with sickle cell anemia. *Obstet. Gynecol.* **48,** 22s-24s.

Perkins, R. P. (1976b). The significance of glucose-6-phosphate dehydrogenase deficiency in pregnancy. *Am. J. Obstet. Gynecol.* **125,** 215-223.

Perrine, R. P., and John, P. (1974). Pregnancy in sickle cell anemia in a Caucasian group. *Am. J. Obstet. Gynecol.* **118,** 29-33.

Peterson, J. A., and Weiss, M. C. (1972). Expression of differentiated functions in hepatoma cell hybrids: Induction of mouse albumin production in rat hepatoma mouse fibroblast hybrids. *Proc. Natl. Acad. Sci. U.S.A.* **69,** 571-575.

Piomelli, S., and Corash, L. (1976). Hereditary hemolytic anemia due to enzyme defect of glycolysis. *Adv. Hum. Genet.* **6,** 165–240.

Pritchard, J. A., Scott, D. E., Whalley, P. J., Cunningham, F. G., and Mason, R. A. (1973). The effects of maternal sickle cell hemoglobinopathies and sickle cell trait on reproductive performance. *Am. J. Obstet. Gynecol.* **117,** 662–670.

Ramirez, F., O'Donnell, J. V., Marks, P. A., Bank, A., Musumeci, S., Schilero, G., Pizzarelli, G., Russo, G., Luppis, B., and Gumbino, R. (1976). Abnormal or absent β m-RNA in β^0 Ferrara and gene deletion in $\delta\beta$-thalassemia. *Nature (London)* **263,** 471–475.

Ricks, P. (1968). Further experience with exchange transfusion in sickle cell anemia and pregnancy. *Am. J. Obstet. Gynecol.* **100,** 1087–1091.

Rivers, G. L., Robbons, A. B., and Schwarts, S. O. (1961). Sickle cell hemoglobin: A clinical study. *Blood* **18,** 385–397.

Rowley, P. T. (1976). The diagnosis of beta-thalassemia trait: A review. *Am. J. Hematol.* **1,** 129–137.

Rucknagel, D. L., and Winter, W. P. (1974). Duplication of structural genes for hemoglobin α and β chains in man. *Ann. N.Y. Acad. Sci.* **241,** 80.

Schmidt, R. M. (1973). Laboratory diagnosis of hemoglobinopathies. *J. Am. Med. Assoc.* **224,** 1276–1280.

Schmidt, R. M., Moo-Pen, W., *et al.* (1975). "Advanced Lab Methods of Hb Detection," 2nd ed., DHEW Publ. No. (CDC) 75-8296, pp. 1–218. US Govt. Printing Office, Washington, D.C.

Schroeder, W. A., Huisman, T. H. J., Shelton, J. B., Kleibauer, E. F., Dozy, A. M., and Robberson, B. (1968). Evidence for multiple structural genes for the γ-chain of human fetal hemoglobin. *Proc. Natl. Acad. Sci. U.S.A.* **60,** 537–544.

Shine, I., and Lal, S. (1977). A strategy to detect β-thalassemia minor. *Lancet* **1,** 692–694.

Silverstein, E., Roadman, C., Byers, R. H., and Kitay, D. Z. (1974). Hematologic problems in pregnancy. III. Glucose-6-phosphate dehydrogenase deficiency. *J. Reprod. Med* **12,** 153–158.

Singer, K., Chernoff, A. L., and Singer, L. (1951). Studies on abnormal hemoglobins. *Blood* **6,** 429–435.

Stamatoyannopoulos, G., Fessas, P., and Papayannopoulou, T. (1969). F-thalassemia: A study of thirty-one families with simple heterozygotes and combinations of F-thalassemia with A_2-thalassemia. *Am. J. Med.* **47,** 194–208.

Tanaka, T. R., and Paglia, D. E. (1971). Pyruvate kinase deficiency. *Semin. Hematol.* **8,** 367–396.

Thomas, E. D., Lochte, H. L., Greenough, W. B., and Wales, M. (1960). In vitro synthesis of foetal and adult haemoglobin by foetal haematopoietic tissues. *Nature (London)* **185,** 396–397.

Vergnes, H., and Clerc, A. (1968). Erythrocyte enzyme activity in pregnancy. *Lancet* **2,** 834.

Walker, J., and Turnbull, E. P. N. (1955). Haemoglobin and red cells in the human fetus. III. Foetal and adult haemoglobin. *Arch. Dis. Child.* **30,** 111–116.

Weatherall, D. J., and Clegg, J. B. (1972). "The Thalassemia Syndromes," 2nd ed. Blackwell, Oxford.

Weatherall, D. J., Clegg, J. B., and Wong, H. B. (1970). The haemoglobin constitution of infants with the haemoglobin Barts' hydrops foetalis syndrome. *Br. J. Haematol.* **18,** 357–367.

Weiss, H. J. (1963). Hereditary elliptocytosis with hemolytic anemia. *Am. J. Med.* **35,** 455–466.

Went, L. N., Anderson, M., MacIver, J. E., and Dixon, H. G. (1960). Sickle cell disease in pregnancy. *Lancet* **2,** 516–521.

Whalley, P. J., Pritchard, J. A., and Richards, J. R. (1963). Sickle cell trait and pregnancy. *J. Am. Med. Assoc.* **186,** 1132–1135.

Whalley, P. J., Martin, F. G., and Pritchard, J. A. (1964). Sickle cell trait and urinary tract infections. *J. Am. Med. Assoc.* **189,** 903–906.

4

Genetic Disorders of Hemostasis

CONNIE H. MILLER AND MARGARET W. HILGARTNER

I. INTRODUCTION

Genetic defects in the hemostatic mechanism result in a multitude of diseases with a wide variety of clinical and laboratory manifestations. Some may pose

Copyright © 1981 by Academic Press, Inc.
All rights of reproduction in any form reserved.
ISBN 0-12-630940-X

life-threatening risks to mother or baby during pregnancy and delivery, and the application of such diagnostic techniques as amniocentesis and fetal blood sampling can add to the dangers. In the X-linked hemophilias, the severity of disease in affected males may cause the potential problems of hemophilia carriers to be overlooked. Other coagulation and platelet disorders affect women as well as men due to their autosomal inheritance patterns. With the exception of von Willebrand disease (which may be as common as classical hemophilia), these disorders are rare. However, they may present significant problems because of the specialized diagnostic and treatment measures that they require.

This chapter deals with the hereditary disorders of plasma coagulation factors and inherited defects in platelet function and number. Those syndromes in which the hemostatic defect plays only a small part are discussed more fully in other chapters; therefore, only the hematologic abnormalities will be described here. For each disorder the clinical features and laboratory findings are described. The inheritance pattern is stated in accordance with principles of classification discussed below. Information regarding any special risks of pregnancy and delivery to affected women is mentioned, although few data are available for the rare disorders. While bleeding disorders in the newborn infant have been well described, information on the fetus is sketchy. Prenatal diagnosis has been reported for only a few disorders; therefore, discussions of its feasibility in other disorders are purely speculative. The general mode of therapy for each disorder is given but for details of treatment procedures and their risks, the reader is referred to the standard works of Hilgartner (1976), Williams *et al.* (1977), and Smith (1979).

A. The Normal Hemostatic Mechanism

The process of hemostasis is considered today to be a surface phenomenon involving the blood vessel, the platelet, and the plasma coagulation factors. When a blood vessel is severed, the exposed endothelium and collagen fibers attract circulating platelets which adhere to the vessel, aggregate, and activate each other to form the platelet plug. Substances released from the vessel wall and agents released from the aggregated platelets interact with the plasma coagulation factors to activate the intrinsic and extrinsic coagulation systems. The intrinsic or intravascular system, which includes factors V, VIII, IX, X, XI, and XII, and the extrinsic or extravascular system, including tissue factor and factors V, VII, and X, culminate in the conversion of prothrombin (F.II) to the enzyme thrombin. Thrombin converts fibrinogen (F.I) into the fibrin monomer by splitting off two fibrinopeptides. The fibrin monomers then coalesce to form a polymer which is stabilized by factor XIII into the firm fibrin clot. Within the clot are incorporated substances that lead to the formation of the enzyme plasmin, which eventually will lyse the clot as the vessel repairs itself. The precursors of both thrombin and plasmin are kept in balance by a variety of natural inhibitors, including anti-

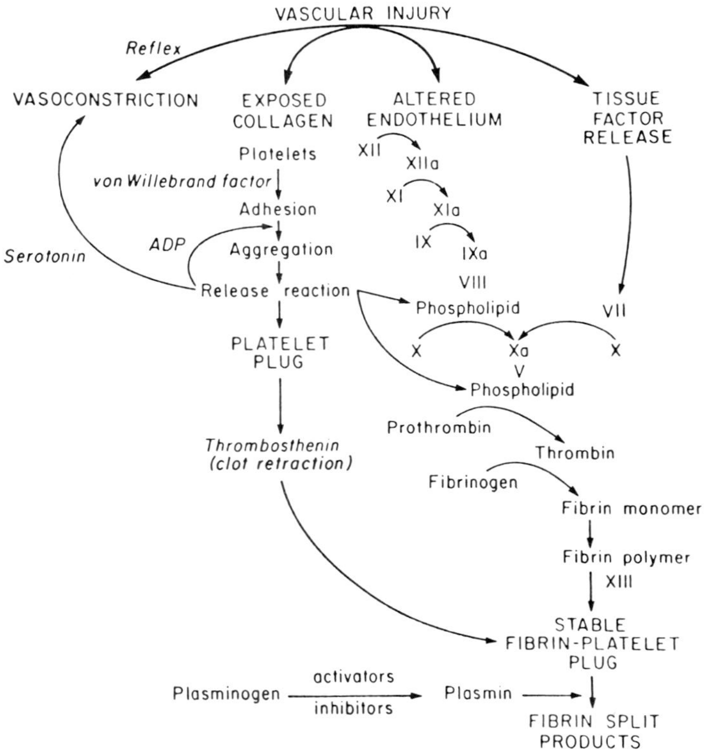

Fig. 1. The hemostatic mechanism From Abildgaard (1979).

thrombins and antiplasmins. Figure 1 illustrates the current concept of the hemostatic mechanism as diagrammed by Abildgaard (1979). An abnormality in the function of any of the integral parts of the scheme may result in a bleeding problem or, in some cases, in thrombosis. This chapter will focus on the bleeding diseases which comprise the vast majority of the genetic disorders of hemostasis.

B. Diagnosis of Clotting Disorders

An accurate diagnosis of the bleeding disorder is required for optimum management of the patient and for appropriate genetic counseling. The mechanism for reaching a diagnosis will be briefly reviewed.

A careful history of the patient and her family is the most valuable tool in screening for both major and minor bleeding disorders. Information must be gathered to determine whether the problem is inherited or acquired and whether the history suggests a vascular, platelet, or coagulation defect, or some combination of defects. Specific questions must be asked since general questions about bleeding often fail to elicit the necessary information. The signs to look for include neonatal bleeding from delivery trauma, from circumcision, or from the umbilicus; easy bruising or prolonged bleeding from minor injuries; oozing gums; deep hematoma formation from intramuscular injection or immunization; nosebleeds (particularly those that are spontaneous and/or difficult to stop); bleeding into joints or any history of painful swollen joints; excessive bleeding following surgery, dental extractions, or childbirth; menorrhagia; hematuria; gastrointestinal bleeding; and hemorrhagic skin lesions. A complete pedigree must be taken and examined for consanguinity as well as for the appearance of any of the above symptoms in any relative. A drug history is vital. The use of some drugs may influence the severity of bleeding in the patient or may cause an acquired bleeding disorder which mimics an inherited one. Drugs, particularly aspirin and aspirin-containing products, may influence laboratory tests as much as two weeks after their use. They are of particular importance if platelet aggregation studies are to be performed. The presence of underlying disease states which may give rise to bleeding should be evaluated, especially the possibility of systemic lupus erythematosus, nephrosis, or hypothyroidism. Unusual drug exposure or systemic disease is a strong possibility if bleeding manifestations are acute with no lifelong history of bleeding.

The physical examination may add a great many clues about the etiology of a bleeding diathesis in addition to assessing the general state of health of the patient. Particular attention should be paid to the skin, mucous membranes, soft tissues, and joints for signs of bleeding or residual damage. Congenital anomalies should also be noted. Such entities as Ehlers-Danlos syndrome, hereditary platelet disorders, mild or severe hemophilia, and von Willebrand disease must be kept in mind.

The history and physical examination can guide one to the efficient use of the laboratory. Laboratory tests should be selected and ordered with an established impression as to the diagnosis, rather than to initiate the diagnostic process. The four screening tests to be carried out are (1) a platelet count; (2) bleeding time; (3) a prothrombin time (PT), which tests the extrinsic and common coagulation pathways; and (4) an activated partial thromboplastin time (PTT), which assesses the intrinsic and common coagulation pathways. The platelet phase of hemostasis requires an adequate number of platelets and normal platelet function. The finding of thrombocytopenia should initiate a search for its underlying cause. A prolonged bleeding time in the presence of a normal platelet count requires further laboratory investigation. Its most likely cause, if aspirin ingestion is ruled out, is von Willebrand disease, the diagnosis of which is discussed in Sec. II, B. If no evidence for von Willebrand disease can be obtained, additional platelet function studies are indicated. Diagnostic tests for disorders of platelet function are discussed by Roberts and Kroncke (1971).

An abnormality in either the PT or PTT suggests a coagulation factor deficiency and calls for specific factor assays. If only the PT is abnormal, a deficiency of factor VII is suggested. When the PTT is abnormal but the PT is normal, factors VIII, IX, and XI should be tested if the patient has a bleeding tendency. This combination of results may also indicate a deficiency of factor XII or one of the other factors recently described which are usually associated with an asymptomatic deficiency state. Specific factor assays are performed by testing the ability of the patient's plasma to correct the clotting abnormality in a plasma known to be deficient in the specific factor being measured. Each assay requires the availability of plasma either artificially depleted of the factor or from a patient severely deficient in it. Therefore, not all hospitals maintain the capability to do these tests routinely. In addition, accurate coagulation tests demand scrupulous methods of collection, preparation, and storage of blood samples as well as some degree of technical skill. The laboratory which performs such tests infrequently cannot always be expected to produce reliable results. For specific testing procedures, the reader is referred to several basic texts (Biggs, 1976; Smith, 1979; Williams *et al.*, 1977).

Normal results in the four screening tests can usually be considered to exclude a major bleeding disorder. However, if the patient has a significant history of bleeding or if family members have an undiagnosed bleeding disorder known to have caused major bleeding, more extensive testing is called for. These screening tests do not detect the rare deficiency of factor XIII, described below. They also may not detect von Willebrand disease. For purposes of genetic counseling, it is often necessary to test other affected family members to establish the diagnosis before evaluating the patient for the carrier state, since partial deficiencies of the factors may be difficult to interpret.

Diagnostic tests performed in infants, particularly the newborn, must take into account the immaturity of the coagulation mechanism in the normal newborn.

C. Hemostasis in the Normal Newborn

Hemostasis is suboptimal in the normal newborn because of quantitative and qualitative differences from the adult. Numerous studies of neonatal coagulation indicate a rather consistent pattern of deficits relative to adult values in screening tests and the activities of some coagulation factors (Table I). These deficiencies, termed *physiologic,* are due to immaturity of the liver enzyme system which produces all of the clotting factors except factor VIII and to the neonatal deficiency of vitamin K, necessary for the production of factors II, VII, IX, and X. The levels fall during the first 3 days of life and then rise to normal adult levels within 2 weeks, although for factors IX and XI there may be a lag period of as long as 9 months in some infants. The prekallikriens and high molecular weight kininogens important for the activation of the clotting and complement systems

TABLE I
Comparison of Test Results in the Normal Adult, Infant, and Fetus[a]

	Adult	Term infant	Preterm infant 32–36 weeks	Preterm infant 27–31 weeks	Fetus 10–15 weeks
Screening tests:					
PT (seconds)	< 13.0	16 (13–20)[b]	17 (12–21)	23	—
PTT, activated (seconds)	< 40.0	55 ± 10[c]	70	—[d]	—
Bleeding time, Ivy (minutes)	4 ± 1.5	4 ± 1.5	4 ± 1.5	—	—
Platelet count ($10^3/mm^3$)	300 ± 50	310 ± 68	290 ± 70	275 ± 60	Platelets present
Coagulation factors:					
Fibrinogen (mg/100 ml)	315 ± 60	246 ± 55	226 ± 70	270 ± 140	120
F.II (%)	100	45 ± 15	35 ± 12	30 ± 10	—
F.V (%)	100	98 ± 40	91 ± 23	72 ± 25	81
F.VII (%)	100	56 ± 16	39 ± 14	32 ± 15	18
F.VIII (%)	100	105 ± 35	98 ± 40	70 ± 30	—
F.IX (%)	100	28 ± 8	—	27 ± 10	—
F.XI (%)	100	30	—	—	—
F.XII (%)	100	51	30	—	—
F.XIII (%)	100	100	100	100	—

[a] Adapted from Hathaway (1975).
[b] Mean (range).
[c] Mean ± standard deviation.
[d] —, No data available.

are also low in the newborn. Qualitative differences include the presence of a "fetal" fibrinogen (Witt *et al.*, 1969) and defects in platelet adhesion and aggregation (Hilgartner, 1968; Mull and Hathaway, 1970).

Differences from the adult are even greater in the fetus and in premature infants of various gestational ages (Table I). This area has been well summarized by Hathaway (1975).

D. Hemostasis in the Normal Pregnant Patient

A normal woman tested during pregnancy may show plasma coagulation factor levels quite different from those that she would display in the nonpregnant state. Kasper and colleagues (1964) and Nilsson and Kullander (1967) have measured various factors over the course of pregnancy in normal women. Their studies agreed that factor VIII and fibrinogen levels increase during pregnancy, whereas factor V shows little change. They differed over the effect on factor IX, with Nilsson and Kullander (1967) finding no change but Kasper *et al.* (1964) observing striking increases. Kasper's group also found increases in factors VII and X, and confirmed earlier studies that showed little change in factor II levels. Other investigators have found factor XI (Hilgartner and Smith, 1965; Nossel *et al.*, 1966) and factor XIII (Coopland *et al.*, 1969) to decrease during pregnancy, and factor XII to increase (Biland and Duckert, 1973). These changes generally begin in the second trimester and progress to term. They reverse at varying rates after delivery. Whether such changes also occur in women affected with coagulation disorders is not clear. What little information is available is noted in each section. However, it is important to be aware of this potential source of error in diagnosis and carrier detection during pregnancy.

E. Inheritance Patterns

The terms used in reports of coagulation disorders to describe their inheritance patterns are sometimes confusing. Many authors have used *incompletely dominant* or *incompletely recessive* to indicate that the disease does not follow a simple easily recognized pattern. Both terms may be unclear unless their usage is defined. Incomplete dominance may have either of two meanings. It may imply that the disorder follows a dominant pattern but that some heterozygotes are not affected; or it may indicate that all heterozygotes are intermediate between the normal and affected homozygotes (Levitan and Montagu, 1971). The first usage may perhaps be better described as dominance with decreased penetrance, and the second corresponds to *co-dominance*. Incomplete recessiveness usually means that some heterozygous carriers manifest the disorder. In some cases, the confusion seems to have arisen because a laboratory abnormality without clinical symptoms was considered to represent manifestation of the disease. As in other genetic diseases, the presence of a detectable carrier state does not make the recessiveness "incomplete" or make the disorder dominant. The inheritance pattern may be unclear in other reports because of genetic heterogeneity of the disorder discussed, i.e., families with different inheritance patterns may have been lumped together.

For purposes of this chapter, only the terms *dominant* and *recessive* have been used. A disorder has been classified as dominant or recessive according to whether the gene must be present in heterozygous or homozygous form to bring about the appearance of the disease, which is itself defined as the presence of sufficient debility to require medical attention (Kirkman, 1972). This is in accordance with the classification of McKusick (1978). In the description of each disorder, evidence of incomplete penetrance and reports of more than one inheritance pattern will be noted. It should be remembered that when few families have been described with a given disorder, each family encountered must be carefully studied to determine whether it concurs with the inheritance pattern previously reported.

II. DISORDERS OF THE PLASMA COAGULATION FACTORS

A. Hemophilia A and B

Hemophilia A (classical hemophilia or factor VIII deficiency) and hemophilia B (Christmas disease or factor IX deficiency) are similar in clinical appearance and mode of inheritance, and may be differentiated only through laboratory techniques. They will be described together, with the term *hemophilia* used to indicate both disorders and the disease specified when there are differences between the two.

Patients with hemophilia may be classified as having severe, moderate, or mild disease according to the level of the factor present. Those patients having less than 1% of the normal level of factor VIII or factor IX have severe disease with bleeding which is often spontaneous, or without known trauma. Patients with moderate disease have 1–5% of the factor and bleed with minor trauma. Mildly affected hemophiliacs have 6–50% of the factor present and bleed only with severe trauma or surgical challenge. Clinical manifestations usually parallel the laboratory findings, although there are individual differences. The severity of the disease is usually constant within a family, with related patients having similar laboratory and clinical manifestations. This leads to the hypothesis that the differences are due to different alleles at the X-chromosome locus.

1. Clinical Features

a. Severe Hemophilia. Severe hemophilia may present at birth with prolonged bleeding from the cord or the navel, cephalohematoma, or bleeding from the circumcision site. Diagnosis may be made on cord blood if less than 1% F.VIII or F.IX is present. Hemorrhagic symptoms have been reported during the first week of life in one-third to one-half of cases (Hartmann and Diamond, 1957; Schulman, 1962), with the majority due to circumcision. The reason for so few problems among affected newborns is not clear since maternal F.VIII does not cross the placenta (Strauss, 1965). The liberation of tissue thromboplastin by the pressure of the circumcision clamp probably prevents more significant bleeding from that procedure.

Occasionally, infants may have oozing when teeth erupt but the most obvious problems begin to appear as the toddler learns to walk. He may develop unusual hematomas from falls. Gum or tongue lacerations and a split frenulum are also common. With the increased activity of childhood, hemarthrosis and deep muscle hematomas develop.

Hemarthrosis is by far the most common and most significant clinical feature of severe hemophilia. Bleeding within the joint usually follows trauma in early childhood; but after two or more hemarthroses in the same joint, bleeding may occur without provocation. Once bleeding has occurred within a joint, deterioration begins and may be progressive. In the past, all patients with severe hemophilia had chronic arthropathy by age 10 years (Jordan, 1958). This should not occur today if therapy is instituted at the first sign of bleeding. While 75% of today's adult hemophiliacs have chronic arthropathy, only 10–15% of adolescents have such residual damage. The philosophy of care has changed radically over the past 10 years. The prognosis of the disease is different today and continues to improve with advances in therapy.

b. Moderate Hemophilia. Moderate hemophilia is often not diagnosed in the neonatal period. Infants with factor levels greater than 1% rarely bleed, even with circumcision (Schulman, 1962; Strauss, 1965). Although laboratory diagnosis can be made at birth in F.VIII-deficient infants, moderate F.IX deficiency may be more difficult to detect due to the physiologic low level of F.IX in the normal neonate. The F.IX assay must often be repeated at 6–9 months of age to verify the diagnosis.

Moderate hemophilia results in bleeding from mild trauma, and spontaneous bleeding is usually not present. However, repeated bleeding into the same joint can initiate chronic hemarthrosis similar to that experienced by severely affected patients. Advances in treatment have greatly improved the outlook for patients with this form of hemophilia.

c. Mild Hemophilia. Mild hemophilia is often not diagnosed until significant challenge to the patient's hemostatic mechanism reveals the defect. Screening tests may not detect the abnormality. Bleeding may occur following dental extraction, surgery, or severe trauma. Patients with mild disease rarely, if ever, have hemarthrosis and find that the disorder does not appreciably interfere with their daily life.

2. Genetics

Descriptions of hemophilia are among the oldest known mentions of genetic diseases. The Talmud of the fifth century describes the decree of Rabbi Judah in the second century that if two sons of a woman died following circumcision, the third must not be circumcised. Later statements in the Talmud and Rabbinic writings show a remarkable understanding of the X-linked inheritance pattern, including the concepts that sons of sisters of a carrier may be at risk as will sons of the same woman by different husbands (Rosner, 1969).

Today, it is recognized that a female relative of a hemophiliac may be identified with certainty as a carrier if she falls into one of the following categories: (1) she is the daughter of a hemophilic man, (2) she is the mother of more than one hemophilic son, or (3) she is the mother of a hemophiliac and has at least one other relative with hemophilia. The risk that a woman who meets one of these requirements will have a hemophilic son is 1 in 4, or 25%, for each pregnancy. Each of her sons will have a 50% chance of being affected, and each of her daughters will have a 50% chance of being a carrier.

Female relatives who may or may not be carriers include (1) the mother of a single hemophilic son with no other affected relative (the son may have hemophilia due to a new mutation on his X chromosome not present in his mother), (2) the sisters of a hemophilic male, and (3) the aunts, cousins, and nieces of an affected male who are related through maternal ties. It has been estimated that there are, on the average, five to six women at risk for transmitting hemophilia for each affected male in the population (World Health Organization, 1977).

3. Carrier Detection

Tremendous advances have been made in the last few years in the detection of carriers for both hemophilia A and hemophilia B. This area has been recently reviewed by Graham (1978). It has long been known that carriers for hemophilia A have a wide range of F.VIII clotting activity levels (Graham *et al.*, 1953). The levels can be shown to produce a normal distribution with a mean level of approximately 50%, as compared to the normal mean of 100%. The considerable overlap between the range for carriers and the range for normal women made carrier detection difficult until the advent of a new test, developed by Zimmerman *et al.* (1971). This immunologic test quantitates the antigen that is detected by a rabbit antibody to F.VIII, the F.VIII-related antigen. This antigen is present in normal amounts in hemophiliacs, indicating the presence of a nonfunctional

F.VIII-related entity. Testing for carriers involves comparing the clotting activity and antigenic activity to look for a discrepancy which might indicate the production of nonfunctional F.VIII. Because of the wide variability in both normal and carrier groups in these two tests, and the significant age effects that are often seen, statistical methods such as discriminant analysis are initially necessary to provide a basis for evaluating the results of the laboratory tests. A committee of the World Health Organization (1977) has recently reviewed laboratory and statistical methods for detecting carriers of both hemophilia A and hemophilia B. Tests for the carrier state for hemophilia A are available at most centers which provide comprehensive care for a large number of hemophilic patients.

Carrier detection in hemophilia B has not been developed to such an advanced stage. The range of F.IX clotting activity in carriers of hemophilia B is also wide, forming a distribution similar to that for hemophilia A carriers (Barrow *et al.*, 1960). However, there appears to be heterogeneity among patients with F.IX deficiency, as shown by Roberts *et al.* (1968). Factor IX antigen, or cross-reacting material (CRM), detected by a human or rabbit antibody may be at normal levels (CRM+), absent (CRM−), or reduced (CRM red). In addition, a prothrombin time test using bovine brain thromboplastin may be normal or prolonged. Kasper *et al.* (1977) and Graham (1978) have described the pitfalls in establishing carrier tests for hemophilia B. If immunologic tests for F.IX antigen are to be used, it appears necessary to establish separate standards for each type of hemophilia B. However, if discrimination is based on F.IX coagulant activity alone, all types can be pooled and a single method used for evaluation. Few laboratories at present have available tests for F.IX antigen. The data for establishing carrier testing based on F.IX coagulant activity can be collected if a sufficient number of known carriers is available, but the actual availability of the test across the country is limited.

4. *Pregnancy*

Testing for the carrier state for either hemophilia A or B cannot currently be reliably performed during pregnancy. There is at present a lack of information regarding the changes occurring in both coagulant and antigenic activities as pregnancy progresses. Factor VIII coagulant activity has been found to rise in normal women during pregnancy, with a gradual increase beginning in the second trimester and approaching two to three times normal levels by term (Kasper *et al.*, 1964; Nilsson and Kullander, 1967). Factor VIII-related antigen also rises during pregnancy, but whether the relationship between the two activities remains constant is uncertain (Bennett and Ratnoff, 1972a; Van Royen and ten Cate, 1973). Graham (1978) indicated that separate reference groups may be needed to establish testing statistics for possible carriers who are pregnant.

The question of whether or not factor IX levels increase during pregnancy seems to be unanswered, with Nilsson and Kullander (1967) finding no change in

15 women followed throughout pregnancy and Kasper and colleagues (1964) finding striking increases in the third trimester in nine subjects. There are apparently no reports of F.IX antigen measured during pregnancy.

5. *Symptomatic Carriers*

Nonpregnant carriers of hemophilia, as a group, have an average level of the deficient factor of 50%. In the heterozygous condition for an autosomal disease, this would be the predicted level for each individual within the group. However, the product of a gene located on the X chromosome is subject to the process known as *lyonization*, or X-chromosome inactivation, described by Mary Lyon in 1962. Because of this phenomenon, only one X chromosome is active in each cell of the female. The other X chromosome becomes the x-chromatin (Barr body). The choice as to which X chromosome is active occurs early in embryonic development, when only a few cells are present, and is fixed for each descendant of those original cells. In a heterozygote for hemophilia, this random choice early in development determines how many cells will be producing normal clotting factor and how many will be producing the defective product. This process accounts for the wide range of levels seen in carriers. Since the values approximate a normal distribution, the number of carriers above the mean of 50% and the number below it should equal. If the level below which bleeding symptoms may appear is taken to be 40%, it can be shown from the data of Elston *et al.* (1976) that approximately 28% of hemophilia A carriers may be expected to have bleeding symptoms. Graham and colleagues (1975a) have postulated that even more carriers of hemophilia B than hemophilia A should have low levels.

The majority of these symptomatic women have levels in the range of the mild hemophiliac and may exhibit similar clinical problems, such as bleeding with surgery, trauma, or dental work. Spontaneous or induced abortion and dilatation and curettage may also be dangerous for these women. Due to the rise in factor levels experienced by most carriers during pregnancy, a history of the absence of bleeding with childbirth should not be taken as assurance that adequate hemostasis will exist for surgical procedures in early pregnancy or in the nonpregnant state (Kasper *et al.*, 1964).

6. *Hemophilic Females*

In a few rare cases, women present with clinical and laboratory findings almost indistinguishable from those of men with hemophilia. These hemophilic females pose a number of problems both from a clinical and a genetic point of view. Each must be carefully studied on an individual basis in order to provide the best clinical care for her and appropriate genetic counseling for her and her family.

Kerr in 1965 found 60 case reports of women with hemophilia A but not all of the patients had been carefully characterized. There have been several recent reports in which the patients and their families have been more thoroughly studied (Lusher *et al.*, 1969; Czapek *et al.*, 1974; Graham *et al.*, 1975b, 1976). Kerr (1965) also mentioned two reports of women with hemophilia B. Subsequent reports include those of Rozman *et al.* (1967), Strauss and Olson (1969), Révész *et al.* (1972), Neuschatz and Necheles (1973), and Rust *et al.* (1975).

Barrow and Graham (1974) have discussed the genetic basis for hemophilia in women. Most of the possible explanations apply equally well to hemophilia A and hemophilia B. The four major states leading to female hemophilia each have several mechanisms by which they can occur. Some of these can be ruled out by careful study, but others are indistinguishable at present. The possible explanations are as follows.

1. The patient is homozygous for hemophilia, having the hemophilia gene on both X chromosomes, because:
   a. She has a hemophilic father and her mother is a carrier.
   b. A mutation occurred on each of her X chromosomes (the likelihood of this rare occurrence is about 1 in 4 billion).
   c. She received one hemophilia gene from a carrier mother or a hemophilic father, and a mutation occurred on her other X chromosome.
2. The patient is a hemophilia carrier in whom the normal gene does not work properly because:
   a. She has a chromosome abnormality that causes the X chromosome bearing the normal gene to be preferentially inactivated, i.e., the x-chromatin (Barr body) always contains the X chromosome with the normal gene. This can be due to an X-autosome translocation, or a structurally abnormal X chromosome.
   b. She is at the extreme end of the distribution of factor levels found in carriers due to random choice as to which X chromosome is inactivated. By chance, all of the normal genes are inactive in the cells that are supposed to be producing the factor.
3. The patient has only one X chromosome, that one bearing the hemophilia gene, due to:
   a. Turner syndrome (45,X Karyotype).
   b. Testicular feminization syndrome (46,XY Karyotype).
   c. Sex chromosome mosaicism.
4. She has a disease other than hemophilia, such as:
   a. von Willebrand disease.
   b. An unusual form of hemophilia, inherited in a different pattern which affects both males and females.
   c. An acquired inhibitor to factor VIII or factor IX.

Determining the basis for hemophilia in a particular woman usually requires chromosomal analysis, laboratory testing to exclude other diseases and inhibitors, and examination of her parents for hemophilia and the carrier state. The two possibilities that cannot be distinguished are random X-chromosome inactivation and a single new mutation in a carrier, which Graham has speculated are about equally likely to occur. If the woman has a slightly higher level of factor VIII or factor IX than her male hemophilic relatives, random inactivation is more likely. The presence of the carrier state rather than homozygosity would be confirmed if the woman gave birth to a normal son.

The clinical features found in hemophilic females vary, although most have experienced hemarthrosis if their factor levels are less than 5%. Menorrhagia and postpartum bleeding may occur but are not invariable. If a chromosome abnormality is present, concomitant treatment for its effects may be necessary.

Genetic counseling can be complex, particularly if a specific genetic diagnosis cannot be made. In almost every instance, however, the patient herself would be at risk for producing a hemophilic son. Her parents also could have an affected son and, unless the basis for the defect was clear, could not be assured with certainty that they would not produce another hemophilic daughter.

7. *Treatment*

The carrier female with a low factor VIII or IX level must be treated for surgical procedures in the same manner as a mild hemophiliac. The woman whose factor level does not rise over 40% of normal in the last trimester of pregnancy may need to receive factor for delivery and during the puerperium. Cryoprecipitate is the product of choice because of lower hepatitis risk, but its use requires laboratory monitoring by a knowledgeable hematologist.

The occasional severely affected female must receive the same treatment as a male hemophiliac, including control of hemarthroses to prevent degenerative joint disease.

8. *Prenatal Diagnosis*

Prenatal diagnostic tests for hemophilia have evolved through several stages. The report of Riis and Fuchs, in 1960, of the first amniocenteses to be performed for any genetic risk involved two hemophilia carriers. In these early procedures, the amniotic fluid cells were stained for sex chromatin. More recently, the culture and karyotyping of amniotic fluid cells have been used to determine fetal sex. However, no biochemical evidence for the presence or absence of hemophilia has yet been found in tissue culture (Jaffe *et al.*, 1973; Blecher *et al.*, 1978). Edgell *et al.* (1978) described the use of linkage in prenatal diagnosis, as well as in carrier detection. The close linkage of hemophilia A and glucose-6-phosphate dehydrogenase (G-6-PD) on the X chromosome permits the tracking of the hemophilia gene by following a G-6-PD allele, which can be detected in amniotic fluid cell culture. This method may be useful in subpopulations which are polymorphic for G-6-PD, such as the U.S. black population; the method requires that the mother be heterozygous for G-6-PD alleles as well as for hemophilia, and that additional pedigree data be available.

The safety of amniocentesis in hemophilia carriers was questioned by Riis and Fuchs (1966), who observed three spontaneous abortions and two premature deliveries among 18 pregnancies of hemophilia carriers. Cederbaum and colleagues (1971) described a spontaneous abortion 20 hours after an unsuccessful amniocentesis in a carrier with 36% factor VIII. Dilatation and curettage resulted in severe hemorrhage in this patient, who had previously undergone successful amniocentesis and abortion. After observing complications in two of their carrier patients, Spiro and Lubs (1976) surveyed 131 centers at which amniocentesis is performed, 112 of them in the United States. Only one-third of the centers had performed amniocentesis on a hemophilia carrier, and of the 11,819 taps reported, 75 (or 0.64%) were for hemophilia. The factor VIII or factor IX level of the mother was known in only one-third of the cases. Of those whose factor VIII level was known, 22% had less than 40% factor VIII. This survey found the frequency of fetal deaths among carriers to be not significantly different from that of all reported amniocenteses. Bleeding problems occurred during or after the pregnancies of nine carriers (12%) among those tested, including the case described by Cederbaum *et al.* (1971) and two pregnancies of one patient. Complications included bleeding after the tap, after cesarean section, and after induced abortion. Monitoring of factor VIII or IX levels before and after the procedures could undoubtedly reduce the risk of unexpected bleeding and allow special preparations such as pretreatment or the procuring of factor replacement materials to be made if necessary.

Reports by Mahoney and colleagues (1978) and Firshein *et al.* (1979) describe the successful prenatal diagnosis of hemophilia A using fetal blood samples obtained by fetoscopy at 18–20 weeks gestation. This test was made possible by the development of an immunoradiometric assay for that portion of factor VIII responsible for clotting activity, as reported by Peake and Bloom (1978). In contrast to the factor VIII-related antigen detected by rabbit antibody, this antigen is recognized by a human factor VIII antibody, or inhibitor, and is undetectable in severe hemophiliacs. The advantage of this measurement over clotting tests is that it can be performed on plasma contaminated with amniotic fluid and/or serum. The diagnosis of a fetus with hemophilia B, using factor IX antigen levels has been reported by Holmberg and colleagues (1980). The safety of fetoscopy in hemophilia carriers and the danger of blood sampling in a potentially hemophilic fetus remain to be explored.

9. Risks to Fetus and Newborn

Little is known about the hemophilic fetus during pregnancy. There appears to be little risk during gestation other than that posed by bleeding problems in the mother. Delivery risk appears to be low. Problems due to birth trauma occur rarely in the uncomplicated delivery of the child with severe hemophilia, but intracranial bleeding, cephalohematoma, and "hidden" bleeding have been reported (Rausen and Diamond, 1961). When a male child known to be at risk for hemophilia is born, it is wise to obtain cord blood for factor VIII or IX assays or at least a PTT which should be extremely prolonged in a severely affected child. A good quality sample may be obtained by venipuncture of the cord prior to separation. Failure to obtain blood at birth may cause a delay of several months in making the diagnosis until a sample of sufficient volume and quality for testing can be obtained. Circumcision should generally be delayed until after testing is completed. The coagulation status of the newborn and clinical features in the infant have been discussed in previous sections.

B. von Willebrand Disease

von Willebrand disease (VWD) is an hereditary hemorrhagic disorder affecting both men and women. Although it exhibits a deficiency of the antihemophilic factor, F.VIII, it can be distinguished from classical hemophilia by means of differences in clinical course, inheritance pattern, and laboratory findings. The disorder was first described by Erik von Willebrand, a Finnish physician, in 1926. His interest was aroused by a family in which five of seven daughters had bled to death at an early age.

The gene for VWD is apparently pleiotropic, exhibiting its effect in both a decrease in F.VIII and a defect in platelet function. Diagnostic criteria include the following laboratory abnormalities: (1) prolonged bleeding time, (2) decreased F.VIII coagulant activity, (3) decreased Willebrand factor (the plasma factor necessary for proper platelet function in tests of platelet adherence to glass surfaces and platelet aggregation in the presence of the antibiotic ristocetin), and (4) decreased F.VIII-related antigen. In contrast, classical hemophilia and its carrier state are characterized by a deficiency of F.VIII coagulant activity alone. The most striking feature of von Willebrand disease may be its variability. Not all laboratory abnormalities are present in each patient, and the constellation of findings may vary within the same patient over time.

Diagnosis of VWD may be difficult in an isolated patient. An abnormal result on any one of the four diagnostic tests described above may indicate VWD, but other disorders must be ruled out, including mild hemophilia and intrinsic platelet defects. Repeated testing is sometimes useful and it may be helpful to test other family members, particularly any reported to have bleeding problems, in hope of clarifying the diagnosis. Within a family known to be transmitting VWD, an abnormality on one test is strong evidence for heterozygosity. Miller and colleagues (1979) have recently discussed the variability of the disease and the efficacies of various diagnostic methods.

1. Clinical Features

The bleeding experienced by patients with von Willebrand disease usually involves the skin and mucous membranes, in contrast to the deep tissue and joint hemorrhages common in hemophilia. Affected individuals often have a history of easy bruising, epistaxis (lessening at puberty), gum bleeding, prolonged oozing from minor wounds, and bleeding after dental extraction or surgery. Hemarthrosis is rare, but it may occur in patients with very low F.VIII levels, usually less than 1%.

Menorrhagia is a serious problem for many women with VWD, and postpartum hemorrhage may also occur. Family histories may include women who bled to death at menarche or after childbirth in the days before adequate treatment was available. It is also not unusual, however, to encounter a woman in the middle years who is unaware of her disorder. In some families nosebleeds and heavy menstrual periods are so common that they are not recognized as abnormal by the family members. Other affected individuals may actually experience no abnormal bleeding until challenged by surgery or trauma.

2. Genetics

"Classic" von Willebrand disease exhibits autosomal dominant inheritance with a wide variability of expression. Since the disease is autosomal rather than X-linked, both men and women may be affected. Children with an affected parent of either sex have a 50% chance of inheriting the mutant gene; however, penetrance is low. When individuals known to have transmitted the disease were examined (Miller *et al.*, 1979), 42% had no detectable laboratory abnormality. Discriminant analysis could detect 81% of the heterozygotes, but it is unlikely to be practical for routine use.

In addition to the autosomal dominant "classic" VWD, there may be a recessive form characterized by greater severity in the affected individuals. These patients have extremely prolonged bleeding times and almost undetectable levels of F.VIII coagulant activity, F.VIII-related antigen, and Willebrand factor. Their parents are usually asymptomatic but can sometimes be diagnosed as heterozygotes by statistical evaluation of their laboratory data (Veltkamp and van Tilburg, 1973). The consanguinity present in some of these cases (Italian Working Group, 1977) has made homozygosity seem likely. A study of one such family caused Miller *et al.* (1979) to postulate that two severely affected patients were homozygous for the dominant form of VWD. Although the disorder was not apparent in the parents of these individuals, it was exhibited in classic form by other relatives. Therefore, the distinction between the two types may not be clear.

3. Pregnancy

The response of women with von Willebrand disease to pregnancy is variable. In 1973, Noller and colleagues surveyed the literature for reports of pregnancy in von Willebrand disease patients. They found eight cases with sufficient information provided for assessing changes during pregnancy, and added four cases of their own. Nine of the 12 patients had significant increases in F.VIII coagulant activity during pregnancy. Those whose levels did not increase had prepregnancy levels of less than 5% F.VIII. In two patients the bleeding time failed to shorten as well. One of these hemorrhaged during labor, as well as postpartum, and the other was treated prophylacticly. Even a rise to two or three times baseline F.VIII coagulant activity may not be sufficient if the patient has a very low level. Of 17 patients whose status was known at term, four hemorrhaged postpartum. Three of the four had levels below 20% and bled 5–18 days after delivery. The patient with a normal F.VIII at term began bleeding on day 25, presumably after her level had returned to baseline. The rapid return to prepregnancy F.VIII activity levels has been noted by several authors (Noller *et al.,* 1973; Strauss and Diamond, 1963) and constitutes the greatest danger to these women. Careful follow-up with repeated measurements of F.VIII activity as well as bleeding time or Willebrand factor is indicated. Treatment of choice is cryoprecipitate, since it may correct the bleeding time as well as the F.VIII level and poses the least risk of hepatitis transmission. As in the hemophilia carrier, a history of normal delivery should not be taken as assurance of adequate hemostasis for surgical procedures in early pregnancy or in the nonpregnant patient. In addition, a previous uncomplicated delivery does not exclude the possibility of a problem later. There are reports of women who have bled with one delivery but not with another (Noller *et al.,* 1973).

4. Prenatal Diagnosis

Prenatal diagnosis of severe von Willebrand disease has been reported, by Hoyer and colleagues (1979). For moderate von Willebrand's disease, careful establishment of control data would be necessary, since most affected individuals would be expected to have a relative reduction in the factors measured rather than a complete absence.

5. Risks to Fetus and Newborn

There is no information at present about the effect of VWD on the fetus. Risks would seem to be largely due to maternal bleeding. The consanguineous parents of one severely affected patient were noted to have produced four spontaneous abortuses and two stillborn infants, but the deaths could not be directly attributed to VWD.

The infant with von Willebrand disease usually has no problems during delivery. The incidence of bleeding does not appear to be greater than that in hemophilia. One infant born at the New York Hospital into a family with moderately severe VWD had sufficient intracranial bleeding to cause a neurologic defect. If circumcision is desired for an infant born into a family with von Willebrand disease, the presence or absence of the disease must be determined before this surgical procedure may be safely done.

6. Treatment

Treatment is seldom necessary for those with mild disease, except at the time of surgical procedures or childbirth. Severely affected patients require more frequent therapy. Fresh frozen plasma and/or cryoprecipitate are the products of choice to replace both F.VIII and the Willebrand factor. Oral contraceptives are useful for control of menorrhagia. The antifibrinolytic agent Amicar (EACA) is also used for that purpose, as well as for control of mouth bleeding.

C. Disorders of Fibrinogen (Factor I)

1. Afibrinogenemia

Congenital afibrinogenemia is a rare disorder due to almost complete absence of fibrinogen from the blood. Both males and females are affected, and more than 100 cases have been reported (Mammen, 1974).

a. Clinical Features. The condition may present in the newborn as hematomas from the trauma of delivery, melena, hematemesis, and bleeding from the umbilicus. Symptoms appearing later may include bleeding from the skin and mucous membranes, such as epistaxis and gastrointestinal bleeding, as well as bleeding into muscles, joints, and the central nervous system. Menstrual bleeding may be normal or prolonged. Surgical procedures may be followed by delayed bleeding.

b. Laboratory Findings. In the homozygote, fibrinogen appears to be completely absent by precipitation and coagulation techniques, while immunologic methods may detect trace amounts (5 mg or less). Since the final substrate for formation of the clot is missing, all screening tests are abnormal, including whole blood clotting time, PT, PTT, and thrombin clotting time. Occasionally slight decreases in other clotting factors have been noted and tests of platelet function such as bleeding time, adhesion, and aggregation are sometimes abnormal. *In vitro* correction can usually be obtained by addition of normal amounts of fibrinogen (Weiss and Rogers, 1971).

c. Genetics. The disorder appears to be autosomal recessive. Heterozygotes have lower than normal fibrinogen levels (hypofibrinogenemia) but are asymptomatic. The original case of Rabe and Salomon (1920) was the product of a consanguineous marriage, as are more than one-half of the reported cases (Egbring *et al.,* 1971).

d. Pregnancy. While fibrinogen levels have been reported to rise in normal women during pregnancy (Kasper *et al.*, 1964; Nilsson and Kullander, 1967), there are no reports of pregnancy in affected women.

The mother of an affected child would be expected to be heterozygous and, therefore, to have no bleeding diathesis herself. However, she would have a 1 in 4 chance of bearing another affected child if the father is the same.

e. Prenatal Diagnosis. Prenatal diagnosis of afibrinogenemia has not been reported but may be possible by fetal blood sampling. Fibrinogen is reported to be detectable in fetal blood prior to 15 weeks gestation (Hathaway, 1975).

f. Fetus and Newborn. No information is available on effects of this disorder on the fetus. The disorder is detectable at birth. Clinical features in the newborn have been discussed above.

g. Treatment. Bleeding episodes are controlled by replacement of fibrinogen with fresh frozen plasma, cryoprecipitate, or concentrates of fibrinogen (Cohn fraction I).

2. *Dysfibrinogenemia*

The qualitative abnormalities of the fibrinogen molecule are called dysfibrinogenemias. Patients homozygous for the abnormality have normal amounts of plasma fibrinogen, which fails to form fibrin in the presence of the enzyme thrombin. Heterozygotes have both normal and abnormal fibrinogen.

Twenty-six families with dysfibrinogenemia have been described with a total of 103 patients (Mammen, 1974). As in the hemoglobinopathies, each of the abnormal fibrinogens has been named according to the city in which it was found (Beck, 1964). The functional abnormality has been characterized in 24 of the 26 families and may be identified as to the phase of fibrin formation in which it appears. Six of the abnormal fibrinogens have defects in the first, or proteolytic, phase in which thrombin cleaves fibrinopeptide A from the A-α chain and fibrinopeptide B from the B-β chain to give fibrin monomers; 17 have defects in the polymerization phase when the fibrin monomers aggregate to form the fibrin network (Mammen, 1974). The molecular defect has thus far been ascertained only in fibrinogen Detroit, in which there is a single amino acid substitution (Blömback *et al.*, 1968).

a. Clinical Features. The dysfibrinogenemias are not uniform in symptomatology. Only 13 of the 26 reported families have exhibited a bleeding disorder. The presentation is usually mild, characterized by easy bruising, prolonged bleeding from minor trauma, wound dehiscence, and menorrhagia. Three families have exhibited thromboses (Menache, 1970).

b. Laboratory Findings. Quantitation of the abnormal fibrinogens varies with the method used. Precipitation and immunologic methods give normal values, while tests dependent on clotting in the presence of thrombin will show abnormal results. The degree of abnormality depends upon the defect present in the family and whether the patient is homozygous or heterozygous. The PT, PTT, and whole blood clotting time may be normal or abnormal, but the thrombin time and reptilase time will always be abnormal.

c. Genetics. The inheritance pattern of the defects is autosomal, with equal distribution in males and females. Most individuals reported have normal fibrinogen present in addition to the abnormal molecule, suggesting that they are heterozygous. Only in the family with fibrinogen Detroit (Mammen *et al.*, 1969) has the homozygous state been clearly demonstrated.

d. Pregnancy and the Fetus. There is no information on the effect of these abnormalities on the pregnant woman or the fetus, or on the possibility of prenatal diagnosis.

e. Treatment. The need and efficacy of therapy for the mild symptoms of dysfibrinogenemia are not known.

D. Disorders of Prothrombin (Factor II)

1. Hypoprothrombinemia

Hereditary deficiency of prothrombin is extremely rare. As of 1971, only 21 proven cases were present in the world literature (Girolami, 1971). Early reports often included cases subsequently found to have deficiencies of other factors, expecially factors V, VII, and X.

The disorder appears to be due to lack of production of the prothrombin protein; however, the defect is not complete since most homozygous patients can produce 10–15% of the normal amount.

a. Clinical Features. The clinical disorder is mild, including mucous membrane bleeding, easy bruising, menorrhagia, and bleeding with dental extractions. Hemarthroses have not been reported.

b. Laboratory Findings. A prolonged PT in the presence of normal levels of factors V, VII, X, and fibrinogen is suggestive of the disorder. However, the PT may be nearly normal in less severe cases. Diagnosis can be made by a two-stage prothrombin assay.

c. Genetics. The inheritance pattern is autosomal recessive, with heterozygotes having few or no symptoms.

d. Pregnancy. No change has been found in prothrombin levels during pregnancy in normal women (Kasper *et al.*, 1964) but there are no reports of pregnancy in affected women.

The mother of an affected child would be expected to be heterozygous and therefore to have no bleeding problems herself. However, she would have a 1 in 4 chance of bearing another affected child by the same father.

e. Prenatal Diagnosis. Since levels of prothrombin are low among normal infants at birth, it is unlikely that its measurement in fetal blood could provide a clear diagnosis.

f. Fetus and Newborn. No information is available on the effects of this disorder on the fetus. As with other prothrombin complex factors (factors VII, IX, and X), this deficiency may present with bleeding after the first day of life that does not respond to vitamin K. Diagnosis may be difficult due to the low levels often found in normal newborns.

g. Treatment. Treatment is usually not necessary, but replacement of the missing factor can be accomplished with fresh frozen plasma or a prothrombin complex concentrate. Vitamin K is of no therapeutic value.

2. *Dysprothrombinemia*

Two families with quantitatively normal but nonfunctional prothrombin have been described. Prothrombin Cardeza (Shapiro *et al.*, 1969) and prothrombin Barcelona (Josso *et al.*, 1971) can be detected by immunological techniques but have low levels of clotting activity. Prothrombin Barcelona has been observed in both the heterozygous and homozygous states, whereas prothrombin Cardeza has been found only in heterozygotes.

E. Factor V Deficiency

A deficiency of factor V (proaccelerin) was first described in 1942 by Owren, who termed it *parahemophilia*. It has since been described in more than 80 cases (Melliger and Duckert, 1971). The nature of the disease may be failure of production of the protein, since antisera against human F.V are not neutralized by plasma from homozygotes (Feinstein *et al.*, 1970; Fratantoni *et al.*, 1972).

1. *Clinical Features*

Factor V deficiency results in a mild bleeding disorder characterized by bleeding from the mucous membranes and into the skin, including epistaxis and gastrointestinal bleeding. More severe bleeding may occur with surgery or trauma. Hemarthrosis is unusual. Menorrhagia may be quite severe.

2. *Laboratory Findings*

The PT, PTT, and whole blood clotting time are prolonged. The definitive diagnosis can be made with a specific F.V assay, using plasma either artificially depleted of the factor or obtained from a patient known to have the defect. A prolonged bleeding time has also been observed in some F.V-deficient patients (Rizza, 1976).

3. *Genetics*

The disease is transmitted as an autosomal recessive. Heterozygotes have 35–50% F.V and are usually asymptomatic.

4. *Pregnancy*

Levels of F.V show little change during pregnancy in normal women (Kasper *et al.*, 1964; Nilsson and Kullander, 1967). There is little information on pregnancy in affected women.

Pregnancy in a heterozygote would not be expected to result in abnormal bleeding if the woman was previously asymptomatic. She would have a 1 in 4 chance of producing a second affected child by the same father.

5. *Prenatal Diagnosis*

Prenatal diagnosis might be possible by fetal blood sampling, since F.V has been reported to be detectable in fetal blood prior to 15 weeks gestation (Hathaway, 1975).

6. *Fetus and Newborn*

No information is available on effects of this disorder on the fetus. It is detectable at birth, and clinical features in the newborn would be similar to those seen in hemorrhagic disease of the newborn.

7. *Treatment*

Bleeding episodes can be treated only with fresh frozen plasma.

F. Factor VII Deficiency

Deficiency of F.VII (also known as proconvertin, stable factor, or serum prothrombin conversion accelerator) was first described by Alexander and colleagues (1951). This rare disorder affects both males and females, and has been reported in more than 58 cases. Some patients presumably have a nonfunctional F.VII molecule, since their plasma can neutralize a rabbit antibody prepared by immunization with a crude F.VII preparation (Goodnight *et al.*, 1971).

1. *Clinical Features*

Factor VII deficiency may result in epistaxis, gastrointestinal bleeding, and skin, muscle, or intracranial bleeding. Hemarthrosis is rare but has been reported. Menstrual bleeding may be very severe.

2. *Laboratory Findings*

The disorder can usually be detected by the presence of a prolonged PT and a normal PTT. A specific assay for F.VII will confirm the diagnosis.

3. *Genetics*

Kerr (1965) reviewed the reported pedigrees of F.VII deficiency and found them to be consistent with autosomal recessive inheritance. Heterozygotes have intermediate levels of the factor but are generally asymptomatic. Earlier confusion regarding the inheritance pattern may have arisen because some homozygous individuals have measurable levels of F.VII.

4. Pregnancy

Levels of F.VII increase during pregnancy in normal women (Kasper *et al.*, 1964; Owen *et al.*, 1964). There is little information on pregnancy in affected women. Risks of delivery to a heterozygous woman can best be judged by examining her history and those of other heterozygotes in the family. She would have a 1 in 4 chance of producing an affected child if the father is also heterozygous.

5. Prenatal Diagnosis

Since F.VII is low in fetal life and at birth, prenatal diagnosis would be difficult by present methods.

6. Fetus and Newborn

The effects of this disorder on the fetus are unknown. Detection at birth may be difficult due to low levels of all of the prothrombin complex factors. However, bleeding that does not respond to vitamin K may be due to this disorder. Fatal intracranial hemorrhage due to birth trauma has been reported (Rabiner *et al.*, 1960). Umbilical bleeding may also be present.

7. Treatment

Treatment may be necessary for severe bleeding episodes, for surgical procedures, or for menstrual bleeding. Fresh frozen plasma is the best source of F.VII, although prothrombin complex concentrates contain appreciable amounts and may be of value.

G. Factor X Deficiency

Factor X (Stuart-Prower factor) deficiency is a very rare coagulation disorder affecting both males and females. It was described by Telfer *et al.* (1956) and Hougie *et al.* (1957) in patients originally thought to have F.VII deficiency. About 20 cases have been reported in the literature. CRM+, CRM−, and CRM-reduced variants of the F.X molecule have been observed (Denson *et al.*, 1970).

1. Clinical Features

Factor X deficiency causes a moderately severe bleeding diathesis with epistaxis, hematomas, and occasionally hemarthrosis. Severe bleeding may follow surgery or trauma. Menstrual bleeding may be quite severe.

2. *Laboratory Findings*

Both PT and PTT will be prolonged, as will the whole blood clotting time, prothrombin consumption, and thromboplastin generation. A specific assay for F.X will confirm the diagnosis.

3. *Genetics*

The disorder demonstrates autosomal recessive inheritance. Homozygotes have 5% or less F.X. Heterozygotes have 30–60% F.X and are usually asymptomatic; however, heterozygotes with levels in the lower range may have bleeding problems with surgery or severe trauma.

4. *Pregnancy*

Levels of F.X rise during pregnancy in normal women (Kasper *et al.*, 1964). Nonetheless, Rizza (1976) reported that a patient with a F.X level of 8% bled excessively after childbirth. Heterozygotes might experience bleeding as well if the F.X level remains low.

5. *Prenatal Diagnosis*

As with other prothrombin complex factors (II, VII, IX), prenatal diagnosis would be difficult due to low levels in the fetus and newborn.

6. *Fetus and Newborn*

There is no information about the effects of this disorder on the fetus. Detection at birth may be difficult due to the low levels expected in the normal neonate. Bleeding which does not respond to vitamin K may be due to this disorder.

7. *Treatment*

Fresh frozen plasma is the treatment of choice for most bleeding episodes. Prothrombin complex concentrates may also be useful.

H. Factor XI Deficiency

Deficiency of F.XI (plasma thromboplastin antecedent, PTA) affects both males and females and is most common in persons of Jewish ancestry. The disorder is apparently due to a deficiency of the protein, since antigen is detectable only in proportion to F.XI activity.

1. *Clinical Features*

The bleeding manifestations of F.XI deficiency are usually mild, with epistaxis and menorrhagia the most common problems. The disorder is frequently detected by screening tests prior to surgery or because of bleeding following minor surgery or dental procedure. Hemarthrosis is unusual.

2. *Laboratory Findings*

F.XI deficiency exhibits abnormal PTT and prothrombin consumption. The whole blood clotting time may or may not be prolonged. A specific assay for F.XI using either congenitally deficient or artificially depleted plasma will confirm the diagnosis.

3. *Genetics*

The disorder shows autosomal recessive inheritance. Homozygotes have less than 20% F.XI, while heterozygotes have 30–65% and are generally asymptomatic.

4. *Pregnancy*

In normal women, F.XI may drop to 60–70% of the normal level by the last trimester of pregnancy (Hilgartner and Smith, 1965; Nossel *et al.*, 1966). Affected women have been reported to bleed at delivery (Czapek, 1973).

A heterozygote would not be expected to have bleeding problems at delivery unless her F.XI dropped below 20%. Offspring of two heterozygotes would have a 1 in 4 chance of being affected.

5. *Prenatal Diagnosis*

Since the level of F.XI is low at birth, diagnosis during fetal life might be difficult.

6. *Fetus and Newborn*

Effects of this disorder on the fetus are unknown but would presumably be slight due to the mild nature of the disorder.

Diagnosis at birth may be difficult due to the low F.XI levels sometimes seen in the normal neonate. The levels rise slowly over the first 60 days of life, with some infants requiring 9 months to reach the normal adult level (Hilgartner and Smith, 1965). In affected infants, the disease may present in the newborn period with bleeding from circumcision or from trauma.

7. *Treatment*

Fresh frozen plasma is the usual mode of therapy for this disorder.

I. Factor XII Deficiency

Deficiency of F.XII (Hageman factor) is a laboratory abnormality not usually associated with hemorrhagic symptoms. It was first observed by Ratnoff and Colopy (1955) during a routine preoperative screening on a patient named

Hageman. It has been described in over 100 cases but may be more common, since it is usually detected by chance.

1. Clinical Features

Bleeding with this disorder, even in patients with no detectable F.XII, is extremely unusual. Neither spontaneous nor postsurgical bleeding occurs. There have been rare reports of women having excessive bleeding with childbirth or dental extraction, but it is difficult to ascribe the bleeding solely to F.XII deficiency.

Egeberg (1970) reported four Norwegian families in which members with F.XII levels about one-half of normal showed a slight bleeding tendency and an increased incidence of cerebral apoplexy. The disorder in these families appears to differ considerably from that described above and has not been commonly encountered.

2. Laboratory Findings

The PTT, prothrombin consumption, whole blood clotting time, and thromboplastin generation time are abnormal, while the prothrombin time is normal. A specific F.XII assay confirms the diagnosis.

3. Genetics

The inheritance pattern in most families appears to be autosomal recessive, with homozygotes having less than 1% of the normal level of F.XII. Heterozygotes usually have 25–60% of the normal level and can be detected only by a specific F.XII assay (Ratnoff and Steinberg, 1962).

Bennett and colleagues (1972b) have reported a family which appears to demonstrate autosomal dominant inheritance. A male and a female sibling with less than 1% F.XII each produced offspring with less than 1% F.XII by unrelated spouses with normal F.XII levels. However, some other family members also displayed the intermediate F.XII levels expected of heterozygotes for the recessive form, indicating incomplete penetrance of the trait if it is truly dominant.

4. Pregnancy

There are conflicting statements on whether F.XII increases or decreases during pregnancy in normal women (Biland and Duckert 1973; World Health Organization, 1972).

Pregnancy and delivery in affected women would not be expected to produce abnormal bleeding.

5. Fetus and Newborn

Factor XII levels are low in the newborn but reach adult levels by 14 days of age (Kurkcuoglu and McElfresh, 1960). There would be no need for prenatal diagnosis, since the deficiency causes no symptoms.

6. Treatment

Therapy is not usually necessary for this disorder.

J. Factor XIII Deficiency

Factor XIII (fibrin stabilizing factor) is necessary for the cross-linkage of fibrin monomers to produce a stable clot. Its absence causes the formation of a soft clot that readily breaks apart, resulting in delayed bleeding occurring after hemostasis is thought to have been attained. The disorder was first described in man by Duckert *et al.* (1960) and is quite rare.

1. Clinical Features

The characteristic pattern of the disease is apparently normal clotting following trauma, with bleeding occurring 24–36 hours later and lasting for several days. Wounds heal slowly, producing large scars. Hemorrhage usually does not occur without trauma, but intracranial and gastrointestinal bleeding have been observed. Hemarthrosis may occur but without residual crippling.

2. Laboratory Findings

All routine coagulation tests are normal in this disorder. The deficiency should be suspected when the clot appears shaggy and friable. The diagnosis is confirmed if the fibrin clot obtained from recalcified plasma is rapidly soluble in 5 *M* urea or 1% monochloracetic acid. Factor XIII may be quantitated by means of a fluorescence assay.

3. Genetics

The disorder is usually inherited as an autosomal recessive trait. Consanguinity has been observed in many of the families reported. Decreased levels of the factor are present in the plasma of some heterozygotes (Lorand *et al.*, 1970).

Ratnoff and Steinberg (1968) proposed that in some families in which only men were affected, the data were most compatible with X-linked inheritance. This matter is still unsettled and both inheritance patterns must be considered possible (McDonagh *et al.*, 1974).

4. Pregnancy

Factor XIII has been reported to fall during pregnancy in normal women (Coopland *et al.*, 1969). In affected women the disorder may present as postpartum hemorrhage (Czapek, 1973). Pregnancy in an affected woman has also been interrupted by a spontaneous abortion attributed to severe decidual bleeding (Fisher *et al.*, 1966). Heterozygous women would not be expected to have bleeding problems.

5. *Prenatal Diagnosis*

Little is known about fetal levels of F.XIII, so the likelihood of prenatal diagnosis is uncertain.

6. *Fetus and Newborn*

Risks to the fetus of an affected woman may include that of spontaneous abortion as discussed above. Special dangers to the affected fetus are not known.

Factor XIII levels in the normal newborn are markedly low. They reach adult levels by 1 month of age. Diagnosis may be made only if the level is extremely low. Umbilical bleeding is the most common symptom of the disorder. Easy bruising and intracranial bleeding may also occur.

7. *Treatment*

The accepted mode of therapy is fresh or fresh frozen plasma. A lyophilized concentrate is available in Germany but not in the United States.

K. Other Factor Deficiencies

1. *Fletcher Factor*

Deficiency of Fletcher factor results in a laboratory abnormality but does not cause a bleeding diathesis. Although the deficient plasma lacks a prekallikrein which activates the coagulation pathway (through F.XII), the fibrinolytic system, and chemotactic activity (Kaplan and Austen, 1975), no related symptoms have been found in the patients described so far.

a. Laboratory Findings. Laboratory findings include prolonged PTT, thromboplastin generation time, and whole blood clotting time. However, the PTT shortens after incubation for 10 minutes. A specific assay may be performed using plasma from a patient known to be deficient in the factor.

Only five families, four of them Black, have been reported with this deficiency. The inheritance pattern appears to be autosomal recessive with homozygotes having a level of 1% or less and heterozygotes 40–72% of the normal level of the factor.

b. Treatment. Treatment for this disorder is unnecessary.

2. *Fitzgerald Factor*

Deficiency of a high molecular weight kininogen associated with the fibrinolytic pathways has been described in four families: Fitzgerald (Saito *et al.,* 1975), Flaujeac (Lacombe *et al.,* 1975), Reid (Lutcher, 1976), and Williams (Colman *et al.,* 1975). They are apparently deficient in the same protein (Wuepper *et al.,* 1975). The deficiency is not associated with a disease state but results in a prolonged PTT, often detected on routine screening. Little else is known about this disorder.

III. HEREDITARY PLATELET DISORDERS

Hereditary platelet disorders may be divided into three categories: (1) those disorders associated with a qualitative platelet abnormality, (2) those disorders in which the platelets are intrinsically normal but function abnormally because of another inherited defect (e.g., von Willebrand disease), and (3) those disorders with hereditary thrombocytopenia or decreased platelet number. Congenital platelet disorders usually result in a life-long history of easy bruisability and mucous membrane bleeding; however, they present a wide spectrum of severity and in some disorders an adult may present with a post-traumatic bleeding episode as the first manifestation of abnormality. Routine screening tests should be normal except for a low platelet count, a prolonged bleeding time, or abnormal clot retraction. If platelet numbers are normal, a functional defect is suggested. This may be either an inherent platelet defect or the deficiency of a plasma factor necessary for normal platelet function. For in-depth discussions, the reader is referred to recent reviews of platelet function and physiology (Hardisty, 1974; Weiss, 1975; Lusher and Barnhart, 1977).

A. Qualitative Platelet Defects

In accordance with the classification of Weiss (1975), these thrombocytopathies are defined as defects of platelet adhesion, the first phase of aggregation, and the second phase of aggregation, including the release phenomenon. Two disorders, the Bernard-Soulier syndrome and von Willebrand disease, exhibit defects in platelet adhesion. Since the von Willebrand disease platelets are qualitatively normal, that disease will be discussed in a separate category. Glanzmann thrombasthenia, which may actually be a heterogeneous group of disorders, results from defective first phase aggregation. Two groups of disorders result from abnormal second phase aggregation: the storage pool diseases and the "aspirin-like" defects.

1. Defects in Adhesion

a. Bernard–Soulier syndrome. This disease was first described by Bernard and Soulier (1948) as a congenital bleeding disorder characterized by giant platelets, with a prolonged bleeding time and normal or near normal platelet count. The defective adhesion in this disorder appears to be due to the absence of a glycoprotein complex on the platelet surface which serves as a receptor for the von Willebrand factor (Nachman, 1977; Phillips, 1977).

1. Clinical features. This syndrome results in moderate to severe membrane bleeding and excessive bruising which may begin in infancy. Symptoms may include epistaxis, gum or gastrointestinal bleeding, menorrhagia, and excessive bleeding following dental extraction, superficial cuts, or surgery.

2. Laboratory findings. The characteristic laboratory feature is giant platelets ranging in size from 2.5 to 8 μm in diameter, approximately the size of a red cell. Platelet counts may fluctuate and may occasionally be low. However, the bleeding is out of proportion to the thrombocytopenia. The bleeding time is prolonged. Aggregation of the patient's platelet-rich plasma is abnormal with ristocetin, but normal in response to other agents. The abnormal aggregation is not corrected by the addition of normal plasma. The patient's plasma contains normal amounts of Willebrand factor as measured in an assay using normal fresh or fixed platelets.

3. Genetics. Bernard–Soulier syndrome is inherited as an autosomal recessive trait. Carriers may also have some giant platelets but are asymptomatic (Cullum *et al.*, 1967).

4. Pregnancy. Risk to an affected mother at delivery may be considerable if symptoms have been severe. Normal platelet transfusions may be required for hemostasis at delivery and in the postpartum period.

5. Prenatal diagnosis. Prenatal diagnosis by detection of giant platelets in fetal blood should be possible.

6. Fetus and newborn. The affected fetus would presumably be at little risk. The newborn infant could have any of the symptoms described above.

7. Treatment. The bleeding from this disorder is not alleviated by corticosteroid therapy and should be controlled by transfusion of normal platelets.

b. Macrothrombocytopathy Associated with Alport Syndrome. Epstein and colleagues (1972) described two families in which Alport syndrome of nephritis and deafness is associated with macrothrombocytes having the typical platelet dysfunction of Bernard–Soulier syndrome. Bleeding symptoms and occasional thrombocytopenia are also present. The mode of inheritance of this trait appears to be autosomal dominant. Additional families have since been reported with no evidence of independent existence of either Alport syndrome or the platelet defect (Ekstein *et al.*, 1975; Bernheim *et al.*, 1976). Bleeding symptoms must be controlled, as in Bernard–Soulier syndrome, by transfusion of normal platelets.

2. Defects in the First Phase of Aggregation

Thrombasthenia (Glanzmann Thrombasthenia). Thrombasthenia was first described by Glanzmann in 1918 as a heterogeneous group of disorders characterized by normal or near normal platelet numbers but incomplete or absent clot retraction. It also exhibits a prolonged bleeding time and life-long chronic nonthrombocytopenic purpura. The heterogeneity of the disorder has been supported by the finding of several different abnormalities in small numbers of patients: decreased surface glycoproteins IIb and IIIa, decreased actinomyosin (the platelet protein responsible for clot retraction), and enzymatic abnormalities in the glycolytic pathway (Lusher and Barnhart, 1977).

1. Clinical features. The disorder is variable in severity. It usually presents in early infancy as generalized purpura. Other hemorrhagic manifestations are generally mild to moderate. Epistaxis, gum oozing, petechiae, and ecchymosis are common in children and adults. Severe bleeding is usually traumatic in origin and fatal complications have been reported. Menorrhagia may be a major problem in affected women, sometimes requiring hormonal suppression of menses (Cronberg and Nilsson, 1968; Vinazzer, 1974). Bleeding manifestations have been reported to decrease in severity with age (Lusher and Barnhart, 1977).

2. Laboratory findings. The characteristic laboratory findings include a normal platelet count, prolonged bleeding time, decreased or absent clot retraction and poor or absent glass bead platelet adhesion. The main defect appears in platelet aggregation. There is no spontaneous aggregation on a slide and no aggregation to any concentration of ADP, thrombin, epinephrine, or collagen. The platelets do, however, aggregate with ristocetin and bovine factor VIII. The release phenomenon is apparently normal.

3. Genetics. The disorder is almost always transmitted as an autosomal recessive. Consanguinity has been present in several reported families (Caen *et al.*, 1966). Heterozygotes are asymptomatic and usually have normal platelet function. Detection of heterozygotes using an antibody from a multiply transfused patient has been described in four families (Levy *et al.*, 1976). A few heterozygotes have also been reported to show abnormal clot retraction (Karaca and Nilsson, 1970; Papayannis and Israëls, 1970).

4. Pregnancy. Bleeding symptoms may continue during pregnancy, and the risk for an affected mother at delivery may be considerable. Profuse bleeding at childbirth has been described (Cronberg and Nilsson, 1968). Cesarean section has been successfully performed using platelet concentrates (Vinazzer, 1974).

5. Prenatal diagnosis. Prenatal diagnosis could perhaps be performed if fetal platelets could be tested for abnormal function.

6. Fetus and newborn. Nothing is known about the disorder in the fetus. The newborn infant may have purpura and ecchymoses of moderate severity. Other clinical features described above may appear soon after birth.

7. Treatment. Treatment is usually not necessary for minor bleeding, although nasal packing may be necessary for epistaxis and anovulatory therapy for menorrhagia. At the time of surgical procedures and dental extractions, transfusion of normal platelet concentrates is usually necessary; however, this therapy may prove a problem if used frequently. It should be reserved for serious bleeding, because the development of isoantibodies to transfused platelets is common.

3. *Defects in the Second Phase of Aggregation*

a. Platelet Storage Pool Diseases. The defect in these diseases originates in the megakaryocyte and seems to be in the platelet dense granules which are absent or fail to package ADP, ATP, serotonin, and calcium properly. The heterogeneous group includes both genetically determined and acquired disorders (Weiss *et al.*, 1977). Clinically the patients are similar, having easy bruisability, mucous membrane bleeding, and a prolonged bleeding time. In the laboratory the abnormality appears as the absence of a second wave of aggregation in response to those agents that trigger the release of endogenous ADP, such as 2 m*M* ADP. The lack of ADP in the dense granules results in a high ATP/ADP ratio, the distinguishing feature of the storage pool diseases. In contrast, the "aspirin-like" defects have a normal ATP/ADP ratio. Other laboratory abnormalities can include decreased adhesion, decreased platelet factor 3 availability, and impaired prostaglandin production. The following disorders have storage pool defects.

1. Storage pool deficiency. This is characterized by the functional platelet abnormalities described above, occurring in otherwise normal individuals without associated congenital abnormalities. A decrease in platelet-dense granules correlates with the impaired release reaction. The disease is autosomal dominant in transmission with mild clinical manifestations that should be of little risk to mother or infant. Should severe bleeding occur, treatment would be with transfusion of normal platelets.

2. Hermansky-Pudlak syndrome. This syndrome is the combination of albinism, a bleeding diathesis with prolonged bleeding time, and an unusual accumulation in bone marrow macrophages of a ceroid-like pigment as first described by Hermansky and Pudlak (1959). The bleeding diathesis has since been found to be due to platelet storage pool deficiency, and the albinism is of the oculocutaneous tyrosinase-positive type.

This syndrome is a rare autosomal recessive disorder described in only 50 cases to date. The hemorrhagic manifestations are usually mild but may magnify bleeding from other causes, as in our case at the New York Hospital, in which bleeding from rectal biopsy and ulcerative colitis was severe and was controlled only by transfusion of normal platelets.

Risk of bleeding in the mother at delivery may be present, requiring transfusion of normal platelets. Risk for the infant would be possible but unusual. Prenatal diagnosis on the basis of albinism or the platelet defect might be possible.

3. Wiskott-Aldrich syndrome (WAS). This is a disease in which platelets have been characterized as deficient in serotonin and the nonmetabolic adenine nucleotides (Grottum *et al.*, 1969). It might, therefore, qualify for inclusion in this group of disorders. The platelets are small and have an increased ratio of ATP/ADP. This disorder, characterized by the triad of eczema, recurrent infections, and thrombocytopenia, will be discussed at length under the group of hereditary thrombocytopenias.

4. Chediak-Higashi syndrome. This is a rare disorder characterized by the triad of partial oculocutaneous albinism, recurrent infections, and hemorrhagic tendency. Mental retardation and intermittent febrile episodes may also be present. The albinism is due to impaired melanin packaging, the infections due to abnormal phagocytic function of neutrophils and macrophages, and the bleeding tendency due to storage pool deficiency. Bleeding time is prolonged, and second phase aggregation is abnormal.

The clinical manifestations of the disorder are variable. A lymphoma-like "accelerated phase" has been described in early childhood, with widespread histiocytic infiltration, pancytopenia, hepatosplenomegaly, and early death due to infection and/or bleeding (Dent *et al.*, 1966; Buchanan and Handin, 1976). Chemotherapy has been useful in this phase. The mean age of death is 9 years, but in those patients who survive, the symptoms tend to abate with age.

This disorder is transmitted as an autosomal recessive trait and has also been described in animals (Windhorst and Padgett, 1973). Consanguinity is present in over one-half of the reported cases, and experimental breeding in animal models has confirmed the inheritance pattern. The carrier state has not been identified using any of the abnormalities. Little information is available on pregnancy in affected women.

Prenatal diagnosis might be possible by detection of the platelet defect, the albinism, or the abnormal leukocytes.

b. "Aspirin-Like" Defects. This second group of abnormalities characterized by defects in the second phase of aggregation are so named because they resemble the defect caused by the drug aspirin. When aspirin is ingested by normal subjects, collagen-induced platelet aggregation and the second phase aggregation with ADP and epinephrine are eliminated *in vitro* and *in vivo,* and the subjects have a prolonged bleeding time. Aspirin interferes with dense-granule release of ATP, ADP, and serotonin. The ATP consumption is diminished and prostaglandins are not produced, with a resultant lack of the second phase of aggregation response. The mechanism of drug action appears to be inhibition of production of thromboxane, the basic platelet reactant which stimulates the release reaction.

Although the acquired defects of aspirin ingestion are far more common, the congenital disorders of impaired second phase aggregation have the same functional defects and a mild hemorrhagic tendency. The platelets of these patients contain normal amounts of ADP but fail to release it. These disorders can be distinguished from the storage pool deficiencies only by assay of platelet ADP. Some patients have been reported also to have a deficiency of platelet factor 3 and/or platelet factor 4. The disorder has been reported in patients without associated abnormalities, but the majority of the patients have connective tissue disorders.

1. "Aspirin-like" defect without associated abnormalities. This defect has been reported in only five cases (Holmsen and Weiss, 1970). The disorder described by O'Brien (1967) and termed *Portsmouth syndrome* is probably the same. The bleeding tendency is mild, and the transmission is autosomal dominant. Prenatal diagnosis might be possible, but the disorder probably presents no risk to mother or infant.

2. Ehlers-Danlos syndrome. This syndrome is characterized by hyperextensibility of the joints, hyperelasticity of the skin, poor wound healing, and easy bruising. The disorder is usually transmitted as an autosomal dominant trait and has variable expressivity of each of the components. The hemorrhagic tendency is usually mild in character with a range of clinical symptoms from easy bruising to surgical bleeding. Laboratory findings range from only a positive tourniquet test to a prolonged bleeding time, defective platelet factor 3 release and defective aggregation with ADP, collagen, and epinephrine (Onel *et al.*, 1973). The pathogenesis of the defect has been ascribed to the abnormal collagen by one author (Hardisty, 1974). The risk to an affected mother of excessive bleeding at delivery would depend on the severity of her hemorrhagic manifestations. If previous symptoms had been severe, normal platelets might be necessary. However, most patients have only mild symptoms.

3. Marfan syndrome. This syndrome which is characterized by arachnodactyly, loose joints and ligaments, ectopia lentis, and dilatation of the aorta, sometimes includes a hemorrhagic tendency. However, this is exceedingly rare and not a major feature of the disorder.

4. Osteogenesis imperfecta. This disease is characterized by frequent fractures, blue sclerae, and osteoporosis. The metabolic defect affects leukocytes and platelets as well as bone, and a mild bleeding diathesis may be present (Estes, 1972; Hathaway *et al.*, 1972). In the laboratory, defective aggregation with ADP, collagen, and epinephrine and defective platelet factor 3 release are seen. The hemorrhagic tendency is not a major feature of the disorder.

For further discussion of the above three diseases see the chapter on connective tissue disorders.

B. Disorders of Platelet Function Due to an Inherited Plasma Defect

This category includes two diseases in which the platelets are intrinsically normal but function abnormally due to deficiency of a plasma factor.

1. Afibrinogenemia

This disorder is discussed fully in Section II under Plasma Coagulation Factor Disorders. It may exhibit a prolonged bleeding time and abnormal platelet adhesion and aggregation which are corrected *in vitro* by the addition of fibrinogen (Inceman *et al.*, 1966; Weiss and Rogers, 1971).

2. *von Willebrand Disease*

This disorder has also been discussed in Section II. The platelets are normal morphologically and when placed in normal plasma, function normally. The deficient plasma factor, called Willebrand factor or ristocetin cofactor, is necessary for platelet adhesion to glass beads and for normal aggregation by the antibiotic ristocetin. It may be measured by the ability of the patient's plasma to aggregate normal platelets when ristocetin is introduced.

The Willebrand factor is synthesized by endothelial cells (Jaffe *et al.*, 1974) and supports adhesion of platelets to the subendothelium of the vessel wall (Tschopp *et al.*, 1974). It has been proposed that the protein acts as a bridge between the platelet membrane and the vessel wall (Gralnick *et al.*, 1977).

C. Hereditary Thrombocytopenias

Several rare forms of hereditary thrombocytopenia have been described. Those which involve an intrinsic platelet defect have already been discussed.

1. *Bernard–Soulier Syndrome*

This giant platelet syndrome results from an abnormal platelet membrane component and has been described above. Many patients have some degree of thrombocytopenia resulting from a shortened platelet life span.

2. *May–Hegglin Anomaly*

This very rare hematologic disorder is usually discovered by chance when a patient has a peripheral blood film examined for an unrelated reason. The disorder is characterized by giant platelets, a variable degree of thrombocytopenia, and basophilic inclusions known as Döhle bodies within leukocytes. The giant platelets may be as large as 20 μm in length but are functionally normal. Thrombocytopenia may be present, absent, or intermittent.

The majority of patients are asymptomatic; however, purpura, increased bruisability, and minor epistaxis have occasionally been reported. The disorder is transmitted as an autosomal dominant and could be diagnosed prenatally in fetal blood. However, potential risk to the affected individual is negligible.

3. *Macrothrombocytopathy Associated with Alport Syndrome*

This association of Bernard-Soulier-like giant platelets with nephritis and deafness has already been discussed. It is mentioned here because of the thrombocytopenia occasionally seen in some patients.

4. *Thrombocytopenia with Absent Radius (TAR Syndrome)*

This disorder is characterized by thrombocytopenic purpura and bilateral absent radii. All patients have additional skeletal abnormalities, and one-third have congenital heart defects. The syndrome must be distinguished from Fanconi anemia, which it resembles. The major differences, noted in Table II, are the age at onset of the hematologic abnormalities, the severity of the findings, the associated congenital anomalies, and the chromosome abnormalities. The disorder has been thoroughly reviewed by Hall *et al.* (1969).

a. Clinical Features. Hemorrhagic manifestations include petechiae, ecchymoses, and mucous membrane bleeding beginning in early infancy. In one-half of the cases reviewed by Hall *et al.* (1969), symptoms appeared within the first week of life. Mortality is high in the first year. However, beyond infancy the prognosis improves, with those patients surviving into adulthood having a lessening of symptoms. At menarche platelet counts rise to low normal levels, and menorrhagia is not usually a problem.

b. Laboratory Findings. At birth megakaryocytes in the bone marrow are decreased or absent, and platelet counts vary between 10,000 and 60,000/mm^3.

TABLE II
Comparison of Fanconi Anemia (FA) and Thrombocytopenia with Absent Radii (TAR)

	FA	TAR
Inheritance	Autosomal recessive	Autosomal recessive
Male:female ratio	2:1	1:2
Skeletal deformities	66%	100%
Carpal abnormalities	38%	Rare
Absent radii:five fingers	0%	100%
Microcephaly	40%	Rare
Nonmusculoskeletal congenital anomalies	Frequent	Rare
Hyperpigmentation	77%	0%
Renal abnormalities	40%	Rare
Chromosome breaks in lymphocytes	Increased	None
Mental retardation	17%	7%
Onset of hematologic signs	5–10 years	Birth to early infancy
Blood findings	Pancytopenia	Thrombocytopenia and leukemoid reactions, ± blood loss anemia
Marrow findings	Panmyelopathy	Decreased or defective megakaryocytes
Prognosis	Poor	Relatively good after first year of life

Both increase with age. Platelets are thought to have normal function and survival, but some cases have been reported to have a functional platelet defect similar to storage pool disease (Zahavi *et al.*, 1977). A leukemoid reaction is frequently present in early infancy with a total leukocyte count as high as 100,000/mm^3.

c. Genetics. The TAR syndrome is transmitted as an autosomal recessive. The apparently unrelated defects are the result of a single highly pleiotropic gene. The basic defect is unknown.

d. Pregnancy and Prenatal Diagnosis. The risk for an affected mother at delivery should be minimal but would depend on the platelet count. Prenatal diagnosis has been performed by roentgenographic detection of the absent radii (Omenn *et al.*, 1978).

e. Fetus and Newborn. An affected infant could be at great risk for intracranial bleeding at the time of delivery. If the diagnosis is suspected, delivery by cesarean section might be planned. Treatment with platelet concentrates soon after delivery should be considered for the affected infant.

f. Treatment. Bleeding symptoms can be controlled only with transfusions of normal platelets. Corticosteroids and splenectomy are not helpful.

5. *Wiskott-Aldrich Syndrome (WAS)*

Wiskott-Aldrich syndrome is characterized by the triad of thrombocytopenia, recurrent infections, and eczema in male children. Most affected children die in the first few years of life from infection or lymphoreticular malignancy. Wiskott-Aldrich syndrome is best known as an immunodeficiency disorder, but hemorrhage can be quite severe in some cases. The immunologic defect is thought to be in B-lymphocyte function, with T-cell dysfunction secondary (Lawton and Cooper, 1977). The hemorrhagic defect is not well characterized, but thrombocytopenia and abnormally small platelets have been consistently described. The thrombocytopenia, apparently, results from a shortened platelet life span, with a turnover rate as much as eight times normal (Baldini, 1972). One would expect large platelets in the face of a shortened life span, if platelet production is normal. However, WAS platelets are one-third normal size (Baldini, 1972). This, in combination with the striking morphologic changes in megakaryocytes observed by Pearson *et al.* (1966), suggests abnormal thrombopoiesis. The combination of thrombocytopenia and small platelets is unique to this disorder (Murphy, 1972).

a. Clinical Features. Wiskott-Aldrich syndrome presents in males within the first year of life with the onset of thrombocytopenic purpura, atopic dermatitis, and frequent viral and bacterial infections. Epistaxis, bloody diarrhea, chronic otitis media, and anemia from blood loss also occur. Petechiae and variable hepatosplenomegaly may be found.

b. Laboratory Findings. Significant thrombocytopenia, anemia, and low IgM are usually found. Platelets are approximately one-third normal size and have a half-life of 24 hours. Studies have shown that normal platelets have a normal life span in WAS patients, while WAS platelets have a short survival in a normal individual. Several authors have noted platelet phagocytosis by macrophages and reticulum cells in the marrow and have suggested that the reticuloendothelial system recognizes these platelets as abnormal (Grottum *et al.*, 1969; White and Gerrard, 1976). In addition, WAS platelets have a functional abnormality of second phase aggregation, as discussed earlier.

c. Genetics. This syndrome is an X-linked recessive trait. No hypothesis about the basic defect has satisfactorily explained both the immunologic and hematologic components of the disease (Krivit and Good, 1959; Rosen, 1971).

d. Pregnancy. No problems are expected in the carrier mother at delivery; however, she has a 1 in 4 chance of producing an affected son.

e. Prenatal Diagnosis. Prenatal determination of fetal sex has allowed the abortion of males, who have a 50% chance of being affected. M. S. Golbus (unpublished observation) has suggested that the disorder might be detected in fetal blood by analysis of platelet size.

f. Fetus and Newborn. Effects on the fetus have not been described. The disease is not usually evident in the newborn, although the platelet abnormality may be detectable.

g. Treatment. Therapy is usually directed against the immunodeficiency, with red cells used as support for the anemia. Some improvement has been seen with transfer factor. Steroids and splenectomy are contraindicated. Transfusion with normal platelets for bleeding problems is the treatment of choice. If a tissue match for bone marrow transplantation is possible, that experimental procedure is providing promising results.

6. Hereditary Thrombocytopenia Resembling ITP

Hereditary thrombocytopenia without other associated anomalies is very rare and must be distinguished from idiopathic thrombocytopenic purpura (ITP), which is acquired. The peripheral smear shows absent platelets with no other abnormality, and the bone marrow shows normal or increased megakaryocytes. Bleeding begins early in life, in contrast to ITP which is usually seen later. Symptoms include petechiae, bruising, and mucous membrane bleeding. Platelet kinetics are helpful in distinguishing the two disorders. In the hereditary form the half-life is usually normal, whereas in ITP it is markedly shortened (Murphy, 1972). A family history of thrombocytopenia confirms the diagnosis. Both autosomal dominant (Seip, 1963) and autosomal recessive (Murphy *et al.*, 1972) forms have been reported. The bleeding manifestations are usually mild and do not warrant treatment.

IV. CONCLUSION

The pregnant patient with a disorder of hemostasis should be thoroughly studied prior to delivery and should be delivered at a hospital where adequate laboratory facilities, a trained hematologist, and replacement therapy appropriate for her condition can be maintained. The family of a patient with a bleeding disorder requires the provision of comprehensive diagnostic and treatment services, which should include genetic counseling and evaluation of the feasibility of prenatal diagnosis if the severity of the disorder warrants.

REFERENCES

Abildgaard, C. F. (1979). Coagulation disorders. *In* "Haematology of Infancy and Childhood" (D. G. Nathan and F. A. Oski, eds.), Saunders, Philadelphia, Pennsylvania.

Alexander, B., Goldstein, R., Landwehr, G., and Cook, C. D. (1951). Congenital SPCA deficiency: A hitherto unrecognized coagulation defect with hemorrhage rectified by serum and serum fractions. *J. Clin. Invest.* **30,** 576.

Baldini, M. G. (1972). Nature of the platelet defect in the Wiskott-Aldrich syndrome. *Ann. N.Y. Acad. Sci.* **201,** 437-444.

Barrow, E. M., and Graham, J. B. (1974). Blood coagulation factor VIII (antihemophilic factor): With comments on von Willebrand's disease and Christmas disease. *Physiol. Rev.* **54,** 23-74.

Barrow, E. M., Bullock, W. R., and Graham, J. B. (1960). A study of the carrier state for plasma thromboplastin component (PTC, Christmas Factor) deficiency, utilizing a new assay procedure. *J. Lab. Clin. Med.* **55,** 936-945.

Beck, E. A. (1964). Abnormal fibrinogen (fibrinogen "Baltimore") as a cause of a familial hemorrhagic disorder. *Blood* **24,** 853-854.

Bennett, B., and Ratnoff, O. D. (1972a). Changes in antihemophilic factor (AHF, factor VIII) procoagulant activity and AHF-like antigen in normal pregnancy, and following exercise and pneumoencephalography. *J. Lab. Clin. Med.* **80,** 256-263.

Bennett, B., Ratnoff, O. D., Holt, J. B., and Roberts, H. R. (1972b). Hageman trait (factor XII deficiency): A probable second genotype, inherited as an autosomal dominant characteristic. *Blood* **40,** 412.

Bernard, J., and Soulier, J. (1948). Sur une nouvelle variste de dystrophie thrombocytaire hémorragipare congénitale. *Sem. Hop.* **24,** 3217-3233.

Bernheim, J., Dechavanne, M., Bryon, P. A., Lagarde, M., Colon, S., Pozet, N., and Traeger, J. (1976). Thrombocytopenia, macrothombocytopathia, nephritis and deafness. *Am. J. Med.* **61,** 145-150.

Biggs, R., ed. (1976). "Human Blood Coagulation, Haemostasis and Thrombosis," 2nd ed. Blackwell, Oxford.

Biland, L., and Duckert, F. (1973). Coagulation factors of the newborn and his mother. *Thromb. Diath. Haemorrh.* **29;** 644-651.

Blecher, T. E., Westby, J. C., and Thompson, M. J. (1978). Synthesis of procoagulant antihaemophilic factor *in vitro*. *Lancet* **2,** 1333-1336.

Blombäck, M., Blombäck, B., Mammen, E. F., and Prasad, A. S. (1968). Fibrinogen Detroit: A molecular defect in the N-terminal disulfide knot of human fibrinogen? *Nature (London)* **218,** 134-137.

Buchanan, G. R., and Handin, R. I. (1976). Platelet function in the Chediak-Higashi syndrome. *Blood* **47;** 941-948.

Caen, J. P. (1972). Glanzmann's thrombasthenia. *Clin. Haematol.* **1,** 383-392.

Caen, J. P., Castaldi, P. A., Leclerc, J. C., Inceman, S., Larrieu, M. J., Probst, M., Bernard, J. (1966). Congenital bleeding disorders with long bleeding time and normal platelet count. I. Glanzmann's thrombasthenia (report of fifteen patients). *Am. J. Med.* **41,** 4-26.

Cederbaum, S. D., Holzman, G. B., and Sparkes, R. S. (1971). "Spontaneous" abortion and haemorrhage following attempted amniocentesis in a carrier of haemophilia A. *Lancet* **2,** 429-430.

Colman, R. W., Bagdasarian, A., Talamo, R. C., Scott, C. F., Seavey, M., Guimareaes, J. A., Pierce, J. V., and Kaplan, A. P. (1975). Williams trait. Human kininogen deficiency with diminished levels of plasminogen proactivator and prekallikrein associated with abnormalities of the Hageman factor-dependent pathways. *J. Clin. Invest.* **56,** 1650-1662.

Coopland, A., Alkjaersig, N., and Fletcher, A. P. (1969). Reduction in plasma factor XIII (fibrin stabilizing factor) concentration during pregnancy. *J. Lab. Clin. Med.* **73,** 144.

Cronberg, S., and Nilsson, I.M. (1968). Investigations in a family with thrombasthenia of moderately severe type with 16 affected family members. *Scand. J. Haematol.* **5,** 17-25.

Cullum, C., Cooney, D. P., and Schrier, S. L. (1967). Familial thrombocytopenic thrombocytopathy. *Br. J. Haematol.* **13,** 147.

Czapek, E. E. (1973). Coagulation problems. *Int. Anesthesiol. Clin.* **11,** 175-201.

Czapek, E. E., Hoyer, L. W., and Schwartz, A. D. (1974). Hemophilia A in a female: Use of factor VIII antigen levels as a diagnostic aid. *J. Pediatr.* **84,** 485-489.

Denson, K. W. E., Lurie, A., DeCataldo, F., and Mannucci, P. M. (1970). The factor X defect: The recognition of abnormal forms of factor X. *Br. J. Haematol.* **18,** 309.

Dent, P. B., Fish, L. A., White, J. G., and Good, R. A. (1966). Chediak-Higashi syndrome: Observation on the nature of the associated malignancy. *Lab. Invest.* **15,** 1634-1642.

Duckert, F., Jung, E., Shmerling, D. H. (1960). A hitherto undescribed congenital haemorrhagic diathesis probably due to fibrin stabilizing factor deficiency. *Thromb. Diath. Haemorrh.* **5,** 179.

Edgell, C. J., Kirkman, H. N., Clemons, E., Buchanan, P. D., and Miller, C. H. (1978). Prenatal diagnosis by linkage: Hemophilia A and polymorphic glucose-6-phosphate dehydrogenase. *Am. J. Hum. Genet.* **30,** 80-84.

Egbring, R., Andrassy, K., Egli, H., and Meyer-Lindenberg, J. (1971). Diagnostiche und therapeutische Probleme bei congenitaler afibrinogenämie. *Blut* **22,** 175-201.

Egeberg, O. (1970). Factor XII defect and hemorrhage: Evidence for a new type of hereditary hemostatic disorder. *Thromb. Diath. Haemorrh.* **23,** 432-440.

Ekstein, J. D., Filip, D. J., Watts, J. C. (1975). Hereditary thrombocytopenia, deafness and renal disease. *Ann. Intern. Med.* **82,** 639-645.

Elston, R. C., Graham, J. B., Miller, C. H., Reisner, H. M., and Bouma, B. N. (1976). Probabilistic classification of hemophilia A carriers by discriminant analysis. *Thromb. Res.* **8,** 683-695.

Epstein, M. A., Sahed, M. A., Piel, C. F., Goodman, J. R., Berrfield, M. R., Kushner, J. H., and Ablin, R. A. (1972). Hereditary macrothrombocytopathia, nephritis and deafness. *Am. J. Med.* **52,** 299-310.

Estes, S. (1972). Platelet abnormalities in heritable disorders of connective tissue. *Ann. N. Y. Acad. Sci.* **201,** 445-450.

Feinstein, D. I., Rapaport, S. J., McGehee, W. G. *et al.* (1970). Factor V anticoagulants: Clinical biochemical immunological observations. *J. Clin. Invest.* **49,** 1578.

Firshein, S. I., Hoyer, L. W., Lazarchick, J., Forget, B. G., Hobbins, J. C., Clyne, L. P., Pitlick, F. A., Muir, W. A., Merkatz, I. R., Mahoney, M. J. (1979). Prenatal diagnosis of classical hemophilia. *New Engl. J. Med.* **300,** 937-941.

Fisher, S., Rikover, M., and Noar, J. (1966). Factor XIII deficiency with severe hemorrhagic diathesis. *Blood* **28,** 34.

Fratantoni, J. C., Hilgartner, M. W., and Nachman, R. L. (1972). Nature of the defect in congenital factor V deficiency study in a patient with an acquired circulatory anticoagulant. *Blood* **39,** 751.

Girolami, A. (1971). The hereditary transmission of congenital 'true' hypoprothombinaemia. *Br. J. Haematol.* **21,** 695–703.

Glanzmann, E. (1918). Hereditäre hämorrhagische Thrombasthenie. Ein Beitrag zur Pathologie der Blutplättchen. *Jahrb. Kinderheilkd.* **88,** 1–12.

Goodnight, S. H., Feinstein, D. I., Osterud, B., and Rapaport, S. I. (1971). Factor VII antibody-neutralizing material in hereditary and acquired factor VII deficiency. *Blood* **38,** 1.

Graham, J. B. (1978). Genotype assignment (carrier detection) in the hemophilias. *Clin. Haematol.* **8,** 115–145.

Graham, J. B., McLendon, W. W., and Brinkhous, K. M. (1953). Mild hemophilia: An allelic form of the disease. *Am. J. Med. Sci.* **225,** 46–53.

Graham, J. B., Barrow, E. S., and Elston, R. C. (1975a). Lyonization in hemophilia: A cause of error in direct detection of heterozygous carriers. *Ann. N. Y. Acad. Sci.* **240,** 141.

Graham, J. B., Barrow, E. S., Roberts, H. R., Webster, W. P., Blatt, P. M., Buchanan, P., Cederbaum, A. I., Allain, J. P., Barrett, D. A., and Gralnick, H. R. (1975b). Dominant inheritance of hemophilia A in three generations of women. *Blood* **46,** 175–188.

Graham, J. B., Miller, C. H., Reisner, H. M., Elston, R. C., and Olive, J. A. (1976). The phenotypic range of hemophilia A carriers. *Am. J. Hum. Genet.* **28,** 482–488.

Gralnick, H. R., Sultan, Y., and Coller, B. S. (1977). Von Willebrand's disease. Combined qualitative and quantitative abnormalities. *N. Engl. J. Med.* **296,** 1024–1030.

Grottum, K. A., Hovig, I., Holmsen, H., Abrahamsen, A. F., Jeremie, M., and Seip, M. (1969). Wiskott-Aldrich syndrome: Qualitative defects and short platelet survival. *Br. J. Haematol.* **17,** 373–388.

Hall, J. G., Levin, J., Kuhn, J. P., Ottenheimer, E. J., vanBerkum, K. A. P., and McKusick, V. A. (1969). Thrombocytopenia with absent radius (TAR). *Medicine (Baltimore)* **48,** 411–439.

Hardisty, R. M. (1974). Hereditary disorders of platelet function. *In* "Platelets" (O. N. Ulutin, ed.), Int. Congr. Ser. 357, pp. 201–207. Excerpta Med. Found., Amsterdam.

Hartmann, J. R., and Diamond, L. K. (1957). Hemophilia and related hemorrhagic disorders. *Practitioner* **178,** 179.

Hathaway, W. E. (1975). The bleeding newborn. *Semin. Hematol.* **12,** 175–188.

Hathaway, W. E., Solomons, O. C., and Ott, J. E. (1972). Platelet function and pyrophosphates in osteogenesis imperfecta. *Blood* **39,** 500–509.

Hermansky, F., and Pudlak, P. (1959). Albinism associated with hemorrhagic diathesis and unusual pigmented reticular cells in the bone marrow: Report of two cases with histochemical studies. *Blood* **14,** 162–169.

Hilgartner, M. W. (1968). Transient functional thrombasthenia in the newborn. *Soc. Pediatr. Res., 38th Annu. Meet.* p. 120.

Hilgartner, M. W., ed. (1976). Hemophilia in children. *Prog. Pediatr. Hematol.-Oncol.* **1.**

Hilgartner, M. W., and Smith, C. H. (1965). Plasma thromboplastin antecedent (factor XI) in the neonate. *J. Pediatr.* **66,** 747.

Holmberg, L., Gustavii, B., Cordesius, E., Kristofferson, A. C., Ljung, R., Lofberg, L., Stromberg, P., Nilsson, I. M. (1980). Prenatal diagnosis of hemophilia B, by an immunoradiometric assay of Factor IX. *Blood* **56,** 397–401.

Holmsen, H., and Weiss, J. H. (1970). Hereditary defect in the platelet release reaction caused by a deficiency in the storage pool of platelet adenine nucleotides. *Br. J. Haematol.* **19,** 643–649.

Hougie, C., Barrow, E. M., and Graham, J. B. (1957). Stuart clotting defect. I. Segregation of an hereditary hemorrhagic state from the heterogeneous group heretofore called "stable factor" (SPCA, proconvertin, factor VII deficiency). *J. Clin. Invest.* **36,** 485.

Hoyer, L. W., Lindsten, J., Blomback, M., Hagenfeldt, L., Cordesius, E., Stromberg, P., Gustavii, B. (1979). Prenatal evaluation of a fetus at risk for severe von Willebrand's (Letter). *Lancet* **2,** 191–192.

Inceman, S., Caen, J., and Bernard, J. (1966). Aggregation, adhesion, and viscous metamorphosis of platelets in congenital fibrinogen deficiencies. *J. Lab. Clin. Med.* **68,** 21–32.

Italian Working Group (1977). Spectrum of von Willebrand's disease: A study of 100 cases. *Br. J. Haematol.* **35,** 97–108.

Jaffe, E. A., Hoyer, L. W., and Nachman, R. L. (1973). Synthesis of antihemophilic factor antigen by cultured human endothelial cells. *J. Clin. Invest.* **52,** 2757–2764.

Jaffe, E. A., Hoyer, L. W., and Nachman, R. L. (1974). Synthesis of von Willebrand factor by cultured human endothelial cells. *Proc. Natl. Acad. Sci. U.S.A.* **71,** 1906–1909.

Jordan, H. H. (1958). "Hemophilic Arthropathies." Thomas, Springfield, Illinois.

Josso, F., deSanchez, J. M., Lavergne, J. M., Menache, D., and Soulier, J. P. (1971). Congenital abnormality of the prothrombin molecule (Factor II) in four siblings: Prothrombin Barcelona. *Blood* **38,** 9.

Kaplan, A. P., and Austen, K. F. (1975). Activation and control mechanism of Hageman factor-dependent pathways of coagulation, fibrinolysis and kinin generation and their contribution to the inflammatory process. *J. Allergy Clin. Immunol.* **56,** 491.

Karaca, M., and Nilsson, I. M. (1970). Diagnosis of mild thrombasthenia. *Lancet* **2,** 1084–1085.

Kasper, C. K., Hoag, M. S., Aggelar, P. M., and Stone, S. (1964). Blood clotting factors in pregnancy; Factor VIII concentration in normal and AHF-deficient women. *Obstet. Gynecol.* **24,** 242.

Kasper, C. K., Osterud, B., Minami, J. Y., Shonick, W., and Rapaport, S. I. (1977). Hemophilia B: Characterization of genetic variants and detection of carriers. *Blood* **50,** 351–366.

Kerr, C. B. (1965). Genetics of human blood coagulation. *J. Med. Genet.* **2,** 254–303.

Kirkman, H. N. (1972). Enzyme defects. *Prog. Med. Genet.* **8,** 125–168.

Krivit, W., and Good, R. (1959). Aldrich's syndrome (thrombocytopenia, eczema and infection in infants): Studies of the defense mechanism. *Am. J. Dis. Child.* **97,** 137–153.

Kurkcouglu, M., and McElfresh, A. E. (1960). The Hageman factor: Determination of its concentration during the neonatal period and presentation of a case of Hageman factor deficiency. *J. Pediatr.* **57,** 61.

Lacombe, M. J., Varet, B., and Levy, J. P. (1975). A hitherto undescribed plasma factor acting at the contact phase of blood coagulation (Flaujeac factor): Case report and coagulation studies. *Blood* **46,** 761–768.

Lawton, A. R., and Cooper, M. D. (1977). Immune deficiency diseases. *In* "Harrison's Principles of Internal Medicine" (G. W. Thorn, R. D. Adams, E. Braunwald, K. J. Isselbacher, and R. G. Petersdorf, eds.), pp. 402–409. McGraw-Hill, New York.

Levitan, M., and Montagu, A. (1971). "Textbook of Human Genetics," pp. 186–187. Oxford Univ. Press, London and New York.

Levy, J. M., Stoll, C., Gardea, A., and Bigel, P. (1976). Clinique et génétique dans la thrombasthénie de Glanzmann. *Nouv. Rev. Fr. Hematol.* **16,** 399–406.

Lorand, L., Urayama, T., Atencio, A., and Hsia, D. Y. (1970). Inheritance of deficiency of fibrin stabilizing factor (factor XIII). *Am. J. Hum. Genet.* **22,** 89.

Lusher, J. M., and Barnhart, M. I. (1977). Congenital disorders affecting platelets. *Sem. Thromb. Haemostasis* **4,** 123–186.

Lusher, J. M., Zuelzer, W. W., and Evans, R. K. (1969). Hemophilia A in chromosomal female subjects. *J. Pediatr.* **74,** 265–271.

Lutcher, C. L. (1976). Reid trait: A new expression of high molecular weight kininogen (HMW-kininogen) deficiency. *Clin. Res.* **24,** 47a (abstr.).

Lyon, M. F. (1962). Sex chromatin and gene action in the mammalian X-chromosome. *Am. J. Hum. Genet.* **14,** 135–148.
McDonagh, J., McDonagh, R. P., Myllyla, G., and Ikkala, E. (1974). Factor XIII deficiency: A genetic study of two affected kindreds in Finland. *Blood* **43,** 327–332.
McKusick, V. A. (1978). "Mendelian Inheritance in Man: Catalogs of Autosomal Dominant, Autosomal Recessive and X-linked Phenotypes," 5th ed. Johns Hopkins Univ. Press, Baltimore, Maryland.
Mahoney, M. J., Hoyer, L., Firshein, S., Lazarchick, J., Forget, B., and Hobbins, J. (1978). Prenatal diagnosis of hemophilia. *Am. J. Hum. Genet.* **30,** 59a (abstr.).
Mammen, E. F. (1974). Congenital abnormalities of the fibrinogen molecule. *Sem. Thromb. Haemostasis* **1,** 184.
Mammen, E. F., Prasad, A. S., and Barnhart, M. I. (1969). Congenital dysfibrinogenemia: Fibrinogin Detroit. *J. Clin. Invest.* **48,** 235.
Melliger, E. J., and Duckert, F. (1971). Major surgery in a subject with factor V deficiency: Cholecystectomy in a parahaemophilic woman and review of the literature. *Thromb. Diath. Haemorrh.* **25,** 438–446.
Menache, D. (1970). Congenitally abnormal fibrinogen. *Thromb. Diath. Haemorrh., Suppl.* **39,** 307.
Miller, C. H., Graham, J. B., Goldin, L. R., and Elston, R. C. (1979). Genetics of classic von Willebrand's disease. I. Phenotypic variation within families. II. Assignment of the heterozygous genotype (diagnosis) by discriminant analysis. *Blood* **54,** 117–145.
Mull, M. M., and Hathaway, W. E. (1970). Altered platelet function in newborns. *Pediatr. Res.* **4,** 229.
Murphy, S. (1972). Hereditary thrombocytopenia. *Clin. Haematol.* **1,** 359.
Murphy, S., Oski, F. A., Naiman, J. L. (1972). Platelet size and kinetics in hereditary and acquired thrombocytopenia. *N. Engl. J. Med.* **286,** 499–504.
Nachman, R. (1977). Von Willebrand's disease and the molecular pathology of hemostasis. *N. Engl. J. Med.* **296,** 1059–1060.
Neuschatz, J., and Necheles, T. F. (1973). Hemophilia B in a phenotypically normal girl with XX(ring)XO mosaicism. *Acta Haematol.* **49,** 108–113.
Nilsson, I. M., and Kullander, S. (1967). Coagulation and fibrinolytic studies during pregnancy. *Acta Obstet. Gynecol. Scand.* **46,** 273–285.
Noller, K. N., Bowie, E. J. W., Kempers, R. D., and Owen, C. A., Jr. (1973). Von Willebrand's disease in pregnancy. *Obstet. Gynecol.* **41,** 865–872.
Nossel, H. L., Niemetz, J., Mibashan, R. S., and Schulze, W. G. (1966). The measurement of factor XI (plasma thromboplastin antecedent). Diagnosis and therapy of the congenital deficiency state. *Br. J. Haematol.* **12,** 133.
O'Brien, J. R. (1967). Platelets: Portsmouth syndrome? *Lancet* **2,** 258.
Omenn, G. S., Hall, J. G., Graham, C. B., and Karp, L. E. (1978). The use of radiographic visualization for prenatal diagnosis. *Birth Defects, Orig. Artic. Ser.* **13** (3D), 217–229.
Onel, D., Ulutin, S. B., and Ulutin, O. N. (1973). Platelet defect in a case of Ehlers-Danlos syndrome. *Acta Haematol.* **50,** 238–244.
Owen, C. A., Amundsen, M. A., Thompson, J. H., Jr., Spittell, J. A., Bowie, E. J. W., Stilwell, G. G., Hewlett, J. S., Mills, S. D., Sauer, W. G., and Gage, R. P. (1964). Congenital deficiency of factor VII (hypoconvertinemia). *Am. J. Med.* **37,** 71.
Papayannis, A. G., and Israëls, M. C. G. (1970). Glanzmann's disease and trait. *Lancet* **2,** 44.
Peake, I. R., and Bloom, A. L. (1978). Immunoradiometric assay of procoagulant factor-VIII antigen in plasma and serum and its reduction in haemophilia. Preliminary studies on adult and fetal blood. *Lancet* **1,** 473–475.
Pearson, H. A., Shulman, N. R., Oski, F. A., *et al.* (1966). Platelet survival in Wiskott-Aldrich syndrome. *J. Pediat.* **68,** 754.

Phillips, D. R. (1977). Platelet membrane glycoproteins in normal and genetically abnormal platelets. *Thromb. Haemostasis* **38,** 232.

Rabe, F., and Salomon, E. (1920). Über Faserstoffmangel im Blut bei einem Falle von Haemophilie. *Dtsch. Arch. Klin. Med.* **132,** 240–244.

Rabiner, S. F., Winick, M., and Smith, C. H. (1960). Congenital deficiency of factor VII associated with hemorrhagic disease of the newborn. *Pediatrics* **25,** 101.

Ratnoff, O. D., and Colopy, J. E. (1955). A familial hemorrhagic trait associated with a deficiency of a clot-promoting fraction of plasma. *J. Clin. Invest.* **34,** 602.

Ratnoff, O. D., and Steinberg, A. G. (1962). Further studies on the inheritance of Hageman trait. *J. Lab. Clin. Med.* **59,** 980.

Ratnoff, O. D., and Steinberg, A. G. (1968). Inheritance of fibrin stabilizing factor. *Lancet* **1,** 25.

Rausen, A. R., and Diamond, L. K. (1961). Enclosed hemorrhage and neonatal jaundice. *Am. J. Dis. Child.* **101,** 164.

Révész, T., Schuler, D., Goldschmidt, B., and Elodi, S. (1972). Christmas disease in one of a pair of monozygous twin girls, possibly the effect of lyonization. *J. Med. Genet.* **9,** 396–400.

Riis, P., and Fuchs, F. (1960). Antenatal determination of foetal sex in prevention of hereditary diseases. *Lancet* **2,** 180–182.

Riis, P., and Fuchs, F. (1966). Sex chromatin and antenatal sex diagnosis. *In* "The Sex Chromatin" (K. L. Moore, ed.), pp. 220–228. Saunders, Philadelphia, Pennsylvania.

Rizza, C. R. (1976). The clinical features of clotting factor deficiencies. *In* "Human Blood Coagulation, Haemostasis and Thrombosis" (R. Biggs, ed.), 2nd ed., pp. 231–248. Blackwell, Oxford.

Roberts, H. R., Kroncke, F. G., Jr. (1971). Tests of platelet activity: Application to clinical diagnosis. *Monogr. Pathol.* **11,** 365–386.

Roberts, H. R., Grizzle, J. E., McLester, W. D., and Penick, G. D. (1968). Genetic variants of hemophilia B: Detection by means of a specific PTC inhibitor. *J. Clin. Invest.* **47,** 360–365.

Rosen, F. S. (1971). The thymus gland and the immune deficiency syndrome. *In* "Immunological Diseases" (H. Samter, ed.), 2nd ed., pp. 472–498. Little, Brown, Boston, Massachusetts.

Rosner, F. (1969). Hemophilia in the Talmud and Rabbinic writings. *Ann. Intern. Med.* **70,** 833–837.

Rozman, C., Castillo, R., Ribas Mundo, M., and Suros, J. (1967). Christmas disease in a girl with female karyotype. *Acta Haematol.* **37,** 217–223.

Rust, L. A., Goodnight, S. H., Freeman, R. K., and Johnson, C. S. (1975). Pregnancy and delivery in a woman with hemophilia B. *Obstet. Gynecol.* **46,** 483–486.

Saito, H., Ratnoff, O. D., Waldmann, R., and Abraham, J. P. (1975). Fitzgerald trait. *J. Clin. Invest.* **55,** 1082–1089.

Schulman, I. (1962). Pediatric aspects of the mild hemophiliac. *Med. Clin. North Am.* **46,** 93.

Seip, M. (1963). Hereditary hypoplastic thrombocytopenia. *Acta Paediatr. Scand.* **52,** 370–376.

Shapiro, S. S., Martinez, J., and Hobburn, R. R. (1969). Congenital dysprothrombinemia; an inherited structural disorder of human prothrombin. *J. Clin. Invest.* **48,** 2251.

Smith, C. H. (1979). "Blood Diseases of Infancy and Childhood." Mosby, St. Louis, Missouri.

Spiro, R., and Lubs, M. L. (1976). Survey of amniocentesis for fetal sex determination in hemophilia carriers. *Clin. Genet.* **10,** 337–342.

Strauss, H. S. (1965). Clinical pathological conference. *J. Pediatr.* **66,** 443.

Strauss, H. S., and Diamond, L. K. (1963). Elevation of factor VIII (AHF) during pregnancy in normal persons and in a patient with VWD. *N. Engl. J. Med.* **269,** 1251–1252.

Strauss, H. S., and Olson, S. L. (1969). Hemophilia B in a female. *Pediatrics* **44,** 268–271.

Telfer, T. P., Denson, K. W., and Wright, D. R. (1956). A "new" coagulation defect. *Br. J. Haematol.* **2,** 308.

Tschopp, T. B., Weiss, H. J., and Baumgartner, H. R. (1974). Decreased adhesion of platelets to subendothelium in von Willebrand's disease. *J. Lab. Clin. Med.* **83,** 296–300.

Van Royen, E. A., and ten Cate, J. W. (1973). Antigen/biological activity ratio for factor VIII in late pregnancy. *Lancet* **2,** 449–450.

Veltkamp, J. J., and van Tilburg, N. H. (1973). Detection of heterozygotes for recessive von Willebrand's disease by the assay of antihemophilic-factor-like antigen. *N. Engl. J. Med.* **289,** 882–885.

Vinazzer, H. (1974). Familial Glanzmann's disease—observations during pregnancy and delivery. *In* "Platelets" (O. N. Ulutin, ed.), Int. Congr. Ser. 357; pp. 255–258. Excerpta Med. Found., Amsterdam.

von Willebrand, E. A. (1926). Hereditär pseudohamofili. *Fin. Laekaresaellsk. Handl.* **68,** 87–112.

Weiss, H. J. (1975). Platelet physiology and abnormalities of platelet function. *N. Engl. J. Med.* **293,** 531–541, 580–588.

Weiss, H. J., and Rogers, J. (1971). Fibrinogen and platelets in the primary arrest of bleeding. Studies in 2 patients with congenital afibrinogenemia. *N. Engl. J. Med.* **285,** 369.

Weiss, H. J., Lages, B. A., Witte, L. D., Kaplan, K. L., Goodman, D. S., Nossel, H. L., and Baumgartner, H. R. (1977). Storage pool disease: Evidence for clinical and biochemical heterogeneity. *Thromb. Haemostasis* **38,** 3.

White, J. G., and Gerrard, J. M. (1976). Ultrastructural features of abnormal blood platelets. *Am. J. Pathol.* **83,** 590–614.

Williams, W. J., Beutler, E., Erslev, A. J., and Rundles, R. W. (1977). "Hematology" 2nd ed. McGraw-Hill, New York.

Windhorst, D. B., and Padgett, G. (1973). The Chediak-Higashi syndrome and the homologous trait in animals. *J. Invest. Dermatol.* **60,** 529–537.

Witt, I., Müller, H., and Kunzer, W. (1969). Evidence for the existence of foetal fibrinogen. *Thromb. Diath. Haemorrh.* **22,** 101–109.

World Health Organization (1972). Inherited blood clotting disorders, *W.H.O., Tech. Rep. Ser.* **504.**

World Health Organization (1977). Methods for the detection of haemophilia carriers: A memorandum. *Bull. W.H.O.* **55,** 675–702.

Wuepper, K. D., Miller, D. R., and Lacombe, M. J. (1975). Flaujeac trait. Deficiency of human plasma kininogen. *J. Clin. Invest.* **56,** 1663–1672.

Zahavi J., Gale, R., and Sacks, Z. (1977). Storage pool disease of platelets in an infant with thrombocytopenic absent radii syndrome simulating Fanconi anemia. *Thromb. Haemostasis* **38,** 283.

Zimmerman, T. S., Ratnoff, O. D., and Littell, A. S. (1971). Detection of carriers of classic hemophilia using an immunologic assay for antihemophilic factor (factor VIII). *J. Clin. Invest.* **50,** 255–258.

5

Rh and Other Blood Group Incompatibilities

JOHN T. QUEENAN

The deleterious effect of blood group incompatibilites on pregnancies has been known for decades. In 1932, Diamond, Blackfan, and Baty reported that hydrops fetalis, icterus gravis neonatorum, and anicteric anemia of the newborn were manifestations of the same disease process (Diamond *et al.*, 1932). They noted there were no cases of fetal–maternal incompatibility of the ABO blood groups in their patients. In 1939, Levine and Stetson described an atypical agglutinin in the

Copyright © 1981 by Academic Press, Inc.
All rights of reproduction in any form reserved.
ISBN 0-12-630940-X

blood of a woman who had just given birth to a stillborn, macerated fetus and who subsequently incurred a transfusion reaction when transfused with apparently compatible blood (Levine and Stetson, 1939). They postulated that maternal immunization was a result of a fetal antigen inherited from the father and absent in the mother. In 1940, Lansteiner and Wiener discovered the Rh antigen on erythrocytes, making possible the investigation of Rh incompatibility (Lansteiner and Wiener, 1940).

Blood group incompatibilities in pregnancy can cause a spectrum of effects in the fetus, ranging from no effect to intrauterine death. Rh and ABO incompatibilities are the best-known examples but other blood group incompatibilities also cause significant clinical disease. Although genetic counseling is frequently initiated only after identification of an affected individual, the genetics of Rh immunization and other blood group incompatibilities make these disorders ideal for genetic counseling. There is a very high level of predictability of outcome in the Rh-immunized pregnancy. The ability to screen for gene products and the ability to detect abnormal antibodies also makes blood group incompatibilities ideal for genetic screening.

In 1956, Bevis recognized the relationship between increased blood pigments in the amniotic fluid (AF) and fetuses with erythroblastosis fetalis (EBF) (Bevis, 1956). He demonstrated that the spectrophotometric analysis of the AF provides a method of predicting the severity of the EBF, and established the value of serial AF determinations in treating the disease. Shortly thereafter, several medical centers throughout the world developed Rh clinics that became the forerunners of the modern high risk pregnancy clinic. Because the patients at risk could be identified by antibody screening, the appropriate patients could be referred for specialized care. By allowing the Rh-negative unimmunized patients to attend the regular clinic and the Rh-negative immunized patients to attend the special clinic, the physicians were able to concentrate the increased level of care on the appropriate high risk group. The AF spectrophotometric analysis provided the clinician with the additional information concerning the severity of the disease in the fetus.

Blood group antigens are generally inherited as simple Mendelian dominant characters, detectable in single dose. Because the 22 autosomes are paired, an individual is either heterozygous or homozygous for the alleles at these loci. The specific autosomes on which various blood group genes are located is often not known.

I. Rh INCOMPATIBILITY

In 1940 Lansteiner and Wiener discovered the Rh antigen on human erythrocytes. They noted that 85% of the human erythrocytes reacted with an antibody produced by immunizing rabbits with Rhesus monkey erythrocytes.

The nomenclature of the Rh system has been the arena of considerable dispute. The Fisher-Race (CDE) and Wiener (Hr Rh) systems are the two best known. The Fisher-Race system will be used in this chapter since it is easily understood and it describes the antigens present and their inter-relationships.

A. The Rh Locus

The actual nature of the human Rh locus is not well defined. The genes occupy a short segment on chromosome 1. This segment constitutes a single genetic region containing three loci. Each loci (C, D, and E) is a genetic determinant which is expressed as an alternative form of antigen.

1. Number of Alleles

These antigens are alleles or alternate forms of the *C, D,* and *E* loci. There are three at the *C* locus (*C, c,* and C^{w}), three at the *D* locus (*D, d,* and D^{u}), and two at the *E* locus (*E* and *e*).

In the Rh system, the offspring inherits three antigens from each parent. The offspring may inherit *CDe* from one parent and *cde* from the other. When that individual has offspring these same combinations of three antigens will be passed to the offspring. This can be explained by the fact that there is a very close linkage of the three loci for the genes determining *C* or *c, D* or *d,* and *E* or *e*.

By determining the strength of erythrocyte agglutination by a specific antibody technique, it is often possible to predict whether a particular blood group gene is present on one or both homologous chromosomes on the basis of dosage effect. However, in certain systems, e.g., the Rh system, the serological reactivity of an antigen varies depending on the presence or absence of another antigen. For example, if the genetic determinants of the Rh antigens *C, D* are on the same chromosome (i.e. in the cis position) the activity of the *D* is weakened. Trans-position effects of *C* and *D* can also occur. For instance, erythrocytes of the genotype (*CDe*/*Cde*) react very poorly with anti-D. But another individual having the *CDe* gene combined with cde reacts very strongly with the same anti-D antibody.

2. Genotyping

If an individual is Rh-positive, the specific Rh genotype can be determined with a high degree of probability. The possible genes and their alleles are *C, D, E, c, d* and *e*. A Rh-negative individual lacks the Rh_0 (*D*) gene and must have d on each homologous chromosome. There is no antiserum against the d antigen; thus, the genotype must be determined by testing for five out of six factors (*C, D, E, c, e*) for which antisera are available. Therefore, the determination of a genotype is calculated on the basis of frequency of occurence of a genotype.

In Table I, examples of Rh genotyping for Rh-positive individuals are pre-

TABLE I
Rh Genotyping of Rh-Positive Individuals

Genes present					Best guess genotype and frequency in %		
C	*D*	*E*	*c*	*e*	*Most likely*	*Next guess*	*Least likely*
+	+	o	+	+	*CDe/cde* 32.6	*CDe/cDe* 2.2	*cDe/Cde* 0.1
+	+	o	o	+	*CDe/CDe* 17.7	*CDe/Cde* 0.8	—
+	+	+	+	+	*CDe/cDE* 11.9	*CDe/cdE* 0.9	*cDE/Cde* 0.3
o	+	+	+	+	*cDE/cde* 11.0	*cDE/cDe* 0.7	*cDe/cdE* 0.1
o	+	+	+	o	*cDE/cDE* 2.0	*cDE/cdE* 0.3	—
o	+	o	+	+	*cDe/cde* 2.0	*cDe/cDe* 0.1	—
+	+	+	o	+	*CDE/CDe* 0.2	—	—
+	+	+	+	o	*CDE/cDE* 0.1	—	—
+	+	+	o	o	*CDE/CDE* <0.0	—	—

sented, with the individual gene configurations on the left and the most likely guess of this specific genotype on the right. The genotypes are presented in the order of frequency of appearance in the population. The frequencies are expressed in percentages after each genotype. The percentages have been rounded off. Genotypes occurring with frequencies $<1{:}10^{-5}$ are omitted from Table I.

For example, suppose testing indicates that a Rh-positive father has *C, D, c,* and *e* but no *E*. The most likely genotype is *CDe/cde,* i.e., heterozygous for the Rh antigen (*D*). This occurs in 32% of the population. He could have genotype *CDE/cDe,* but this only occurs with a frequency of 2.2%. Even less likely, he could have genotype *cDe/Cde,* however, this only occurs with a frequency of 0.1%.

Occasionally, the need arises to genotype an Rh-negative individual. Most Rh-negative individuals have the genotype *cde/cde*. As shown in Table II this occurs in 15.1% of the population of Caucasians of European decent.

It is possible for an Rh-positive patient to become immunized to *C, E, c,* or *e* antigens. Most often this occurs as the result of transfusion. For instance, an Rh-

TABLE II
Rh Genotyping of Rh-Negative Individuals

Genes present					Best guess genotype and frequency in %	
C	*D*	*E*	*c*	*e*	Most likely	Next likely
o	o	o	+	+	*cde/cde* 15.1	—
o	o	+	+	+	*cdE/cde* 0.9	—
+	o	o	+	+	*Cde/cde* 0.8	—
+	o	+	+	+	*Cde/cdE* 0.02	*CdE/cde* 0.004
o	o	+	+	o	*cdE/cdE* 0.01	—

positive mother, with a genotype *cDE/cDE*, (2.0% of the population) may become immunized to the *C* antigen because it is absent in her blood. The father's genotype should be done to determine if he had the *C* antigen. If he is Rh-negative and genotyping reveals that he has *C, c,* and *e* but no *D* or *E*, his genotype is *Cde/cde* which occurs with a frequency of 0.8%. In this instance, he is heterozygous for the *C* antigen and 50% of the offspring can be expected to have EBF.

Rh immunization has very predictable effects upon pregnancy. These will be considered in terms of maternal and fetal interactions.

B. Effects on Mother

The process of becoming immunized to a blood group antigen is not associated with symptoms in the mother. Common situations causing immunization to a blood group antigen are transplacental hemorrhage during pregnancy, significant exposure of the mother to fetal erythrocytes following a delivery, exposure to fetal erythrocytes due to a spontaneous or induced abortion, or, in rare instances, the result of transfusion of incompatible blood. In recent years, Rh immunization has been observed in "mainline" narcotic addicts. The method of immunization is believed to be exposure to Rh-positive blood by sharing syringes.

The appearance of antibodies can be detected by screening the maternal serum. IgM antibodies usually appear at the onset of immunization, followed by the appearance of IgG antibodies. There are no symptoms or untoward effects due to this immunization evident to the mother. Indeed, further exposure to Rh-positive blood may cause an anamnestic rise in her antibody titer, with the titer reaching a very high level. Even during the anamnestic response, the mother experiences no signs or symptoms.

Even during pregnancy the Rh-immunized patient who carries a Rh-positive fetus usually has no untoward signs or symptoms other than those of a normal unimmunized pregnancy. If the immunization is believed to be a threat to the fetus, the mother will need to undergo serial amniocenteses and be delivered of her affected fetus at 38 weeks gestation or earlier. At no time, however, will she experience signs or symptoms as a result of her affected fetus. If the mother has been incorrectly typed as Rh-positive and no follow-up antibody screening is performed, this absence of maternal awareness that a hemolytic process is occurring *in utero* may allow the EBF to go undetected until the mother notes a loss of fetal activity. This situation may also occur with an Rh-positive mother who develop anti-c or anti-e antibodies.

In the severest form of EBF when the fetus becomes markedly hydropic, the mother may develop rapid weight gain, marked edema, proteinuria, and hypertension. The maternal serum proteins become markedly decreased (Queenan, 1978). The "Maternal syndrome" or the "mirror syndrome" is ominous because it generally presages fetal death.

C. Effects on Fetus and Neonate

If the mother becomes immunized to the Rh antigen, she eventually develops IgG antibodies. If the fetus inherits the *Rh* gene from the father, the fetus will automatically be affected by EBF. EBF may range from a very mild form characterized only by a positive cord blood direct antiglobulin (Coombs) test, the development of elevated reticulocyte count, and mild hyperbilirubinemia in the neonate, to a most severe form in which the fetus dies during the second trimester. Fortunately, most Rh EBF is not so severe.

1. Mild Erythroblastosis Fetalis

Mild EBF is generally associated with a low antibody titer in the maternal serum. Prediction that the disease is mild can be confirmed if the laboratory performs accurate antibody titers. Even though the fetus does well clinically, the process of destruction of erythrocytes due to an antigen–antibody reaction continues *in utero*. The fetus responds to this hemolytic process with increased erythrocyte production. At birth, the only evidence of hemolytic anemia may be a positive cord blood direct antiglobulin (Coombs) test and an elevated reticulocyte count due to increased erythropoiesis. The cord blood hemoglobin usually is above 14 g/100 ml; it may even be normal. In the mildly affected neonate at birth, the liver and spleen may be slightly enlarged. The neonate is usually not pale and almost never jaundiced. Hydrops fetalis is not present. The neonatal course over the next 24 hours may be a mild hyperbilirubinemia that responds well to ultraviolet phototherapy. However, even mildly affected neonates may have persistent hemolysis due to the presence of anti-D antibodies, which have a half-life of 28 days. Furthermore, if the neonate is premature with less developed conjugating function, and unable to tolerate hyperbilirubinemia safely, then even the mildly affected newborn will require exchange transfusions to prevent kernicterus.

2. Moderate Erythroblastosis Fetalis

The moderately affected neonate has a cord blood hemoglobin of 10–14 g/100 ml. The cord blood direct antiglobulin (Coombs) test is positive, and a higher level of maternally derived anti-D antibodies is present. Until these antibodies are dissipated, either by natural degradation or by exchange transfusions, the antibodies will continue to react with the erythrocytes produced by the neonate. Because the hemolytic anemia may be severe, extramedullary hematopoiesis begins during fetal life and may be pronounced at birth. This may be manifested by hepatosplenomegaly and the increased number of nucleated erythrocytes and markedly elevated reticulocyte count in the peripheral blood.

3. Severe Erythroblastosis Fetalis

In the severely affected pregnancy, the fetus has pronounced extramedullary erythropoiesis as it attempts to counteract the rampant hemolytic anemia. In the extreme form, the fetus may be able to produce erythrocytes adequately and stay alive. In other instances, however, the fetus is unable to keep up with the antigen–antibody reaction occurring even as the erythrocytes are formed in the bone marrow, liver, or spleen. The fetus mounts an all out effort for survival against an overwhelming hemolytic anemia, e.g., extramedullary hematopoiesis in such unlikely organs as the kidney, lung, and heart. The cord blood hemoglobin is below 10g/100 ml.

In this form of EBF, the fetus frequently becomes edematous. If this process is mild, it is described as prehydropic edema. If it is extreme, there is marked anasarca affecting all fetal tissue. On amniography the markedly hydropic fetus creates a classic picture (Fig. 1). Although amniography has been almost completely replaced by ultrasound scanning, this picture demonstrates a fetus in breech presentation with massive scalp edema (arrows), marked soft tissue edema, and no evidence of swallowing as noted by the absence of opacified amniotic fluid in the gastrointestinal tract. Figure 2 is an example of ultrasound scanning which has replaced amniography. This demonstrates a transverse scan of the fetal abdomen with the arrows pointing to early ascites. The placenta, superior to the fetus, is normal in thickness and does not yet show evidence of hydropic change.

If the fetus is born alive, the hematological problems are marked anemia in the presence of high output, congestive heart failure, decreased platelets, and WBC. If the neonate is hydropic, the serum albumin is markedly reduced. Hypoalbuminemia creates a special problem for the neonatologist. Serum bilirubin determinations only indicate the levels in the intravascular compartment. Bilirubin diffuses throughout the interstitial and intracellular fluid. Albumin binds bilirubin very effectively; thus, hypoalbuminemia makes the exchange transfusion less effective in removing bilirubin. Salt-poor albumin may be administered prior to an exchange transfusion to increase the efficiency of bilirubin removal. If the fetus is hydropic, the neonatalogist must first administer packed red blood cells because administration of albumin could lead to an increased intravascular volume and severe cardiac failure.

It has been reported that hydropic neonates are not hypervolemic at birth and may have no evidence of cardiomegaly. However, if born alive and kept alive, affected neonates rapidly develop cardiomegaly and congestive heart failure. *In utero* as the severity of hemolysis increases in the fetus, erythropoietin levels rise. The extramedullary erythropoiesis becomes markedly pronounced. Islets of extramedullary erythropoiesis increase in the liver, and portal venous obstruction develops with increased portal and umbilical venous pressure. Ascites develops with increased hepatic erythropoiesis. The intrahepatic blood flow becomes obstructed. Distortion of hepatic cords and interference with the blood supply causes damage to the liver cells. The serum albumin decreases, the fetus becomes hypoproteinemic and develops generalized anasarca producing the classic picture of hydrops fetalis.

Regardless of the severity of the immunization, the Rh-positive fetus gener-

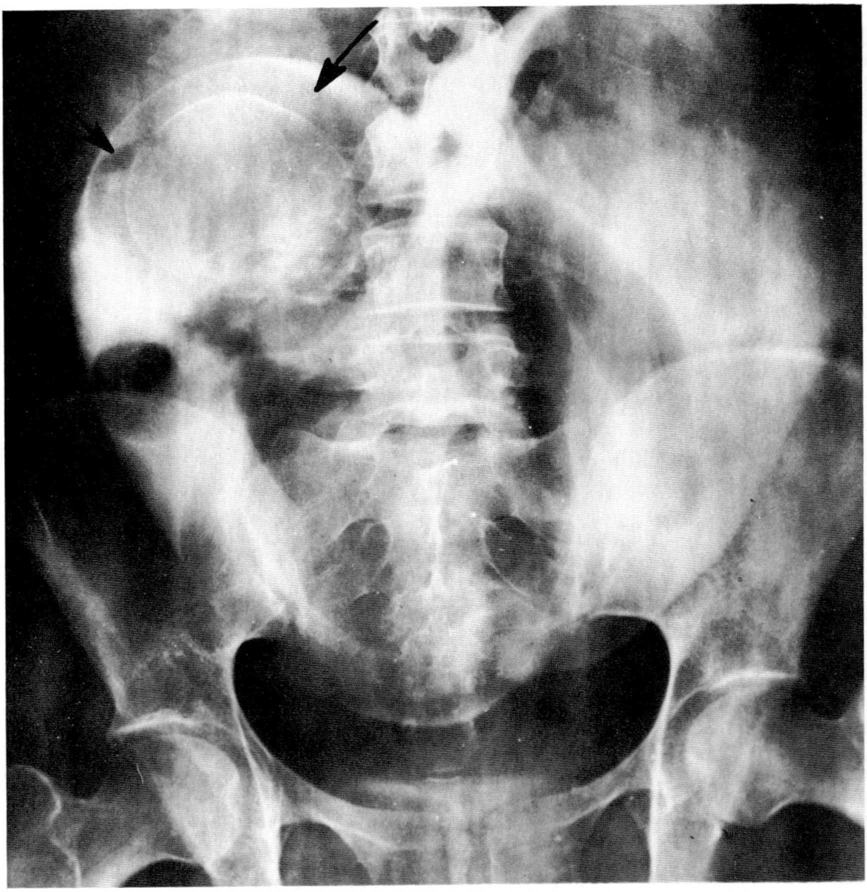

Fig. 1. Amniogram of a fetus in the breech presentation, demonstrating massive scalp edema (arrows). No opacification is noted in the fetal gastrointestinal tract. (From Queenan, 1968.)

ally does not die prior to the seventeenth or eighteenth week of gestation. This is fascinating because destruction of erythrocytes begins much earlier. Perhaps the fetus is able to derive nutrition and respiratory exchange by other means during the development of the circulatory system.

D. Antepartum Detection and Clinical Management

Prenatal office or clinic visits usually include a variety of maternal serum screening. In all instances a blood group and Rh type must be determined. Rh-negative patients should be screened for Rh antibodies. If the patient has a

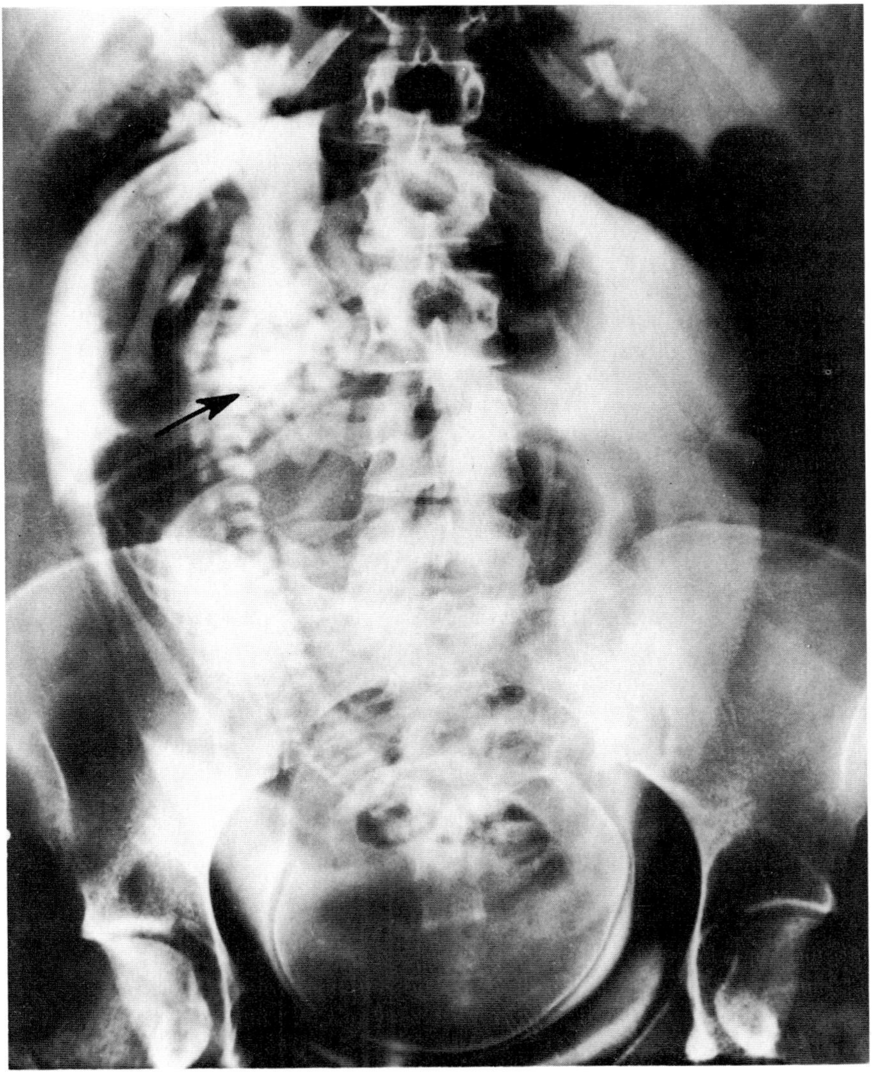

Fig. 2. Transverse ultrasound scan of fetal abdomen with arrow pointing to early fetal ascites.

history of cesarean sections, blood transfusions, or induced abortions, she should be screened for irregular antibodies, even if Rh-positive.

Table III shows the frequencies with which antibodies of the Rh system are detected in the obstetrical patient (Queenan *et al.*, 1969). D is the strongest antigen therefore, it is not surprising that anti-D is the most common antibody, occurring with a relative frequency of 100. The next most common are anti-C and anti-E, each of which occurs with a relative frequency of 10. Next in frequency is anti-c, 4, and finally anit-e with a frequency of 1.

These are all immune antibodies capable of producing EBF if the fetus has the corresponding antigen. They also can cause transfusion reactions.

More than one antibody can occur simultaneously. If anti-D is detected, an anti-C or anti-E may also be present, particularly if a patient has had a transfusion (Queenan *et al.*, 1969). Anti-D may be detected only in Rh-negative individuals, but anti-C and anti-E may be detected in Rh-negative or positive individuals. These will be discussed in detail later in this chapter.

If the patient is Rh-negative, she will require repeat screening for Rh antibodies at 24 weeks gestation and every 4 weeks thereafter. If the husband is Rh-negative it is best not to assume paternity but to screen *all* Rh-negative patients on a routine basis. This may avoid a tragedy due to breakdown in office routine or if a question about the paternity develops later.

Antibody screening is performed on serum by an indirect antiglobulin (Coombs) technqiue. Some laboratories screen for IgM antibodies, which usually appear with an initial immunization and subsequently disappear. The reactivity of the serum in an albumin medium may also be tested. However, the test that most closely correlates with the severity of disease is the indirect antiglobulin (Coombs) test, a technique performed by testing a serum for the presence of antibodies. In the first stage of the test, antigen-positive erythrocytes become coated during the period of incubation with the individual's serum, if the serum contains the corresponding antibodies. Red blood cell coating is detected by washing the erythrocytes and adding antiglobulin (Coombs) reagent and testing for agglutination. Results are reported as a titer. The titer strength starts with undiluted, 1:1, 1:2, and so on, up to a titer as high as 1:1024 or greater. In addition, a scoring test based upon the degree of agglutination is available. The reliability of titers in laboratories with a small case load may be questioned; however, in an experienced laboratory, valuable information may be obtained by the indirect antiglobulin (Coombs) method.

One dilemma develops in trying to determine whether the fetus is Rh-positive or Rh-negative. If an anti-D antibody is detected on routine prenatal screening and the father is Rh-positive, it becomes important to determine whether the father is heterozygous or homozygous. A heterozygous father has an equal chance of siring Rh-negative and Rh-positive offspring. If he is homozygous, all babies will be Rh-positive.

TABLE III
Rh-Hr System[a]

Antibody	Relative incidence of detection	Natural	Immune	Mode of action			Transfusion reaction	Percent of blood compatible
				Saline	Albumin	AHG		
Anti-D	100	Unlikely	Yes	Occasional	Yes	Yes	Probable	15
Anti-C	10	Unlikely	Yes	Occasional	Yes	Yes	Probable	30
Anti-E	10	Unlikely	Yes	Freq	Yes	Yes	Probable	70
Anti-c	4	Unlikely	Yes	Occasional	Yes	Yes	Probable	20
Anti-e	1	Unlikely	Yes	Occasional	Yes	Yes	Probable	3

[a]In order of frequency of antibody detection.

The *first immunized pregnancy* is one in which the antibody screen is negative at the initial visit but becomes positive later in pregnancy. If the screening is accurate, the antibody develops in response to the fetal erythrocyte antigen in the current pregnancy. A chance for error occurs if the initial antibody screening is erroneously reported to be negative, but the patient is already immunized from a prior pregnancy. Fortunately, this sort of error is rare. Generally, if the initial antibody screen is negative and the patient subsequently develops a detectable antibody, the fetus is Rh-positive.

When a maternal anti-D antibody is present during a pregnancy in which the fetus is Rh-positive, the neonate will be affected with EBF. Because the Rh antigen will be present on all the fetal erythrocytes and the IgG anti-D antibodies cross the placenta to cause an antigen–antibody reaction, the predictability of this occurrence is 100%. If the fetus is Rh-negative, no disease will occur because the IgG anti-D antibodies that cross the placenta will have no antigen on the fetal erythrocytes with which to react. In this case the stimulus for Anti-D antibodies may have been a prior pregnancy or transfusion.

In any *subsequent immunized pregnancies* the Rh status of the fetus is not known. The level of antibody may rise, fall, or remain the same without any related significance (Queenan, 1977). In subsequent immunized pregnancies the clinician thus cannot rely on the antibody titer to predict fetal status; however, the strength of the antibody as determined by the antibody titer should have a general correlation with clinical disease.

According to data based upon 444 patients on whom indirect antiglobulin (Coombs) tests were performed prior to AF analysis, Allen *et al.* (1954) concluded that a patient with no prior history of a stillborn or a hydropic baby and with an antibody titer of 1:32 or less has a 96% chance of having her baby alive *in utero* at 37 weeks gestation. If her titer is 1:64 or greater, the chance is 83%. If the mother has a history of a stillborn or hydropic baby, the chance is only 33%. If performed accurately, the indirect antiglobulin (Coombs) test can give the clinician very valuable information.

If a patient begins a pregnancy with a titer of 1:64 and a similar titer is found throughout the remainder of pregnancy, then the genotype of the Rh-positive father is very important in trying to determine if the fetus is Rh-positive or Rh-negative. If genotyping the father reveals that he is homozygous Rh-positive the fetus is almost definitely Rh-positive. If the father is heterozygous for the D antigen, there is a 50% chance of the fetus being Rh-positive and a 50% chance of it being Rh-negative.

1. Amniotic Fluid Analysis

The blood Rh antibody titer indicates the degree of maternal immunization. In the *first immunized pregnancy,* it may be predictive of the fetal condition and the pregnancy outcome. However, once a high Rh antibody titer has been reached or during a *subsequent immunized* pregnancy, the fetal condition and pregnancy outcome can be completely independent of the antibody level in the maternal blood. For instance, if the maternal serum antibody level is high, i.e., 1:512 and the paternal genotype is heterozygous, the fetus could be Rh-negative. In the severely immunized patient or in the subsequent immunized pregnancy serum antibody titers are not useful. AF analysis is the only means of determining the fetal condition and predicting the pregnancy outcome.

2. *The Amniocentesis*

Prior to the initial amniocentesis an ultrasound scan should be performed. By visualizing the fetus, placenta, and AF, the operator can avoid trauma to the placenta and fetus and locate pockets of AF. The risks of transplacental hemorrhage, i.e., further fetal anemia and increased maternal immunization, must be avoided. The possibility of sampling AF from only one sac in unsuspected twins is another reason for doing an ultrasound scan. Figure 3 is a transverse scan of an anteriorly implanted placenta. The arrows point to the chorionic plate. Note the sonolucent area representing a pocket of AF on the patients left.

The ultrasound scan should be done prior to each subsequent amniocentesis to locate the exact position of the placenta, the fetus, and pockets of AF.

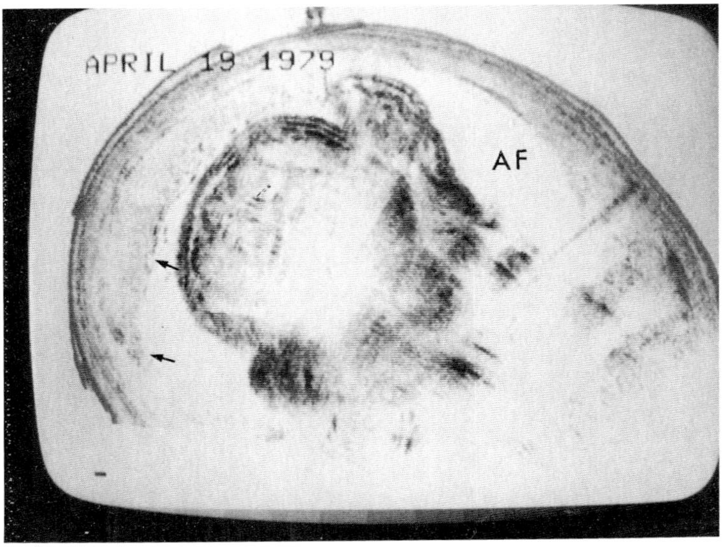

Fig. 3. Transverse ultrasound scan of a right anterior to right lateral placenta. Small arrows point to chorionic plate. The sonolucent area represents a pocket of amniotic fluid (AF).

Ideally the amniocentesis is performed in the area of the fetal small parts as shown in Fig. 4. However, if the placenta is located at that site, the area behind the fetal neck may be used (see Fig. 5). The head and neck must be protected with the operator's hand during the insertion of the needle to avoid trauma to the cord or fetus. A third method is to displace the fetal head upwards in the uterus. One hand holds the fetal head up, the other hand inserts the needle suprapubically into the amniotic sac. The objections to this technique are that it is manipulative; there may be a slightly higher incidence of premature rupture of membranes, and the needle must be passed through an area that is difficult to render sterile by a prep.

3. Amniotic Fluid Bilirubin

Once AF is obtained, it should be centrifuged immediately to remove any particulate matter. Oxyhemoglobin contamination will cause a major peak at 412, and minor peaks at 540 and 575 millimicrons (see Fig. 6). The AF must be protected from sunlight which causes a false low spectrophotometric bilirubin determination.

The AF is scanned on a spectrophotometer to determine the bilirubin content. All patients have small amounts of bilirubin in early pregnancy. The bilirubin concentration rises until about 22–26 weeks gestation, then it decreases to term. This decrease with increasing gestation, is such a constant factor that Mandelbaum reported that it could be used to determine fetal maturity (Mandelbaum and

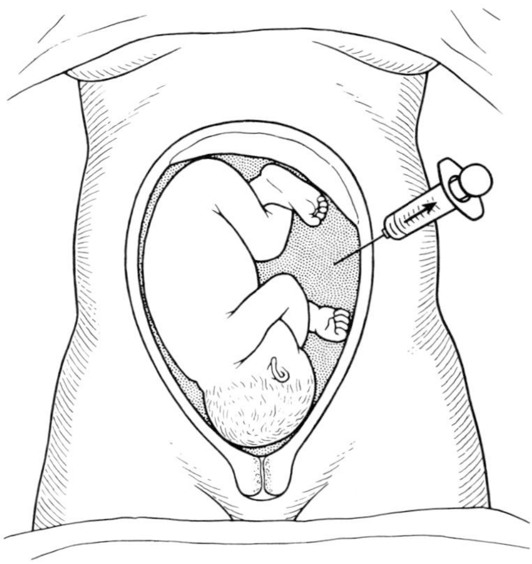

Fig. 4. Diagram demonstrating the technique of amniocentesis in the area of the fetal small parts. (From Queenan, 1966.)

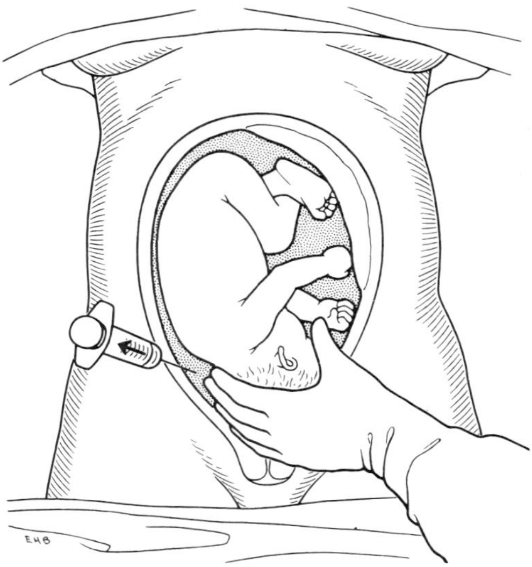

Fig. 5. Diagram demonstrating the technique of amniocentesis in the area posterior to the fetal neck. Operator's hand is shown shielding the fetus so that the needle cannot injure it. (From Queenan, 1966.)

Robinson, 1966). He showed that the AF bilirubin content generally decreases or disappears as the fetus reaches term.

The spectrophotometric scan should be done on a continuous recording spectrophotometer. The AF is generally scanned from 350 to 750 nm. If no bilirubin is present, the scan will nearly parallel the zero baseline with a slightly increased absorbance in the lower wavelengths as seen in Fig. 7. When the AF contains bilirubin, the absorbance starts to increase at 375 millimicrons reaching a peak at 450 millimicrons and returns to normal at 525 nm as shown in Fig. 8.

Figure 9 shows serial AF determinations on a patient at 27 weeks, 32 weeks, and 35 weeks gestation. Note that the AF bilirubin level decreases with increasing gestation.

4. Methods of Interpretation

Liley showed that AF bilirubin values placed on a three zone graph would predict the outcome of pregnancy (Liley, 1963) (see Fig. 10). Generally, his method requires two AF values being placed on the three zone graph in order to predict the outcome of the pregnancy. In many instances such a prediction is adequate for clinical purposes. But occasionally, monitoring the condition of the fetus is necessary.

Numerous other investigators proposed methods of AF analysis in the Rh-immunized pregnancy. At Georgetown University School of Medicine, we use

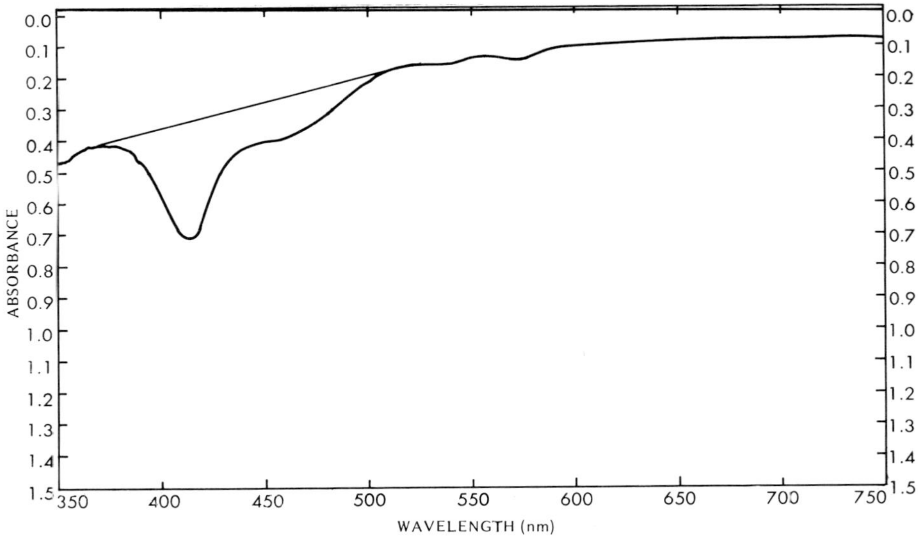

Fig. 6. Spectrophotometric scan of normal amniotic fluid at 37 weeks gestation. Fluid was contaminated with oxyhemoglobin. Note characteristic peak at 412 millimicrons associated with smaller peaks at 540 and 575 millimicrons. (From Queenan, 1975.)

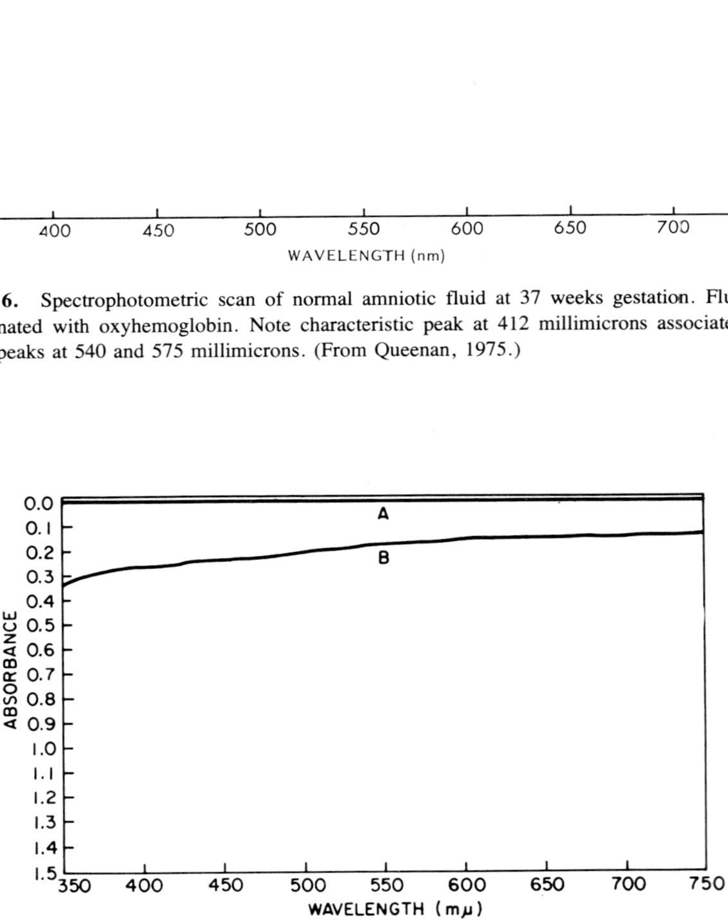

Fig. 7. Spectrophotometric scan of (A) distilled water and (B) normal amniotic fluid. Note that normal amniotic fluid approximates a straight line, roughly paralleling the distilled water scan, with a slightly greater absorbance in the shorter wavelengths. (From Queenan, 1971.)

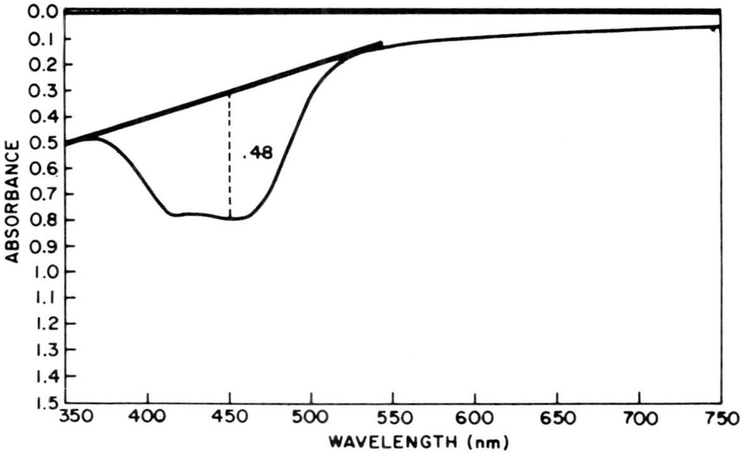

Fig. 8. Spectrophotometric scan of amniotic fluid from a severely immunized Rh-negative patient. The amniotic fluid scan has a large peak at 450 millimicrons. An arbitrary (heavy) line is drawn where the scan would have traced if there were no bilirubin. The difference is measured by the upright (broken) line at 450 millimicrons which in this case is 0.48 absorbance units. (From Queenan, 1971.)

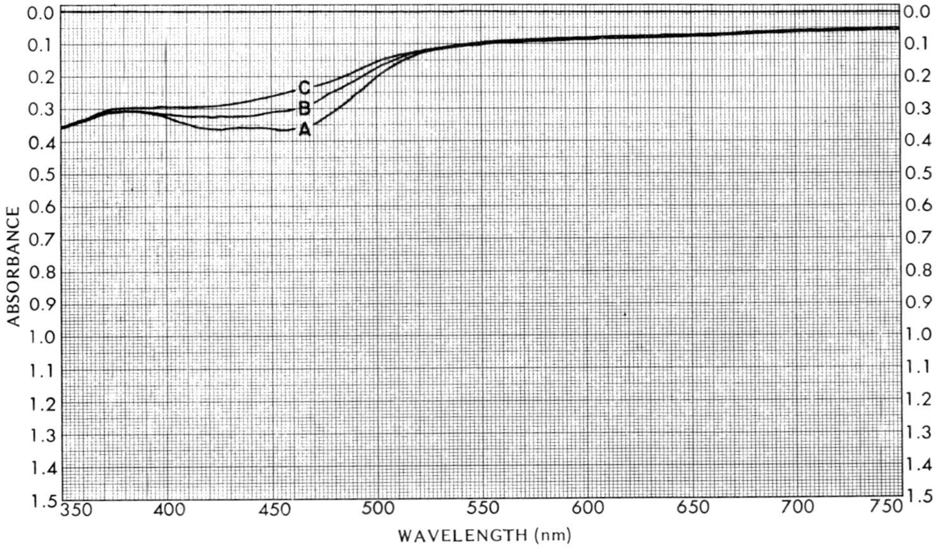

Fig. 9. Three spectrophotometric scan of amniotic fluid at 27 (A), 32 (B), and 35 (C) weeks of gestation. Note the decrease in deviation from normal with increasing maturity. (From Queenan, 1977.)

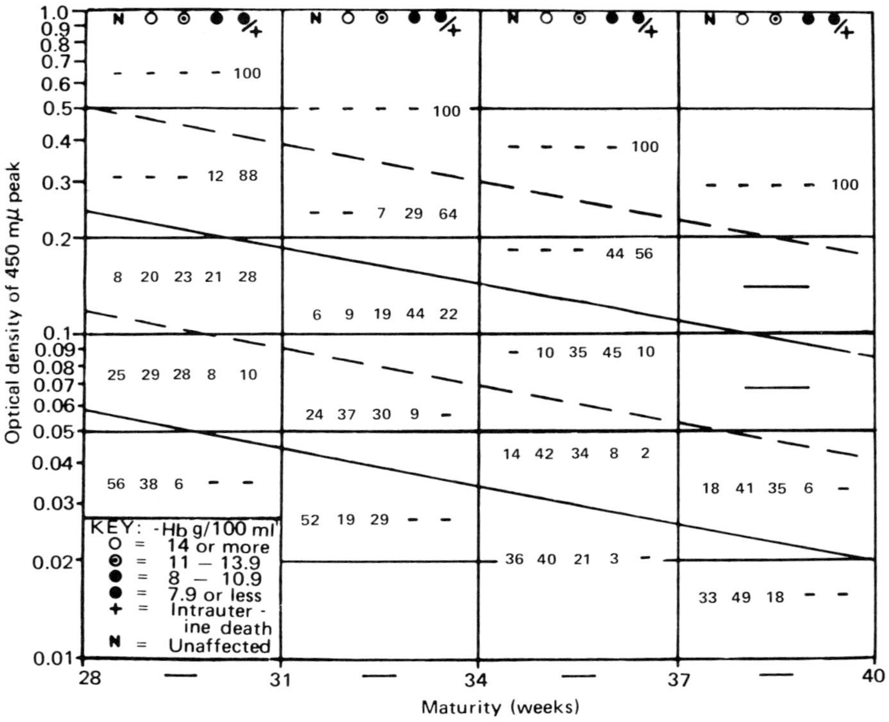

Fig. 10. Liley graph divided into three prognostic zones. The lower zone included Rh-negative and mildly affected fetuses. The upper zone included severely affected fetuses. The middle zone included two types of fetuses; those that were mildly affected or those that had severe disease. The graph indicates the percentage probability of the various grades of affliction for the peak size in a single specimen. (From Liley, 1963.)

the AF bilirubin trend to determine the severity of the disease in the fetus (Queenan, 1977). If the trend of bilirubin is decreasing on serial determinations, the fetus may be Rh-negative, unaffected, or have mild or moderate EBF. If the bilirubin trend is horizontal or rising, the fetus is Rh-positive and may be severely involved or may die *in utero*. The AF trend is an excellent predictor of the outcome of pregnancy. More importantly, the AF bilirubin trend monitors the condition of the fetus. It can be used to determine the week to week management of the Rh-immunized pregnancy.

5. *Clinical Application*

The AF bilirubin trend method is based on the principle that the bilirubin trend is extremely reliable in predicting the outcome of pregnancy in the Rh-immunized patient. The basic approach may be summarized:

1. The initial amniocentesis is performed at 28–29 weeks gestation unless the obstetric history or antibody titer indicates earlier AF evaluation.
2. Amniocentesis is repeated 1- 3-week intervals based on the level of AF bilirubin.
3. Serial AF bilirubin values are studied to determine the trend.
   A. If the trend is falling, the fetus is safe until a scheduled delivery at 38 weeks gestation (Fig. 11A).
   B. If the trend is rising an IUT is indicated between 25 and 32 weeks (Fig. 11C).
   C. If the trend is rising or horizontal, after 32 weeks gestation, delivery is indicated (Fig. 11B).
4. AF maturity studies and ultrasound BPD are used in conjunction with AF bilirubin values to determine the optimal time to deliver (Queenan, 1977).

Amniotic fluid analyses, when properly used, are essentially 100% effective in detecting the fetus that will die *in utero*. The main source of error is the patient

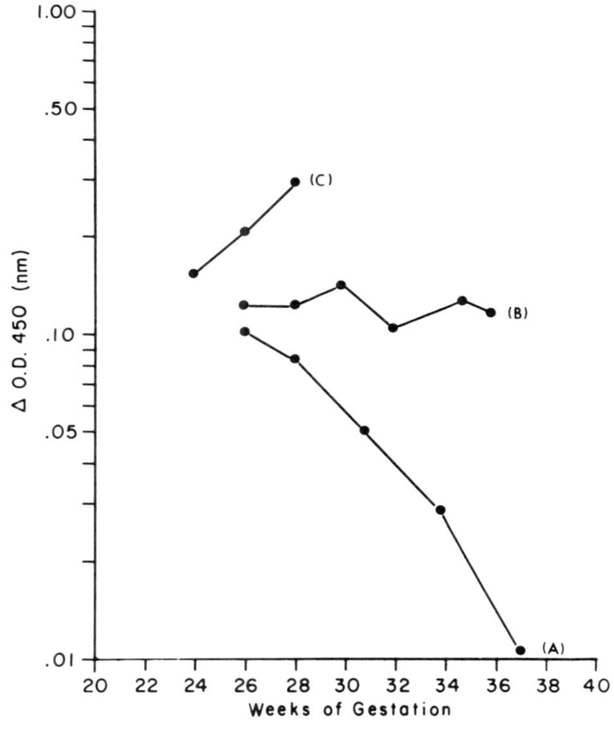

Fig. 11. Examples of the three amniotic fluid bilirubin trends. If the trend is decreasing (A), the fetus may be Rh-negative, unaffected, or have mild or moderate EBF. If the trend is horizontal (B) or rising (C), the fetus is Rh-positive and may be severely affected or may die *in utero*.

with polyhydramnios. This condition may be diagnosed clinically and can be confirmed by ultrasound scanning. It can cause a seriously low bilirubin determination, giving a false sense of security while the fetus deteriorates.

Since the various methods of management for the Rh patient are beyond the scope of this chapter, the reader is referred to "Modern Management of the Rh Problem" (Queenan, 1977).

E. Rh-Immune Prophylaxis

Considerable information exists concerning the development of Rh immunization. An individual may become immunized to a blood group antigen in numerous ways. The mechanisms of becoming immunized are as follows.

1. Incompatible Blood Transfusion

The incidence of an Rh-negative individual becoming immunized after receiving 500 ml of Rh-positive blood is approximately 95%. Although not all individuals will demonstrate antibodies following a transfusion, a subsequent challenge by the Rh antigen will evoke an antibody response in almost 100%.

2. Intramuscular Injection of Rh-Positive Blood

Several decades ago it was accepted practice to inject children exposed to infectious diseases with blood from individuals who had known cases of viral disease in order to offer them a dose of antibodies. Unfortunately, it was not until many years later that it was learned that this passive immunization had caused some of these individuals to become immunized to the Rh or other blood group factors. Obviously, this practice is contraindicated and has not been done over the last two decades.

3. Transplacental Hemorrhage during Pregnancy

Throughout pregnancy fetal erythrocytes may be detected in the maternal circulation. The Kleihauer-Betke stain technique provides an excellent means of quantitating the magnitude of the transplacental hemorrhage (TPH). The frequency of TPHs increases throughout pregnancy, such that fetal erythrocytes are present in a detectable amount in approximately 70% of patients at term (Clayton *et al.*, 1966). The magnitude of the TPH also increases throughout the pregnancy (Woodrow and Donohoe, 1968). Generally, few or no fetal cells are detectable in the first trimester, whereas TPH of 0.04 ml are relatively common at term (Wimhöfer, 1962).

4. Transplacental Hemorrhage at Delivery

Wimhöfer *et al.* (1962) showed that the frequency and magnitude of TPHs are increased postpartum. Queenan *et al.* (1962) showed that blood in the uterine cavity immediately postpartum enters the maternal circulation. Thus, procedures like traumatic manual removal of the placenta, which causes escape of fetal blood, will allow this blood to enter the maternal circulation. Woodrow and Donohoe (1968) showed that the size of the TPH correlates with the likelihood of immunization, but, immunizations were found even in patients in whom no fetal red cells could be detected. The major time of risk of Rh immunization appears to be the late third trimester and at delivery.

5. *Protection from Immunization*

In 1909 Theobold Smith showed that an antigen which would cause an antibody response in a laboratory animal would not do so if injected in that same animal with an excess of corresponding antibodies. This principle became the rationale for modern Rh-immune prophylaxis.

Not all Rh-negative individuals are at the same risk for Rh immunization. Levine (1943) showed that Rh-negative patients were less likely to become immunized if their fetus was ABO incompatible with the mother. Additionally, Stern *et al.* (1961) had shown that Rh-positive red blood cells that were incompatible in the ABO blood group would not immunize an Rh-positive individual.

6. *Rh Immunoglobulin*

Simultaneously and independently, Clark, Woodrow, and Finn (see Finn *et al.*, 1961) in Liverpool and Gorman, Freda, and Pollack (1964) in New York City embarked on paths of research which culminated in the Rh-prophylaxis system used today. Freda *et al.* (1964) developed an Rh immunoglobulin which could be given intramuscularly. They recruited nine Rh-negative group 0 male volunteers at Sing Sing prison. Once a month, on five successive occasions, they administered 2 ml of Rh-positive whole blood with 5 ml of a potent Rh immunoglobulin before each injection of blood. The rationale for this experiment was based on Theobald's principle that an antibody administered in excess will protect the individual against subsequent injections of the corresponding antigen. Five of the volunteers were used as controls, receiving 2 ml of Rh-positive whole blood but receiving no Rh immunoglobulin. None of the four treated men produced antibodies whereas four out of five of the controls developed antibodies.

In a second experiment by these investigators, 27 volunteers were given an intravenous injection of 10 ml of Rh-positive blood and 3 days later 4 ml of Rh immunoglobulin was given to 14 of them. The 13 others were controls receiving no Rh immunoglobulin. In this series, none of the protected individuals developed Rh antibodies, whereas 6 of the 13 controls did.

7. *Clinical Use*

Subsequent to these prisoner experiments, the Rh immunoglobulin was tested in Rh-negative mothers who delivered Rh-positive infants, who were compatible to the mother in the ABO blood group. The Rh immunoglobulin was demonstrated to be highly effective in preventing Rh immunization when administered within 72 hours postpartum. Today, the administration of Rh-immune prophylaxis has become a routine procedure following the exposure of an Rh-negative patient to any Rh antigenic stimulation. The indications for administration of Rh-immune globulin to an Rh-negative individual are (1) spontaneous abortion, (2) induced abortion, (3) amniocentesis, (4) ectopic pregnancy, (5) premature delivery, (6) term delivery, and (7) transfusion with incompatible blood.

The risk of Rh-immunization has been studied. Approximate levels of risk are as follows: (1) spontaneous abortion, 3.5%; (2) induced abortion, 5.5%; (3) first term pregnancy (up to the time of delivery, 1.2%; (4) subsequent pregnancies, 10%.

Since the Rh-immune prophylaxis is passive immunization by the administration of antibody, the procedure must be carried out every time the patient is exposed to Rh-positive antigens.

The question of antepartum administration has been raised as a means of preventing the 1–2% of patients who might become immunized during their first pregnancy. Although superficially this appears attractive, certain considerations make this economically unfeasible.

1. 40% of Rh-negative mothers will deliver Rh-negative infants.
2. Only 2% of patients are immunized by the first delivery.
3. Patients tend to have smaller families so that even if the patient becomes immunized, with good obstetric care she should be able to have her family.
4. The administration of Rh immunoglobulin should be effective against the Rh-positve fetal erythrocytes circulating in the maternal bloodstream. If the fetus is Rh-positive, the Rh immunoglobulin will be rapidly removed from the maternal blood stream as it combines with the red blood cells of the fetus. Therefore, even if Rh-immune prophylaxis were administered, for instance at 28 weeks gestation, it would fail to protect against a TPH at 30, 32, or 34 weeks gestation because the prophylaxis would be removed from the circulation.

The 72-hour interval for intramuscular administration of Rh immunoglobulin postpartum is an arbitrary time. It was derived from the protocol used for studies at Sing Sing. A patient who fails to receive Rh-immune prophylaxis within the 72-hour period, may still have the product administered at a later time.

Occasionally, the patient who is a candidate for Rh-immune prophylaxis will have what appears to be a minimal immunization. The level of antibodies may be so low that it is barely detectable by the laboratory. If there is any question, it is better to administer the Rh immunoglobulin because it can do no harm and it may be helpful.

The so-called crossmatch performed prior to the administration of Rh immunoglobulin is actually a fetal red blood cell screen. If the crossmatch is positive, it indicates a significant population of Rh-positve fetal erythrocytes in the maternal circulation. In this instance, more than one ampule of Rh immunoglobulin should be administered. The best way to determine when an adequate amount of Rh-immune prophylaxis has been administered is to do an indirect Coombs test. If it is positive, it indicates that an antibody excess exists and that the patient has been adequately protected.

II. ABO INCOMPATIBILITY

A. ABO Loci

The ABO system comprises three major alleles. The phenotypes are recognized by reactions with anti-A and anti-B antibodies. The usual phenotypes are A, B, AB, and O. The possible genotypes are *A/A*, *A/O*, *B/B*, *B/O*, *A/B*, and *O/O*. The phenotype and genotype *O/O* are detectable by the absence of reaction to anti-A and anti-B antibody. It is not possible to differentiate between *A/A* and *A/O* or between *B/B* and *B/O* without doing studies on other family members.

The genes for determining the ABO blood groups occupy the corresponding loci on a pair of autosomes. For instance, one of the parents might transmit the gene responsible for the A antigen and the other gene responsible for group O. The offspring would be heterozygous *A/O* but would type as group A because A is dominant over O.

When an individual has an A or B antigen she invariably lacks the corresponding antibody. When she lacks the antigen, she has the antibody. For example, an individual with A/A will have anti-B but no anti-A. An O/O individual will have anti-A and anti-B.

B. Effects on Mother

The A and B antigens are not confined to human erythrocytes. They are found in a wide variety of gram-negative bacteria which are normally present in the gastrointestinal tract. The A and B antigens are found on all somatic cells of the body with the exception of the nervous system. 80% of individuals will secrete the antigens in soluble form in tears, saliva, urine, and sweat. This represents an enormous antigenic mass which will react with maternal isoagglutinins which can enter the fetal circulation.

Although most individuals do not have anti-A or anti-B antibodies at birth, they develop these due to the antigenic stimulus of the A and B antigens in the gastrointestinal tract bacteria. The isoagglutinins generally appear within the first 6 months of life in individuals who lack the specific antigen. There are no signs or symptoms that accompany this isoimmunization. The antibodies are present for the lifetime of the individual. No prior exposure to blood is necessary for an immunization to A or B antigens. Therefore, unlike Rh-EBF, approximately 40% of cases of ABO-EBF occur in the first pregnancy.

C. Effects on Fetus and Neonate

ABO incompatibility occurs more than twice as often as incompatibility due to Rh. The problem can occur without prior exposure to blood and is commonly seen in the first pregnancy. The major difference is that ABO incompatibility usually causes mild neonatal disease with mild jaundice during the first 24 hours. Intrauterine death is extremely rare, if it occurs at all.

ABO-EBF is often so mild that it goes undetected. The newborn affected infant rarely has pronounced anemia, although it may be possible to detect increased erythropoiesis by an elevated reticulocyte count, the presence of nucleated erythrocytes and micropherocytosis.

The cord blood bilirubin is never significantly elevated. However, it can rise sharply over the first 24 hours of life. The maximum levels of bilirubin are frequently observed on the second or third days of life and thereafter return to normal. Occasionally, the disease is more pronounced and marked hyperbilirubinemia occurs. But, usually less than 10% of neonates need phototherapy or exchange transfusions. If the hemolysis is persistent and severe, exchange transfusions are essential to avoid risk of kernicterus.

The cord blood direct antiglobulin (Coombs) test is generally positive. However, the reaction may be weak.

The anti-A and anti-B antibodies produced by group O individuals are predominantly IgG whereas the anti-A of group B individuals and the anti-B of group A individuals are predominantly IgM. This explains why most ABO-EBF occurs in group A or B infants born to group O mothers.

Since intrauterine death almost never occurs, and since ABO-EBF is generally a mild disease, amniocentesis is not indicated. The experience of using anti-A or anti-B antibody titers as an index of severity in the prenatal period has been unrewarding. Since the fetus does not die *in utero,* the clinical problem may be managed by careful evaluation and therapy starting at birth. Because only a few neonates develop severe disease, the outlook for the neonate is usually excellent. In the setting of rampant and persistent hemolysis, aggressive management is essential to prevent the disaster of brain damage.

III. OTHER BLOOD GROUP INCOMPATIBILITIES

There are over 100 blood group antigens. More than 30 of these are capable of causing EBF. Only those with clinical significance will be considered in this chapter. Table IV lists the frequency of detection of irregular antibodies.

TABLE IV
Irregular Antibodies[a]

Antibody	Relative incidence of detection	Natural	Immune	Mode of action: Saline	Albumin	AHG	Transfusion reaction	Percent of blood compatible
Nonspecific	28	Probable	Probable	—	—	—	Possible	—
Anti-Kell	10	Unlikely	Yes	Occasional	Occasional	Yes	Probable	90
Anti-Le^b	10	Yes	Unlikely	Yes	Yes	Yes	Probable	20 (40 B)
Anti-Le^a	7	Yes	Unlikely	Yes	Yes	Yes	Probable	78 (82 B)
Anti-I	5	Yes	Unlikely	Yes	Yes	Unlikely	Unlikely	<0.1
Anti-P_1	5	Yes	Unlikely	Occasional	Occasional	Yes	Probable	<0.1
Anti-M	5	Yes	Unlikely	Yes	Yes	Occ	Unlikely	22
Anti-Fya	4	Unlikely	Yes	Unlikely	Unlikely	Yes	Probable	30 (89 B)
Anti-Jk^a	2	Unlikely	Yes	Unlikely	Unlikely	Yes	Probable	25
Anti-S	2	Unlikely	Yes	Occasional	Occasional	Yes	Probable	45
Anti-V	2	Unlikely	Yes	Unlikely	Unlikely	Yes	Probable	100 (82 B)

[a]In order of frequency of antibody detection.

A. The Kell System

The Kell antigen, (K) is a strong antigen easily capable of immunizing a K-negative individual. Nine percent of the white population are K-positive, having the genotype *KK* or *Kk*. Kell immunization can cause EBF. Only 0.2% have the genotype *KK*. If an individual is Kell-negative, the genotype is *kk* or Cellano-positive.

Naturally occuring anti-K is extremely rare. The best test for anti-K is the indirect antiglobulin (Coombs) test. In almost all instances, anti-K is formed as an immune antibody. When detected, the patient commonly has a history of having received blood transfusions.

Two other pairs of antithetrical antigens Kp^a and Kp^b and Js^a and Js^b are inherited together with K and k. This is similar to the Rh system wherein certain combinations of these antigens are common and others are very rare.

B. The Lewis System

Two Lewis antigens, Le^a and Le^b are present on secreted glycoprotein molecules which are only secondarily absorbed onto the erythrocyte. This unique property makes it possible to change the phenotype of Lewis (a+, b−) blood by injecting the erythrocytes into the circulation of a Lewis (a−, b+) individual.

The Lewis antibody is encountered relatively commonly in the obstetrical population. The important clinical feature of the Lewis system is that the antibodies are usually IgM and, therefore, do not cross the placenta. Since they rarely cross the placenta, the likelihood of Lewis-EBF is extremely small. This is further supported by the fact that during fetal life, the Lewis gene is not very active. At birth, the erythrocyte reactions with anti-Le^a are very weak and reactions with anti-Le^b cannot be detected for several weeks. The Lewis antigen becomes progressively absorbed onto the newborn's erythrocytes during the neonatal period.

If a patient did develop an IgG anti-Lewis antibody which would cross the placenta, it is unlikely that the reaction with the Lewis-positive fetal erythrocytes would be strong enough to cause clinical disease. Therefore, the Lewis system is not important from the standpoint of EBF. It does pose a problem to the immunized mother, however, because it may be difficult to find compatible blood should she need a transfusion.

C. The I System

The erythrocytes of almost all healthy adults carry the antigen I. Only rarely is an individual I-negative. Anti-I was first described as a potent cold autoagglutinin in the serum of a patient with acquired hemolytic anemia. Since the antibodies are generally IgM, they do not cause clinical EBF. But potent anti-Is are common in pregnant women and cause delays in crossmatching blood in the mother.

D. The P System

Approximately 74% of individuals have the Plantigen or are P-positive, and 26% are P-negative. Individuals who are P-negative frequently have anti-P_1 in their serum as a cold agglutinin. There is considerable variation in the strength of the P antigen. Anti-P and anti-P_1 is capable of causing a transfusion reaction. But, since anti-P is usually reactive only at low temperatures and is an IgM, it is doubtful that it poses a serious threat of EBF.

E. The MNS System

The antigens M and N are inherited in a manner consistent with two allelic genes. Some evidence indicates that N may be a precursor of M, perhaps analogous to H as a precursor of A and B. The anti-M is capable of causing EBF but it is unlikely to cause a transfusion reaction.

F. The Duffy (Fy) System

Approximately 66% of individuals are Fy (a+) belonging to the genotype Fy^aFy^a or Fy^aFy^b. Testing with anti-Fy^b serum will differentiate these genotypes.

In Blacks, 68% have Fy(a-b-), a combination not been found in Whites. Most instances of anti-Fy^a are immune antibodies. Only rarely does anti-Fy^a occur as a natural occurring antibody. Since anti-Fy^a develops as an IgG, it can cause EBF.

G. The Kidd (JK) System

75% of people possess Jk^a antigen. 25% have Jk^a [phenotype JK (a+,b−)] and 50% have Jk^a/Jk^b [phenotype Jk (a+,b+)]. The remaining 25% of the population have Jk^b and Jk^b [phenotype Jk^b (a−,b+)]. The Kidd immunization can cause EBF. Since the antigen is not as strong as the Rh or Kell antigen, it is most commonly seen in patients who have had a history of blood transfusion. The antibodies are best detected by an indirect antiglobulin (Coombs) test.

IV. CONCLUSION

The incidence of Rh immunization is decreasing markedly due to the Rh immunoglobulin. However, because administration constitutes passive immunization there will always be failures. Rh immunization will remain as an infrequent but clinically significant problem. The incidence of ABO incompatibility will remain the same. The importance of early detection must be emphasized in view of the trend towards early discharge from the hospital. The incidence of immunization to irregular antibodies will probably increase slightly due to an increase in blood transfusions and induced abortions. Since many of these immunizations can cause clinically significant disease, aggressiveness in detection should be encouraged in order to provide optimal antepartum care for the fetus at risk.

REFERENCES

Allen, F. H., Diamond, L. K., and Jones, A. R. (1954). *N. Engl. J. Med.* **251,** 453–459.

Bevis, D. C. A. (1956). *J. Obstet. Gynaecol. Commonw.* **63,** 68–75.

Clayton, E. M., Feldhaus, W., and Phythyon, J. M. (1966). *Obstet. Gynecol.* **28,** 194–197.

Diamond, L. K., Blackfan, K. D., and Baty, J. M. (1932). *J. Pediatr.* **1,** 269–309.

Finn, R., Clarke, C. A., Donohoe, W. T. A., McConnell, R. B., Sheppard, P. M., Lehane, D., and Kulke, W. (1961). *Br. Med. J.* **1,** 1486–1490.

Freda, V. J., Gorman, J. G., and Pollack, W. (1964). *Transfusion* **4,** 26–32.

Gorman, J. G., Freda, V. J., and Pollack, W. (1964). *Proc. Congr. Int. Soc. Hematol. 9th, 1962* Vol. 2, p. 545.

Lansteiner, K., and Wiener, A. S. (1940). *Proc. Soc. Exp. Biol. Med.* **43,** 223.

Levine, P. (1943). *J. Hered.* **34,** 71–80.

Levine, P., and Stetson, R. E. (1939). *J. Am. Med. Assoc.* **113,** 126–127.

Liley, A. W. (1963). *Am. J. Obstet. Gynecol.* **86,** 485–494.

Mandelbaum, B., and Robinson, A. R. (1966). *Obstet. Gynecol.* **28,** 118–120.

Queenan, J. T. (1977). "Modern Management of the Rh Problem." Harper and Row, Hagerstown, Maryland.

Queenan, J. T. (1978). *In* "Amniotic Fluid—Research and Clinical Application" (D. V. T. Fairweather and T. K. A. Eskes, eds), 2nd ed., pp. 187–208. Excerpta Med. Found., Amsterdam.

Queenan, J. T., Landesman, R., Nakamoto, M., and Wilson, K. H. (1962). *Obstet. Gynecol.* **20,** 774–780.

Queenan, J. T., Smith, B. D., Haber, J. M., Jeffrey, J., and Gadow, E. C. (1969). *Obstet. Gynecol.* **34,** 767–771.

6

Cardiovascular Defects

SHERMAN ELIAS AND RUTH M. YANAGI

Because of advances in medical management and surgical correction, an increasing number of women with congenital heart disease (CHD) reach childbearing age. Accordingly, geneticists and obstetricians will be called upon not only to manage the pregnancies of women who themselves have CHD, but also to counsel families regarding genetic risks. In our discussion of the relationship between CHD and pregnancy we shall consider the following areas: (1) epidemiology, (2) etiologies of various disorders, (3) drug teratogenicity, and (4) prenatal diagnosis. For each disorder we shall consider effects of pregnancy upon (1) women with the disorder and (2) their offspring.

Copyright © 1981 by Academic Press, Inc.
All rights of reproduction in any form reserved.
ISBN 0-12-630940-X

I. EPIDEMIOLOGY AND MORTALITY STATISTICS

A. Incidence

In a recent report, Hoffman and Christianson (1978) attempted to ascertain the incidence of CHD in a large cohort of births. The authors point out several important deficiencies in previous studies. Spuriously low rates have been reported in studies which have (1) followed up children for only a short time after birth and (2) used registry data which would detect surviving infants but not include infants who died shortly after birth. Even studies that gave relatively higher rates might still underestimate frequencies if the initial diagnosis is not made by a cardiologist, thus missing minimal lesions. In addition, few attempts had been made to determine the frequency of false negative diagnosis. Finally, stillborn fetuses, who are known to have a high rate of cardiac malformations, are usually not evaluated. In order to minimize such biases, Hoffman and Christianseon (1978) studied a large cohort of births in the San Francisco–East Bay region. Their experimental design allowed derivation of information concerning stillborn fetuses and long-term follow-up of infants for adequate assessment of cardiac status.

The study group was conducted among members of a prepaid medical care program (Kaiser Foundation Health Plan) over the enrollment period of 1959–1966. There were 20,504 pregnant women entered into the study, with pregnancies terminating in 19,044 livebirths and 458 stillbirths. 163 liveborn children were diagnosed as having definite CHD and an additional 31 children were regarded as having possible CHD. The crude incidence rate was 8.8/1000 live births for definite CHD and 10.4/1000 for both definite and possible CHD. The incidence rate of CHD was 7.9% among all stillborn fetuses subjected to autopsy and 10.2% among those in which the autopsy report was evaluated as being sufficiently detailed to determine the presence of CHD. Approximately 30% of children with CHD had anomalies of other organ systems. Fifty children (2.63/1000 livebirths) were diagnosed as having a chromosomal abnormality and of these, 30% had CHD.

B. Maternal Age

Although maternal age has been shown to correlate with CHD associated with syndromes due to chromosomal nondisjunction, the evidence for such parental age association in CHD per se is not well defined. In a study from British Columbia, Renwick *et al.* (1964) used data collected from the Registry for Handicapped Children and Adults and the Division of Vital Statistics. They determined that twice as many babies with CHD (excluding Down syndrome) were born to mothers over age 39 years as would be expected on the basis of the maternal age distribution for all livebirths in the population. Caution must be used in interpreting these data because (1) all cardiovascular malformations were grouped as a single entity, (2) paternal age was not considered, and (3) birth order was not analyzed. Among mothers ages 40–45 years, Campbell (1965) found a statistically significant increased risk for offspring with tetralogy of Fallot (excluding Down syndrome). In addition, there was an increased risk of ventricular septal defect in mothers ages 35–39 years who were having their sixth or more child. In general, however, it has been concluded that the parental age is of little or no etiologic significance in CHD (Nora and Nora, 1978a).

C. Congenital Heart Disease and Maternal Mortality

Hibbard (1975) reviewed the maternal mortality in California during the 9-year period 1960–1968. Excluding trauma unrelated to pregnancy, there was 1362 maternal deaths associated with 3,194,000 reportable births. Seventy-seven deaths were attributable to cardiac disease, most commonly rheumatic heart disease (28 deaths). Eighteen women were reported to have died from complications of CHD. Other cardiac diseases responsible for maternal deaths included idiopathic myocarditis, coronary heart disease, lupus erythematosis with cardiomyopathy, pericarditis, ventricular arrhythmia, pulmonary hypertension, congestive failure, and pheochromocytoma.

Copeland *et al.* (1963) ascertained data from five state maternal mortality committees and found only 21 cases of congenital heart disease in 2684 maternal deaths. There was no attempt to correlate these deaths with other causes of maternal death or with the total number of births in these states during the 6 to 11 years these committees were in operation.

Kahler (1975) reviewed five different series comprising 490 patients with CHD who underwent 1135 pregnancies. There were six maternal deaths in this group; four occurred in patients with the Eisenmenger syndrome (two with ventricular septal defect and two with patent ductus arteriosus), one in a patient with tetralogy of Fallot and one in a patient with "discrete" subaortic stenosis complicated by subacute bacterial endocarditis.

From the studies cited above it may be concluded that CHD has not been a major cause of maternal mortality over the past several decades. However, over this same period of time there have been important advances in cardiovascular surgery. Women who heretofore would have died earlier in life are now surviving into the reproductive years as result of major cardiac surgery. Pregnancy may be expected to impose substantial risks to such women, and thereby increase the proportion of overall maternal mortality as result of CHD.

D. Effects of Maternal Congenital Heart Disease on the Fetus

The risk to fetal well being appears to increase in direct proportion to the severity of cardiac disease complicating pregnancy. In 1948 Bunim and Rubricicius reported 30% perinatal mortality rate among mothers with congestive heart failure, 9% perinatal loss with "compensated" heart disease, and 7% loss among patients with normal hearts. In a study from the Ohio State University Hospital, Copeland *et al.* (1963) reviewed the case histories of 125 women with CHD who had a total of 372 pregnancies. There were 77 (20.7%) fetal losses. Eleven losses occurred from 20 weeks gestation to 27 days after birth, yielding a perinatal mortality rate of 3% ("about average"). The highest risk group appeared to be those patients with pulmonary hypertension (defined as pulmonary artery pressure greater than 60 mm Hg) or maternal cyanosis. Thirty-three of the 77 fetal losses occurred in the presence of these two cardiovascular complications.

Schaefer *et al.* (1968) reviewed 120 pregnancies in 51 patients with CHD delivered at the New York Lying-In Hospital. Twenty-four pregnancies (20%) ended in spontaneous abortion. Among the 96 infants, 14 (14.6%) were premature by weight, and 5 (5.2%) were stillborn. The 14.6% incidence of prematurity was significantly higher than 7.3% in the New York Lying-In Hospital population.

Finally, Kahler (1975) reviewed the data from five different series and ascertained 490 patients with congenital heart disease who underwent 1135 pregnancies. The fetal loss for the entire grouped series was 214, for a fetal mortality rate of 18.9%.

II. ETIOLOGIES OF CONGENITAL HEART DISEASE

Congenital heart disease may be categorized into four major etiologic categories: (1) chromosomal disorders—5%; (2) single gene mutations—3%; (3) environmental factors (e.g., rubella, thalidomide)—1-2%; (4) genetic-environmental interaction (polygenic/multifactorial inheritance)—90% (Nora and Nora, 1978a).

A. Chromosome Abnormalities Associated with Cardiovascular Defects

Chromosomal abnormalities have been estimated to account for approximately 5% of congenital cardiac malformations (Nora, 1977). Such malformations associated with chromosomal abnormalities invariably exist as part of a multisystem malformation syndrome rather than as an isolated defect (e.g., ventricular septal defect only).

Anders *et al.* (1965) examined 156 patients with CHD, excluding Down syndrome and Turner syndrome, and found no major chromosomal abnormalities. Five of these patients were found to have "variations in size" of certain chromosomes, particularly chromosome 16 (4 cases). German *et al.* (1966) performed cytogenetic studies of 35 families in each of which multiple members had cardiac anomalies. With the exception of nonconsistent autosomal variants, there were no aberrations.

In 1970, Dahl studied 100 patients with congenital cardiovascular disease. Twelve patients were trisomy G (presumably trisomy 21), four patients were 45,X and one patient with associated "severe extracardiac malformations" had a partial trisomy of a group C chromosome and translocation of C/D. An additional patient had either an "isochromosomy" or a pericentric inversion of the X chromosome which was also found in his brother, father, and paternal grandfather, none of whom had CHD. The remaining 82 patients had normal chromosomal complements.

Thus, screening patients with congenital heart defects for chromosomal abnormalities is not of particular value unless the defect coexists with other malformations. With the advent of newer methods of chromosome banding, a number of familial malformation syndromes involving the cardiovascular system which have heretofore been categorized as of "unknown etiology" will undoubtedly be recognized as chromosomal in origin.

1. Autosomal Trisomies

Warkany *et al.* (1966) reviewed the relative frequency of various cardiac malformations associated with trisomies 21, 18, or D (or 13–15). In reviewing four necropsy series, they reported congenital heart defects in 52% (140/272) of children with Down syndrome. The single most common lesion was ventricular septal defect (68/272); however, if the atrial defects (i.e., atrial septal defect, ostium primum, and atrioventricular communis) are combined, they were the most common malformation (total of 112/272, or 64%). Of the 140 autopsied children with Down's syndrome, 64 (45%) had only a single cardiac malformation, whereas 76 (55%) had combined lesions. In reviewing 84 necropsy cases of trisomy 18, all of the children except one were found to have structural cardiac defects. Ventricular septal defect was present in 64 (74%) and patent ductus arterious in 47 (56%), and other various defects in small percentages. Of 32 autopsied cases in trisomy D (13–15), 27 (84%) had cardiac malformations, most often ventricular septal defect with patent ductus arteriosus.

Finally, Nora and Nora (1978a) have recently summarized their experience of the incidence of CHD in liveborn infants with various chromosomal abnormalities (Table I).

TABLE I
Congenital Heart Diseases (CHD) in Selected Chromosomal Aberrations[a]

Population studied	Incidence of CHD (%)	Most common lesions[b] 1	2	3
General population	1	VSD	PDA	ASD
Trisomy 21	50	VSD or AV canal	ASD	PDA
Trisomy 18	99+	VSD	PDA	PS
Trisomy 13	90	VSD	PDA	Dex
Trisomy 22	67	ASD	VSD	PDA
Partial trisomy 22 (cat-eye)	40	Complex TAPVR	VSD	ASD
4p-	40	ASD	VSD	PDA
5p- (cri-du-chat)	20	VSD	PDA	ASD
Trisomy 8 (mosaic)	50	VSD	ASD	PDA
Trisomy 9 (mosaic)	50	VSD	Coarc	DORV
13q-	25	VSD		
+14q-	50	PDA	ASD	Tet
18q-	50	VSD		
XO Turner	35	Coarc	AS	ASD
XXXXY	14	PDA	ASD	ARCA

[a]From Nora and Nora (1978b).

[b]ARCA, Anomalous right coronary artery; Dex, dextroversion; DORV, double outlet right ventricle; VSD, ventricular septal defect; PDA, patent ductus arteriosus; ASD, atrial septal defect; TAPVR, total anomalous pulmonary venous return.

2. *Sex Chromosomes*

The association of cardiac anomalies with sex chromosome abnormalities is considerably lower than that of the autosomal trisomies. For example, only 10–16% of 45, X individuals have a cardiovascular anomaly with coarctation of the aorta and ventricular septal defect being the most frequent (Simpson, 1976). Rosenthal (1972) has suggested that individuals with Klinefelter syndrome (47,XXY) also have an increased likelihood of having cardiac defects; however, no specific risk figures are availabe. Among defects associated with Klinefelter's syndrome are tetralogy of Fallot, patent ductus arteriosus, atrial septal defects, and ventricular septal defects (Rosenthal, 1972). It has also been suggested that the frequency and severity of malformations is related to the degree of X-chromosome polysomy (Grumbach, 1968). For example, congenital heart defects, mental retardation, skeletal anomalies, and hypogonadism are all features associated with the 48,XXXXY complement (Karsh *et al.*, 1975).

3. Genetic Counseling for Chromosomal Abnormalities Associated with CHD

The incidence of a chromosomal abnormality is about 1/200 newborns (Hook and Hamerton, 1977). Situations known to be associated with an increased likelihood of having offspring with a chromosomal abnormality include (1) advanced maternal age; (2) a parent with a balanced translocation, inversion, or other chromosomal abnormality; and (3) a previous child with a chromosomal abnormality, particularly autosomal trisomy. Thus, when congenital heart defects are associated with chromosomal abnormalities, amniocentesis for prenatal diagnosis should be considered in subsequent pregnancies. The risk is 1% if the proband was trisomic and 2–10% if a parent has a balanced translocation, but probably not increased if the proband has a *de novo* rearrangement.

4. Obstetrical Considerations

Limited data are available regarding the outcome of pregnancies in which a mother has both an unbalanced chromosomal abnormality and CHD (Mullins *et al.*, 1960; Sharrer *et al.*, 1975). In the rare instance that a woman with an unbalanced chromosomal complement associated with a congenital cardiac defect becomes pregnant, the authors recommend the following: (1) because most of these individuals are mentally retarded, abortion, and sterilization should be considered; (2) both the patient and her family must be counseled regarding the high risk of a chromosomally abnormal fetus and amniocentesis offered for prenatal diagnosis if the pregnancy is to be continued; and (3) regardless of the maternal chromosomal abnormality, such pregnancies should be managed in a fashion similar to that utilized for chromosomally normal women with CHD.

B. Mendelian Disorders

Nora (1971) estimated that about 2% of congenital heart lesions are caused by single mutant genes. Analogous to the chromosomal abnormalities, most congenital heart defects attributed to single gene mutations are also associated with abnormalities of other organ systems and comprise recognizable syndromes. There are, however, several noteworthy exceptions in which single gene mutations apparently result in abnormalities limited to the heart, and these will be discussed first.

1. Mendelian Disorders with Expression Limited to the Heart

a. Asymmetric Septal Hypertrophy. Asymmetric septal hypertrophy (ASH), an autosomal dominant disorder, is a distinct cardiomyopathy showing a spectrum of clinical manifestations. Affected individuals may be totally asymptomatic or may show marked left ventricular outflow obstruction and mitral regurgitation. The more severe form is most commonly designated as idiopathic hypertrophic subaortic stenosis (IHSS). Asymmetric septal hypertrophy, detected by echocardiography, is generally defined by a ventricular system that is disproportionately thickened compared to the posterior–basal left ventricular free wall (Henry *et al.,* 1973). When outflow obstruction occurs, it appears to be the result of systolic narrowing of the left ventricular outflow tract caused by apposition of the anterior leaflet of the mitral valve with the hypertrophied ventricular septum (Henry *et al.,* 1972). In addition, necropsy studies have revealed that papillary muscles of patients with IHSS often are displaced by this grossly thickened septum (Roberts, 1973).

b. Genetics. Probably the earliest report of familial transmission of IHSS was by Davies (1952). He described a family in which five sibs (four females, one male) and the child (male) of one of these sibs were affected by congenital subaortic stenosis determined by electrocardiographic, radiographic, and physical diagnostic criteria. Subsequently, a number of such reports were published with individuals of both sexes affected in multiple generations (Bercu *et al.,* 1958; Pare *et al.,* 1961; Wood *et al.,* 1962; Kariv and Solomon, 1962; Braunwald *et al.,* 1964; Horick *et al.,* 1966). In 1973 Henry *et al.* showed that ASH is a characteristic abnormality present in all patients falling into the IHSS spectrum. These investigators subsequently studied the inheritance of IHSS in 30 families, using ASH as the disease marker. In 28 cases at least one first-degree relative of the index case also had ASH. Of 105 first-degree relatives, 49 (46%) were determined to have ASH by echocardiography, autopsy, or cardiac catheterization. Forty-six percent of the 63 males had ASH and 45% of the females were affected. They concluded that "most, and probably all patients with ASH have a genetic defect that is transmitted as an autosomal dominant trait with a high degree of penetrance." In addition, they emphasize that ASH is the fundamental anatomic defect in IHSS and that its clinical spectrum ranges from asymptomatic persons to patients with typical symptomatic IHSS (Clark *et al.,* 1973).

c. Obstetrical Considerations. In 1968 Turner *et al.* reported their experience with 13 pregnancies in nine women affected with IHSS. In five of these nine patients other relatives were affected with IHSS; however, neither detailed pedigree information nor documentation was given. Initially, these investigators advocated elective cesarean section because they felt the hazards of tachycardia, uncontrolled bleeding, or expulsive efforts associated with delivery per vagina, were unacceptable. After gaining additional experience, they subsequently allowed patients to deliver per vagina using the following protocol: (1) prophylactic antibiotics to prevent bacterial endocarditis, a complication for which these patients are at increased risk; (2) administration of propranolol during labor; (3) blood available in case of hemorrhage; (4) positioning the patient in the lateral recumbant position to avoid uterine compression of the inferior vena cava; (5) shortening the second stage of labor, presumably by forceps delivery to avoid expulsive efforts; and (6) administration of ergometrine at the end of the second stage to minimize postpartum bleeding. In their series, all mothers and infants did well.

Kolibash *et al.* (1975) reported a patient with IHSS who near term manifested severe symptoms (Class IV, New York Heart Classification). The patient survived her pregnancy, labor, and delivery and subsequently returned to a Class II status. The authors suggest that significant worsening of symptoms is the rule in pregnant women with IHSS as result of (1) increased blood volume; (2) decreased systemic arterial resistance accompanying the increase in cardiac output characteristic of the latter half of pregnancy; (3) inferior vena cava obstruction by the uterus when patients are supine; and (4) increased adrenergic activity during labor and delivery.

Recently Datta *et al.* (1978) described a woman affected with IHSS who was successfully treated with propranolol throughout her pregnancy. They cautioned that the effects of propranolol during pregnancy may include (1) excessive uterine activity; (2) adverse maternal cardiovascular effects such as decrease in myocardial contractility, cardiac output, heart rate, oxygen consumption, an increase in left ventricular and diastolic pressure; and (3) intrauterine growth retardation as result of decreased uterine blood flow, increased myometrial tone, and placental insufficiency. They advocate either local blockage or light anesthesia for delivery per vagina rather than regional anesthesia. If cesarean section becomes necessary, light general anesthesia is preferable. Finally, the neonate must be carefully observed for bradycardia as result of β-adrenergic blockade and hypoglycemia (the presumed mechanism being inhibition of epinephrine-stimulated glycolysis).

d. Antenatal Diagnosis. Since ASH is inherited in autosomal dominant fashion, the risk for each offspring of an affected parent being similarly affected is 50%. Antenatal diagnosis of ASH might be possible by ultrasonographic studies of the fetal heart demonstrating ventricular septal hypertrophy using the posterior-basal left ventricle as a standard of comparison. Further experience with fetal echocardiography would be required before applying such techniques on a clinical basis (see Section IV).

2. *Cardiac Conduction System Abnormalities*

A number of families have been reported in which cardiac conduction system abnormalities appear to be the result of single gene mutations. Among the conditions in which autosomal dominant inheritance has been suggested are right bundle branch blocks (Combrink *et al.*, 1962), isolated complete right bundle branch block (Esscher *et al.*, 1975), and complete heart block (Gazes *et al.*, 1965). Families showing autosomal dominant inheritance of atrial septal defect (fossa ovalis type) with prolonged atrioventricular conduction might be included within this category (Bizarro, 1970; Pease *et al.*, 1976).

In addition, reports of multiple affected sibs with condition abnormalities suggest an autosomal recessive form (Verel and Chandrasekhar, 1970; Husson *et al.*, 1973). Although there are undoubtedly Mendelian disorders of the conduction system which may result from either (1) developmental abnormalities of the conduction system itself or (2) the secondary effects of abnormal blood or nerve supply to the system, the high prevalence and marked heterogeneity of conduction defects makes genetic delineation difficult in most cases.

a. Obstetrical Considerations. Evidence that cardiac conduction defects, without coexisting anomalies, are an infrequent cause of maternal mortality is indirect. Hibbard (1975) reviewed 77 maternal deaths attributed to cardiac disease among 3,194,000 reportable births in California during the 9-year period 1960–1968. Of these 77 cases, two were the result of ventricular arrhythmia, both of which were determined to be "idiopathic" in origin. However, conduction disorders are not infrequently encountered during pregnancy. In most cases these disturbances are minor and self-limiting (Ueland, 1977). Nevertheless, in certain situations arrhythmias may be life-threatening, especially the occurrence of atrial fibrillation. In such cases, cardioversion has been utilized with no ostensible adverse effects to the fetus (Schroeder and Harrison, 1971). Ueland (1977) has also used quinidine during pregnancy without problems. If other family members are affected with cardiac conduction abnormalities, both the mother and the neonate should be carefully evaluated for similar abnormalities. Finally, congenital heart block in the fetus has been reported during the third trimester (Harris, 1952). Newer methodology in fetal electrocardiography and phonocardiography hold promise for increasing the reliability in establishing the diagnosis of conduction defects antenatally (see Section IV).

3. Hypoplastic Left Heart Syndrome

In 1958 Noonan and Nadas proposed the term *hypoplastic left heart syndrome (HLHS)* to describe infants with an obstructive lesion on the left side of the heart, with hypoplasty of the left ventricle and right ventricular hypertrophy. The left ventricle varies from a vestigial structure with coexisting aortic and mitral valve atresia to a lesser form of the syndrome characterized by a hypoplastic aortic arch and only a relative underdevelopment of the left ventricle. The mortality of the complete syndrome continues to be at or close to 100% with or without surgical treatment.

a. Genetics. Although McKusick (1978) continues to catalogue HLHS as an autosomal recessive disorder, considerable evidence now indicates that this disorder is inherited in polygenic/multifactorial fashion. In 1971 Shokeir reported five independent families from Saskatchewan with a total of 13 affected individuals. Consanguinity was demonstrated among the parents of three of the sibships. Their initial impression was that HLHS was inherited in autosomal recessive fashion. Subsequently, Holmes *et al.* (1974) reported that the inheritance of the HLHS was most consistent with polygenic/multifactorial inheritance. In a follow-up study, Brownell and Shokeir (1976) investigated the families of 64 cases with HLHS and determined that the recurrence risk among later sibs was 2%, a figure consistent with polygenic inheritance. Nevertheless, McKusick (1978) believes that a subgroup with autosomal recessive inheritance cannot be totally excluded.

b. Obstetrical Considerations. Since individuals with the HLHS generally do not survive beyond the neonatal period, there have been no pregnancies in affected women. Prenatal diagnosis should be possible by ultrasonographic studies of the fetal heart.

4. X-Linked Congenital Heart Disease

Monteleone and Fagan (1969) reported a family in which four males were affected with congenital mitral and aortic insufficiency; a fifth male had congenital mitral stenosis only; a sixth male was said by history to have congenital heart disease. The affected males comprised all males in three generations of this family. The authors concluded that the involvement of only males in this family could best be explained by X-linked recessive inheritance; however, additional reports would be necessary to confirm this observation. Finally, fetal echocardiography might prove useful for antenatal diagnosis in such families.

C. Mendelian Disorders II: Syndromes with Associated Cardiovascular Defects

In numerous Mendelian conditions cardiovascular defects are part of a syndrome (Tables II, III, and IV).

Several important principles should be emphasized. First, most syndromes exhibit varied expressivity. That is, the severity of cardiovascular involvement among individuals affected with a given syndrome (e.g., the Marfan syndrome) may range from mild to severe. As a rule, there is less variability in gene expression among members of a single kindred than between different family groups. Such observations suggest that gene expression depends at least partly upon modifying genes, which would differ more between than within families. Second, some individuals who almost certainly have the appropriate genotype may not exhibit any detectable phenotypic abnormalities (i.e., nonpenetrance). The degree of penetrance of a mutant gene also depends upon the criteria used to search for its presence. In many cases close examination of such individuals will reveal minimal expression of an autosomal dominant disorder which would otherwise remain unnoticed (e.g., detection of café-au-lait spots in person at risk for neurofibromatosis). Finally, in certain metabolic disorders the heart per se is

TABLE II
Autosomal Dominant Disorders Involving the Cardiovascular System

Condition	Major abnormalities	Cardiovascular defects
Holt-Oram syndrome	Skeletal anomalies: hypoplasia of thumbs; hypoplasia of radius; occasionally phocomelia	Atrial septal defect (ADS); ventricular septal defect (VSD)
Multiple lentigines syndrome	Multiple lentigines; hypertelorism; deafness	Pulmonic stenosis
Noonan syndrome	Short statures; webbed neck; low posterior hair line; low set ears; cubitus valgus; cryptorchidism	Pulmonic stenosis; VSD; ASD
Waardenburg syndrome	Lateral displacement of medial canthi; broad and high nasal bridge; partial albinism; occasional deafness	VSD
Tuberous sclerosis	Adenoma sebaceum; seizures; mental retardation	Myocardial rhabdomyoma; cardiomyopathy
Craniofacial dysostosis (Crouzon syndrome)	Carniosynotosis; shallow orbits with exopthalmos; maxillary hypoplasia	Coarctation of the aorta; patent ductus arteriosus (PDA)
Ehlers-Danlos syndrome (several forms)	Dermal hyperelasticity and joint laxity; skin fragility	Aortic dissection
Idiopathic hypertrophic subaortic stenosis	Cardiac defects as noted	Left ventricular hypertrophy; hypertrophy of ventricular septum with obstructed outflow
Marfan syndrome	Arachnodactyly; joint hyperextensibility; ectopia lentis	Aortic dilatation with regurgitation; dissecting aortic aneurysm; mitral regurgitation
Mandibulofacial dysostosis (Treacher-Collins syndrome)	Downward slant of palpebral fissures; lower eyelid coloboma; hypoplasia of zygomatid arches; dysplastic auricles; deafness; micrognathia	PDA; VSD; ASD
Neurofibromatosis syndrome (Von Recklinghausen disease)	Multiple neurofibromata; café-au-lait spots	Neurofibromata of the heart; arrhythmias
Osteogenesis imperfecta type II	Skeletal fragility; blue sclerae; odontogenesis imperfecta	Aortic regurgitation
Ventricular fibrillation with prolonged QT interval	Syncope; sudden death	Conduction abnormalities; prolonged QT interval; ventricular fibrillation

structurally normal; however, an abnormal deposition of substrate within the myocardium may secondarily cause clinical disease (e.g., glycogen accumulation in Pompe disease).

Of special note is the frequent involvement of the heart in disorders of connective tissue. See Chapter 2 for additional discussion of the Ehlers-Danlos syndrome, the Marfan syndrome, pseudoxanthomas elasticum, osteogenesis imperfecta, neurofibromatosis, thanatophoric dwarfism, the Conradi syndrome, and the Ellis-van Creveld syndrome.

Obstetrical Considerations

Important considerations regarding the Mendelian disorders associated with cardiovascular defects and their relationships to pregnancy include:

1. Appropriate genetic counseling should be given to couples at risk concerning the relatively high probability of inheriting the mutant gene. Such risks are relatively straightforward, e.g., 50% for autosomal dominant disorders and 25% for autosomal recessive disorders; however, because of varied expression and penetrance the likelihood of a child having a clinically significant disorder is often less.

2. Prenatal diagnosis should be offered whenever feasible (e.g., certain glycogen storage diseases and mucopolysaccaridosis are associated with cardiomyopathies, and are also amenable to prenatal diagnosis through amniotic fluid cell studies).

3. The physiologic changes of pregnancy adversely effect the maternal cardiovascular system. For example, in the Marfan syndrome the hemodynamic alterations of pregnancy may predispose to rupture of an aortic aneurysm or can lead to aortic ring dilitation and aortic valve involvement resulting in regurgitation and left ventricular failure (Elias and Burkowitz, 1976) (see also Chapter 2).

4. In addition to cardiovascular abnormalities, defects in other organ systems may also pose problems to either the pregnant woman or her fetus. For example, patients with sickle cell anemia may develop myocardial hemosiderosis, cardiomegaly, mitral insufficiency, and occasionally pericarditis. During pregnancy affected women are particularly prone to sickling crises, pneumonia, urinary tract infections, toxemia, and pulmonary embolism. These complications have led to maternal mortality rates varying from 0 to 25% (Perkins, 1971). Horger (1974) has postulated that the concominant intrauterine hypoxia may be the cause

TABLE III
X-Linked Disorders Involving the Cardiovascular System

Syndrome	Associated anomalies	Cardiovascular defects
Aase syndrome	Triphalangeal thumb; congenital anemia	Ventricular septal defect
Mucopolysaccharidosis type II (Hunter syndrome)	Sulfioduronide sulfatase deficiency; coarse facies; short stature; mental deficiency; hepatosplenomegaly	Coronary occlusive disease; valvular disease
Duchenne muscular dystrophy	Pseudohypertrophic muscle weakness; predominantly proximal	Myocardiopathy
Ehlers-Danlos type V	Hyperextensible skin; bruising tendency	Mitral valve incompetance
Focal dermal hypoplasia (Goltz syndrome)	Skin atrophy; digital, oral, and ocular anomalies; telangectasia	Aortic stenosis; atrial septal defect

TABLE IV
Autosomal Recessive Disorders Involving the Cardiovascular System

Condition	Major abnormalities	Cardiovascular defects
Mucopolysaccharidosis I (Hurler syndrome)	Growth; mental retardation, course facies; cornea opacification; depressed nasal bridge; hepatosplenomegaly	Thickened coronary vessels and mitral values
Kartagener syndrome	Situs inversus; bronchiectasis; sinusitis	Dextrocardia; cardiac septation anomalies
Limb reduction–ichthyosis syndrome	Unilateral hypomelia; unilateral skin hypoplasia	Partial to complete absence of cardiac septum; occasional truncus arteriosis; valve defects
Mucolipidosis III (Pseudo–Hurler polydystrophy)	Course facies; corneal opacity; mild retardation; stiff joints, short stature	Aortic valve disease with regurgitation
Acrocephalopolysyndactyly type III (Carpenter syndrome)	Acrocephaly; polydactyly and syndactyly; lateral displacement of inner canthi	Patent ductus arteriosus (PDA); ventricular septal defect (VSD); transposition of the great vessels
Chondroectodermal dysplasia (Ellis–van Creveld syndrome)	Short distal extremities; polydactyly; nail hypoplasia; occasional mental retardation	Single atrium; atrial septal defect (ASD)
Dysencephalia splanchnocystica (Meckel–Gruber syndrome)	Encephalocele; polydactyly; polycystic kidneys	PDA; septal defectus; coarctation of aorta; PS
Pseudothalidomide syndrome (Roberts syndrome)	Hypomelia; midfacial defects; severe growth deficiency; profound mental retardation	ASD
Thanatophoric dwarfism syndrome	Severe growth deficiency; short extremities; large cranium; flat vertebrae; narrow short ribs; occasional brain anomalies	PDA
Conradi syndrome	Short limbed dwarfism; ankylosis; lenticular cataracts; depressed nasal bridge	VSD; PDA
Glycogen storage disease II (Pompe disease)	α-1,4-Glucosidase deficiency; profound hypotonia; muscle firmness	Myocardiopathy with extensive glycogen deposition; ECG shows large QRS complexes and shortened P-R interval
Glycogen storage disease III (Forbes disease)	Amylo-1,6-glucosidase deficiency; hepatosplenomegaly; growth retardation	Myocardiopathy; polysaccharide accumulation in heart; nonspecific ECG changes
Friedreich ataxia	Spinocerebellar degeneration	Myocardiopathy; arrhythmias

Lawrence–Moon–Biedl syndrome	Obesity; mental retardation; polydactyly; syndactyly; retinitis pigmentosa; genital hypoplasia; hypogonadism	VSD; tetralogy of Fallot
Cockayne syndrome	Precocious senile appearance; photophobia; retinal degeneration; deafness; short stature; mental retardation	Premature atherosclerotic cardiovascular disease
Cardioauditory syndrome	Sensorneural deafness; syncopal episodes; sudden death	ECG abnormalities including large T waves and prolonged QT intervals indicating abnormal ventricular repolarization
Mucopolysaccharidosis IV (Morquio syndrome)	Chondroitin-6-sulfate *N*-acetyl glucosamine-4-sulfate-sulfatase deficiency; severe growth retardation; mild coarse facies; corneal opacification; short anteverted nose; widely spaced teeth; hepatosplenomegaly	Aortic regurgitation; coronary vessel disease
Mucopolysaccharidosis VI (Maroteaux–Lamy syndrome)	Arylsulfatase B deficiency; short stature; coarse facies; corneal opacification; thick nostrils and lips; hepatosplenomegaly	Aortic stenosis
Pseudoxanthoma elasticum syndrome	Systemic degeneration of elastic fibers; yellow discoloration of skin, especially at flexure sites; retinal degeneration with angiod streaks	Endocardial thickening; medial arterial degeneration
Heredopathia atactica polyneuritiformis (Refsum disease)	Retinitis pigmentosa; polyneuropathy; cerebellar ataxia	Atrioventricular conduction defects
Sickle cell disease	Intravascular sickling; undernourished appearance; short trunk; narrow hips and shoulders; chronic skin ulceration; especially about the ankles; splenic fibrosis in adults; hepatomegaly	Cardiomegaly; mitral insufficiency
Smith–Lemli–Opitz syndrome	Anteverted nostrils; ptosis of eyelids; syndactyly of second and third toes; hypospadias; cryptorchidism	VSD; PDA
Radial asplasia–thrombocytopenia syndrome (TAR syndrome)	Thrombocytopenia; bilateral absence of radius	Tetralogy of Fallot; ASD; dextrocardia
β-Thalassemia (Cooley's anemia)	Suppressed β-chain synthesis; severe anemia	Congestive heart failure secondary to anemia

of both the increased rate of stillbirths and the increased incidence of low birth weight infants.

D. Polygenic/Multifactorial Inheritance for Cardiovascular Defects

In most cases of congenital cardiovascular defects, the proportion of affected relatives is greater than in the general population; however, the number of affected individuals in such families is less than would be expected with simple Mendelian inheritance. In addition, no chromosomal anomalies are present. The most common model put forward to explain such family patterns is that of polygenic/multifactorial inheritance.

Among the important implications of the polygenic/multifactorial model are the following: (1) the recurrence risk increases as the number of affected first-degree relatives increases; (2) when a sex difference in the population incidence of a disorder exists, the relatives of the less frequently affected sex will be more frequently affected; (3) the more severe a malformation, the higher the recurrence risk; and (4) the recurrence risk in a first-degree relative of an affected individual may be approximated by the square root of the population frequency ($\sqrt{p}$) (Edwards, 1960).

To support the hypothesis that most common forms of CHD are the result of polygenic/multifactorial inheritance, Nora (1971) randomly surveyed 1478 patients with 13 types of congenital heart lesions and the number of sibs in each category who were also affected with congenital heart defects (Table V). They compared this "empiric risk" for recurrence in sibs to the "expected" recurrence risk as estimated by the square root of the population frequency ($\sqrt{p}$) for each defect (Edwards, 1960). The frequency of congenital heart disease in the population was approximated by combining the results of National Institutes of Health (Mitchell *et al.*, 1969) and the Kaiser-Berkley (Yerushalmy, 1969) surveys. The empiric recurrence risk for each malformation closely approximates the expected recurrence risk (in the range of 1–4%). The higher the frequency of the heart disease in the population, the higher the frequency of the disorders in sibs. Nora and Nora (1976) also calculated the recurrence risk among 1120 children where a parent was affected with one of seven congenital heart defects. Again, the recurrence risk was determined to be in the range of 2 to 4%, similar to the recurrence risk among sibs. The more common the defect, the higher the frequency for recurrence in offspring. For purposes of genetic counseling, Nora and Fraser (1974) emphasized that although predicted and observed recurrence risks are in close agreement, it is preferable to use empiric recurrence risks when available.

To illustrate the hereditary predisposition for congenital heart defects in various "family types," Nora (1971) set forth a model of hypothetical distribution

TABLE V
Observed and Expected Recurrence Risks in Sibs in 1478 Probands with Congenital Heart Lesions[a]

Anomaly	Probands	Affected sibs		
		No.	%	Exp. ($\sqrt{p}$)
VSD	212	24/543	4.4	5.0
PDA	204	17/505	3.4	3.5
Tetralogy	157	9/338	2.7	3.2
ASD	152	11/342	3.2	3.2
PS	146	10/345	2.9	2.9
AS	135	7/317	2.2	2.1
Coarctation	128	5/272	1.8	2.4
Transposition	103	4/209	1.9	2.2
A-V canal	73	4/151	2.6	2.0
Tricuspid atresia	51	1/96	1.0	1.4
Ebstein anomaly	42	1/96	1.1	0.7
Truncus	41	1/86	1.2	0.7
Pulmonic stenosis	34	1/77	1.3	1.0
Total	1478	95/3376		

[a] From Nora (1971).

curves based upon polygenic/multifactorial inheritance (Fig. 1). This model assumes that (1) individuals must be predisposed to cardiovascular maldevelopment; (2) a genetic predisposition to react adversely to the environmental trigger exists; and (3) the environmental insult must occur during the vulnerable period of cardiac development.

Although the Nora model of cardiovascular defects based on polygenic/multifactorial inheritance is useful for counseling patients regarding recurrence risks, several important shortcomings should be noted. For example, the data used to estimate the population frequency for various heart defects are primarily derived from only two surveys, one national (Mitchell *et al.*, 1969) and one regional (San Francisco-East Bay area) (Yerushalmy, 1969). On the other hand, the empiric data from which the relatives of affected propositi are drawn come from Texas, Wisconsin, and Canada (Nora *et al.*, 1970). Thus the population used to derive incidence is not necessarily comparable to that of relatives chosen for the study. More appropriately, the population frequency would be determined from the same georgraphical area as the study group. In addition, stillborn fetuses of relatives, who are known to have a high rate of cardiac malformations, were not considered in their data.

Finally, support for polygenic/multifactorial inheritance of common cardiac malformations may be found in animal studies. Mulvihill (1972) reviewed the summaries of all dogs seen in 10 veterinary school clinics (70,401 animals); 282

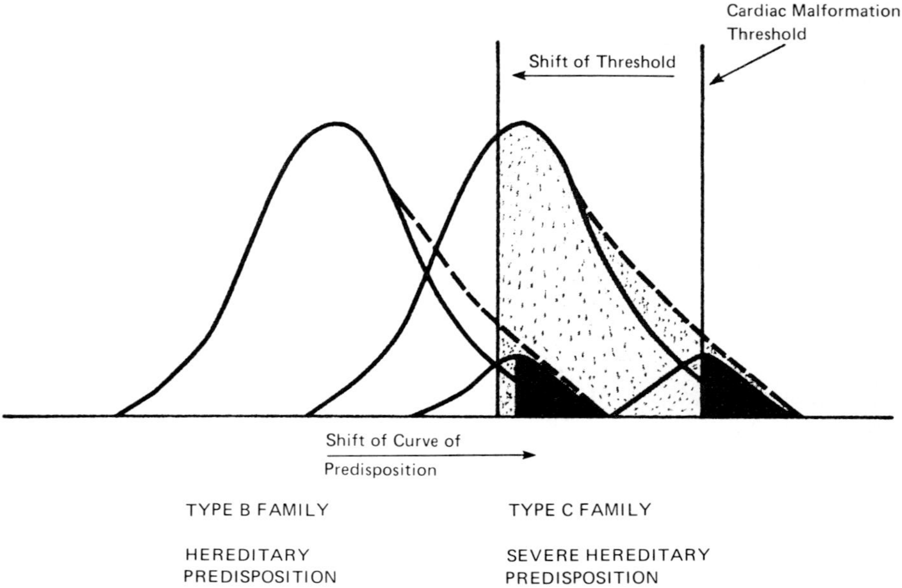

Fig. 1. Hypothetical quasi-continuous and discontinuous models for multifactorial inheritance of congenital heart diseases. In the typical low-risk (type B) family, the threshold of malformation may be crossed by a symmetrical curve of polygenic predisposition or by a slightly skewed curve in which there is a population with mendelian risks (blackened area). In the high-risk type C family, the normally distributed curve of predisposition moves to the right, and the threshold to the left, accounting for all of the risk in the quasi-continuous models of Falconer and Edwards. Only in the presence of significant dominance may Morton's model be appreciated as accounting for a small portion of families at risk. Although the overall empiric recurrence-risk figures are the same in all three models, the risks in a given family must be specifically identified as high or low risk on the basis of the individual pedigree. (From Nora and Nora, 1978b.)

had CHD. Purebred dogs had three times the frequency of cardiovascular defects as crossbreeds and certain tended to have a single type of lesion. Thus, animal data add evidence that common congenital heart defects have a genetic basis which results not from single gene mutations but is consistent with multifactorial inheritance (Nora, 1971). Nevertheless, genetic heterogeneity undoubtedly exists and certain forms of CHD are probably the result of single mutant genes.

Specific Malformations

a. Communications between Systemic and Pulmonary Circuits with Predominantly Left to Right Shunts. There are four major defects which result in communication between systemic and pulmonary circuits and result in left to right shunts: (1) atrial septal defects, (2) ventricular septal defects, (3) patent ductus arteriosus, and (4) endocardial cushion defects. A brief description of each defect will be given initially. Since these defects produce similar physiologic alterations they shall be considered collectively with respect to clinical manifestations.

1. Atrial septal defects. There are five different types of interatrial communications. The secondum (fossa ovalis) atrial septal defect results from abnormal development of the septum secondum. The sinus venosus defect, located superior to the fossa ovalis, is in close proximity to the ostium of the superior vena cava, and is frequently associated with an anomalous venous connection from the upper lobe of the right lung. The ostium primum atrial septal defect (partial endocardial cushions and its adjacent tissues) lies low in the atrial septum, and is usually associated with a cleft mitral valve. Finally, there is the rare common atrium defect in which the atrial septum is either absent or rudimentary.

2. Ventricular septal defects. Ventricular septal defects (VSD) vary with respect to size and location. The most common form is the membranous or infracristal defect, located near the aortic valve. In the supracristal or subpulmonary form the defect lies anteriorly, just below the pulmonary valve. Abnormal development of the endocardial cushions may result in a VSD located both above and below the septum. Finally, VSD, may occur in the muscular portion of the ventricular septum, and are frequently multiple.

3. Patent ductus arteriosus. The ductus arteriosus normally begins to constrict a few hours after delivery, followed by a more gradual constriction to produce final functional closure by 1–8 days (Lees and Sunderland, 1977). Failure of the ductus arteriosus to constrict may result in persistence of fetal circulation with blood being diverted from the right ventricle into the aorta.

4. Endocardial cushion defects. Abnormal development of the endocardial cushions may be partial, resulting in an ostium primum atrial septal defect, or complete resulting in additional significant deficiency of the intraventricular septum, producing a common atrioventricular canal. The atrioventricular valves are usually malformed, deficient, or abnormally attached to the ventricular septum.

5. Obstetrical considerations. The most common findings in isolated atrial septal defect (ASD), ventricular septal defect (VSD), and patent ductus arteriosus (PDA) are (1) left to right cardiovascular shunting due to the pulmonary vascular resistance being lower than systemic pressure and (2) lower right ventricular and pulmonary artery pressures than left ventricular and aortic pressures. Unless there are additional associated anomalies (e.g., pulmonary stenosis) shunting is predominantly left to right in endocardial cushion defects.

During pregnancy the most frequent complications associated with left to right cardiovascular shunting are congestive heart failure and bacterial endocarditis (specifically, in PDA and VSD). In 1960, Mendelson reviewed the outcome of 402 pregnancies associated with PDA, ASD, and VSD that were reported from 1938 to 1960. Nine women underwent surgical ligation of the PDA prior to pregnancy, and 22 underwent the procedure during pregnancy. There were no maternal deaths and only one fetal loss in each group. In two cases of ASD surgically repaired (one prior to pregnancy and one during), both mothers and fetuses survived.

In 1967, Mortenson and Ellsworth reviewed 227 pregnancies in 96 women with left to right cardiovascular shunts before and after surgical repair. The cardiovascular lesions included ASD, VSD, PDA, atrioventricular canal, "balanced" tetralogy of Fallot, and anomalous pulmonary venous connections. Fifty-four women had 134 pregnancies before surgical repair: 49 women had uncomplicated pregnancies; 26 women had cardiovascular complication (e.g., congestive failure); 59 pregnancies resulted in spontaneous abortion or stillbirth. Forty-two women had 93 pregnancies following surgical repair: 83 were uncomplicated, and 1 was associated with a cardiovascular complication; 9 pregnancies resulted in spontaneous abortion or stillbirth. Finally, Meffert and Stansel (1968) and Collins *et al.* (1977) have reported successful surgical repair of ASD and PDA during pregnancy with subsequent delivery of normal newborns.

In patients with large left to right shunts and pulmonary artery hypertension, narrowing and obliterative pulmonary vascular disease occurs as result of medial hypertrophy, intimal thickening, and prominence of the elastic membrane; therefore, shunting becomes bidirectional and subsequently reversed, the so-called Eisenmenger syndrome. Ultimately pulmonary hypertension, cyanosis, polycythemia and clubbing develop, and the patient may go into congestive heart failure.

During normal pregnancy there is a hormonally mediated decrease in both peripheral and pulmonary vascular resistences (Metcalfe and Ueland, 1974; Burch, 1977). By contrast, patients with the Eisenmenger syndrome have high pulmonary vascular resistance in combination with decreased systemic resistance, thus leading to a progression in symptomatology. Accordingly, pregnancy in such patients is particularly dangerous. Jones and Howitt (1965) reported 26 pregnancies in 16 patients with the Eisenmenger syndrome; five of these resulted in a maternal death. Many of the maternal deaths occur either during parturition or during the first postpartum week (Howitt, 1971). Neilson *et al.* (1970, 1971) concluded that prior diagnosis of the Eisenmenger syndrome is a strong contraindication for allowing pregnancy to continue to term. Thor-Björn and Werkö (1974) have stated that once started, termination of pregnancy may carry as great a risk as continuation; however, the basis for such a statement appears tenuous.

b. Coarctation of the Aorta. In coarctation of the aorta (CA) the lumen of the aorta narrows, usually near insertion of the ductus arteriosus or its vestige. Coarctation of the aorta occurs more commonly in males (M:F = 2:1) and comprises about 5% of CHD (Nora and Nora, 1978a). The estimated incidence of CA in females is 1:300 to 1:1000 (Pritchard, 1953).

Bonnet (1903) originally classified the defect into an "infantile" form with a long diffuse narrowing of the aorta and an "adult" form where the coarctation was characterized by an abrupt short isthmic narrowing. This chronologic classification was subsequently found to be unsatisfactory because some infants had the "adult" form and, conversely, some adults had the "infantile" form. Various other classifications have been proposed, including that of Johnson *et al.* (1951) based on the position of the ductus in relation to the coarctation (preductal or postductal) and whether the ductus is patent or closed. The classification of Nadas and Fyler (1972) describes the coarctation in terms of a long "preductal" lesion between the left subclavian artery and the ductus, and the localized "post-ductal" lesion at or beyond the insertion of the ductus.

Obstetrical considerations. Hamilton and Stewart (1933) reported the first recognized case of CA associated with pregnancy. Labor and delivery were described as uncomplicated. In 1940 Mendelson reported three cases of CA associated with pregnancy and reviewed an additional 26 cases in the literature. Five of the 26 women died during pregnancy or in the immediate postpartum period (two of aortic rupture, one of cardiac failure, one of a cerebrovascular accident, and one of endocarditis); 10 patients surviving pregnancy had residual deleterious effects. Mendelson concluded that women with CA should be discouraged from becoming pregnant and urged contraception, sterilization, and early therapeutic abortion. If pregnancies were permitted to continue, he recommended elective cesarean section at term. Pritchard (1953) reviewed 77 cases of CA and pregnancy and reported two additional cases. Ten percent of these women died either during pregnancy or in the early puerperium. Causes of death included aortic rupture (four cases), heart failure (two cases), and cerebrovascular accident (one case). The material mortality rate was not decreased among cases delivered by cesarean section (7.1%) as compared to the group allowed to deliver per vagina (6.5%). Pritchard, therefore, concluded that cesarean section should be limited to obstetrical indications. In addition, he felt that therapeutic abortion was infrequently indicated in these patients. In 1960, Mendelson expanded his earlier survey of CA associated with pregnancy and ascertained an additional nine maternal deaths. Thus, in a total of 200 women with uncorrected CA there were four maternal deaths (7%) and a cardiovascular complication rate of 40%. He continued to support his original recommendation regarding therapeutic abortion in these patients. Recently, Ueland (1978) has recommended that women with uncomplicated CA may be allowed to labor using either systemic analgesia or conduction anesthesia. The second stage of labor should be shortened by forceps delivery to reduce expulsive efforts. Patients should be given prophylactic antibiotics during labor and at least 48 hours following delivery. There are several reports of surgical repair of CA during pregnancy with successful maternal and fetal outcomes (Collins *et al.*, 1968; Barash *et al.*, 1975). However, the general consensus has been that surgical repair during pregnancy should be avoided if possible. Women who have had surgical correction of uncomplicated CA appear to tolerate pregnancy well.

c. Tetralogy of Fallot. The four components of this malformation are (1) ventricular septal defect; (2) obstruction to right ventricular outflow; (3) overriding of the aorta; and (4) right ventricular hypertrophy. Tetralogy of Fallot (TF) constitutes about 10% of all forms of congenital heart disease and is the anomaly most frequently responsible for cyanosis in patients over 1 year of age (Kahler, 1975). The lesion responsible for the most significant hemodynamic disturbance is pulmonary stenosis, the severity of which determines the degree of right to left shunting and the amount of blood flow to the lungs. Patients with extensive lesions usually do not survive infancy or childhood without surgical correction, but in the mildest forms they may reach the fourth decade (Guntheroth and Kawabori, 1977). Palliative surgery consists of anastomosing the systemic and pulmonary circulations to increase pulmonary blood flow.

Obstetrical considerations. Ueland (1978) delineated the following pregnancy-associated cardiovascular changes as being particularly hazardous to the pregnant woman with TF: (1) reduction in systemic peripheral vascular resistance and (2) diminution in venous return from such factors as supine hypotension, anesthesia during labor and for delivery, excessive blood loss at delivery, and ambulation in the early puerperium with pooling of blood in the legs, presumably predisposing to thrombophlebitis.

Mendelson (1960) reviewed 26 pregnant women with TF collected from the New York Lying-In Hospital and from the literature. Ten of the 26 women developed severe congestive heart failure. In this series there were three maternal mortalities (one at 26 weeks gestation; two during the postpartum period) and no fetal losses. In 1974 Batson reported 27 pregnancies in eight women with TF with no maternal deaths. In this series there were eight premature deliveries, three stillbirths, and 11 abortions. In combining the data from five studies, Kahler (1975) tabulated 22 pregnancies in 12 patients with TF. The fetal loss rate was 31.8%, and one maternal death was reported.

Despite these published risks, most authors agree that TF is not necessarily a contraindication to pregnancy. Indeed, there have now been several reports of successful pregnancies in women who have undergone complete repair of TF (Loh and Tan, 1971; Lewis *et al.,* 1972; Ralstin and Dunn, 1976). However, Ueland (1978) cautions that in those women who have had only palliative surgery (i.e., shunting procedures or pulmonary valvotomy) pregnancy represents a continued risk requiring close medical supervision.

d. Valvular Aortic Stenosis. Campbell 1968b, 1973) estimated that valvular aortic stenosis (VAS) comprises approximately 5% of congenital heart lesions. In this condition the valve can have a single cusp or fusion of the three cusps. With the unicusp valve, there is about 90% incidence of associated aortic insufficiency; with the bicusp valve, this complication is rare (Falcone *et al.,* 1971). The majority of patients with VAS do well during the first three decades; however, deteterious hemodynamic changes are often progressive during these asymptomatic decades (Campbell, 1968a).

Obstetrical considerations. Relatively few cases of VAS in association with pregnancy have been reported. Arias and Pineda (1978) recently reported such a patient, and reviewed an additional 23 cases in the English language literature. Among these 24 women were four maternal deaths (maternal mortality rate of 17%) and 40 pregnancies with 13 fetal losses (fetal mortality rate of 32%). The authors concluded (1) termination of pregnancy is usually unnecessary, (2) surgical replacement of the aortic valve during pregnancy may be dangerous to both mother and fetus, (3) patients with severe aortic stenosis should be hospitalized during the duration of pregnancy, and (4) delivery per vagina is preferred unless obstetric complications dictate cesarean section. The authors felt that there was insufficient experience regarding the safest anesthesia either for delivery per vagina or cesarean section. They suggested caution in using conduction anesthesia because of sympathetic blockade with loss of peripheral vascular resistance which could result in a postenotic vascular resistance drop, an increase in the valvular gradient, and sudden cardiac failure. On the other hand, they noted frequent fluctuations in heart rate and cardiac output during intubation and extubation, which could be dangerous in these women with labile hemodynamics. In their own case they used local anesthesia for cesarean section.

e. Ebstein Anomaly. This abnormality consists of a downward displacement of an abnormal tricuspid valve into the right ventricular cavity which is reduced in size, and the right atrial chamber is correspondingly enlarged. Consequently, the effective output from the right side of the heart is decreased because of the small size of the functionary right ventricle. The severity of the disorder depends on the extent of malattachment and the structural anomalies of the tricuspid valve (Nadas and Fyler, 1972). Associated anomalies are common, interatrial communication being most frequently encountered. Although survival beyond the sixth decade has been reported, most authors report the mean age of death in the second decade (Van Mierop *et al.*, 1977).

Obstetrical considerations. Only a few reports of pregnancy associated with Ebstein's anomaly have been published. In 1960, Mendelson reported three pregnancies in two women with Ebstein's anomaly. In one instance, severe cardiac failure developed during the woman's second pregnancy. From a 10-year record review at the Ohio State University Hospital, Copeland *et al.* (1963) ascertained 125 women with CHD who had 372 pregnancies. Two of the three women with Ebstein's anomaly were cyanotic; however, there were no maternal deaths. These three women had a total of six pregnancies, two of which resulted in fetal losses (cause not stated). In 1970, Neilson *et al.* reported an additional patient with Ebstein's anomaly and cyanosis; she had three uncomplicated pregnancies.

As suggested by Littler (1970) women with Ebstein's anomaly should be carefully evaluated before advising against pregnancy or its termination because a "successful outcome may be achieved." The possible hazards during pregnancy are (1) congestive heart failure; (2) arrhythmias; (3) embolization; and (4) subacute endocarditis. The presence of cyanosis appears to increase the incidence of spontaneous abortions.

f. Two Chambered Heart. Seeds and Cefalo (1977) reported a patient with a single atrium, a single ventricle, and one atrioventricular valve. The pregnancy was complicated by intractable hemoptysis; however, the patient survived. Apparently labor and delivery were uncomplicated.

g. Tricuspid Atresia. In tricuspid atresia there is no direct communication between the right atrium and the right ventricle because the right atrioventricular valve is atretic. The right atrial blood is shunted through either the foramen ovale or an atrial defect into the left atrium, where complete mixture with pulmonary venous blood occurs.

Obstetrical considerations. Collins *et al.* (1977) reported a case of pregnancy associated with tricuspid atresia and reviewed two other cases reported in the literature. Only one infant survived in a total of nine pregnancies. Among the maternal complications reported are (1) cardiac failure; (2) pulmonary emboli; (3) arterial embolization with grand mal seizure; and (4) premature labor. The authors conclude that therapeutic abortion is indicated and sterilization should be encouraged. Should pregnancy occur, they recommend immediate heparinization to decrease emboli, and avoidance of conduction anesthesia because of the danger of shunt reversal and intractable heart failure.

III. DRUGS AND CARDIOVASCULAR DEFECTS

Several drugs have been implicated in producing cardiac malformations in offspring of women exposed during pregnancy. Essential components to the production of such defects by teratogens are (1) genetic predisposition to maldevelopment; (2) genetic predisposition to react adversely to a specific agent; (3) dose effect; and (4) exposure during a vulnerable period of development. A few drugs have been shown to carry a considerable risk of disturbing organogenesis, thereby producing congenital malformations. Other drugs may produce these results in a small but significant number of individuals. Table VI provides a selected list of drugs which have been associated with cardiovascular teratogenicity.

One drug which deserves special consideration is coumarin, an agent commonly used for long-term anticoagulation. With increasing frequency, women who require such therapy are becoming pregnant (e.g., following reparative cardiac surgery with placement of prosthetic valves). Teratogenic effects specifi-

TABLE VI
Possible Cardiovascular Teratogens[a]

Drug	Most common malformations
Alcohol	Ventricular septal defect (VSD); patent ductur arterosis (PDA); atrial septal defect (ASD)
Amphetamines	VSD; PDA; ASD; transposition of great arteries (TGA)
Anticonvulsants	
Hydantoin	Pulmonary stenosis (PS); aortic stenosis (AS); coarctation of aorta; PDA
Trimethadione	TGA; tetralogy of Fallot (TF); hypoplastic left heart
Lithium	Ebstein's anomaly; tricuspid atresia; ASD
Progestagens	VSD; TGA; TF
Thalidomide	TF; ASD; complex lesions

[a] Modified from Nora and Nora (1978a). The validity of many of these teratogens, as well as their risks, are not known.

cally identified with coumarin exposure during pregnancy have been recognized (Holzgreve *et al.,* 1976; Pauli *et al.,* 1977). First trimester exposure has been associated with nasal hypoplasia, stippled epiphyses, and possibly mental retardation. Second trimester exposure appears to result in mental retardation, microcephaly, and optic atrophy. Whether the teratogenic effect is mediated through multiple small hemorrhages or some other means is not yet known. Recently Hall *et al.* (1980) reviewed the published cases in which coumarin derivatives or heparin were administered during pregnancy. Of 418 reported cases, one-sixth resulted in abnormal liveborn, one-sixth resulted in abortion or stillbirth, and two-thirds in apparently normal infants. In the past it was recommended that pregnant women who require long-term anticoagulation be taught self-administration of heparin (Hill and Pearson, 1971). However, among the 135 published cases of heparin administration during pregnancy reviewed by Hall *et al.* (1980), the infants in one-eighth were stillborn, in one-fifth premature (a third of whom died), and, again at most, in two-thirds apparently normal. Because of the substantial risks of both the coumarin derivatives and heparin, and the inherent risks of pregnancy complicated by the indications for anticoagulation, Hall *et al.* suggested that pregnancy prevention is usually indicated. However, if pregnancy does occur, a normal outcome may be expected in about two-thirds of the pregnancies regardless of the anticoagulant utilized.

IV. PRENATAL DIAGNOSIS OF ISOLATED CONGENITAL HEART DEFECTS

With recent technical advances in ultrasonography (e.g., M-mode, grey-scale, and real-time) the prenatal diagnosis of polygenic heart disease may be possible. Patel and Goldberg (1976) diagnosed a three-to-one congenital heart block during the third trimester by means of an M-mode motion type of ultrasonic display showing a reduced fetal heart rate of approximately 60 beats per minute. With the information provided by ultrasound, a pediatric cardiac surgeon was available at delivery for insertion of a pacemaker. Using similar methods, J. C. Hobbins (Yale University, New Haven, Connecticut, personal communications, 1979) has been successful in establishing the third trimester prenatal diagnosis of (1) tricuspid atresia with hypoplastic right ventricle; (2) isolated hypoplastic right ventricle; and (3) first degree heart block in a fetus of a mother who subsequently was diagnosed as having lupus erythematosis. In addition, those investigators are prospectively evaluating pregnancies where mothers have been exposed to progestagens during the first trimester, specifically looking for transposition of the great vessels.

Finally, Hamilton *et al.* (1977) suspected a congenital heart defect in a 32-week-fetus using simultaneous fetal phonocardiographic and abdominal signals. The infant was subsequently shown to have a complex form of cyanotic heart disease consisting of a common atrium, complete atrioventricular canal, double outlet right ventricle, valvular and infundibular pulmonic stenosis, aneurysmal right aortic arch with mirror-image branching, and interrupted left-sided vena cava with azygous continuation. Such electronic fetal monitoring systems may further expand our capabilities for prenatal diagnosis of CHD in the future. However, at present such techniques must still be regarded in the stage of applied research and reliability cannot be stated.

REFERENCES

Anders, J. M., Moores, E. C., and Emanuel, R. (1965). Chromosome studies in 156 patients with congenital heart disease. *Br. Heart J.* **27,** 756–762.

Arias, F., and Pineda, J. (1978). Aortic stenosis and pregnancy. *J. Reprod. Med.* **20,** 229–231.

Barash, P. G., Hobbins, J. C., Hook, R., Stansel, H. C., Whittemore, R., and Hehre, F. W. (1975). Management of coarctation of the aorta during pregnancy. *J. Thorac. Cardiovasc. Surg.* **69,** 781–784.

Batson, G. A. (1974). Cyanotic congenital heart disease and pregnancy. *J. Obstet. Gynaecol. Br. Commonw.* **81,** 549–553.

Bercu, B. A., Dietteret, G. A., Danforth, W. H., Pund, E. E., Jr., Ahlvin, R. C., and Belliveau, R. R. (1958). Pseudoaortic stenosis produced by ventricular hypertrophy. *Am. J. Med.* **25,** 814–818.

Bizarro, R. O., Callahan, J. A., Feldt, R. H., Kurland, L. T., Gordon, H., and Brandenburg, R. O. (1970). Familial atrial septal defect with prolonged atrioventricular conduction: A syndrome showing the autosomal dominant pattern of inheritance. *Circulation* **41,** 677–683.

Bonnet, L. M. (1903). Sur la lésion dite sténose congénitale de l'aorte dans la région de l'isthme. *Rev. Med. (Paris)* **23,** 108–126.

Braunwald, E., Costas, T., Lambrew, S., Rockoff, D., Ross, J., Jr., and Morrow, A. G. (1964). Idiopathic hypertrophic subaortic stenosis: A description of the disease based upon an analysis of 64 patients. *Circulation* **30** Suppl. 4, 1–119.

Brownell, L. G., and Shokeir, M. H. K. (1976). Inheritance of hypoplastic left heart syndrome (HLHS): Further observations. *Clin. Genet.* **9,** 245–249.

Bunim, J. J., and Rubricuis, J. (1948). The determination of the prognosis of pregnancy in rheumatic heart disease. *Am. Heart J.* **35,** 282–297.

Burch, G. E. (1977). Heart disease and pregnancy. *Am. Heart J.* **93,** 104–116.

Campbell, M. (1965). Causes of malformations of the heart. *Br. Med. J.* **2,** 895–904.

Campbell, M. (1968a). The natural history of congenital aortic stenosis. *Br. Heart J.* **30,** 514–526.

Campbell, M. (1968b). The incidence and later distribution of malformations of the heart. *In* "Pediatric Cardiology" (H. Watson, ed.), pp. 71–83. Lloyd-Luke, London.

Campbell, M. (1973). Incidence of cardiac malformations at birth and later, and neonatal mortality. *Br. Heart J.* **35,** 189–200.

Clark, C. E., Henry, W. L., and Epstein, S. C. (1973). Familial prevalence of idiopathic subaortic stenosis. *N. Engl. J. Med.* **289,** 709–714.

Collins, H. A., Daniel, R. A., and Scott, H. W., Jr. (1968). Cardiac surgery during pregnancy. *Ann. Thorac. Surg.* **5,** 300–309.

Collins, M. L., Leal, J., and Thompson, N. J. (1977). Tricuspid atresia and pregnancy. *Obstet. Gynecol.* **50,** Suppl., 725–735.

Combrink, J. M., Davis, W. H., and Snyman, H. W. (1962). Familial bundle branch block. *Am. Heart J.* **64,** 397–400.

Copeland, W. E., Wooley, C. F., Ryan, J. M., Runco, V., and Levin, H. S. (1963). Pregnancy and congenital heart disease. *Am. J. Obstet. Gynecol.* **86,** 107–110.

Dahl, G. (1970). Chromosomal conditions in congenital heart disease. *Acta Paediatr. Scand.* **59,** 65–73.

Datta, S., Kitzmiller, J. L., Ostheimer, G. W., and Schoenbaum, S. C. (1978). Propranolol and parturition. *Obstet. Gynecol.* **51,** 577–581.

Davies, L. G. (1952). A familial heart disease. *Br. Heart J.* **14,** 206–212.

Edwards, J. H. (1960). Simulation of Mendelism. *Acta Genet. Stat. Med.* **10,** 63–70.

Elias, S., and Burkowitz, R. L. (1976). The Martan syndrome and pregnancy. *Obstet. Gynecol.* **47,** 358–361.

Esscher, E., Hardell, L. I., and Michaelsson, M. (1975). Familial, isolated, complete right bundle-branch block. *Br. Heart J.* **37,** 745–747.

Falcone, M. W., Roberts, W. C., Morrow, A. G., and Perloff, J. K. (1971). Congenital aortic stenosis resulting from unicommisural valve: Clinical and anatomic features in twenty-one adult patients. *Circulation* **44,** 272–280.

Gazes, P. C., Culler, R. M., Taber, E., and Kelley, T. E. (1965). Congenital familial cardiac conduction defects. *Circulation* **32,** 32–34.

German, J., Ehlers, K. H., and Engle, M. A. (1966). Familial congenital heart disease. II. Chromosomal studies. *Circulation* **34,** 517–523.

Grumbach, M. M. (1968). The reproductive system. *In* "The Biologic Basis of Pediatric Practice" (R. E. Cooke, ed.), pp. 1058–1082. McGraw-Hill, New York.

Guntheroth, W. G., and Kawabori, I. (1977). Ted-Rad of Fallot. *In* "Heart Disease in Infants and

Children'' (A. J. Moss, F. H. Adams, and G. C. Emmanouilides, eds.), 2nd ed., pp. 276–289. Williams & Wilkins, Baltimore, Maryland.

Hall, J. G., Paul, R. M., and Wilson, K. M. (1980). Maternal and fetal sequelae of anticoagulation during pregnancy. *Am. J. Med.* **68,** 122–140.

Hamilton, B. E., and Stewart, C. C. (1933). Four living adults with coarctation of the aorta. *N. Engl. J. Med.* **209,** 958–959.

Hamilton, L. A., Fisher, E., Horn, C., DuBrow, I., and Vidyasagar, D. (1977). A new prenatal cardiac diagnostic device for congenital heart disease. *Obstet. Gynecol.* **50,** 491–494.

Harris, J. W. S. (1952). A case of congenital heart block recognized at the thirty-eighth week of pregnancy. *J. Obstet. Gynaecol. Br. Commonw.* **59,** 404–405.

Henry, W. L., Clark, C. E., Glancy, D. L., and Epstein, S. E. (1972). Non-invasive (echocardiographic) measurement of the left ventricular outflow gradient in IHSS. *Circulation* **46,** Suppl. II, 62 (abstr.).

Henry, W. L., Clark, C. E., and Epstein, S. E. (1973). Asymmetric septal hypertrophy: Echocardiographic identification of the pathognomonic anatomic abnormality of IHSS. *Circulation* **47,** 225–233.

Hibbard, L. T. (1975). Maternal mortality due to cardiac disease. *Clin. Obstet. Gynecol.* **18,** 27–36.

Hill, W. C., and Pearson, J. W. (1971). Outpatient intravenous heparin therapy for antepartum iliofemoral thrombophlebitis. *Obstet. Gynecol.* **37,** 785–789.

Hoffman, J. I. E., and Christianson, R. (1978). Congenital heart disease in a cohort of 19,502 births with long-term follow-up. *Am. J. Cardiol.* **42,** 641–647.

Holmes, L. B., Rose, V., Child, A. H., and Kratzer, W. (1974). Commentary on the inheritance of the hypoplastic left heart syndrome. Clinical delineation of birth defects. XVI. Urinary system and others. *Birth Defects, Orig. Artic. Ser.* **10**(4), 228–230.

Holzgreve, W., Carey, J. C., and Hall, B. D. (1976). Warfarin-induced fetal abnormalities. *Lancet* **2,** 914–915.

Hook, E. B., and Hamerton, J. L. (1977). The frequency of chromosome abnormalities in consecutive newborn studies—differences between studies—results by sex and by severity of phenotypic involvement. *In* ''Population Cytogenetics'' (E. B. Hook and I. H. Porter, eds.), pp. 63–79. Academic Press, New York.

Horger, E. O. (1974). Hemoglobinopathies in pregnancy. *Clin. Obstet. Gynecol.* **17,** 127–162.

Horkick, L., Petkovich, N. J., and Boton, C. F. (1966). Idiopathic hypertrophic subvalvular stenosis. A study of a family involving four generations. Clinical, hemodynamic and pathologic observations. *Am. J. Cardiol.* **17,** 411–418.

Howitt, G. (1971). Heart disease and pregnancy. *Practitioner* **206,** 765–780.

Husson, G. S., Blackman, M. S., Rogers, M. C., Bharati, S., and Lev, M. (1973). Familial congenital bundle branch system disease. *Am. J. Cardiol.* **32,** 365–369.

Johnson, A. L., Ferencz, C., Wiglesworth, F. W., and McRae, D. L. (1951). Coarctation of the aorta complicated by patency of the ductus arteriosus: Physiologic considerations in the classification of coarctation of the aorta. *Circulation* **4,** 242–250.

Jones, A. M., and Howitt, G. (1965). Eisenmenger syndrome in pregnancy. *Br. Med. J.* **1,** 1627–1631.

Kahler, R. L. (1975). Cardiac disease. *In* ''Medical Complications During Pregnancy'' (N. Burrow and T. F. Ferris, eds.), pp. 105–149. Saunders, Philadelphia, Philadelphia.

Kariv, I., and Solomon, M. (1962). Familial cardiomyopathy. *Proc. Tel. Hashomer Hosp.* **1,** 53–57.

Karsh, R. B., Knapp, R. F., Nora, J. J., Wolfe, R. R., and Robinson, A. (1975). Congenital heart disease in 49,XXXXY. *Pediatrics* **56,** 462–464.

Kolibash, A. J., Ruiz, D. E., and Lewis, R. P. (1975). Idiopathic hypertrophic subaortic stenosis in pregnancy. *Ann. Intern. Med.* **82,** 791–794.

Lees, M. H., and Sunderland, C. O. (1977). Diseases of the cardiovascular system. *In* ''Neonatal-

Perinatal Medicine, Diseases of the Fetus and Newborn'' (R. E. Behrman, ed.), 2nd ed., pp. 446–574. Mosby, St. Louis, Missouri.

Lewis, B. S., Rogers, N. M. A., and Gotsman, M. S. (1972). Successful pregnancy after repair of Gallot's tetralogy. *S. Afr. Med. J.* **46,** 934–936.

Littler, W. A. (1970). Successful pregnancy in a patient with Ebstein's anomaly. *Br. Heart J.* **32,** 711–713.

Loh, T. F., and Tan, N. C. (1971). Fallot's tetrology and pregnancy. A report of a successful pregnancy after complete repair. *Med. J. Aust.* **2,** 141–145.

McKusick, V. A. (1978). ''Mendelian Inheritance in Man,'' 5th ed., p. 552. Johns Hopkins Univ. Press, Baltimore, Maryland.

Meffert, W. G., and Stansel, H. C. (1968). Open heart surgery during pregnancy. *Am. J. Obstet. Gynecol.* **102,** 1116–1120.

Mendelson, C. L. (1940). Pregnancy and coarctation of the aorta. *Am. J. Obstet. Gynecol.* **39,** 1014–1021.

Mendelson, C. L. (1960). Congenital heart disease. *In* ''Cardiac Disease in Pregnancy: Medical Care, Cardiovascular Surgery and Obstetrics Management as Related to Material and Fetal Welfare'' (C. E. Heaton, ed.), pp. 122–175. Davis, Philadelphia, Pennsylvania.

Metcalfe, J., and Ueland, K. (1974). Maternal cardiovascular adjustments to pregnancy. *Prog. Cardiovasc. Dis.* **16,** 336–374.

Mitchell, S. C., Korones, S. B., and Berendes, H. W. (1969). Incidence of congenital heart disease in 54,033 births. *Circulation* **40,** Suppl. III, 148 (abstr.).

Monteleone, P. L., and Fagan, L. F. (1969). Possible X-linked congenital heart disease. *Circulation* **39,** 611–614.

Mortensen, J. D., and Ellsworth, H. S. (1967). Pregnancy before and after surgical correction of left-to-right cardiovascular shunts. *Obstet. Gynecol.* **29,** 241–243.

Mullins, D. H., Estrada, W. R., and Gready, T. G. (1960). Pregnancy in an adult mongoloid female. *Obstet. Gynecol.* **15,** 781–783.

Mulvihill, J. J. (1972). Comments on the epidemiology of congenital heart disease in dogs. *Birth Defects, Orig. Artic. Ser.* **8** (5), 175–177.

Nadas, A. S., and Fyler, D. C. (1972). ''Pediatric Cardiology,'' 3rd ed. Saunders, Philadelphia, Pennsylvania.

Neilson, G., Galea, E. G., and Blunt, A. (1970). Congenital heart disease and pregnancy. *Med. J. Aust.* **1,** 1086–1088.

Neilson, G., Galea, E. G., and Blunt, A. (1971). Eisenmenger's syndrome and pregnancy. *Med. J. Aust.* **1,** 431–434.

Noonan, J. A., and Nadas, A. S. (1958). The hypoplastic left heart syndrome. *Pediatr. Clin. North Am.* **5,** 1029–1056.

Nora, J. J. (1971). Etiologic factors in congenital heart disease. *Pediatr. Clin. North Am.* **18,** 1059–1074.

Nora, J. J. (1977). Etiologic aspects of congenital heart disease. *In* ''Heart Disease in Infants, Children, and Adolescents'' (A. J. Moss, F. H. Adams, and G. C. Emmanouilides, eds.), 2nd ed., pp. 3–11. Williams & Wilkins, Baltimore, Maryland.

Nora, J. J., and Fraser, F. C. (1974). Cardiovascular disease. *In* ''Medical Genetics: Principles and Practice,'' pp. 331–345. Lea & Febiger, Philadelphia, Pennsylvania.

Nora, J. J., and Nora, A. H. (1976). Recurrence risks in children having one parent with a congenital heart disease. *Circulation* **53,** 701–702.

Nora, J. J., and Nora, A. H. (1978a). ''Genetics and Counseling in Cardiovascular Diseases.'' Thomas, Springfield, Illinois.

Nora, J. J., and Nora, A. H. (1978b). The evolution of specific genetic and environmental counseling in congenital heart disease. *Circulation* **57,** 205–213.

Nora, J. J., McGill, C. W., and McNamara, D. G. (1970). Emperic recurrence risks in common and uncommon congenital heart lesions. *Teratology* **3,** 325–330.

Pare, J. A. P., Fraser, R. G., Pirozynski, W. J., Shanks, J. A., and Stubington, D. (1961). Hereditary cardiovascular dysplasia. A form of familial cardiomyopathy. *Am. J. Med.* **31,** 37–62.

Patel, J., and Goldberg, B. B. (1976). Prenatal genetic diagnosis and ultrasound. *Clin. Obstet. Gynecol.* **19,** 893–907.

Pauli, R. M., Hall, J. G., and Shaul, W. L. (1977). Spectrum of ultra-uterine effects of Warfarin. *In* "Fifth International Conference on Birth Defects" (J. W. Littlefield, F. J. G. Ebling, and I. W. Henderson, eds.), Abstr. Pap., P. 68. Excerpta Med. Found., Amsterdam.

Pease, W. E., Nordenberg, A., and Ladda, R. L. (1976). Familial atrial septal defect with prolonged atrioventricular conduction. *Circulation* **53,** 759–762.

Perkins, R. P. (1971). Inherited disorders of hemoglobin synthesis and pregnancy. *Am. J. Obstet. Gynecol.* **111,** 120–159.

Pritchard, J. A. (1953). Coarctation of the aorta and pregnancy. *Obstet. & Gynecol. Surv.* **8,** 775–791.

Ralstin, J. H., and Dunn, M. (1976). Pregnancies after surgical correction of tetrology of Fallot. *J. Am. Med. Assoc.* **235,** 2627–2628.

Renwick, D. H. G., Miller, J. R., and Paterson, D. (1964). Estimates of incidence and prevalence of Monglolism and of congenital heart diesease in British Columbia. *Can. Med. Assoc. J.* **91,** 365–371.

Roberts, W. C. (1973). Valvular, subvalvular, and supravalvular aortic stenosis. *Cardiovasc. Clin.* **5,** 104–126.

Rosenthal, A. (1972). Cardiovascular malformations in Klinefelter's syndrome; Report of three cases. *J. Pediatr.* **80,** 471–743.

Schaefer, G., Arditi, L. I., Solomon, H. A., and Ringland, J. E. (1968). Congenital heart disease and pregnancy. *Clin. Obstet. Gynecol.* **11,** 1048–1063.

Scharrer, S., Stengel-Rutkowski, S., Rodewald-Rudescu, A., Erdlen, E., and Zang, K. D. (1975). Reproduction in a female patient with Down's syndrome. *Humangenetik* **26,** 207–214.

Schroeder, J. S., and Harrison, D. C. (1971). Repeated cardioversion during pregnancy: Treatment of refractory paroxysmal atrial tachycardia during three successive pregnancies. *Am. J. Cardiol.* **27,** 445–446.

Seeds, J. W., and Cefalo, R. C. (1977). Pregnancy with congenital heart disease: A two-chambered heart. *Am. J. Obstet. Gynecol.* **127,** 213–214.

Shokeir, M. H. K. (1971). Hypoplastic left heart syndrome: An autosomal recessive disorder. *Clin. Genet.* **2,** 7–14.

Simpson, J. L. (1976). "Disorders of Sexual Differentiation: Etiology and Clinical Delineation." Academic Press, New York.

Thor-Björn, C., and Werkö, L. (1974). Management of heart disease in pregnancy. *Prog. Cardiovasc. Dis.* **16,** 407–419.

Turner, G. M., Oakley, C. M., and Dixon, H. G. (1968). Management of pregnancy complicated by hypertrophic obstructive cardiomyopathy. *Br. Med. J.* **4,** 281–284.

Uchida, I. A. (1978). Familial occurrence of congenital heart disease. *In* "Heart Disease in Infancy and Childhood" (J. D. Keith and R. D. Rowe, eds.), 3rd ed., pp. 153–159. Macmillan, New York.

Ueland, K. (1977). Pregnancy and cardiovascular disease. *Med. Clin. North Am.* **61,** 17–41.

Ueland, K. (1978). Cardiovascular diseases complicating pregnancy. *Clin. Obstet. Gynecol.* **21,** 429–442.

Van Mierop, L. H. S., Schiebler, G. L., and Victorica, B. E. (1977). Anomalies of the tricuspid valve resulting in stenosis or incompetence. *In* "Heart Disease in Infants, Children and

Adolescents" (A. J. Moss, F. H. Adams, and G. C. Emmanouiledes, eds.), pp. 262–276. Williams & Wilkins, Baltimore, Maryland.

Verel, D., and Chandrasekhar, K. P. (1970). Congenital complete heart block in siblings. *Cardiology* **55,** 242–248.

Warkany, J., Passarge, E., and Smith, L. B. (1966). Congenital malformations in autosomal trisomy syndromes. *Am. J. Dis. Child.* **112,** 502–517.

Wood, R. S., Taylor, W. J., Wheat, M. W., Jr., and Schriebler, G. L. (1962). Muscular subaortic stenosis in childhood: Report of occurrence in three siblings. *Pediatrics* **30,** 749–758.

Yerushalmy, J. (1969). The child health and development studies. Berkley and Oakland, California-description and selected findings. *Exerpta Med. Found. Int. Congr. Ser.* **191,** 20 (abstr.).

7

Respiratory System Disorders

SHERMAN ELIAS

A number of genetic disorders involve the respiratory system, either primarily or as part of a multisystem syndrome. The obstetrician may be called upon to (1) provide medical care for a woman who is herself affected with such a disorder or (2) offer genetic counseling and, if available, prenatal diagnosis to couples at risk for having affected offspring. This chapter will consider genetic disorders of the respiratory system and their relationships to pregnancy.

I. STRUCTURAL ABNORMALITIES OF THE RESPIRATORY SYSTEM

The structural abnormalities of the respiratory system may be categorized into those defects primarily involving the respiratory tract alone (trachea, bronchi, and lung parenchyma) and those in which the respiratory tract is part of a malformation syndrome involving other organs.

Copyright © 1981 by Academic Press, Inc.
All rights of reproduction in any form reserved.
ISBN 0-12-630940-X

A. Defects Primarily Involving the Respiratory Tract

1. Tracheoesophageal Fistula

The incidence of esophageal atresia with or without tracheoesophageal fistula (TEF) is about 1 in 3000 births (Cudmore, 1978). At least seven different anatomical varieties of TEF have been described (Waterson *et al.*, 1963). About one-half of the infants with TEF have other anomalies with renal, cardiac, and rectal defects being most common. Recently the so-called VATER association (an acronym for vertebral, anal, tracheoesophageal, radial, and renal anomalies) has been recognized (Quan and Smith, 1973). Infants with this disorder frequently have only one umbilical artery further emphasizing the obstetric principle that the umbilical cord should be carefully inspected.

Dennis *et al.* (1973) concluded that most cases of esophageal atresia (with or without tracheal fistula) appear to be inherited in polygenic/multifactorial fashion. Familial cases have been reported including multiple affected sibs, concordance in monozygotic twins, and one kindred with affected individuals in three generations (Grieve and McDermott, 1939; Copleman *et al.*, 1950; Sloan and Haight, 1956; Woolley *et al.*, 1961; Dennis *et al.*, 1973). Genetic heterogeneity cannot be excluded and Mendelian inheritance is suggested in rare families. There have been several reports of women who have had successful repair of TEF and subsequently delivered infants who were similarly affected (Dennis *et al.*, 1973). It is therefore recommended that a nasogastric tube be passed to confirm the continuity of the esophagus in infants who have a first degree relative with either esophageal atresia or TEF.

Cozzi and Wilkinson (1975) reviewed 142 pregnancies in which fetuses were affected with either esophageal atresia or TEF. Twenty-one percent were preterm with a mean weight of 2 kg, 19% had intrauterine growth retardation with a mean weight of 2.23 kg, and the remaining 60% were born at term and weighed more than 2.5 kg. In 44 patients (30%) there were other severe malformations, most commonly anorectal abnormalities. Respiratory complications occurred in 41 of the 142 patients (29%). Pneumonia was more frequently diagnosed in term infants, and among these there was a significantly higher incidence in the small-for-dates infants as compared to those of normal birth weight. Of the 142 patients, five died before surgery could be performed. Of six patients with TEF (without esophageal atresia) the fistula was divided and all did well. Of the remaining 131 infants (TEF with esophageal atresia), 90 underwent primary surgical repair with division of the fistula and anastomosis of the esophagus. Mortality was similar for all babies treated by primary repair. In 41 cases the operation was done in stages, usually because of poor general health or because the gap between segments of the esophagus was too wide for primary anastomosis. Among this group the mortality rate was significantly higher in preterm infants as compared to term infants. There was no statistically significant difference in the mortality rate between term infants weighing more than 2.5 kg as compared to term infants of low birth weight (less than 2.5 kg).

There is a definite interrelationship between maternal polyhydraminos and fetal lesions which interfere with normal swallowing and alimentation. In reviewing 220 cases of alimentary tract obstruction at the Columbus Children's Hospital, Lloyd and Clatworth (1958) found that 7 of 53 infants (13.9%) with esophageal atresia were delivered of mothers with polyhydramnios. Duenhoelter *et al.* (1976) estimated that 1 of 15 healthy pregnant women with polyhydramnios carries a fetus with proximal gastrointestinal obstruction. The presence of polyhydraminos which is not readily explained (i.e., associated with diabetes mellitus, erythroblastosis fetalis, or multiple gestation) should, therefore, alert the physician to the possibility of a congenital malformation, especially of the upper gastrointestinal tract. Theoretically TEF associated with esophageal atresia should be antenatally diagnosable by amniography, at least in the third trimester of pregnancy. Such a diagnosis would enable corrective surgery to be performed soon after birth and reduce the likelihood of pulmonary aspiration and subsequent pneumonia. Finally, Pitkin and Reynolds (1975) have demonstrated that amniotic fluid protein is swallowed by the term monkey fetus and hydrolyzed in the gut into its component amino acids which are then absorbed and utilized in fetal protein synthesis. Thus, it has been suggested that the association between esophageal anomalies and impaired fetal growth may be due to the inability of the fetus to utilize swallowed amniotic fluid protein as a source of nutritional nitrogen requirements.

2. *Tracheobronchiomegaly (Mournier–Kuhn Syndrome)*

In 1932 Mournier–Kuhn first described tracheobronchiomegaly (dilatation of the trachea and main stem bronchi), often associated with diverticula and sacculations between cartilaginous rings. Clinical manifestations include a loud whooping chronic cough with poor ability to raise secretions. Expiratory collapse of the poorly supported airways leads to repeated attacks of bronchitis and pneumonitis (Himalstein and Gallagher, 1973). Roentgenographic findings include deep corrugations produced by redundant tissue between cartilaginous rings and a widened tracheal air column.

The basic lesion causing the enlarged diameter of the trachea and bronchi is unknown. Johnston and Green (1965) have proposed that tracheobronchiomegaly is inherited in autosomal recessive fashion. They reported five cases of the disorder, two of which were sibs whose parents were from the same county in Georgia, suggesting consanguinity.

The pregnant woman with tracheobronchiomegaly should be managed similarly to patients with bronchiectasis (see below). Finally, no method for prenatal diagnosis is currently available.

3. *Bronchiectasis*

Most familial cases of bronchiectasis are associated with other systemic abnormalities (e.g., cystic fibrosis). However, Danielson *et al.* (1967) reported an interesting family in which four of five sibs were affected with middle lobe bronchiectasis of early onset (all prior to age 17 years). No specific abnormalities of the tracheobronchial tree or immunoglobulins were demonstrated in any of these children. The mother, father, and oldest son were reported to have no significant history of respiratory disease. An autosomal recessive mode of inheritance was suggested.

Haynes (1969) estimated the incidence of bronchiectasis during pregnancy to be between 1 in 2500 to 1 in 3200. He observed that some women with long-standing bronchiectasis show increased dyspnea and cyanosis as gestation nears completion, although such cases are the exception. Most women will experience little or no clinically significant changes as result of an uncomplicated pregnancy. However, chronic bronchiectasis may result in nonspecific systemic debilitation of the pregnant patient. Haynes recommended frequent postural drainage and antibiotic therapy during episodes of acute infection. When necessary, ephedrine or aminophylline may be administered for bronchial spasm. During labor and delivery, agents which promote inspissation of bronchial secretions (e.g., atropine, scopolamine) or suppress the cough reflex (e.g., morphine, meperidine) should be avoided. Local infiltration, pudendal block, or conduction anesthesia are preferable to general anesthesia, and forceps should be utilized to shorten the second stage of labor. Finally, postpartum sterilization should be considered in patients with advanced disease.

4. *Bronchomalacia (Williams–Campbell Syndrome)*

In 1960 Williams and Campbell described an unusual type of bronchiestasis associated with generalized bronchomalacia of the bronchial tree. The clinical features are suggestive of asthmatic bronchitis or cystic fibrosis with chronic cough, wheezing, febrile attacks, and emphysema. There appears to be a considerable variation with respect to clinical severity. Some individuals will manifest symptoms of respiratory distress, whereas others will show no evidence of the disorder until about 1 year of age. Death as result of infection or severe pulmonary compromise may occur during the newborn period; however, individuals have survived into the second decade.

Wayne and Taussig (1976) reported two sibs who developed respiratory symptoms at birth and were both diagnosed as having the Williams-Campbell syndrome. Agosti *et al.* (1974) reported four of five affected sibs with generalized bronchomalacia. Although both parents originated from the same small village in Italy and had the same family name, consanguinity was not established with certainty. The clinical presentations of the cases reported by Agosti *et al.* were more severe than the typical case of Williams and Campbell syndrome suggesting heterogeneity of bronchial flaccidity conditions. Because of the high probability of consanguinity and multiple affected sibs, they assumed the condition to be inherited as a ''recessive character'' in their reported family.

Treatment of the pregnant woman with bronchomalacia would be similar effects to that of bronchiectasis (see above). Prenatal diagnosis is currently unavailable.

5. *Familial Spontaneous Pneumothorax*

The heritability of pulmonary bullae is uncertain and genetic heterogeneity probably exists. The condition occurs more frequently in young men than in young women with a ratio of 5:1 (Najafi and Guzman, 1977). Undoubtedly, the majority of affected individuals remain undiagnosed unless the bulla ruptures to produce ''spontaneous'' pneumothorax, or the defect is diagnosed at the time of chest roentgenography. A number of familial cases with multiple affected sibs (suggesting autosomal recessive inheritance) have been reported (Leites and Tannenbaum, 1960; Gibson, 1977). Families in which a parent and offspring were affected suggest that an autosomal dominant form of this condition may exist (Boyd, 1957; Leman and Dines, 1973).

Spontaneous pneumothorax may occur as a complication of various disease processes (e.g., tuberculosis emphysema); however, it is most commonly seen in young people as result of rupture of a bulla. Clinical symptoms vary greatly depending upon the degree and speed of functional loss of the lung. In some cases the symptoms are minimal, whereas in others there is an acute onset of dyspnea and unilateral chest pain. If a patient already has impaired pulmonary function, even a small pneumothorax may cause severe respiratory distress. Tension pneumothorax may occur with mediastinal displacement and, if unrelivened, may cause death as result of asphyxia. The diagnosis may be confirmed by chest roentgenogram or tomography. This condition has rarely been associated with pregnancy and only a few cases have been reported (Brantley *et al.*, 1961; Vance, 1968; Najafi and Guzman, 1977). Although small collections of air may resorb spontaneously, an intercostal tube with water-seal drainage is usually regarded as an important emergency treatment. In some cases, thorocotomy to excise the lesion, reinflation of the lung, and obliteration of the pleural space may be necessary. Most patients do well with appropriate management.

6. *Familial Idiopathic Pulmonary Fibrosis (Hamman–Rich Syndrome)*

In 1944, Hamman and Rich reported four patients with an unusual form of pneumonia characterized by progressive fibrosis of both lungs. The disease was accompanied by progressive hypoxia and ultimately right ventricular failure, and

TABLE I
Respiratory System Involvement in Selected Genetic Disorders

Syndrome	Respiratory tract malformation or disorder	Major associated abnormalities	Mode of inheritance
Kartagener syndrome	Sinusitis; bronchiectasis	Situs inversus	AR
Marfan syndrome	Interstitial parenchymal disease with ''honeycombing''; diffuse and apical bullous emphysema; anomalous lobation of lungs; bronchial stenosis; bronchogenic cysts; bronchomalacia	Arachnodactyly with hyperextensibility; ectopia lentis; aortic dilatation	AD
Achondroplasia	Supernumerary lobes of lung	Disproportionate short-limb dwarfism; low nasal bridge; caudal narrowing of spinal canal	AD
Asphyxiating thoracic dystrophy (Jeune syndrome)	Narrow thorax; pulmonary hypoplasia	Hypoplastic iliac wings; occasional polydactyly; occasional chronic nephritis	AR
Hypospadias–dysphagia syndrome (G syndrome)	Pulmonary hypoplasia; lack of pulmonary segmentation; pulmonary aplasia; tracheoesophageal fistula	Hypospadias in males; imperferate anus; dysphagia	AD
Bilateral renal agenesis	Pulmonary hypoplasia	Oligohydramnios; ''Potter facies'' including deep, semicircular infraorbital skin fold, broad	Most commonly P/M

		flattened nose, small receding mandible, low-set ears deficient in cartilage; contraction deformities of extremities; undescended testes	
Rubinstein–Taybi syndrome	Azygous lobe of lung	Developmental retardation; broad terminal phalanges of thumbs and halluces; hypoplastic maxilla; undescended testes	Sporadic P/M
Rowley–Rosenberg syndrome	Atelectasis; recurrent pulmonary infections	Growth retardation; poor muscular development; scanty adipose tissue; right ventricular hypertrophy; pulmonary hypertension	AR
Tuberous sclerosis	Cystic alveolar degeneration; hamartoma of lung; bronchogenic cysts; diffuse interstitial fibrosis	Adenoma sebaceum; seizures; mental retardation; phakomata of retina	AD
Neurofibromatosis (Von Recklinghausen disease)	Fibrosing alveolitis; bronchial obstructive disease	Café-au-lait spots; multiple neurofibromata	AD
Hereditary hemorrhagic telangectasia (Osler–Rendu–Weber syndrome)	Arteriovenous fistulas of lung	Epistaxes; multiple telangectasia	AD
Dermatorrhexis cutis hyperelastica (Ehlers–Danlos syndrome)	Spontaneous rupture of the lung with mediastinal emphysema and/or pneumothorax; bronchiectasis; ectasia of the lung; subplural blebs	Hyperextensibility of skin; joint laxity; poor wound healing	AD

AD, Autosomal dominant; AR, autosomal recessive; P/M, polygenic/multifactorial.

in each case was fatal within a year of onset. Since this original description, a number of patients who more or less fit this clinical description have been reported. Although most cases do not appear to be familial, an autosomal dominant form of interstitial pulmonary fibrosis has been recognized (Swaye *et al.*, 1969). This disorder rarely affects pregnant women. Should the diagnosis be made, treatment is similar to that of bronchiectasis (see above).

7. *Pulmonary Alveolar Microlithiasis*

This condition is characterized by multiple microcalculi located in the alveoli. Histopathologic studies of the microcalculi show calcium carbonate or calcium phosphate (Goodman, 1973). Chest roentgenograms show granular to small nodular calcium densities distributed diffusely and uniformly throughout the lungs, especially in the lower lobes. Clinical manifestations include dyspnea, cyanosis, and, occasionally, cor pulmonale. This condition is inherited in autosomal recessive fashion and appears to be more common in people of Spanish ancestry (Caffrey ane Altman, 1965; Lopez-Areal *et al.*, 1965). There have been no reports or pregnancy in an affected individual. Presumably management would be the same as in the nonpregnant state, namely, symptomatic treatment with oxygen, expectorants, and antibiotics.

B. Syndromes Involving the Respiratory Tract

There are numerous genetic disorders in which the respiratory tract may be involved as part of a multisystem syndrome. These include both Mendelian and polygenic/multifactorial disorders (Table I) as well as chromosomal abnormalities (Table II). In most cases there is impaired reproductive fitness and affected women rarely become pregnant. In those disorders in which pregnancies have been reported there is essentially no information that deals with respiratory

TABLE II
Respiratory Tract Involvement in Various Cytogenetic Disorders

Chromosomal abnormality	Respiratory system disorders occasionally encountered
Del(5p)	Atrophy of vocal cords, hypoplasia of epiglottis; laryngomelacia; falciform plicae in the meso-arytenoid space
Trisomy 13	Pulmonary stenosis; situs inversus of lungs
Trisomy 18	Poorly developed thorax; tracheo-esophageal fistula
Trisomy 21	Lower IgE with resultant increased susceptibility to respiratory infection
47,XXY (Klinefelter syndrome)	Bronchiectasis; pulmonary fibrosis

problems per se. Undoubtedly this is because the other clinical manifestations of these disorders have more important clinical significances. Some of these genetic disorders are considered in greater detail in other chapters.

II. FUNCTIONAL ABNORMALITIES OF THE RESPIRATORY SYSTEM

Among the heritable disorders of pulmonary function, three will be discussed because of their important relationships to pregnancy. These include (1) cystic fibrosis, (2) α-1-antitrypsin deficiency, and (3) bronchial asthma.

A. Cystic Fibrosis

1. Genetics

Cystic fibrosis (CF) is the most common lethal genetic disease of Caucasians with an estimated incidence between 1:1900 to 1:3700 in the United States (Merritt *et al.*, 1962; Steinberg and Brown, 1960). The genetic evidence is compelling that this disorder is the result of an autosomal recessive gene; however, the possibility of genetic heterogeneity exists (Bearn, 1972; Rao and Nadler, 1974). In contradistinction to other rare autosomal recessive disorders, consanguinity is rarely encountered due to the high heterozygote frequency, estimated to be about 3% in Caucasians (Steinberg and Brown, 1960). As yet no clearly discernible heterozygote advantage has been identified. There has been no chromosomal assignment for the CF gene(s), nor any linkage to blood groups or major HLA haplotypes (Goodchild *et al.*, 1976; Polymenidis *et al.*, 1973). Finally, despite numerous attempts, the basis metabolic defect(s) in CF remain as yet unknown. The reader is referred to an excellent review by Nadler *et al.* (1978) for an in-depth perspective of recent hypotheses.

2. Clinical Manifestations

Most clinically significant manifestations of CF are the result of obstruction of organ ducts by secretions of abnormally high viscosity. Involvement of the lungs is the hallmark of this disorder with 90% of patients eventually succumbing from pulmonary insufficiency (Nadler *et al.*, 1978). The major lung problem is the end result of mucous obstruction of airways and bacterial or viral infection; however, the specific factors that actually initiates the pathologic process is uncertain. This viscous mucus and infection combination leads to hyperinflation of airways, atelectasis, recurrent pneumonia, bronchitis, bronchiectases, abscess formation, and secondary cardiac decompensation. As the disease progresses, pulmonary function deteriorates and gas exchange becomes markedly impaired. Eventually both hypoxemia and hypercarbia occur. There is wide variation in the severity of CF with approximately 60% of patients dying before age 10 years (Bearn and Danes, 1978). Individuals rarely live beyond age 25 years.

Other manifestations associated with CF include (1) pancreatic insufficiency as result of duct obstruction by eosinophilic secretions, (2) meconium ileus of the newborn, (3) intestinal intussusception, (4) rectal prolapse, (5) hepatic cirrhosis, usually of the biliary type, (6) portal hypertension, (7) cholelithiasis and cholecystitis, (8) salivary gland duct obstruction, and (9) maldevelopment of the mesonephric duct derivatives (e.g., absent vas deferens). The most reliable biochemical abnormality is an elevated concentration of sweat sodium and chloride which is present at birth, persists through life, and is not related to the severity of other clinical problems (Danks *et al.*, 1965; di Sant'Agnese and Talamo, 1967; Steinrud *et al.*, 1974).

3. Obstetrical Considerations

With the advent of antibiotics and modern respiratory therapy techniques, an increasing number of women with CF are reaching the reproductive years. To date, however, relatively few cases of pregnancy in CF patients have been reported. The first description of pregnancy in a woman with CF was by Siegel and Siegel in 1960. This 20-year-old white primigravida was delivered of a normal infant at 34 weeks gestation. The last 2 months of her pregnancy were complicated by progressive deterioration of her pulmonary status, and she died six weeks postpartum of cardiorespiratory failure. The authors concluded that the pregnancy seriously complicated her underlying disease. In 1966 Grant *et al.* reported 10 cases of CF associated with pregnancy. Two cases were observed at the National Institutes of Health, six were ascertained from national cystic fibrosis organizations, and two were reviewed from the literature. In this series five patients suffered no apparent adverse effects as result of pregnancy; in fact, two of these women tolerated two and three pregnancies, respectively, without serious difficulty. In the other five women there was progression of pulmonary disease leading to death in two cases. Those women with lesser pulmonary compromise prior to conception appeared to do better during pregnancy than those with more advanced disease. Eleven of 13 offspring were normal at birth, although three were premature. Of the two that died, one was stillborn and macerated and the other had severe respiratory distress. In a 1975 report at the Cystic Fibrosis "GAP" Conference (Anonymous, 1975) from a national survey of 82 cystic fibrosis centers, adequate information was ascertained to evaluate obstetric outcome in 70 pregnancies. Among these patients there were eight maternal deaths in the period from conception to 4 months post partum. Maternal morbidity was also increased; pulmonary symptoms became worse in 60%, cor pulmonale developed in 14%, and hyperglycemia in 3%. There were 47 viable infants and stillbirths and 18 therapeutic abortions.

Recently, di Sant'Agnese and Davis (1979) reported 75 patients with CF aged 18 to 47 years and reviewed another 232 cases reported in the literature. Among their patients, five women proved fertile; there were three livebirths, one stillbirth, one spontaneous abortion, and three therapeutic abortions. From their experience and review of the literature, the authors concluded that because of the potential adverse effect of pregnancy on pre-existing pulmonary disease there is a significant hazard for both mother and offspring unless the mother's clinical status is good to excellent. Furthermore, there is a higher risk of producing a child with CF (1:40 if the father is of unknown genetic status versus 1:1500–2000 in the general population) and the probability of the child's premature loss of a parent.

The consensus appears to be that such women are indeed at increased risk for both exacerbation of their pulmonary disease as well as premature delivery. Such risks would presumably be minimized by efficient and frequent pulmonary toilet, early treatment of pneumonitis, optimal antepartum nutrition, and antenatal electronic fetal monitoring (e.g., nonstress testing). Sodium balanced may be a problem, particularly in patients with biliary cirrhosis. Intrapartum management should include (1) oxygen administration and intermittent positive pressure breathing; (2) avoidance of high doses of narcotics or barbiturates, which might result in respiratory depression; (3) avoidance of atropine and scopolamine, which may cause inspissation of bronchial secretions; (4) avoidance of inhalation anesthetics; (5) electronic fetal monitoring; and (6) shortening of the second stage of labor by low forceps (Novy *et al.*, 1967; Larsen, 1972). Finally, patients with severe pulmonary compromise and hypoxemia must be counseled prior to pregnancy regarding the potential dangers and sterilization should be considered.

4. *Prenatal Diagnosis*

Despite numerous reports of alterations in many systems (Mangos and Talamo, 1976), there is currently no reliable method for the prenatal diagnosis of CF. In 1968 Danes and Bearn showed that cytoplasmic intravesicular metachromasia was present in skin fibroblasts of individuals with CF as well as obligatory heterozygotes, and it was at one time suggested that this might prove of value in prenatal diagnosis. Subsequently, however, the detection of metachromatic granules in cultivated amniotic fluid cells was shown to be an unreliable method for detecting an affected fetus (Nadler *et al.*, 1969). In 1973 Bowman *et al.* demonstrated that a protein in the media of cultivated amniotic fluid cells from a fetus heterozygous for CF would inhibit cilia action in oyster gills. The heterozygous status of the fetus was inferred since the mother was affected with CF and at 13 months the child was symptom free. This factor was not present in media from normal fetuses. To date, however, the oyster cilia assay has not been perfected to the point where heterozygotes can be reliably distinguished from homozygotes and cannot, therefore, serve as a marker for prenatal diagnosis (Bowman, 1976). Recently, Walsh-Platt and Nadler (1979) have shown markedly reduced MUGB (4-methylumbelliferylguanidinobenzoate) reactivity in the amniotic fluid from two pregnancies which were subsequently delivered of infants with CF. They suggest that this assay may be of potential value for the prenatal diagnosis of CF.

B. α-1-Antitrypsin Deficiency

α-1-Antitrypsin (A-1-AT), a glycoprotein with a molecular weight of about 50,000 (Crawford, 1973) is the major α-1-globulin in human serum and comprises about 4% of the total serum protein level with a normal concentration of approximately 2.2 mg/ml (Talamo, 1975). In addition, A-1-AT is present in most body fluids, including cerebrospinal fluid, pulmonary secretions, saliva, nasal secretions, tears, and mother's milk (Talamo *et al.*, 1972). Its major property is that of an inhibitory factor for a wide spectrum of proteolytic enzymes including elastase and collagenase from leukocytes and macrophages, plasmin, thrombin, collagenase, chymotrypsin, and some bacterial enzymes. Despite its name, however, A-1-AT does not appear to be an important inhibiting factor against trypsin. Thus, A-1-AT serves as a protective mechanism against serious tissue-damaging effects of various enzymes.

1. *Genetics*

By techniques of acid starch gel and immunoelectrophoresis, it has been shown that A-1-AT is inherited by a series of codominant alleles (Fagerhol and Laurell, 1967; Fagerhol and Gedde-Dahl, 1969). There have now been about 25 allelic types of A-1-AT identified on the basis of serum electrophoretic mobility, and each has been classified into a *Pi* (protease inhibitor system) with an assigned letter (Fagerhol and Laurell, 1967). Most persons are homozygotes for Pi^M which, in North America and Central Europe, has a gene frequency of 0.95 (Kueppers, 1973). Other common alleles are designated by the following letters: *B,E,F,G,I,L,M,Mi,N,P,R,S,V,W,X,Z*. The gene locus is designated as *Pi* and the alleles are referred to by superscript capital letters (e.g., Pi^M, Pi^Z, Pi^S, etc.). Accordingly, a heterozygotes would be written as Pi^M/Pi^Z or Pi^{MZ} and correspond to the phenotype MZ. The position of the letter in the alphabet is intended to indicate the approximate electrophoretic mobility with respect to the anode.

2. *Clinical Manifestations*

In 1964 Laurell and Eriksson first reported association between a deficiency of A-1-AT in serum and chronic obstructive pulmonary disease which characteristically develops by the third or fourth decade of life. Subsequent reports confirmed this association, with the vast majority of familial cases being Pi^{ZZ}; however, a few cases have been described with types Pi^{SZ}, Pi^{SS}, or Pi^{PZ}. Kueppers (1978) has estimated that the risk of an individual with Pi^{ZZ} developing chronic obstructive pulmonary disease (COPD) during his or her lifetime is 30 times higher than that of the general population. Clinical manifestations usually begin insiduously, later leading to progressive dyspnea, productive chronic cough, bronchitis, and, later, respiratory failure with cor pulmonale. Pulmonary function studies and radiographic findings are generally compatible with obstructive pulmonary disease. Smoking is considered an important exacerbating factor to the development of chronic bronchitis and emphysema (Kueppers, 1978). Finally, other disorders which have been implicated with certain types of A-1-AT include (1) neonatal or childhood hepatitis and hepatic cirrhosis, (2) neonatal respiratory distress syndrome, (3) childhood glomerulonephritis, and (4) aneuploidy in offspring of affected parents.

3. Obstetrical Considerations

There is little information describing the relationship between A-1-AT deficiency and pregnancy. Giesler *et al.* (1977) reported a 37-year-old primagravida with Pi^{ZZ} and severe COPD who was monitored throughout pregnancy and during the postpartum period. Her pulmonary function studies and arterial blood gases remained relatively unchanged. Ultrasonographic measurements of the fetal biparietal diameter showed progressive intrauterine growth, and serial oxytocin challenge tests begun at 34 weeks gestation remained normal. At 38 weeks gestation she was delivered of a 2320 g female infant by low forceps under continuous lumbar epidural anesthesia. The 5-minute Apgar score was 10. The authors indicate that successful pregnancy outcome is likely provided there is (1) avoidance of smoking, (2) prenatal pulmonary physiotherapy, and (3) aggressive treatment of all pulmonary infections. These recommendations are in general agreement with accepted management for other forms of chronic pulmonary disease associated with pregnancy.

It is now technically feasible to screen large numbers of individuals for A-1-AT deficiency using automated quantitative biochemical analysis (Scheinberg, 1971). Such screening programs would permit early initiation of preventative medicine measures, particularly counseling regarding the hazards of smoking. Finally, prenatal diagnosis by fetoscopically directed fetal blood sampling might also be possible.

C. Bronchial Asthma

Although asthma occurs worldwide, the disorder appears to be more prevelant in urbanized populations. Available evidence suggests an incidence of 0.5–9% with a median of about 3.5% (Smith, 1974). Asthma may have its onset at any age. In some individuals asthma appears to bear a close relationship to atopic dermatitis, hay fever, and other allergic states; and immunologic, infectious, autonomic nervous system, and psychologic factors may all be involved in an individual patient. Furthermore, there is considerable disagreement among allergists and pulmonary physiologists concerning a clearcut definition of asthma. Undoubtedly, this is due to complex interrelationships between asthma and other pulmonary diseases such as emphysema and chronic bronchitis, as well as a lack of understanding of many of the principal genetic and nongenetic factors contributing to the disease (Bias and Marsh, 1977).

1. Genetics

Early reports concerning the genetics of asthma suggested simple Mendelian inheritance (Drinkwater, 1909; Adkinson, 1920). It was not, however, until 1967 that Leigh and Marley first suggested that this disease is inherited in polygenic/multifactorial fashion. This hypothesis gained further support from twin studies. In a questionnaire series of 7000 twin pairs born in Sweden, Edfors-Lubs (1971) found concordance values for monozygotic and dizygotic twin-pairs to be 19% and 4.8% for asthma, 21.4% and 13.7% for allergic rhinitis, and 15.5% and 4.0% for eczema, respectively. Although genetic factors contribute to the predisposition for the development of allergic diseases, they apparently do not dictate which disorder will be manifested by the patient. The reader is referred to a recent review by Bias and Marsh (1977) for a detailed discussion of the immunogenetics of asthma and related allergic states.

2. Clinical Manifestations

Bronchial asthma is characterized by paroxysms of dyspnea, wheezing, and coughing as result of temporary narrowing of the bronchi by smooth muscle spasms, mucosal edema, and tenacious mucus. In most patients asthma occurs in the form of acute attacks lasting for hours or days separated by remissions during which symptoms are reduced or absent. During an attack the affected individual shows varying degrees of tachypnea, cyanosis, hyperinflation of the chest, recruitment of the accessory muscles of respiration, and audible wheezing. The $PaCo_2$ typically remains normal or low because adequate ventilation can usually be maintained to eliminate carbon dioxide. Arterial hypoxemia, however, commonly occurs in patients with moderately severe asthma. Fatigue and dehydration are common features. The term *status asthmaticus* is applied when a severe attack is prolonged, usually beyond 24 hours. In this setting respiratory arrest and death may occur.

3. Obstetrical Considerations

Pregnancy exerts no consistently predictable effect on asthma. In a recent review Weinstein *et al.* (1979) tabulated that about one-third of patients with asthma were reported to improve during pregnancy, one-third had no change in their condition, and one-third became worse. Patients who were severe asthmatics prior to conception tended to have a worse prognosis during pregnancy, as defined by (a) requiring steroid treatment or (b) becoming symptomatic at least weekly. Gordon *et al.* (1970) reviewed the pregnancies from 277 patients with bronchial asthma among 30,861 deliveries investigated in the Collaborative Perinatal Study. Among this cohort of 277 women five deaths occurred related to asthma or its complications. The risk of prematurity among offspring of asthmatics was not significantly increased; however, the perinatal mortality rate was almost twice that of others in the study. Among the 16 severe asthmatic patients, 28% of the pregnancies resulted in perinatal mortalities, 35% in low birth weight infants, and 12.5% in neurologically abnormal infants at 1 year.

Management of the asthmatic patient during pregnancy entails (1) psychogenic support, (2) environmental control, (3) desensitization, (4) avoidance of sensitizing medications (e.g., aspirin, phenobarbital, penicillin), (5) prophylactic antibiotics, and (6) possibly administration of cromolyn sodium (however, this new agent has not yet been fully evaluated for potential teratogenicity) (Kochenour and Lavery, 1976). Acute asthmatic attacks must be promptly treated with appropriate medications (e.g., epinephrine, aminophylline, steroids), hydration, oxygen, nebulized bronchodilators, and, in extreme cases, ventilatory support. During labor administration of narcotics, barbiturates, or other sedatives should be minimized because these compounds tend to depress the respiratory center and the cough reflex. Vaginal delivery is preferred unless cesarean section is indicated for obstetric reasons. Finally, local or conduction anesthesia is preferable to inhalation anesthesia because they do not interfere with respiratory exchange.

REFERENCES

Adkinson, J. (1920). The behavior of bronchial asthma as an inherited character. *Genetics* **5,** 363–418.

Agosti, E., De Filippi, F. R., and Chiussi, F. (1974). Generalized familial bronchomalacia. *Acta Paediatr. Scand.* **63,** 616–618.

Anonymous (1975). "Problems in Reproductive Physiology and Anatomy in Young Adults with Cystic Fibrosis." Report of Cystic Fibrosis GAP Conference, New Orleans, Louisiana.

Bearn, A. G. (1972). Cell culture in inherited disease with some notes on genetic heterogeneity. *N. Engl. J. Med.* **285,** 764–767.

Bearn, A. G., and Danes, S. (1978). Cystic fibrosis. *In* "Genetic Determinants of Pulmonary Disease" (S. D. Litwin, ed.), pp. 75–92. Dekker, New York.

Bias, W. B., and Marsh, D. G. (1977). The genetic base of asthma: Current studies of the genetics of the IgE-mediated immune response in man. *In* "Asthma, Physiology, Immunopharmacology, and Treatment" (L. M. Lichtenstein and K. F. Austen, eds.), pp. 21–34. Academic Press, New York.

Bowman, B. G. (1976). Factors related to cystic fibrosis. *In* "Cystic Fibrosis. Projections into the Future" (J. A. Mangos and R. C. Talamo, eds.), pp. 277–284. Intercontinental Medical Book Corp., New York.

Bowman, B. H., Lockhart, L. H., Herzberg, V. L., Barnett, D. R., Armstrong, D., and Kramer, J.

(1973). Cystic fibrosis: Synthesis of ciliary inhibitor by amniotic cells. *Clin. Genet.* **4,** 461-463.

Boyd, D. H. A. (1957). Familial spontaneous pneumothorax. *Scott. Med. J.* **2,** 220-221.

Brantley, W. M., Del Valle, R. A., and Schoenbucher, A. K. (1961). Bilateral spontaneous pneumothorax complicating pregnancy. *Am. J. Obstet. Gynecol.* **81,** 42-44.

Caffrey, P. R., and Altman, R. S. (1965). Pulmonary alveolar microlithiasis in premature twins. *J. Pediatr.* **66,** 758-763.

Copleman, B., Cannata, B. C., and London, W. (1950). Tracheo-oesophageal anomalies in siblings. *J. Med. Soc. N.J.* **47,** 415.

Cozzi, F., and Wilkinson, A. W. (1975). Low birthweight babies with oesophageal atresia or tracheo-oesophageal fistula. *Arch. Dis. Child.* **50,** 791-795.

Crawford, I. P. (1973). Purification and properties of normal human alpha$_1$-antitrypsin. *Arch. Biochem. Biophys.* **156,** 215-222.

Cudmore, R. E. (1978). Oesophageal atresia and tracheo-oesophageal fistula. *In* "Neonatal Surgery" (P. P. Rickham, J. Lister, and I. M. Irving, eds.), 2nd ed., pp. 189-208. Butterworth, London.

Danes, B. S., and Bearn, A. G. (1968). A genetic cell marker in cystic fibrosis of the pancreas. *Lancet* **1,** 1061-1063.

Danielson, G. K., Hanson, C. W., and Cooper, E. C. (1967). Middle lobe bronchiectasis. Report of an unusual family occurrence. *J. Am. Med. Assoc.* **201,** 605-608.

Danks, D. M., Allan, J., and Anderson, C. M. (1965). A genetic study of fibrocystic disease of the pancreas. *Ann. Hum. Genet.* **28,** 323-356.

Dennis, N. R., Nicholas, J. L., and Kovar, I. (1973). Oesophageal atresia. Three cases in two generations. *Arch. Dis. Child.* **48,** 980-982.

di Sant'Agnese, P. A., and Davis, P. B. (1979). Cystic fibrosis in adults. 75 cases and a review of 232 cases in the literature. *Am. J. Med.* **66,** 121-132.

di Sant'Agnese, P. A., and Talamo, R. C. (1967). Pathogenesis and physiopathology of cystic fibrosis of the pancreas. *N. Engl. J. Med.* **277,** 1287-1308.

Drinkwater, H. (1909). Mendelian heredity in asthma. *Br. Med. J.* **1,** 88.

Duenhoelter, J. H., Santos-Ramos, R., and Rosenfield, C. R. (1976). Prenatal diagnosis of gastrointestinal tract (GI) obstruction. *Obstet. Gynecol.* **47,** 618-620.

Edfors-Lubs, M. L. (1971). Allergy in 7000 twin pairs. *Acta Allergol.* **26,** 249-285.

Fagerhol, M. K., and Gedde-Dahl, T., Jr. (1969). Genetics of the Pi serum types. Family studies of the inherited variants of serum alpha$_1$-antitrypsin. *Hum. Hered.* **19,** 354-359.

Fagerhol, M. K., and Laurell, C.-B. (1967). The polymorphism of "prealbumins" and α_1 antitrypsin in human sera. *Clin. Chim. Acta* **16,** 199-203.

Gibson, G. J. (1977). Familial pneumothoraces and bullae. *Thorax* **32,** 88-90.

Giesler, C. F., Buehler, J. H., and Depp, R. (1977). Alpha$_1$-antitrypsin deficiency. Severe obstructive lung disease and pregnancy. *Obstet. Gynecol.* **49,** 31-34.

Goodchild, M. C., Edwards, J. H., Glenn, K. P., Grindey, C. G., Harris, R., Makintosh, P., and Wentzel, J. (1976). A search for linkage in cystic fibrosis. *J. Med. Genet.* **13,** 417-419.

Goodman, R. M. (1973). The respiratory system. *In* "Clinical Genetics" (A. Sorsby, ed.), 2nd ed., pp. 393-408. Butterworth, London.

Gordon, M., Niswander, K. R., Berendes, H., and Kantor, A. G. (1970). Fetal morbidity following potentially anoxigenic obstetric conditions. VII. Bronchial asthma. *Am. J. Obstet. Gynecol.* **106,** 421-429.

Grand, R. J., Talamo, R. C., di Sant'Agnese, P. A., and Schwartz, R. H. (1966). Pregnancy in cystic fibrosis of the pancreas. *J. Am. Med. Assoc.* **195,** 117-124.

Grieve, J. G., and McDermott, J. G. (1939). Congenital atresia of the oesophagus in two brothers. *Can. Med. Assoc. J.* **41,** 185-186.

Hamman, L., and Rich, A. R. (1944). Acute diffuse interstitial fibrosis of the lungs. *Bull. Johns Hopkins Hosp.* **74,** 177-212.

Haynes, D. M. (1969). "Medical Complications During Pregnancy." McGraw-Hill, New York.

Himalstein, M. R., and Gallagher, J. C. (1973). Tracheobronchomegaly. *Ann. Otol., Rhinol, & Laryngol.* **82,** 223-227.

Johnston, R. F., and Green, R. A. (1965). Tracheobronchomegaly. *Am. Rev. Respir. Dis.* **91,** 35-50.

Kochenour, N. K., and Lavery, J. P. (1976). Managing asthma in the pregnant patient. *Contemp. Obstet./Gynecol.* **7,** 27-30.

Kueppers, F. (1973). Annotation: α-Antitrypsin. *Am. J. Hum. Genet.* **25,** 677-686.

Kueppers, F. (1978). Inherited differences in alpha$_1$-antitrypsin. *In* "Genetic Determinants of Pulmonary Disease" (S. D. Litwin, ed.), pp. 23-74. Decker, New York.

Larsen, J. W. (1972). Cystic fibrosis and pregnancy. *Obstet. Gynecol.* **39,** 880-883.

Laurell, C. B., and Eriksson, S. (1963). The electrophoretic alpha$_1$-antitrypsin deficiency. *Scand. J. Clin. Lab. Invest.* **15,** 132-140.

Leigh, D. E., and Marley, E. (1967). "Bronchial Asthma: A Genetic, Population and Psychiatric Study." Pergamon, Oxford.

Leites, V., and Tannenbaum, E. (1960). Familial spontaneous pneumothorax. *Am. Rev. Respir. Dis.* **82,** 240-241.

Leman, C. B., andDines, D. E. (1973). Treatment of recurrent spontaneous familial pneumothorax. *J. Am. Med. Assoc.* **225,** 1256.

Lloyd, J. R., and Clatworth, H. E. (1958). Hydramnios as an aid to the early diagnosis of congential obstruction of the alimentary tract: A study of the maternal and fetal factors. *Pediatrics* **21,** 903-909.

Lopez-Areal, L., Zumarraga, R., Turner, C. G., Granizo, I. F. M., Vara Cuadrado, F., and Duque Fraile, J. (1965). Microlithiasis alveolar pulmonar familiar e infantil. (Descripcion de cinco casos in dos familias.) *Rev. Clin. Esp.* **97,** 389-395.

Mangos, J. A., and Talamo, R. C. (1976). "Cystic Fibrosis. Projections into the Future." Stratton Intercontinental Med. Book Corp., New York.

Merritt, A. D., Hanna, B. L., Todd, C. W., and Meyers, T. L. (1962). The incidence and mode of inheritance of cystic fibrosis. *J. Lab. Clin. Med.* **60,** 998-999.

Mounier-Kuhn, P. (1932). Dilitation de la trachée: Constations radiographiques et bronchoscopiques. *Lyon Med.* **150,** 106-109.

Nadler, H. L., Swae, M. A., Wodnicki, J. M., and O'Flynn, M. E. (1969). Cultivated amniotic-fluid cells and fibroblasts derived from families with cystic fibrosis. *Lancet* **2,** 84-85.

Nadler, H. L., Rao, G. J. S., and Taussig, L. M. (1978). Cystic fibrosis. *In* "The Metabolic Basis of Inherited Disease" (J. B. Stanbury, J. B. Wyngaarden, and D. S. Fredrickson, eds.), 4th ed., pp. 1683-1710. McGraw-Hill, New York.

Najafi, J. A., and Guzman, L. G. (1977). Spontaneous pneumothorax in labor. *Am. J. Obstet. Gynecol.* **129,** 463-464.

Novy, M. J., Tyler, J. M., Schwachman, H., Easterday, C. L., and Reid, D. E. (1967). Cystic fibrosis and pregnancy. Report of a case, with a study of pulmonary function and arterial blood gases. *Obstet. Gynecol.* **30,** 530-536.

Pitkin, R. M., and Reynolds, W. A. (1975). Fetal ingestion and metabolism of amniotic fluid protein. *Am. J. Obstet. Gynecol.* **123,** 356-363.

Polymenidis, Z., Ludwig, H., and Gotz, M. (1973). Cystic fibrosis and HL-A antigens. *Lancet* **2,** 1452.

Quan, L., and Smith, D. W. (1973). The VATER association. *J. Pediatr.* **82,** 104-107.

Rao, C. J. S., and Nadler, H. L. (1974). Arginine esterase in cystic fibrosis of the pancreas. *Pediatr. Res.* **6,** 684.

Scheinberg, H. (1971). Some problems in the biochemical approach to detecting human mutations. *In* "Monitoring, Birth Defects and Environment: The Problem of Surveillance" (E. B. Hook, D. T. Janerich, and I. H. Porter, eds.), pp. 207–212. Academic Press, New York.

Siegel, B., and Siegel, S. (1960). Pregnancy and delivery in a patient with cystic fibrosis of the pancreas. Report of a case. *Obstet. Gynecol.* **16,** 438–440.

Sloan, H., and Haight, C. (1956). Congenital atresia of the esophagus in brothers. *J. Thorac. Surg.* **32,** 209–215.

Smith, J. M. (1974). Incidence of atopic disease. *Med. Clin. North Am.* **58,** 3–24.

Steinberg, A. G., and Brown, D. C. (1960). On the incidence of cystic fibrosis of the pancreas. *Am. J. Hum. Genet.* **12,** 416–424.

Steinrud, J. S., Winkel, S., and Flensborg, E. W. (1974). Screening for cystic fibrosis with chloride electrode. *Dan. Med. Bull.* **21,** 251–255.

Swaye, P., van Ordstrand, H. S., McCormack, L. J., and Wolpaw, S. E. (1969). Familial Hamman-Rich syndrome; report of 8 cases. *Dis. Chest* **55,** 7–12.

Talamo, R. C. (1975). Basic and clinical aspects of the alpha$_1$-antitrypsin. *Pediatrics* **56,** 91–99.

Talamo, R. C., Langley, C. E., Barber, J. C., Berman, M. B., and Hyslop, N. E. (1972). Distribution and quantitation of alpha$_1$-antitrypsin in dilute biological fluids: Its localized increase in tears of patients with corneal ulceration. *Pediatr. Res.* **6,** 119.

Vance, J. P. (1968). Tension pneumothorax in labor. *Anaesthesia* **23,** 94–97.

Walsh-Platt, M. M., and Nadler, H. L. (1979). Reaction of human amniotic fluid proteases with 4-methyl-umbelliferylguanidinobenzoate (MUGB). *Pediatr. Res.* **13,** 427.

Waterson, D. J., Bonham-Carter, R. E., and Aberdeen, E. (1963). Congenital tracheoesophageal fistula in association with oesophageal atresia. *Lancet* **2,** 55–57.

Wayne, K. S., and Taussig, L. M. (1976). Probable familial congenital bronchiectasis due to cartilage deficiency (Williams-Campbell syndrome). *Am. Rev. Respir. Dis.* **114,** 15–22.

Weinstein, A. M., Dubin, B. D., Podleski, W. K., Spector, S. L., and Forr, R. S. (1979). Asthma and pregnancy. *J. Am. Med. Assoc.* **241,** 1151–1165.

Williams, H., and Campbell, P. (1960). Generalized bronchiectasis associated with deficiency of cartilage in the bronchial tree. *Arch. Dis. Child.* **35,** 182–191.

Woolley, M. M., Chinnock, R. F., and Paul, R. H. (1961). Premature twins with esophageal atresia and tracheoesophageal fistula. *Acta Paediatr. Scand.* **50,** 423–430.

8

Renal Disorders

STANLEY A. MENDOZA,
CLYDE H. BECK, JR.,
AND WILLIAM R. GRISWOLD

I. PREGNANCY AND THE KIDNEY

A variety of genetic diseases which involve the kidney progress to cause decreased renal function. In some cases, renal destruction is so severe that chronic hemodialysis or renal transplantation is required to maintain life. In these patients, the effects of the renal failure on the pregnancy are generally more important than the effects of the specific genetic disease. For this reason, a brief review of the effects of kidney disease, in general, on pregnancy seems appropriate. For more details on this subject, the reader should consult Ferris (1979). In Section II there will be a discussion of the specific genetic diseases involving the kidneys.

Copyright © 1981 by Academic Press, Inc.
All rights of reproduction in any form reserved.
ISBN 0-12-630940-X

A. Effects of Pregnancy on the Normal Kidney

During a normal pregnancy, normal kidneys increase approximately 1 cm in cephalocaudad length. This size change is thought to be due to a 50% increase in blood volume and a 30–40% increase in cardiac output which occurs by the 24th week of gestation (Kerr, 1968). Histologically, the kidneys do not change; however, the smooth muscle of the ureters hypertrophies and the connective tissue undergoes hyperplasia due to the hormonal changes of pregnancy. In addition, there is dilatation and decreased peristalsis of the entire collecting system. Occasionally, there is mild mechanical obstruction of the ureters at the pelvic rim in the last weeks of pregnancy. These changes in the collecting system may not return to normal until 3 months postpartum (Davison and Lindheimer, 1978). The dilatation of the collecting system may lead to a greater than normal incidence of vesicoureteral reflux of urine and urinary tract infections. In addition, the enlarged collecting system makes the precise clinical measurement of renal function more difficult.

During pregnancy, renal blood flow increases by 30–50% and intrarenal vascular resistance decreases. Both of these changes are maximal by mid-pregnancy and then return toward normal. At the same time that renal blood flow increases, there is a 30–40% increase in glomerular filtration rate (GFR). This change persists throughout pregnancy (Sims and Krantz, 1958). Late in pregnancy, serum albumin concentration falls approximately 1 g/dl, offsetting the effect on GFR of the fall in systemic blood pressure to be mentioned below. These glomerular changes result in an increased filtration fraction and a fall in blood urea nitrogen to less than 10 mg/dl and serum creatinine to less than 0.5 mg/dl. Hypouricemia usually occurs with blood values of less than 3 mg/dl being common.

Despite an increase in intravascular volume, there is a fall in blood pressure during pregnancy. Thus, diastolic pressure greater than 75 mm Hg in the second trimester or greater than 85 mm Hg in the third trimester is abnormally high. There is an increase in plasma renin activity and renin substrate, as well as aldosterone excess which is nonsuppressible with salt loading. There is a resistance to angiotensin II perhaps related to increased levels of prostaglandin E (Gant *et al.*, 1973). The increase in prostaglandin levels may be due to increased prostaglandin production by the uterus (Ferris, 1979).

In addition to changes in glomerular function, there are tubular changes leading to glucosuria, aminoaciduria (up to 2 g/day), increased vitamin excretion, a decrease in bicarbonate threshold and mild proteinuria (less than 300 mg/day). Most of these changes are due to increased filtration of the substances without a concomitant increase in the tubular maximum (Ferris, 1979).

B. Pregnancy in Renal Failure

The success of a pregnancy seems to be influenced by the development of intravascular volume expansion and the increased GFR mentioned above. If renal function is impaired by 50%, as in a diabetic for example, these physiologic changes in kidney function will not occur (Younger *et al.*, 1962). It is unusual to complete a normal pregnancy if the serum creatinine of the pregnant woman is greater than 3 mg/dl or blood urea nitrogen greater than 30 mg/dl at the beginning of the childbearing period (Zacur and Mitch, 1977). Many authors feel that pregnancy should not be undertaken if the prepregnant serum creatinine is greater than 1.7 mg/dl (Bear, 1976).

If a woman with chronic renal failure is already pregnant when first medically evaluated, her blood pressure, weight gain, serum uric acid, blood urea nitrogen and 24-hour creatinine clearance, and urinary protein must be followed carefully. If her glomerular filtration rate declines or blood pressure rises in the first few weeks of pregnancy, a permanent decrease in the mother's renal function may ensue. In this setting, there is little hope for a viable infant. In addition to the hypertension of preeclampsia mentioned later, one must search for other treatable causes of decreased renal function in a pregnant female. If a urinary tract infection, ureteral obstruction, electrolyte imbalance or diuretic excess is corrected, renal function may improve and a successful pregnancy may result.

The impact of a pregnancy on chronic glomerular disease and/or nephrotic syndrome is difficult to assess. Older authors differ on the impact of a pregnancy on women with glomerulonephritis. Ferris (1979) has reviewed the series of pregnant women with chronic glomerulonephritis published since 1956. If a woman has chronic glomerular disease at the time of conception, the greatest medical risk is preeclampsia. This occurs with greater frequency and earlier in pregnancy (before the 28th week) in patients with glomerulonephritis. Fetal loss is 18% overall and may be as high as 45% if the patient has associated hypertension and proteinuria. Fetal loss is especially likely if there is significant hypertension before the pregnancy. Loss of renal function in this situation is probably a result of the preeclamptic state. If such cases are well handled medically, the maternal mortality is usually no greater than normal.

Modest hypertension or proteinuria of any cause significantly alters perinatal infant mortality. In a series by Naeye and Freedman (1979), 1339 of 42,241 patients had diastolic blood pressures greater than 85 mm Hg or proteinuria of 1+ or greater. The incidence of placental infarct, abruptio placentae, placental growth retardation, and acute infection of amnionic fluid was greater than normal. These findings and several other miscellaneous abnormalities were associated with the perinatal death rates shown in Table I. In a series by Felding

TABLE I
Effect of Hypertension and/or Proteinuria on Perinatal Death Rate[a]

Group	Perinatal death rate/1000 births
Normal	17
Hypertension only	26
Hypertension plus proteinuria	37.9

[a]After Naeye and Freedman (1979).

(1968a) the perinatal mortality rate was increased from 8% to 31% when the mother was hypertensive. Furthermore, the incidence of prematurity was increased from 12% to 37% in these pregnancies. As a general rule, treatment of hypertension existing before pregnancy or occurring with pregnancy as a result of renal disease is critical. The risk of volume depletion with such treatment is preferable to the risks of uncontrolled hypertension.

Nephrotic syndrome from any cause will not necessarily recur with pregnancy. The course of nephrotic syndrome during pregnancy is unpredictable. The impact on the fetus is quite similar to that of chronic glomerulonephritis. Pre-existing hypertension decreases the chances of a successful pregnancy. Proteinuria may or may not worsen. Usually a pregnancy is successful in the nephrotic woman if she is not azotemic. There is good correlation between the depression of serum albumin and the subnormal birth weight of the infant. As a pregnancy progresses, the already depressed serum albumin in a nephrotic patient will fall even further due to the usual decrease in serum albumin in late pregnancy. This leads to increasing edema. Diuretic treatment of the edematous nephrotic pregnant woman will lead to intravascular volume depletion and decrease uteroplacental perfusion. In addition, it may aggravate the hypercoagulable state and heighten the chances of pulmonary emboli (Gill and Hayslett, 1969; Grunfeld *et al.*, 1973).

The course of inherited diseases leading to nephrolithiasis is unaffected by pregnancy. Kidney stones are, however, one of the most common nonuterine causes of abdominal pain in pregnancy (Coe *et al.*, 1978). Kidney stones and other heritable conditions leading to abnormal urine flow may predispose a woman to pyelonephritis both before and during pregnancy.

Women with chronic interstitial nephritis, of which chronic pyelonephritis is one of the most common causes, have an increased incidence of preeclampsia. Perinatal mortality is increased if women with interstitial nephritis develop hypertension during their pregnancy (Felding, 1968b). The severity of problems associated with chronic pyelonephritis is similar to that of women with glomerular diseases. As mentioned earlier, the urine of a normal pregnant woman is more static and carries nutrients (glucose and amino acids) that facilitate bacterial growth. This increases the risk of a recurrence of urinary tract infection in women with chronic pyelonephritis, as well as increasing the incidence of urinary tract infection in normal pregnant women. Maternal urinary tract infection is significant to fetal development since it causes decreased renal blood flow and has been associated with growth retardation, prematurity, and congenital abnormalities in the infant (Davison and Lindheimer, 1978). In addition, there is an increased incidence of fetal death. A recent study was performed utilizing antibody-coated bacteria in the urine as evidence of renal parenchymal involvement with infection. This study revealed that pregnant women with antibody-coated bacteria in the urine had lower glomerular filtration rates than those with antibody-negative bacteria in the urine or sterile urine (Harris *et al.*, 1976).

C. Pregnancy in Patients on Chronic Hemodialysis

Women receiving hemodialysis treatment for kidney failure have less than one-tenth of normal renal function. With the addition of dialysis their chances of conception remain poor since the treatment does not reverse the anovulation associated with severe azotemia (Section I,B). In addition, libido decreases with chronic renal failure and menses often become irregular or absent (Perez *et al.*, 1978). The largest series of pregnancies in hemodialysis patients includes seven live births (one with multiple congenital anomalies) and 68 abortions of which 36 were spontaneous and 32 were surgically induced (Jacobs *et al.*, 1977). To maximize the possibility of a successful pregnancy in a dialysis patient, careful attention must be paid to fluid balance, blood pressure control, and measurable parameters of adequacy of dialysis such as blood urea nitrogen. Demonstration of abnormalities in female gonadotrophic hormones and associated uterine and ovarian hormones has recently been reported for patients with chronic renal failure (Schalch *et al.*, 1975; Wass *et al.*, 1978).

The decision to have a child despite chronic renal failure must be made by the patient after consideration of many factors. The infant may inherit a genetic disease present in the mother or other family members, may have psychologic problems as a result of the chronic renal failure of the mother and may not have a living mother for the total time necessary to reach adulthood. Indeed, the 5-year survival rate for hemodialyzed women in the childbearing years is approximately 75%. Physicians frequently fail to advise the patient of the gravity of the situation. Davison and Lindheimer (1978) reviewed 75 patients with chronic renal failure who completed 90 pregnancies with live births following successful renal transplantation. One-third of these had no contraceptive advice and 24 of the 90 pregnancies were not diagnosed until the second trimester.

D. Pregnancy after Renal Transplantation

Following successful renal transplantation, sexual function and fertility return to normal with menses usually becoming regular in 4–6 months (Merkatz *et al.*, 1971). Women who become pregnant post-transplant must be carefully observed for hypertension, proteinuria, and the loss of glomerular filtration rate especially late in the pregnancy. A decrease in glomerular filtration rate is often permanent. Despite the pelvic placement of most kidney transplants, vaginal delivery is usually possible. A review of 75 women with 90 pregnancies revealed 71 vaginal deliveries and 19 cesarean sections (Davison and Lindheimer, 1978). Both intravenous pyelography and pelvimetry were helpful in deciding the optimal route of delivery. Fifty-three of these women had living-related donors and 22 received kidneys from cadaveric donors.

Maternal complications of post-transplant pregnancies have included severe hypertension, allograft rejection, and uterine rupture (Farber, 1978). Most infants born to women following transplantation are below average in birth weight. There is a 20% incidence of premature membrane rupture (Nolan *et al.*, 1974). Other infant problems have included steroid-induced hypoglycemia, growth retardation, prematurity, respiratory distress syndrome, hypoadrenalism, and hepatic insufficiency (Papoff *et al.*, 1977; Lower *et al.*, 1971).

Cytomegalovirus (CMV) infects as many as one-third of patients receiving kidney transplants. While babies born to transplanted mothers usually do not have a clinical CMV infection, those with antibodies to CMV have been noted to have lower IQ values and a greater incidence of hearing loss than control infants (Hanshaw *et al.*, 1976).

Immunosuppressive drugs such as azathioprine and corticosteroids used in transplant patients have been shown to cause fetal abnormalities in animals (Fraser and Fainstat, 1951; Rosenkrantz *et al.*, 1967). In humans the incidence of such abnormalities appears to be much less, perhaps not increased over background levels for the two drugs mentioned above. In addition to lymphoid hypoplasia and adrenocorticoid insufficiency in the newborn (Lower *et al.*, 1971), there have been abnormalities in the chromosomes of peripheral lymphocytes of newborn infants from mothers who have had a renal transplant. The chromosomal gaps, fragments, deletions and breaks have all disappeared in 3–30 months. The question of germ cell abnormality and transmission of such defects remains unanswered (Leb *et al.*, 1971), as well as their relationship to the immunosuppressive drugs.

Though women have conceived as soon as 6 weeks post-transplant, it is preferable to wait for transplant stability before adding the stress of pregnancy to the transplanted kidney. The following criteria are useful in deciding whether a woman can safely become pregnant following a renal transplant (Davison *et al.*, 1976):

1. Two years of stable, good health following the renal transplant
2. Stature compatible with good obstetrical outcome
3. Absence of proteinuria
4. Absence of significant hypertension
5. No evidence of graft rejection
6. No pelvicalyceal distention on a recent IVP
7. Serum creatinine less than 2 mg/dl
8. Dose of prednisone 15 mg per day or less and dose of azathioprine no greater than 3 mg/kg per day

II. PREGNANCY IN SPECIFIC GENETIC DISEASES INVOLVING THE KIDNEYS

A. Diseases Primarily Affecting Glomerular Function

1. Hereditary Nephritis (Alport Syndrome)

Hereditary glomerulonephritis may produce a wide spectrum of renal disorders from asymptomatic hematuria to chronic glomerulonephritis and end stage renal disease (Kaufman *et al.*, 1970; Whalen *et al.*, 1961). Bilateral VIII nerve deafness, beginning in childhood as a high tone hearing loss detectable only by audiograms, is often seen in families with this disorder. When familial nephritis and deafness occur in the same kindred, the term *Alport syndrome* is appropriate; in the absence of familial deafness, hereditary nephritis is the best name for this disease. Ocular abnormalities including cataracts, nystagmus, myopia, and lenticonus are observed in a small percentage of patients.

Males, as a general rule, are more severely affected with glomerulonephritis than females (Albert *et al.*, 1969; Chappell and Kelsey, 1960). In men the disease usually progresses to end stage renal failure by the third or fourth decade of life. Most females with the disorder have mild glomerulonephritis manifest by hematuria, proteinuria, mild hypertension, and mild azotemia. There are, however, exceptions and in some families females can develop end stage renal disease.

The inheritance pattern is usually autosomal dominant with variable penetrance (Chazan *et al.*, 1971). In some families an X-linked dominant inheritance has been reported. The disorder does not recur in a renal transplant. Morphologically, a proliferative glomerulonephritis with focal segmental thickening of the glomerular basement membrane has been observed. Foam cells, which were once thought to be pathognomonic for the condition, need not be present; conversely, foam cells may be seen in several nonhereditary renal diseases. Immunofluoresence studies show minimal deposits of immunoglobulin and complement in the glomeruli (Kaufman *et al.*, 1970; Spear *et al.*, 1970). Serum complement levels are characteristically normal.

Pregnancy has been reported to exacerbate the findings of glomerulonephritis. Exacerbation of proteinuria sufficient to produce nephrotic syndrome during pregnancy has been observed (Gill and Hayslett, 1969). Other patients have developed preeclampsia with an increase in hypertension and azotemia during pregnancy. Patients with end stage renal disease have the problems with conception and pregnancy discussed earlier (Sections I,B, C, and D).

Accurate diagnosis of this condition in the parents is imperative for purposes of genetic counseling. Mothers, who have only a mild form of glomerulonephritis, may pass on a mutant allele that produces end stage renal disease in male offspring. More than one parent in our clinic has said they would have planned their family differently if they had been properly counseled in the genetics of this disorder. This has been especially true when the mother has only a mild illness, but male offspring have a severe illness. Accurate diagnosis requires a positive family history, abnormal audiograms, and renal biopsy findings consistent with the disorder. Prenatal diagnosis is not possible at this time, but fetal sex determination could be useful in families in which only males are severely affected. The infant may be affected by the complications of preeclampsia. He rarely shows features of the hereditary nephritis at birth; some features of the disease may develop in childhood.

2. *Fabry Disease*

Fabry disease is an inborn error of metabolism resulting from defective activity of the lysosomal enzyme α-galactosidase A. The X-linked disorder is manifest by dark-red to black angiokeratomata, pain which can be either chronic and associated with paresthesias or episodic with excruciating pain in the palms and soles, corneal opacities and other ocular lesions, cardiac involvement sometimes progressing to myocardial infarction and/or congestive heart failure, and progressive renal failure resulting in uremia (Desnick *et al.*, 1978). In the heterozygous females, the disease is milder. Most affected women have skin and eye lesions. Colley *et al.* (1958) reported that the mother of a patient with Fabry disease had glomerular lesions identical to those of her son and another affected male. In addition, Ropers *et al.* (1977) reported that four of eight heterozygotes had severe clinical signs of the disease and all of these had enzyme levels much less than 50% normal. Two of these women had cardiac and renal involvement. The authors suggested that these findings were compatible with nonrandom (preferential) X-chromosome inactivation.

The disorder is characterized by absence of α-galactosidase A activity in plasma, white blood cells and cultured skin fibroblasts. Detection of heterozygotes can be difficult because there is some overlap between enzyme activity in normal individuals and in heterozygotes. It has been suggested that measuring the ratio of α-galactosidase activity to β-galactosidase activity in plasma or white blood cells may improve heterozygote detection (Spence *et al.*, 1977). Confirmation of the heterozygous state, if necessary, can be done by demonstrating two cell populations with respect to α-galactosidase A activity in cloned fibroblasts (Romeo and Migeon, 1970).

The propriety of pregnancy in a woman heterozygous for this condition is complex. A major consideration is the health of the mother. As noted above, heterozygotes may have minimal involvement with only corneal lesions or may have extensive disease with progressive renal failure. In the latter case, the effects of renal disease on pregnancy discussed above (Sections I,B, C, and D) become important. Genetic counseling is appropriate in this disorder. The heterozygous woman should be informed that, on the average, one-half of her daughters will also be heterozygous and one-half of her sons will be hemizygous (affected). The prospective parents should be aware of the implications of these facts. The mother should be told that prenatal diagnosis of the condition is possible (Brady *et al.*, 1971). The criteria for prenatal diagnosis are absence of α-galactosidase A activity in cultured amniotic fluid fibroblasts and demonstration of a 46, XY karyotype in these cells.

3. Nail–Patella Syndrome

The Nail–Patella syndrome is transmitted as an autosomal dominant disorder; the locus for this syndrome is linked to the gene controlling the ABO blood groups (Jameson *et al.*, 1956). Affected individuals have skeletal abnormalities, fingernail dysplasia and renal disease. Fingernails may be absent or incompletely formed. Absence of the patella, iliac horns, and flexion deformities of the elbows and hips are also observed. Renal dysfunction, seen in 40% of patients, varies from mild proteinuria to nephrotic syndrome and end stage renal disease (Hoyer *et al.*, 1972; Bennett *et al.*, 1973). Bennett *et al.* (1973) found that failure to concentrate the urine was the most common defect. More seriously affected patients have hematuria, proteinuria, hypertension, and azotemia. Serum complement is normal. Renal biopsy shows focal thickening of the glomerular basement membrane in mild cases, although patients with azotemia may also have proliferative glomerulitis and focal interstitial nephritis. Electron microscopic studies have shown a thickened glomerular basement membrane with unique fibrils having the periodicity of collagen (Hoyer *et al.*, 1972).

Two patients with end stage renal disease have been successfully transplanted. A renal biopsy obtained one year after the transplant showed no evidence of recurrent nephropathy (Uranga *et al.*, 1973).

Several patients with mild renal involvement have been reported to have a normal pregnancy (Morgan, 1964; Pussell *et al.*, 1977). Patients with end stage renal disease will have the problems discussed earlier (Sections I,B, C, and D). Birth weights in offspring have been normal. Affected infants will have the skeletal abnormalities at birth. Proteinuria may be absent in the first several years of life and develop later in childhood or young adulthood.

4. Hereditary Nephritis Associated with Deafness and Thrombocytopenia

There have been two reports of the association of hereditary nephritis with deafness and thrombocytopenia (Epstein *et al.*, 1972; Eckstein *et al.*, 1975). This condition appears to be inherited in an autosomal dominant manner. The renal disease in these patients can be severe with 4 of the reported 13 cases becoming uremic between ages 18 and 27 years. There were no pregnancies in the affected patients of Epstein *et al.* (1972). There were 11 pregnancies in 2 women (mother and daughter) reported by Eckstein *et al.* (1975). The only complication was one premature delivery. Neither woman had decreased renal function at the time of the pregnancies but both had significant proteinuria.

B. Diseases Primarily Affecting Tubular Function

1. Renal Tubular Acidosis

a. Type I (Distal). Type I renal tubular acidosis, also called distal renal tubular acidosis, is characterized by an inability to lower the pH of the urine below 6.0 to 6.5 due to defective secretion of hydrogen ion in the distal renal tubule (Morris *et al.*, 1972). The hereditary form of this condition is inherited as an autosomal dominant. There are also several acquired forms of the disease. Patients affected with this condition also have mild hyperparathyroidism, hypercalciuria, hyperphosphaturia, hypokalemia, low urine citrate, and hypophosphatemia (Coe and Firpo, 1975). In untreated cases, nephrocalcinosis can develop and may cause end stage renal disease.

Accurate diagnosis of this condition is essential because the complications of hyperparathyroidism and hypercalciuria can be reversed by alkali therapy; sodium bicarbonate in a dose of 2–5 mEq/kg/day titrated to give a fasting serum bicarbonate concentration greater than 26 mEq/1 is usually adequate. Nephrocalcinosis can be prevented by vigorous bicarbonate therapy. If nephrocalcinosis is present at the time therapy is begun, it does not progress but it is not reversed.

Pregnancy has been reported to exacerbate nephrolithiasis in two cases (Pines and Mudge, 1951; Buckalew *et al.*, 1968). Theoretically, infants could be affected with hypoparathyroidism due to maternal hyperparathyroidism, but this has not been reported in this disorder. However, almost certainly treatment with alkali during pregnancy will be helpful to both mother and fetus. Infants affected with this disorder will have systemic acidosis with failure to acidify the urine.

Vomiting and failure to thrive are the most common presenting symptoms. Prenatal diagnosis is not possible at this time.

b. Type II (Proximal). Defective reabsorption of bicarbonate by the proximal tubule also causes metabolic acidosis. This disorder is called type II renal tubular acidosis. By definition, other functions of the proximal tubule, including reabsorption of glucose and phosphorus, are normal in this condition. Although many cases are sporadic and transient, a family in whom the disease was inherited as an autosomal dominant trait was recently described (Brenes *et al.*, 1977). These patients did not have abnormalities of calcium and phosphorus metabolism; bone X-rays were normal. Pregnancies were uncomplicated in the family studied and newborns were clinically well. The age of onset of the syndrome could not be ascertained from existing data, although one patient became ill in infancy with vomiting and failure to thrive. He was only partially treated with alkali therapy and developed short stature. Siblings of this patient were normal in height even when affected with acidosis.

2. *Renal Glycosuria*

Renal glycosuria is an autosomal dominant condition in which there is defective reabsorption of glucose by the proximal tubule. Patients characteristically show glycosuria with a normal blood sugar. Renal glycosuria must be carefully distinguished from diabetes mellitus. Renal glycosuria is usually a benign condition; however, there are reports of dehydration and ketosis when this condition is complicated by starvation and/or pregnancy (Marble, 1932; Khachadurian and Khachadurian, 1964; Horowitz and Schwartzer, 1955; Krane, 1966). The dehydration presumably results from an osmotic diuresis induced by the urinary glucose. Maternal glomerular filtration rate increases during pregnancy, increasing the quantity of glucose delivered to the renal tubule. Since the individual with renal glycosuria is already reabsorbing glucose maximally, the extra glucose will be excreted in the urine. Adequate fluid intake is essential for the pregnant patient with renal glycosuria in order to avoid dehydration.

3. *Hypophosphatasia*

Hypophosphatasia is an autosomal recessive condition characterized by a low serum alkaline phosphatase concentration, high urinary phosphoethanolamine excretion, and abnormal bone formation which is histologically similar to rickets (Fraser, 1957; Kozlowski *et al.*, 1976). The illness has been subclassified according to age of onset. Many patients present at birth with disease incompatible with life; others present in childhood with rickets. A few patients reach adulthood manifesting only osteomalacia, fragile bones and poor dentition. Three pregnancies have been completed successfully in patients with the adult form of the disease (Bethune and Dent, 1960; Pimstone *et al.*, 1966). In one case delivery was by Cesarean section because of a small pelvis. The offspring were normal.

Prenatal diagnosis has been made in one case on the basis of absence of fetal osseous structures on ultrasound examination, normal amniotic fluid α-fetoprotein (which excluded anencephaly), and absence of alkaline phosphatase in the amniotic fluid (Benzie *et al.*, 1976). Severe osseous deformities were observed in the aborted fetus. The carrier state is characterized by subnormal serum alkaline phosphatase concentrations and high levels of urinary phosphoethanolamine. Determinations of maternal alkaline phosphatase must be made cautiously since the level of this enzyme rises during pregnancy as a result of fetal alkaline phosphatase activity in the maternal circulation.

4. Cystinuria

Cystinuria is a complex disorder characterized by defective reabsorption of lysine, arginine, ornithine, and cystine by the renal tubule and intestine. There is evidence for the existence of three genetically distinct disorders based on differences in the details of the transport of these amino acids by the small intestine (Rosenberg *et al.*, 1966). These biochemical differences are beyond the scope of the present manuscript, but are available in a recent review (Thier and Segal, 1978). Cystinuria is a clinically significant condition because cystine is highly insoluble and patients are plagued by urinary calculi and their complications. Dahlberg *et al.* (1977) reviewed 89 patients with this condition and found considerable morbidity resulting from active stone formation, frequent surgical procedures and multiple urinary tract infections. The urinary tract infections appeared to result from both the calculi and the surgical manipulation. Only two patients developed advanced renal failure.

Therapy for this condition includes high fluid intake to decrease the concentration of cystine in the urine and alkalinization of the urine since the solubility of the amino acid increases dramatically above pH 7.5. Dahlberg *et al.* (1977) reported that fluids plus alkalinization rarely produce dissolution of established stones but did prevent new stone formation. The drug penicillamine forms a mixed disulfide cysteine-penicillamine when reacted with cystine. The mixed disulfide is much more soluble than cystine and stone dissolution frequently results from treatment with penicillamine (Dahlberg *et al.*, 1977).

It has been suggested that the disease is more serious in men than women but Dahlberg *et al.* (1977) found no difference between the sexes. There are three possible effects of pregnancy on a woman with cystinuria. First, if the mother had renal insufficiency the problems discussed earlier (Sections I,B, C, and D) could arise. Second, if she developed a urinary tract infection during pregnancy this could have deleterious effects on both mother and fetus (Section I,B). The existence of renal calculi would be expected to increase the risk of urinary tract infection during gestation. Third, the effect of penicillamine on the fetus is controversial. There have been at least three reports of infants with lax skin and unusual facies born to women taking penicillamine during gestation. (One mother had cystinuria; one had rheumatoid arthritis; the third had Wilson's disease.) Two infants died following intra-abdominal catastrophes in the neonatal period (Mjolnerod *et al.*, 1971; Solomon *et al.*, 1977). The third infant appeared to recover and develop normally (Linares *et al.*, 1979). On the other hand, there have been numerous reports of normal pregnancies in women taking the drug. Lyle (1978) reported on 27 such pregnancies, 8 in women who received the drug throughout gestation and 19 who stopped the drug in early gestation. The only abnormality seen was in one infant with a ventricular septal defect. Scheinberg and Sternlieb (1975) reported 29 normal pregnancies in 18 women receiving penicillamine. In spite of the controversy, it certainly seems appropriate to use penicillamine during pregnancy only if it is absolutely unavoidable.

C. Diseases Affecting Both Tubular and Glomerular Function

1. Cystinosis

a. Nephropathic Cystinosis. Nephropathic cystinosis is a metabolic disorder characterized by the intracellular accumulation of cystine within lysosomes. This disease is inherited as an autosomal recessive trait. Children with nephropathic cystinosis are usually normal until about 6 months of age. They then develop signs of tubular dysfunction including polyuria, aminoaciduria, glycosuria, phosphaturia, and potassium wasting. This Fanconi syndrome is associated with failure to thrive, growth failure, acidosis, and rickets. Although many of the electrolyte abnormalities can be managed medically, the children develop progressive glomerular damage which leads to end stage renal failure, usually before the end of the first decade (Schneider *et al.*, 1978). There is now considerable experience with renal transplantation in these patients (ASC/NIH Renal Transplant Registry, 1975). It is clear that there is no recurrence of the primary disease in the transplanted kidney. On the other hand, transplantation does not correct the primary metabolic defect (Mahoney *et al.*, 1973; Malekzadeh *et al.*, 1977). Although several cystinotic children with functioning renal transplants have reached puberty in the mid- or late-teens, there have been no pregnancies in this group of patients of which we are aware. Intrauterine diagnosis of this condition is possible using pulse labeling of cultured amniotic fluid cells with [^{35}S]cystine (Schneider *et al.*, 1974).

b. Benign Cystinosis. Patients with this variant of cystinosis have elevated intracellular levels of cystine and crystals in the cornea, bone marrow, and white blood cells. They do not develop retinopathy or renal disease (Schneider *et al.*, 1978). There is no reason to expect that pregnancy in women with the condition would be different from normal. Indeed, this was the case in one reported patient who had four uneventful pregnancies (Lietman *et al.*, 1966).

c. Intermediate Cystinosis. This variant of cystinosis is similar to nephropathic cystinosis but the signs of disease appear later, the tubular dysfunction is less severe and the glomerular insufficiency progresses less rapidly. These patients eventually develop end stage renal failure requiring chronic hemodialysis and/or a renal transplant. The effects of renal failure on pregnancy discussed earlier (Sections I,B, C, and D) are relevant to these patients, but to date no pregnancies have been reported.

2. *Hepatolenticular Degeneration (Wilson Disease)*

Hepatolenticular degeneration (HD) is an autosomal recessive trait characterized by neurological dysfunction, liver disease progressing to cirrhosis, Kayser-Fleischer rings in the cornea, and renal dysfunction (Sass-Kortsak and Bearn, 1978). Neurological manifestations include dystonia with spasticity, drooling and rigidity, pseudosclerosis with tremors, and/or psychological abnormalities. The liver involvement is progressive and may produce hepatic failure within weeks of the onset of hepatic abnormalities although slower progression is usual. The Kayser-Fleischer ring in the cornea is pathognomonic for the disease. It can sometimes be seen with the unaided eye but usually a slit lamp is required. There are a number of reports of renal abnormalities in HD. Reynolds *et al.* (1966) reported a single patient with decreased renal plasma flow and glomerular filtration rate. This patient had aminoaciduria, intermittent glycosuria, and hypercalciuria. Renal biopsy showed abnormalities in the proximal tubules by both light and electron microscopy. The authors felt that these findings were compatible with the observed proximal tubular dysfunction. Lew *et. al.* (1970) performed careful renal function studies on 19 patients with HD. Numerous abnormalities were reported. The patients had decreased renal plasma flow and glomerular filtration rate, as reported previously. All untreated patients had increased uric acid clearance and aminoaciduria. Increased urinary excretion of potassium, calcium, phosphate, sodium, and protein was common. Response to ammonium chloride loading was subnormal and three patients had renal tubular acidosis. Mild abnormalities in renal function usually improved with treatment (see below) but patients with significantly decreased renal function did not improve.

Hepatolenticular degeneration is associated with a decrease in serum copper concentration, a decrease in the serum concentration of the copper-containing protein ceruloplasmin, and an increase in urinary copper. Furthermore, the copper content of involved organs such as liver, brain, lens, and kidney is markedly elevated. It has been shown that the disease can be effectively treated by chelation of copper increasing urinary copper excretion. At the present time, penicillamine is the drug of choice for HD. If the drug is started in an asymptomatic patient, symptoms do not develop. If started later, patients improve although extensive disease may not resolve.

There are numerous reports of pregnancy in HD. Successful pregnancies are very uncommon in untreated symptomatic women with HD (Walshe, 1977). In addition, it has been suggested that asymptomatic women with HD may have an unusually high incidence of spontaneous abortion (Klee, 1979). Although the use of penicillamine in pregnancy is controversial (Section II, B,4) it is clear that treatment with the drug allows a woman to become pregnant and is compatible with a normal pregnancy. Scheinberg and Sternlieb (1975) reported 29 normal pregnancies in 18 women receiving penicillamine. Marecek and Graf (1976) reported seven women receiving penicillamine who had eight normal infants. The drug was stopped when pregnancy was discovered in five women and stopped when pregnancy was planned in the other two women. One woman with advanced portal hypertension had an exacerbation of the disease. These authors recommend continuing penicillamine during pregnancy but at reduced doses (250–500 mg daily). Walshe (1977) reported on 15 pregnancies of which 13 were in women on penicillamine. The drug was continued throughout gestation in six pregnancies and stopped in the first trimester in seven others. All infants were normal. Nevertheless, it seems appropriate to be conservative regarding the use of penicillamine during pregnancy in view of the two reported infants with lax skin, unusual facies, and fatal intra-abdominal catastrophes during the neonatal period (Mjolnerod *et al.*, 1971; Solomon *et al.*, 1977) (Section II,B,4).

D. Structural Diseases

1. Polycystic Kidney Disease

There is considerable controversy regarding the classification of polycystic kidney disease. For the purpose of this work, it is sufficient to divide these entities into an infantile form which is inherited as an autosomal recessive trait and an adult form which is inherited as an autosomal dominant trait. Although this division is convenient, it probably is an oversimplification. A more precise classification of polycystic kidney disease will require better understanding of the basic defect in each condition.

a. Infantile Polycystic Kidney Disease. This condition is characterized by cystic dilatation of the collecting ducts involving the renal medulla and cortex. The cysts are lined with columnar or cuboidal epithelium (Lieberman *et al.*, 1971). In addition, there is a consistent hepatic lesion with portal fibrosis and enlargement and distortion of biliary channels (Bernstein, 1978). There is a wide spectrum of severity in this condition. In the most severe form, a newborn infant is found with massive renal enlargement and oliguria. Occasionally, the kidneys are so large that delivery may be complicated. This severe form of infantile polycystic kidney disease is associated with oligohydramnios, flattened facies with low set ears (Potter facies), and respiratory distress. The respiratory distress is partially a result of abdominal distention and partially due to pulmonary hypoplasia. These infants usually die early in life from respiratory failure. Virtually all of the nephrons are cystic. In surviving children, renal enlargement is less extensive. Renal failure may ensue but this is not invariable. In some patients, renal involvement is minimal and progressive hepatic disease dominates the clinical picture. Lieberman *et al.* (1971) regard these latter patients as having a distinct disease, congenital hepatic fibrosis, but this is controversial. Blyth and Ockenden (1971) have proposed that infantile polycystic kidney disease can be divided into four groups depending on the age at presentation, progression and extent of involvement of the kidneys and liver. This division is also controversial but emphasizes that the condition has a similar course within a single family but not necessarily between families. Since survival of affected patients to adult life has been reported, pregnancy is possible but we are unaware of any reports of a successful pregnancy in this condition. At the present time, there is no known method to detect the carrier state. In the severely affected fetus, prenatal diagnosis may be made late in gestation by ultrasonography.

b. Adult Polycystic Kidney Disease. Adult polycystic kidney disease is an autosomal dominant trait with high penetrance. The condition becomes apparent at various ages, most commonly in the third or fourth decade. It has, however, been recognized in the fetus and newborn (Shokier, 1978). In yet other individuals, the disease is not recognized until the eighth or ninth decade. In the United States, adult polycystic kidney disease is the third most common cause of renal failure accounting for approximately 5.1% of end stage renal disease (ASC/NIH Renal Transplant Registry, 1973). The most common presenting complaint is flank pain. The onset of pain usually antedates renal enlargement sufficient to allow the kidneys to be palpated. Increased renal size is usually noted before the development of significant renal insufficiency. Among the complications of adult polycystic kidney disease are hypertension, hematuria, and infection. There have been a number of reports of pregnancies in patients with this condition. It appears that the severity of the disease in the mother at the time of gestation is a critical factor in determining the outcome of the pregnancy. Landesman and Scherr (1956) reported on the results of 64 pregnancies in 28 women with adult polycystic kidney disease. In the 23 women with clinically undetectable disease (designated Group A), there was no increase in the incidence of complications during gestation. In the four women with known adult polycystic kidney disease, hypertension, and normal renal function (Group B), there was an increased incidence of maternal morbidity and fetal mortality. The single pregnancy in a woman with adult polycystic kidney disease and renal insufficiency (Group C) produced a stillborn fetus. Other reports have confirmed a high incidence of maternal death in Group C pregnancies. Furthermore, until recently, there was no report of a live birth from a Group C mother. However, Katz *et al.* (1979) reported the successful pregnancy and delivery of a Group C woman.

It has been suggested that genetic counseling in this condition should include the suggestion of pregnancy at a relatively young age since the disease is likely to progress with time. There are no specific tests permitting reliable detection of individuals with clinically inapparent disease, although ultrasound examination and intravenous pyelography may be abnormal in the absence of symptoms. A prenatal diagnosis has not been made.

2. *Tuberous Sclerosis*

Tuberous sclerosis is a genetic disease usually characterized by the triad of seizures, mental retardation and adenoma sebaceum. It is usually inherited as an autosomal dominant trait although there have been reports of siblings with the condition whose parents were normal (Lowry *et al.*, 1979). No finding is uniformly present in patients with the disease, Pampiglione and Moynahan (1976) reviewed 100 children with tuberous sclerosis. Ninety-eight had some seizure disorder (69 had infantile spasms and 75 had other seizure disorders); 88 had mental retardation (IQ less than 70). Seventy-seven had adenoma sebaceum and a like number had multiple nonpigmented areas of the skin. Lagos and Gomez (1967) reported a series of 71 patients with the disease. In this series, 26 had normal intelligence. In addition to the skin and central nervous system involvement, about 80% of the patients have retinal phakomas, 80% have renal hamartomas, 50% have areas of hyperostosis and 20–40% have cardiac rhabdomyomas (Chalhub, 1976). Seventy-five percent of patients with this disease die by age 20 years, usually as a result of neurological disease (Anderson and Tannen, 1969).

The renal involvement in patients with tuberous sclerosis is variable. Most patients appear to have renal hamartomas. In general, these are asymptomatic, but the renal involvement can be the presenting complaint. Chonko *et al.* (1974) reported three patients who presented with renal disease. One of these had end stage renal failure, one had gross hematuria and one had flank pain. We have treated an infant who presented with bilateral flank masses in the neonatal period. A diagnosis of polycystic kidney disease was made initially but the infant later developed infantile spasms and hypopigmented areas of the skin. Although renal failure is uncommon in these patients, there are a growing number of reports of this complication (Lynne *et al.*, 1979; Nickel *et al.*, 1974; Anderson and Tannen, 1969).

There is a lack of information regarding pregnancy in women with tuberous sclerosis. At least three important considerations exist. The first is genetic counseling. As noted above, the disease is an autosomal dominant and, therefore, the affected mother would have a 50% chance of having an affected fetus. In view of the severity of the disease with its high incidence of seizures, mental retardation, and early death, these facts should be clearly presented to the mother. At this time, prenatal diagnosis is not possible. The second consideration is the possible teratogenic effect of the anticonvulsants which the mother may require. In addition, if the mother has renal insufficiency, the problems discussed above (Section I,B, C, and D) could arise.

REFERENCES

Albert, M. S., Leeming, J. M., and Wigger, H. J. (1969). Familial nephritis associated with the nephrotic syndrome. *Am. J. Dis. Child.* **117,** 153–155.

Anderson, D., and Tannen, R. L. (1969). Tuberous sclerosis and chronic renal failure. *Am. J. Med.* **47,** 163–168.

ASC/NIH Renal Transplant Registry (1973). The 11th report of the human renal transplant registry. *J. Am. Med. Assoc.* **226,** 1197–1204.

ASC/NIH Renal Transplant Registry (1975). Renal transplantation in congenital and metabolic diseases. *J. Am. Med. Assoc.* **232,** 148–153.

Bear, R. A. (1976). Pregnancy in patients with renal disease: A study of 44 cases. *Obstet. Gynecol.* **48,** 13–18.

Bennett, W. M., Musgrave, J. E., Campbell, R. A., Elliot, D., Cox, R., Brooks, R. E., Lovrien, E. W., Beals, R. K., and Porter, G. A. (1973). Nephropathy of the nail-patella syndrome. *Am. J. Med.* **54,** 304–319.

Benzie, R., Doran, T. A., Escoffrey, W., Gardner, H. A., Hoar, D. I., Hunter, A., Malone, R., Miskin, M., and Rudd, N. L. (1976). Prenatal diagnosis of hypophosphatasia. *Birth Defects, Orig. Artic. Ser.* **12** (6), 271–282.

Bernstein, J. (1978). Polycystic disease. *In* "Pediatric Kidney Disease" (C. M. Edelmann, Jr., ed.), Vol. II, pp. 557–570. Little, Brown, Boston, Massachusetts.

Bethune, J. E., and Dent, C. E. (1960). Hypophosphatasia in the adult. *Am. J. Med.* **28,** 615–622.

Blyth, H., and Ockenden, B. G. (1971). Polycystic disease of kidneys and liver presenting in childhood. *J. Med. Genet.* **8,** 257–284.

Brady, R. O., Uhlendorf, B. W., and Jacobson, C. B. (1971). Fabry's disease: antenatal diagnosis. *Science* **172,** 174–175.

Brenes, L. G., Brenes, J. N., and Hernandez, M. M. (1977). Familial proximal renal tubular acidosis. *Am. J. Med.* **63,** 244–252.

Buckalew, V. M., Jr., McCurdy, D. K., Ludwig, G. D., Chaykin, L. B., and Elkinton, J. R. (1968). Incomplete renal tubular acidosis. *Am. J. Med.* **45,** 32–42.

Chalhub, E. G. (1976). Neurocutaneous syndromes in children. *Pediatr. Clin. North Am.* **23,** 499–516.

Chappell, J. A., and Kelsey, W. M. (1960). Hereditary nephritis. *Am. J. Dis. Child.* **99,** 401–407.

Chazan, J. A., Zacks, J., Cohen, J. J., and Garella, S. (1971). Hereditary nephritis. *Am. J. Med.* **50,** 764–771.

Chonko, A. M., Weiss, S. M., Stein, J. H., and Ferris, T. M. (1974). Renal involvement in tuberous sclerosis. *Am. J. Med.* **56,** 124–132.

Coe, F. L., and Firpo, J. J., Jr. (1975). Evidence for mild, reversible hyperparathyroidism in distal renal tubular acidosis. *Arch. Intern. Med.* **135,** 1485–1489.

Coe, F. L., Parks, J. H., and Lindheimer, M. D. (1978). Nephrolithiasis during pregnancy. *N. Engl. J. Med.* **298,** 324–326.

Colley, J. R., Miller, D. L., Hunt, M. S. R., Wallace, H. J., and de Wardener, H. E. (1958). The renal lesion in angioteratoma corpus diffusium. *Br. Med. J.* **1,** 1266–1268.

Dahlberg, P. J., vanden Berg, C. J., Kurtz, S. B., Wilson, D. M., and Smith, L. H. (1977). Clinical features and management of cystinuria. *Mayo Clin. Proc.* **52,** 533–542.

Davison, J. M., Lind, T., and Uldall, P. R. (1976). Planned pregnancy in a renal transplant recipient. *Br. J. Obstet. Gynaecol.* **83,** 518–527.

Davison, R. J., and Lindheimer, M. D. (1978). Renal disease in pregnant women. *Clin. Obstet. Gynecol.* **21,** 411–427.

Desnick, R. J., Klionsky, B., and Sweeley, C. C. (1978). Fabry's disease. *In* "The Metabolic Basis of Inherited Disease'' (J. B. Stanbury, J. B. Wyngaarden, and D. S. Frederickson, eds.), 4th ed., pp. 810-840. McGraw-Hill, New York.

Eckstein, J. D., Feilip, O. J., and Watts, J. C. (1975). Hereditary thrombocytopenia, deafness, and renal disease. *Ann. Intern. Med.* **82,** 639-645.

Epstein, C. J., Sahud, M. A., Piel, C. F., Goodman, J. R., Bernfield, M. R., Kushner, J. H., and Ablin, A. R. (1972). Hereditary macrothrombocytopathia, nephritis, and deafness, *Am. J. Med.* **52,** 299-310.

Farber, M. (1978). Pregnancy and renal transplantation. *Clin. Obstet. Gynecol.* **21,** 931-935.

Felding, C. F. (1968a). The obstetric prognosis in chronic renal disease. *Acta Obstet. Gynecol. Scand.* **47,** 166-172.

Felding, C. F. (1968b). Pregnancy following renal diseases. *Clin. Obstet. Gynecol.* **11,** 579-593.

Ferris, T. F. (1979). Kidney and pregnancy: A review. *In* ''Strauss and Welt's Diseases of the Kidney'' (L. E. Earley and C. W. Gottshalk, eds.), pp. 1321-1356. Little, Brown, Boston, Massachusetts.

Fraser, D. (1957). Hypophosphatasia. *Am. J. Med.* **22,** 730-746.

Fraser, F. C., and Fainstat, T. D. (1951). Production of congenital defects in the offspring of pregnant mice treated with cortisone. *Pediatrics* **8,** 527-533.

Gant, N. F., Daley, G. L., Chand, S., Whalley, P. J., and Macdonald, P. C. (1973). A study of angiotension II pressor response throughout primigravid pregnancy. *J. Clin. Invest.* **52,** 2682-2689.

Gill, G. N., and Hayslett, P. (1969). Hereditary nephritis and pregnancy. *Am. J. Obstet. Gynecol.* **104,** 19-23.

Grunfeld, J. P., Bois, E. P., and Hinglais, N. (1973). Progressive and non-progressive hereditary chronic nephritis. *Kidney Int.* **4,** 216-228.

Hanshaw, J. B., Scheiner, A. P., Moxley, A. W., Gaev, L., Abel, V., and Scheiner, B. (1976). School failure and deafness after ''silent'' congenital cytomegalovirus infection. *N. Engl. J. Med.* **295,** 468-470.

Harris, R. E., Thomas, V. L., and Shelokov, A. (1976). Asymptomatic bacteriuria in pregnancy: Antibody-coated bacteria, renal function, and intrauterine growth retardation. *Am. J. Obstet. Gynecol.* **126,** 20-25.

Horowitz, L., and Schwartzer, S. (1955). Renal glycosuria. *J. Pediatr.* **47,** 634-639.

Hoyer, J. P., Michael, A. F., and Vernier, R. L. (1972). Renal disease in nail-patella syndrome: Clinical morphologic studies. *Kidney Int.* **2,** 231-238.

Jacobs, C., Brunner, F. P., Chantler, C., Donckerwolcke, R. A., Gurland, H. J., Hathaway, R. A., Selwood, N. H., and Wing, A. J. (1977). Combined report on regular dialysis and transplantation in Europe, VII, 1976. *Proc. Eu. Dialysis Transplant. Assoc.* **14,** 3-70.

Jameson, R. J., Lawler, S. D., and Renwick, J. H. (1956). Nail-patella syndrome: Clinical and linkage data on family G. *Ann. Hum. Genet.* **20,** 348-353.

Katz, M., Quagliorello, J., and Young, B. K. (1979). Severe polycystic kidney disease in pregnancy: Report of fetal survival. *Obstet. Gynecol.* **53,** 119-124.

Kaufman, D. B., McIntosh, R. M., Smith, F. G., Jr., and Vernier, R. L. (1970). Diffuse familial nephropathy: A clinicopathological study. *J. Pediatr.* **77,** 37-47.

Kerr, M. G. (1968). Cardiovascular dynamics in pregnancy and labour. *Br. Med. Bull.* **24,** 19-24.

Khachadurian, A. K., and Khachadurian, L. A. (1964). The inheritance of renal glycosuria. *Am. J. Hum. Genet.* **16,** 189-194.

Klee, J. G. (1979) Undiagnosed Wilson's disease as cause of unexplained miscarriage. *Lancet* **2,** 423.

Kozlowski, K., Sutcliffe, J., Barylak, A., Harrington, G., Kemperdick, H., Nolte, K., Rheinwein, H., Thomas, P. S., and Uniecka, W. (1976). Hypophosphatasia. *Pediatr. Radiol.* **5,** 103-117.

Krane, S. M. (1966). Renal glycosuria. *In* ''The Metabolic Basis of Inherited Disease'' (J. B.

Stanbury, J. B. Wyngaarden, and D. S. Fredrickson, eds.), 2nd ed., pp. 1221–1229. McGraw-Hill, New York.

Lagos, J. C., and Gomez, M. R. (1967). Tuberous sclerosis. *Mayo Clin. Proc.* **42,** 26–49.

Landesman, R., and Scherr, L. (1956). Congenital polycystic kidney disease in pregnancy. *Obstet. Gynecol.* **8,** 673–676.

Leb, D. E., Weisskopf, B., and Kunovitz, B. S. (1971). Chromosome aberrations in the child of a kidney transplant recipient. *Arch. Intern. Med.* **128,** 441–444.

Lew, M. L., Strickland, G. T., and Gutman, R. A. (1970). Renal function in Wilson's disease: Response to penicillamine therapy. *Am. J. Med. Sci.* **260,** 381–398.

Lieberman, E., Solinas-Madrigal, L., Gwinn, J. L., Brennan, L. P., Fine, R. N., and Landing, B. H. (1971). Infantile polycystic disease of the kidney and liver. *Medicine (Baltimore)* **50,** 277–318.

Lietman, P. S., Frazier, P. D., Wong, V. G., Shotton, D., and Seegmiller, J. E. (1966). Adult cystinosis—a benign disease. *Am. J. Med.* **40,** 511–517.

Linares, A., Zanranz, J. J., Rodriguez-Alarcon, J., and Diaz-Perez, J. L. (1979). Reversible cutis laxa due to maternal D-penicillamine therapy. *Lancet* **2,** 43.

Lower, G. D., Stevens, L. E., Najarian, J. S., and Reemtsma, K. (1971). Problems from immunosuppressives during pregnancy. *Am. J. Obstet. Gynecol.* **111,** 1120–1121.

Lowry, R. B., Dunn, H. G., and Paris, R. P. (1979). Inheritance of tuberous sclerosis. *Lancet* **1,** 216.

Lyle, W. H. (1978). Penicillamine in pregnancy. *Lancet* **1,** 606–607.

Lynne, C. M., Carrion, H. N., Bakshandeh, K., Nadji, M., Russel, E., and Politano, V. A. (1979). Renal angiomyolipomia; polycystic kidney, and renal cell carcinoma in patient with tuberous sclerosis. *Urology* **14,** 174–176.

Mahoney, C. P., Striker, G. E., Fetterman, G. H., Hickman, R. O., Schneider, J. A., and Marchioro, T. L. (1973). Renal transplantation in childhood cystinosis. *Brith Defects, Orig. Artic. Ser.* **9** (2), 141–147.

Malekzadeh, M. H., Neustein, H. B., Schneider, J. A., Pennisi, A. J., Ettenger, R. B., Uittenbogaart, C. H., Kogut, M. D., and Fine, R. N. (1977). Cadaver renal transplantation in children with cystinosis. *Am. J. Med.* **63,** 525–533.

Marble, A. (1932). Renal glycosuria. *Am. J. Med. Sci.* **183,** 811–827.

Marecek, Z., and Graf, M. (1976). Pregnancy in penicillamine-treated patients with Wilson's disease. *N. Engl. J. Med.* **295,** 841–842.

Merkatz, I. R., Schwartz, G. H., David, D. S., Stenzel, K. H., Riggio, R. R., and Whitsell, J. C. (1971). Resumption of female reproductive function following renal transplantation. *J. Am. Med. Assoc.* **216,** 1749–1754.

Mjolnerod, O. K., Dommerud, S. A., Rasmussen, K., and Gjeruldsen, S. T. (1971). Congenital connective-tissue defect probably due to D-penicillamine treatment in pregnancy. *Lancet* **1,** 673–675.

Morgan, B. E. (1964). Nail-patella syndrome and pregnancy. *Obstet. Gynecol.* **24,** 644–646.

Morris, R. C., Jr., Sebastian, A., and McSherry, E. (1972). Renal acidosis. *Kidney Int.* **1,** 322–340.

Naeye, R. L., and Friedman, E. A. (1979). Causes of perinatal death associated with gestational hyprtension and proteinuria. *Am. J. Obstet. Gynecol.* **133,** 8–10.

Nickel, W. R., Reed, W. B., and Lachman, R. (1974). The kidney in tuberous sclerosis. *Birth Defects, Orig. Artic. Ser.* **10**(4), 80–86.

Nolan, G. H., Sweet, R. L., Laros, R. K., and Roure, C. A. (1974). Renal cadaver transplantation followed by successful pregnancies. *Obstet. Gynecol.* **43,** 732–739.

Pampiglione, G., and Moynahan, E. J. (1976). The tuberous sclerosis syndrome: Clinical and EEG studies in 100 children. *J. Neurol., Neurosurg. Psychiatry* **39,** 666–673.

Papoff, P., Whetham, J. C. G., Katz, A., and de Veber, G. A. (1977). Pregnancy in renal transplant recipients: Report of two successful pregnancies in a patient with impaired renal function. *Can. Med. Assoc. J* **117,** 1288–1295.

Perez, R. J., Lipner, H. Abdulla, N., Cicotto, S., and Abrams, M. (1978). Menstrual dysfunction of patients undergoing chronic hemodialysis. *Obstet. Gynecol.* **51,** 552–555.

Pimstone, B., Eisenberg, E., and Silverman, S. (1966). Hypophasphatasia: Genetics and dental studies. *Ann. Intern. Med.* **65,** 722–729.

Pines, K. L., and Mudge, G. H. (1951). Renal tubular acidosis with osteomalacia. *Am. J. Med.* **11,** 302–311.

Pussell, B. A., Charlesworth, J. A., MacDonald, G. J., and Baker, W. de C. (1977). The nail patella syndrome—a report of a family. *Aust. N.Z. J. Med.* **7,** 20–23.

Reynolds, E. S., Tannen, R. L., and Tyler, H. R. (1966). The renal lesion in Wilson's disease. *Am. J. Med.* **40,** 518–527.

Romeo, G., and Migeon, B. R. (1970). Genetic inactivation of the α-galactosidase locus in carriers of Fabry's disease. *Science* **170,** 180–181.

Ropers, H. H., Wienker, T. F., Grimm, T., Schroetter, K., and Bender, K. (1977). Evidence for preferential X-inactivation in a family with Fabry disease. *Am. J. Hum. Genet.* **29,** 361–370.

Rosenberg, L. E., Downing, S., Durant, J. L., and Segal, S. (1966). Cystinuria: Biochemical evidence for three genetically distinct diseases. *J. Clin. Invest.* **45,** 365–371.

Rosenkrantz, J. G., Githens, J. H., Cox, S. M., and Kellum, D. L. (1967). Azathioprine (Imuran) and pregnancy. *Am. J. Obstet. Gynecol.* **97,** 387–394.

Salant, D. J., Marcus, R. G., Milne, F. J., Meyers, A. M., Botha, J. R., and Myburgh, J. A. (1976). Pregnancy in renal transplant recipients. *S. Afr. Med. J.* **50,** 1288–1290.

Sass-Kortsak, A., and Bearn, A. G. (1978). Hereditary disorders of copper metabolism. *In* "Metabolic Basis of Inherited Disease" (J. B. Stanbury, J. B. Wyngaarden, and D. S. Fredrickson, eds.), 4th ed., pp. 1098–1126. McGraw-Hill, New York.

Schalch, D. S., Gonzalez-Barcena, D., Kastin, A. J., Landa, L., Lee, L. A., Zamora, M. T., and Schally, A. V. (1975). Plasma gonadotropins after administration of LH-releasing hormone in patients with renal or hepatic failure. *J. Clin. Endocrinol. Metab.* **41,** 921–925.

Scheinberg, I. H., and Sternlieb, I. (1975). Pregnancy in penicillamine-treated patients with Wilson's disease. *N. Engl. J. Med.* **293,** 1300–1302.

Schneider, J. A., Verroust, F. M., Kroll, W. A., Garvin, W. A., Horger, E. O., III, Wong, V. G., Spear, G. S., Jacobson, C., Pellett, O. L., and Becker, F. L. A. (1974). Prenatal diagnosis of cystinosis. *N. Engl. J. Med.* **290,** 878–882.

Schneider, J. A., Schulman, J. D., and Seegmiller, J. E. (1978). Cystinosis and the Fanconi syndrome. *In* "Metabolic Basis of Inherited Disease" (J. B. Stanbury, J. B. Wyngaarden, and D. S. Fredrickson, eds.), 4th ed., pp. 1660–1682. McGraw-Hill, New York.

Shokier, M. K. K. (1978). Expression of "adult" polycystic renal disease in the fetus and newborn. *Clin. Genet.* **14,** 61–72.

Sims, E. A. H., and Krantz, K. E. (1958). Serial studies of renal function during pregnancy and the puerperium in normal women. *J. Clin. Invest.* **37,** 1764–1774.

Solomon, L., Abrams, G., Dinner, M., and Berman, L. (1977). Neonatal abnormalities associated with D-penicillamine treatment during pregnancy. *N. Engl. J. Med.* **296,** 54–55.

Spear, G. S., Whitworth, J. M., and Konigsmark, B. W. (1970). Hereditary nephritis with nerve deafness. *Am. J. Med.* **49,** 52–63.

Spence, M. W., Goldbloom, A. L., Burgess, J. K., D'Entremont, D., Riplety, B. A., and Weldon, K. L. (1977). Heterozygote detection in angiokeratoma corporis diffusion (Anderson-Fabry disease). *J. Med. Genet.* **14,** 91–99.

Thier, S. O., and Segal, S. (1978). Cystinuria. *In* "Metabolic Basis of Inherited Disease" (J. B. Stanbury, J. B. Wyngaarden, and D. S. Fredrickson, eds.), 4th ed., pp. 1578–1592. McGraw-Hill, New York.

Uranga, V. M., Simmons, R. L., Hoyer, J. R., Kjellstrand, C. M., Buselmeier, T. J., and Najarian, J. S. (1973). Renal transplantation for the nail patella syndrome. *Am. J. Surg.* **125,** 777–779.

Walshe, J. M. (1977). Pregnancy in Wilson's Disease. *Q. J. Med.* **46,** 73–83.

Wass, V. J., Wass, J. A. H., Reis, L., Edwards, C. R. W., and Ogg, C. S. (1978). Sexual hormone changes underlying menstrual disturbances in haemodialysis. *Proc. Eur. Dial. Transplant Assoc.* **15,** 178–186.

Whalen, R. E., Huang, S., Peschel, E., and McIntosh, H. D. (1961). Hereditary nephropathy, deafness and renal foam cells. *Am. J. Med.* **31,** 171–186.

Younger, D., Rees, S. B., and White, P. (1962). The kidney in pregnancy complicated by diabetes mellitus. *Clin. Obstet. Gynecol.* **5,** 462–481.

Zacur, H. A., and Mitch, W. E. (1977). Renal disease in pregnancy. *Med. Clin. North Am.* **61,** 89–109.

9

Genetic Disorders of the Gastrointestinal Tract

JEROME I. ROTTER AND RICHARD APOSTOL

Copyright © 1981 by Academic Press, Inc.
All rights of reproduction in any form reserved.
ISBN 0-12-630940-X

I. INTRODUCTION

Although the study of the genetics of the gastrointestinal disorders has a long history, it has not received the same degree of interest as that of other organ systems. During this past decade there have been a number of advances. Many of these advances have been so recent that it is simply unknown how they relate to pregnancy. Thus, while we now know much about the genetics of several of the gastrointestinal disorders, and also about the course of some of the more common disorders during pregnancy, in most cases we are ignorant of the interaction of the specific genetic predispositions with the pregnant state.

Our knowledge of the genetics of the gastrointestinal disorders is similar to that of most medical disorders. We recognize many disorders, mostly rare, whose genetic bases and patterns of inheritance are well understood (McKusick, 1978). These include those disorders with simple Mendelian patterns of inheritance and the chromosomal disorders. These rare disorders, even in the aggregate, account for only a relatively small proportion of the total morbidity and mortality due to gastrointestinal disease. Our understanding of the more common gastrointestinal disorders, such as peptic ulcer, chronic gastritis, inflammatory bowel disease, coeliac disease, gallstones, gastrointestinal cancer, and common gastrointestinal malformations, has lagged considerably. While familial aggregation has been appreciated for many years, it has often been unclear whether the aggregation has a genetic basis. Even when it was clear that genetic factors were important, the inheritance has remained ill defined. It is here, in the area of common GI disorders, that perhaps the greatest strides have occurred over the last decade (Rotter *et al.*, 1980; Steinberg *et al.*, 1980).

Ideally, this chapter would describe all those GI disorders in which inheritance plays a role, identify how the pregnant state alters the course of disease for each disorder, explain how the disorder in a woman may influence the pregnancy outcome, and finally describe how a fetus with the disorder may affect the pregnancy and the woman. Unfortunately, this will not be possible primarily for two reasons. First, the majority of these genetic disorders have not been delineated sufficiently so they can be studied as independent entities. Second, as a consequence, any knowledge regarding the maternal–fetal interaction applies to a phenotype, e.g., ulcer, which represents a whole group of disorders, rather than genetically and etiologically distinct diseases. By presenting the current state of knowledge in the area of GI genetics and resulting maternal–infant interactions we hope to encourage more systematic obstetrical investigations as these disorders become genetically better understood, and to encourage the obstetrician to appreciate other GI disorders with yet unrecognized maternal–fetal interactions.

We will first list, in a number of tables, the Mendelian disorders that present with involvement of the gastrointestinal organs. These will serve as an aid to the obstetrician who is faced with genetic counseling questions regarding the potential inheritance of a variety of gastrointestinal disorders. Although these tables are lengthy, evidence for a maternal–fetal interaction is known only for a few of the disorders. We will discuss two of those in which Mendelian inheritance seems most likely and where there is information regarding the interaction in the pregnant state.

We will follow with a discussion of the common gastrointestinal disorders beginning with the recent advances in their genetics. We will in turn present what is known regarding the interaction of pregnancy with each disorder—both the effect of pregnancy on the natural history of the disorder, and the effect of the disorder on the pregnancy outcome.

For a complete review of the genetics of the gastrointestinal disorders or of all gastrointestinal disorders during pregnancy, the interested reader is referred to several review articles and volumes (for GI genetics: McConnell, 1966, 1973, 1976; Bergsma *et al.*, 1972; Passarge, 1979; Rotter *et al.*, 1980; Steinberg *et al.*, 1980; for GI disorders in pregnancy: Barnes, 1974; Haemmerli, 1975; Bynum, 1977; Miller, 1977).

II. GASTROINTESTINAL DISORDERS WITH SIMPLE MENDELIAN INHERITANCE

All geneticists are indebted to McKusick for his periodically updated catalog of Mendelian disorders in man (McKusick, 1978). In Tables I–X, we have extracted from the catalog those disorders that involve the gastrointestinal tract, and listed them by the organ involved or by clinical presentation, along with their pattern of inheritance. Most malformation syndromes without a recognized Mendelian pattern of inheritance are not included nor are the various chromosomal disorders, even though both may involve the gastrointestinal tract as part of their overall pattern of malformations. [The interested reader may consult Smith (1976), Passarge (1979), and the chapter by Simpson in this volume for further details of these latter disorders.]

The format for these tables is presented in the footnotes for Table I. The reader unfamiliar with McKusick's volume is encouraged to skim through a copy. It is crucial for the reader to keep in mind that for those disorders not having an asterisk preceding McKusick's number, the inheritance pattern is not well established and should not be used for genetic counseling before clearly examining the evidence upon which the classification was made.

Although the tables primarily are a listing of Mendelian disorders, the reader will notice that under the inheritance column there are other designations besides *dominant, recessive,* and *sex-linked. Heterogeneity* is listed for those disorders which in actuality are probably more than one entity, a subset of which may be Mendelian (e.g., see the discussion on peptic ulcer in Section IV). *Multifactorial*

TABLE I

Esophagus

Condition[a]	McKusick's[b] listing	Inheritance[c]
Achalasia, familial esophageal	*20,040	Recessive
Behcet syndrome (inflammatory oral lesions)	10,965	?Dominant
Deafness, congenital with vitiligo and achalasia	*22,135	Recessive
Hernia, hiatus	14,240	Complex
Keratosis palmaris et plantaris with esophageal cancer (congenital sliding hiatal hernia, lower esophagus lined by gastric mucosa)	*14,850	Dominant
Microcephaly, hiatus hernia and nephrotic syndrome	*25,130	Recessive
Tracheoesophageal fistula with esophageal atresia	18,990	Heterogeneity, feature of several syndromes
Tracheoesophageal fistula	18,995	Heterogeneity, feature of several syndromes
Vater association[IV][d] (anal atresia, tracheo-esophageal fistula, esophageal atresia)	19,235	Unknown

[a] Material in parentheses indicates gastrointestinal features that are not obvious from the boldface McKusick listing.

[b] Numbers refer to listing of conditions in McKusick (1978). Conditions in the 10,000s are thought to be autosomal dominant; those in 20,000s autosomal recessive; and those in 30,000s sex-linked recessive. The inheritance pattern of these conditions with asterisks are thought by McKusick to be definitive. See text for additional notes.

[c] The stated inheritance patterns with question markers are tentative classifications. Please refer to text for further explanation of the listed inheritance patterns.

[d] Numbered superscripts indicate that this disorder is cross-referenced in the table of the same number, for example in this case, Table IV.

is listed when the disorder appears to fit the traditional multifactorial model (many genes of small effect acting additively, with possibly additional environmental influences). *Complex* is listed when the disorder could be a single entity, but the inheritance is not at the present time known to be due to an abnormal gene at a single locus, nor clearly multifactorial. Those disorders listed as a complex could be heterogeneous, or may be due to abnormal genes at more than one locus.

The terms *heterogeneity, complex,* and *multifactorial*, are applied to those disorders that have been studied in some detail and for which there is strong evidence that they have genetic determinants, and yet the mode of inheritance of this genetic predisposition is unclear. One disorder, coeliac disease, is listed as inherited in a 2-locus pattern. In the discussion of coeliac disease (Section VI) the

TABLE II

Stomach and Duodenum[a]

Condition	McKusick's listing	Inheritance
Amyloidosis IV (peptic ulcer)	*10,510	Dominant
Duodenal ulcer, hyperpepsinogenemic I	*12,685	Dominant
Endocrine adenomatosis, multiple type I (islet cell adenomas, peptic ulcer, duodenal carcinoid)	*13,110	Dominant
Pernicious anemia[8]	17,090	Complex
Pernicious anemia, congenital, due to defect of intrinsic factor[8]	*26,100	Recessive
Pyloric atresia	*26,595	Recessive
Pyloric stenosis, infantile[b]	26,600	Multifactorial, feature of several syndromes
Tremor, nystagmus and duodenal ulcer	*19,031	Dominant

[a] Refer to Table I footnotes for additional footnotes.
[b] See discussion later in this chapter.

evidence for this mode of inheritance will be presented. Numerous gastrointestinal malformations are features of several syndromes; this has been indicated in the tables.

A. Acrodermatitis Enteropathica

Acrodermatitis enteropathica (AE) is an autosomal recessive disorder characterized by skin lesions, diarrhea, and alopecia (Danbolt and Closs, 1942; Dillaha *et al.*, 1953; Vedder, 1956). The skin lesions are usually limited to the extremities and the areas surrounding body orifices. These lesions are most commonly an acute erythematous vesiculo-bullous or pustular dermatitis. They may also become hyperkeratotic plaques. Clinical symptoms become apparent in infancy although usually not until after weaning if the infant is breastfed. Human milk has also been effective in symptom management. When left untreated, the course is progressive with severe failure to thrive and recurrent monilial or bacterial infections. In those patients surviving into childhood and adulthood without the recently reported zinc therapy, additional clinical characteristics include delayed sexual maturation, growth retardation, lethargy, mental depression, photophobia, and ocular abnormalities including blepharitis, conjunctivitis, and corneal opacities (Hambridge *et al.*, 1977).

Diiodohydroxyquin had been the treatment of choice since 1953 (Dillaha *et al.*, 1953) until Moynahan and Barnes (1973) reported remission with dietary zinc supplementation (see also Neldner and Hambridge, 1975; Walravens *et al.*, 1976). Moynahan (1974) has reviewed the early evidence leading to the identifi-

TABLE III

Small Intestine

Condition[a]	McKusick's listing	Inheritance
Analphalipoproteinemia; Tangier disease (intestinal lipid storage)	*20,540	Recessive
Argyrophil myenteric plexus, deficiency of (obstruction, short small intestine, malrotation and pyloric hypertrophy)	*20,793	Recessive
Proneness to appendicitis	10,770	?Dominant
Chloride diarrhea, familial[8b]	*21,470	Recessive
Carcinoid, intestinal (appendiceal, ileum)	11,490	?Dominant
Celiac disease: gluten-sensitive enteropathy[8b]	11,690	Two locus
Cornelia de Lange syndrome (intestinal atresia)	21,790	Unknown
Cutis laxia (diverticula of esophagus, duodenum, and ileum)	*21,900	Recessive
Deafness, nerve type, with mesenteric diverticula of small bowel and progressive neuropathy[8] (malabsorption of fat and intestinal loss of serum protein)	22,140	?Recessive
Diverticulosis of bowel, hernia, retinal detachment (femoral and/or inguinal hernias)	22,333	?Recessive
Duodenal atresia	*22,340	Recessive, heterogeneity, feature of several syndromes
Enteropathy, protein-losing[8] (secondary to hepatic vein stenosis)	*22,630	Recessive
Intestinal atresia, multiple (stomach to anus)	*24,315	Recessive
Intestinal pseudoobstruction (hereditary hollow visceral myopathy)	14,758	?Dominant, heterogeneity
Jejunal atresia; apple peel syndrome[7] (mesentery agenesis)	*24,360	Recessive
Lymphangiectasia, intestinal (intestinal protein loss, small bowel dilated lymphatic spaces, and partial villous atrophy)	*15,280	Dominant
Lymphedema, hereditary I; Nonne–Milroy, or early-onset type (loss of albumin into intestinal tract)	*15,510	Dominant
Megaduodenum and/or megacystis	*15,530	Dominant
Neuromata, mucosal, with endocrine tumors; mucosal neuroma syndrome; multiple endocrine neoplasia, type III[4] (megacolon, colonic diverticula)	*16,230	Dominant
Polyposis coli, juvenile type[4] (hamartomas)	*17,490	Dominant
Polyposis, familial, of entire gastrointestinal tract[4]	17,500	?Dominant

(*continued*)

TABLE III—*Continued*

Condition[a]	McKusick's listing	Inheritance
Polyposis, intestinal, II; Peutz-Jeghers[4] (entire GI tract)	*17,520	Dominant
Polyposis, intestinal, III; Gardner syndrome[4] (colon, stomach, small intestine)	*17,530	Dominant
Polyposis, skin pigmentation, alopecia and fingernail changes; Cronkhite-Canada	17,550	Nongenetic
Regional enteritis[4b]	26,660	Complex
Volvulus of midgut	*19,325	Dominant

[a] Refer to Table I footnotes for table format.
[b] See discussion later in text.

cation of zinc deficiency as playing a role in the etiology at AE. The clinical features of AE are similar to those of nutritional zinc deficiency in animals and man (Hambridge *et al.*, 1977). Optimal growth occurs for AE patients only with zinc therapy (Hambridge *et al.*, 1977). Behavioral changes in infants with AE have been noted within hours of zinc administration with decreased irritability, nervousness, and restlessness (Walravens *et al.*, 1978). An unexplained observation in AE patients are the reports of clinical improvement at puberty in both males and females (Guy, 1927; Epstein and Vedder, 1960; Verburg *et al.*, 1974; Neldner and Hambridge, 1975; Wells and Winkelmann, 1961).

Acrodermatitis enteropathica is established as an autosomal recessive disorder. Early reports showed familial aggregation with multiple affected siblings, sex ratio approximately one, reports of consanguinity, and no parent to offspring transmission (Danbolt and Closs, 1942; Dillaha *et al.*, 1953; Vedder, 1956). There have been unsuccessful attempts to identify the heterozygote state by looking for inclusion bodies in the intestinal Paneth cells (Bohane *et al.*, 1977). Hirsh *et al.* (1976) found lower serum zinc levels in both the parents and siblings of two brothers with AE. They also reported that one of the siblings had what appeared to be a milder form of the disease with excessive drooling and eruptions in the malar area as an infant. In this family the affected boys' hair turned red at the time of disease activity and reverted to its former color in times of control; a finding not elsewhere reported. There are other reports of milder forms of AE suggesting the possibility of either genetic heterogeneity or clinical variability (Tompkins and Livingood, 1969; Wells and Winkelmann, 1961; Piper, 1955). Additionally, clinical AE with normal zinc levels in the patients' serum, skin, or hair has been reported (Portnoy and Molokhia, 1975; Garretts and Molokhia, 1977). Even though these patients had ostensibly normal zinc levels, they improved with the administration of zinc supplementation.

TABLE IV

Large Intestine[a]

Condition	McKusick's listing	Inheritance
Aganglionosis, total intestinal	20,255	?Recessive
Anal anomaly (skin covered, or anteriorly positioned)[b]	*10,556	Dominant
Anal-sacral anomalies[b]	20,550	?Recessive
Anorectal anomalies (stenosis or imperforate)[b]	10,710, 20,750, 30,180	Heterogeneous, feature of several syndromes
Anus, imperforate, with hand, foot, and ear anomalies[b]	*10,748	Dominant
Cancer of colon (cancer family syndrome)	11,450	?Dominant
Cat eye syndrome (imperforate anus)	11,547	?Chromosomal
Diverticulosis of bowel, hernia, retinal detachment (femoral and/or inguinal hernias)	22,333	?Recessive
Enterocolitis (diarrhea, abdominal swelling)	22,615	?Recessive
Hirschsprung disease with ulnar polydactyly, polysyndactyly of the big toes and ventricular septal defect	23,575	?Recessive
Megacolon, aganglionic; Hirschsprung disease[c]	24,920	Multifactorial, feature of several syndromes
Microcolon	25,140	?Recessive
Neuromata, mucosal, with endocrine tumors: mucosal neuroma syndrome: multiple endocrine neoplasia, type III[3] (megacolon, colonic diverticula)	*16,230	Dominant
Polydactyly, imperforate anus, vertebral anomalies	17,410	?Dominant
Polyposis coli, juvenile type[3] (hamartomas)	*17,490	Dominant
Polyposis, familial, of entire gastrointestinal tract[3]	17,500	?Dominant
Polyposis, intestinal, I; familial polyposis of the colon	*17,510	Dominant
Polyposis, intestinal, II; Peutz-Jeghers[3] (entire GI tract)	*17,520	Dominant
Polyposis, intestinal, III; Gardner syndrome (colon, stomach, small intestine)[3]	*17,530	Dominant
Polyposis, intestinal, IV (scattered, discrete)	17,540	?Dominant
Polyposis, intestinal, with multiple exostoses (sigmoid colon)	17,545	?Dominant
Pancreatic insufficiency, combined exocrine[5,8] (imperforate anus)	26,045	?Recessive
Regional enteritis[3]	26,660	Complex
Rieger syndrome (anal stenosis)	*18,050	Dominant
Schinzel syndrome (anal atresia)	18,145	?Dominant
Trypsinogen deficiency[5,8] (imperforate anus)	*27,600	Recessive

(*continued*)

TABLE IV—*Continued*

Condition	McKusick's listing	Inheritance
Turcot syndrome; malignant tumors of the central nervous system associated with familial polyposis of the colon	27,630	?Recessive
Vater assocation[1] (anal atresia, tracheo-esophageal fistula, esophageal atresia)	19,235	Unknown
Waardenburg syndrome (aganglionic megacolon)	*19,350	Dominant

[a] Refer to Table I footnotes for explanation of table format.

[b] There are numerous syndromes that include anorectal malformations as a feature. These are reviewed by Finsky (1978).

[c] See Passarge (1979, 1980) for an extensive review.

The underlying mechanism leading to the zinc-deficient state has remained evasive although it is currently an area of vigorous research. Lombeck *et al.* (1975) demonstrated with absorption studies that there is an actual decrease of zinc absorption in the intestines of AE patients and that excretion of zinc in sweat, urine, or feces is not increased. Intestinal biopsies have shown mucosal abnormalities that resolve after zinc therapy, suggesting that these abnormalities

TABLE V

Pancreas[a]

Condition	McKusick's listing	Inheritance
Cystic fibrosis (biliary cirrhosis, meconium ileus)	*21,970	Recessive
Hyperlipoproteinemia, type I; familial hyperchlomicronemia idiopathic hyperlipemia of Burger–Grutz type, essential familial hyperlipemia (pancreatitis, abdominal pain, hepatosplenomegaly)	*23,860	Recessive
Hypoglycemia with absent pancreatic alpha cells	24,055	?Recessive
Lipase, congenital absence of pancreatic[8]	*24,660	Recessive
Pancreas, annular (causing duodenal stenosis)	16,775	?Dominant
Pancreatic carcinoma	26,035	?Recessive
Pancreatic insufficiency and bone marrow dysfunction; Shwachman syndrome	*26,040	Recessive
Pancreatic insufficiency, combined exocrine[4,8] (imperforate anus)	26,045	?Recessive
Pancreatitis, hereditary (abdominal pain)	*16,780	Dominant
Trypsinogen deficiency[4,8] (imperforate anus)	*27,600	Recessive

[a] Refer to Table I footnotes for explanation of table format.

TABLE VI

Liver and Spleen[a]

Condition	McKusick's listing	Inheritance
Alpers diffuse degeneration of cerebral gray matter (Poliodystrophia cerebri progressiva) with hepatic cirrhosis	*20,370	Recessive
Amyloidosis, familial visceral (hepatosplenomegaly)	*10,520	Dominant
Antitrypsin deficiency of plasma (cirrhosis)	20,740	Recessive, ?heterozygote susceptibility
Asplenia with cardiovascular anomalies; Ivemark syndrome	*20,853	Recessive
Biliary atresia, extrahepatic	21,050	?Recessive
Biliary malformation with renal tubular insufficiency (extrahepatic and intrahepatic biliary hyperplasia)	21,055	?Recessive
Byler disease; fatal intrahepatic cholestasis	*21,160	?Recessive
Cancer, hepatocellular	11,455	?Dominant
Cerebro-hepato-renal; Zellweger syndrome (intrahepatic biliary dysgenesis)	*21,410	Recessive
Ceroid storage disease (ceroid deposits of liver, spleen, and intestinal mucosa)	21,420	?Recessive
Cholestasis and lymphedema	*21,490	Recessive
Cholestasis with peripheral pulmonary stenosis	11,845	?Dominant
Cholesterol ester storage disease of liver	21,500	?Recessive, (allelic to wolman disease)
Cirrhosis, familial	11,890	Heterogeneity
Crigler–Najjar syndrome, type I (jaundice)[b]	*21,880	Dominant
Fatty metamorphosis of viscera (jaundice)	22,810	?Recessive
Giant cell hepatitis, neonatal	*23,110	Recessive
Glycogen storage disease, type IV; brancher deficiency (cirrhosis)	*23,250	Recessive
Hemochromatosis	*14,160 *23,520	Recessive
Hyperbilirubinemia I; Gilbert disease (jaundice)[b]	*14,350	?Dominant
Hyperbilirubinemia, arias type; Crigler–Najjar syndrome, type II (jaundice)[b]	*14,380	?Dominant
Hyperbilirubinemia II; Dubin-Johnson syndrome (hepatomegaly, abdominal pain)[b]	*23,750	Recessive
Hyperbilirubinemia, rotor type[b]	*23,745	Recessive
Hyperbilirubinemia, conjugated, type III	23,755	?Recessive
Hyperbilirubinemia, shunt (jaundice, splenomegaly)	*23,780	Recessive
Hyperbilirubinemia, transient familial neonatal[b]	*23,790	?Recessive
Intrahepatic cholestasis[b]	24,330	?Recessive
Jaundice, familial obstructive, of infancy	*30,860	?Sex-linked
Mediterranean fever, familial[7]	*24,910	Recessive

(*continued*)

TABLE VI—*Continued*

Condition	McKusick's listing	Inheritance
Retino-hepato-endocrinologic syndrome (degenerative liver disease)	26,804	?Recessive
Splenic hypoplasia	*27,140	Recessive
Splenoportal vascular anomalies (splenomegaly)	27,150	?Recessive
Tyrosinemia (cirrhosis, hepatosplenomegaly)	*27,670	Recessive
Wilson disease; hepatolenticular degeneration	*27,790	Recessive
Wolman disease[8] (hepatosplenomegaly, GI xanthomatous changes)	*27,800	Recessive

[a] Refer to Table I footnotes for explanation of table format.
[b] See discussion later in text.

are secondary to the zinc deficiency and are not the primary defect (Kelly *et al.*, 1976; Bohane *et al.*, 1977; Lombeck *et al.*, 1974).

The striking onset of symptoms occurring at weaning has led many investigators to study differences between human breast milk and cows' milk or artificial formulas. Hurley *et al.* (1977) found a zinc binding ligand in human but

TABLE VII

Associated Structures

Condition[a]	McKusick's listing	Inheritance
("Prune belly") abdominal muscles, absence of, with urinary tract abnormality and cryptorchidism	10,010	Heterogeneity
Chemodectoma, intraabdominal, with cutaneous angiolipomas	11,835	?Dominant
Diverticulosis of bowel, hernia, retinal detachment (femoral and/or inguinal hernias)	22,333	?Recessive
EMG (Exomphalos-macroglossia-giantism); Beckwith-Wiedemann (omphalocele)	22,560	Unknown
Hernia, anterior diaphragmatic	30,695	?Sex-linked
Hernia, double inguinal	14,235	Heterogeneity, feature of several syndromes
Jejunal atresia; apple peel syndrome[3] (mesentery agenesis)	*24,360	Recessive
Omphalocele	16,475 25,835	Heterogeneity, feature of several syndromes
Situs inversus viscerum	27,010	?Recessive

[a] Refer to Table I footnotes for explanation of table format.

TABLE VIII

Malabsorption[a]

Condition[b]	McKusick's listing	Inheritance
Abetalipoproteinemia (lipid)	*20,010	Recessive
Acrodermatitis enteropathica (diarrhea, zinc)[c]	*20,110	Recessive
β-Sitosterolemia (increased intestinal absorption of steroids)	21,025	?Recessive
Blue diaper syndrome; familial hypercalcemia with nephrocalcinosis and indicanuria (intestinal tryptophan transport defect)	21,100	?Recessive
Celiac disease; gluten sensitive enteropathy[3c]	11,690	Two locus
Chloride diarrhea, familial[3c]	*21,470	Recessive
Cystinuria (abnormal intestinal transport of all dibasic amino acids)	*22,010	Recessive
Deafness, nerve type, with mesenteric diverticula of small bowel and progressive neuropathy[3] (malabsorption of fat and intestinal loss of serum protein)	22,140	?Recessive
Dibasicaminoacidura I (L-lysine)	12,600	?Dominant
Dibasicaminoacidura, type II (mild malabsorption)	*22,270	Recessive
Disaccharide intolerance, type I; congenital sucrose-isomaltose malabsorption	*22,290	Recessive
Disaccharide intolerance, type II; congenital lactose intolerance	*22,300	Recessive
Disaccharide intolerance, type III; adult lactose deficiency	*22,310	Recessive
Glucose-galactose malabsorption	*23,160	Recessive
Enterokinase deficiency	*22,620	Recessive
Enteropathy, protein-losing[3] (secondary to hepatic vein stenosis)	*22,630	Recessive
Folic acid, transport defect involving	*22,905	Recessive
Glutamate-asparate transport disease	23,165	?Recessive
Hartnup disease (defect of intestinal and renal transport of certain neutral amino acids)	*23,450	Recessive
Hypomagnesemia, primary	24,130	?Recessive
Hypobetalipoproteinemia	14,595	?Dominant
Lipase, congenital absence of pancreatic[5]	*24,660	Recessive
Lipid transport defect of intestine	24,670	?Recessive (allelic to abetalipoproteinemia)
Lysine malabsorption syndrome	24,795	?Recessive
Menkes syndrome; kinky hair disease (copper malabsorption)	*30,940	Sex-linked
Pancreatic insufficiency, combined exocrine[4,5] (imperforate anus)	26,045	?Recessive
Pernicious anemia[2]	17,090	Complex

(*continued*)

TABLE VIII—*Continued*

Condition[b]	McKusick's listing	Inheritance
Pernicious anemia, congenital, due to defect of intrinsic factor[2]	26,100	Recessive
Pernicious anemia, juvenile, due to selective intestinal malabsorption of vitamin B12 with proteinuria; Imerslund syndrome	*26,110	Recessive
Pseudovitamin D deficiency rickets; Vitamin-D-dependent rickets (calcium malabsorption)	*26,470	Recessive
Trypsinogen deficiency[4,5] (imperforate anus)	*27,600	Recessive
Wolman disease[6] (hepatosplenomegaly, xanthomatous GI changes)	*27,800	Recessive
Wolman disease with hypolipoproteinemia and acanthocytosis	27,810	?Recessive

[a] Refer to Table I footnotes for explanation of table format.

[b] For this table, if malabsorbed substance is known, it is listed in parentheses when not implicit in the condition's name.

[c] See discussion in text.

not cow milk. Eckhert *et al.* (1977) showed a difference in the zinc binding ligands in cow and human milk. They suggested that this species difference in zinc-binding ligands may explain why symptoms occur upon weaning. Hurley (1977) hypothesized that the zinc binding ligand in human milk permits or facilitates the absorption of zinc in the neonate until the development of its own intestinal zinc binding ligand occurs. Whether this human ligand enhances ab-

TABLE IX

Bleeding and Rupture[a]

Condition	McKusick's listing	Inheritance
Blue rubber bleb nevus (hemangiomas of GI tract)	*11,220	Dominant
Ehlers-Danlos type IV (bowel)	13,005 22,535	?Heterogeneity
Hemangiomas of small intestine (cavernous)	14,090	?Dominant
Neurofibromatosis (intestines)	*16,220	Dominant
Pseudoxanthoma elasticum (hemmorhage from arterial changes)	*26,480	Recessive
Telangiectasia, hereditary hemorrhagic, of Rendu, Osler, and Weber	*18,730	Dominant

[a] Refer to Table I footnotes for explanation of table format.

TABLE X.

Abdominal Pain[a]

Condition	McKusick's listing	Inheritance
Angiokeratoma, diffuse; Fabry disease; hereditary dystopic lipidosis	*30,150	Sex-linked
Celiac artery stenosis from compression by median arcuate ligament of diaphragm	11,687	?Dominant
Mediterranean fever, familial[7]	*24,910	Recessive
Pelger–Huet-like anomaly and episodic fever with abdominal pain	26,057	?Recessive

[a] Refer to Table I footnotes for explanation of table format.

sorption, actually transports zinc across the intestine prior to development of specific mechanisms, or protects the zinc from chelation with other dietary components is unclear. In infants with AE, the intestinal zinc binding ligand is presumably absent or ineffective. Casey *et al.* (1978, 1979) suggested that the zinc binding ligand in AE patients may be defective in its binding. They compared the binding of zinc to duodenal secretions in normal individuals and AE patients, and found decreased binding in patients with AE.

The specific nature of the zinc binding ligand or ligands defective in AE patients is a matter of debate. Proposals have included the prostaglandin PGE_2 (Evans and Johnson, 1977), citrate (Hurley *et al.,* 1979), and picolinic acid (Krieger and Evans, 1980). The levels of prostaglandin precursors, such as linoleic and arachidonic acid have been reported to be low by some investigators (Neldner *et al.,* 1974; Cash and Berger, 1969; White and Montalvo, 1973), and normal by others (Kayden and Cox, 1973). Citrate is present in cows' milk at only 25% of the level found in human milk (Hurley *et al.,* 1979). Picolinic acid is a metabolite of tryptophan, and its levels have been reported to be both low (Robertson *et al.,* 1973) and normal (Neldner *et al.,* 1974). It is of interest that picolinic acid is structurally similar to diiodohydroxyquin and hence may explain its prior success as a mode of therapy for nearly 30 years. Its path of action has remained unknown, but Hambridge *et al.,* (1977) suggest that its chelating properties may lead to increased zinc absorption in AE patients. Since, as described above, there is great clinical and biochemical variability in AE patients, more than one ligand may be important, with different AE patients having a different biochemical basis for their disorders, i.e. genetic heterogeneity.

Of further interest is the evidence for zinc deficiency as the basis for the lethal milk syndrome of mice (Piletz and Ganschow, 1978). In this disease female mice homozygous for the lethal milk allele are unable to successfully nurse their offspring, and these offspring develop a disease similar to AE. In this disorder, however, unlike in AE, it appears that the defect is not in the infant but in the mother, resulting in defective transport of zinc from the mother's serum to her milk, and subsequent zinc deficiency in the offspring.

Acrodermatitis enteropathica is an intriguing disorder from the perspective of maternal-infant interactions. First is the evidence reviewed above that a factor in the mother's milk will delay the onset of clinical symptoms. In addition, there are reports that there may be an increased likelihood of congenital malformations in infants of mothers with AE. There have been very few reports of pregnancy in AE patients (reviewed by Hambridge *et al.*, 1975). In fact, there are very few reports of individuals with AE reaching adulthood. Now that clinical management of the disorder is increasingly successful, adulthood and pregnancy cases should be less scarce.

To our knowledge, only three women diagnosed as having AE have had published pregnancy histories. The first report is that of Vedder (1956). In his family of five affected siblings, the eldest had four pregnancies. The first resulted in male infant described as "a typical achondroplastic dwarf that made a feeble respiratory effort and died in one-half hour." (This seems somewhat unlikely since neonatal death is unusual for achondroplasia; some other neonatal skeletal abnormality would be more likely.) This infant's birth weight was reported to be five pounds, and four ounces. Her three other pregnancies all resulted in healthy, normal infants. For the last three pregnancies the patient was on diiodohydroxyquin therapy (Epstein and Vedder, 1960). Verburg *et al.* (1974) report a normal pregnancy outcome for woman with a mild form of AE. She received diiodohydroxyquin during the pregnancy. The patient of Neldner and Hambridge (1975) had two pregnancies: one spontaneously terminated at 3 months and the second resulted in a stillborn anencephalic infant at 8 months gestation. There are other reports of adult females with AE, but pregnancy or menstruation information is not provided (Piper, 1955; Tompkins and Livingood, 1969). Hence, of the seven reported pregnancies, one resulted in a near term anencephalic, one led to the birth of an infant with some form of neonatal dwarfism, and one resulted in a first trimester spontaneous termination. Given the extremely small number of pregnancies, nothing definite can be said about pregnancy outcomes for females with AE.

However, a relationship between maternal zinc deficiency and central nervous system defects has been claimed by Sever and Emanuel (1972). This evidence consists essentially of the high frequency of clinically recognizable human zinc deficiency in Egypt and Iran, two countries known to have increased rates of central nervous system malformations. In addition, low maternal zinc levels in mice have led to increased incidence of malformations in offspring, especially CNS defects (Neldner and Hambridge, 1975; Warkany and Petering, 1972).

Of potential importance for the future offspring of AE patients is the observation that even in normal pregnancies, plasma zinc levels have been shown to decline (Halsted and Smith, 1970; Hambridge and Droegemueller, 1974). The pregnant state may thus lower the already subnormal levels in AE females. It has also been noted that the symptoms of AE became more severe at the onset of pregnancy, and remission occurs rapidly in the postpartum period. As AE patients are now surviving and presumably will have increased fertility on modern zinc treatment regimens, it will be of great importance to observe the course of their pregnancies and their outcomes.

B. Congenital Chloride Diarrhea

Congenital chloride diarrhea (CCD) is a rare autosomal recessive disorder with approximately half of the cases reported originating from Finland (Norio *et al.*, 1971; Holmberg *et al.*, 1977a). At birth the infants present with excessive chloride loss in watery diarrhea. In the largest reported series (Holmberg *et al.*, 1977a) all 21 infants were born before term, with 19 being 2 weeks or more premature. In this report the mean birth length was normal for gestation but mean weight was slightly higher, probably due to increased intestinal retention of water. Two-thirds of the infants were born with distended abdomens, giving the false impression of intestinal obstruction (Perheentupa *et al.*, 1965). Without treatment most children die in infancy due to hypokalemia and dehydration. Those surviving without treatment do so by spontaneously balancing their electrolytes but there is evidence of retarded psychomotor development (Holmberg *et al.*, 1977a). Treatment consists of life-long electrolyte and fluid replacement. Renal function and growth are reduced without adequate replacement therapy (Holmberg *et al.*, 1977b). The disorder appears to be a deficiency of active Cl^-/HCO_3^- transport in the ileum and colon with resulting fecal loss of chloride and osmotic diarrhea. There is also impairment of sodium absorption as a consequence of the acidity of the intestinal contents (Holmberg *et al.*, 1975).

As reviewed by Norio *et al.* (1971) and Holmberg *et al.* (1977a), the criteria for autosomal recessive inheritance are more than adequately met; the disease has been reported from consanguinous relationships, horizontal transmission within pedigrees and equal sex distribution is observed. After correction for ascertainment, the proportion of affected siblings is 0.18–0.29 (Norio *et al.*, 1971). Heterozygotes have been found to have normal stools, plasma chloride, potassium bicarbonate, and blood pH (Norio *et al.*, 1971).

The disorder first presents clinically *in utero* with polyhydramnios, suggesting that the diarrhea begins before birth. No meconium is usually found at birth. Electrolyte concentrations in amniotic fluid are normal (Holmberg *et al.*, 1977a). Hence, CCD is an abnormality in which the fetal phenotype affects the pregnancy. There have been no reported successful attempts of prenatal diagnosis, although with a positive family history (specifically, a prior affected infant) the onset of polyhydramnios is a likely indication of CCD in the fetus. To our knowledge, to date there are no reports of pregnancy in affected women, although there is at least one report of an affected male reaching adulthood (Gorden and Levitin, 1973).

III. COMMON GASTROINTESTINAL DISORDERS

Before potential maternal–fetal interactions in a genetic disorder can be identified, the disorder must be genetically well-delineated and potential heterogeneity recognized. This is a crucial consideration for common disorders, including most GI disorders. The genetic study of common diseases, gastrointestinal or otherwise, is bedeviled by a host of problems. These include disease definition, ascertainment, delayed age of onset, high frequency in the population, variable expression, and heterogeneity. The genetic uniqueness of each individual probably underlies individual differences in susceptibility to and expression of different diseases. Most of the common gastrointestinal disorders show familial aggregation and have a higher incidence in monozygotic (identical) than in dizygotic (nonidentical) twins, and yet do not follow a simple Mendelian pattern of inheritance. By studying a disease within and between families, genetic studies often reveal that the disorder under study consists of several distinct diseases that can be then separated by pathophysiology and/or natural history. This important concept of etiologic and genetic heterogeneity has implications not only for pathophysiology, but for diagnosis and treatment as well, since pathogenetically distinct disorders may well differ in their optimal therapy, prevention, and of concern here, potential maternal–fetal interactions. Thus, before maternal–fetal interactions can be successfully investigated, it must be determined whether the disorder under study is in fact one disease. Only when genetically and etiologically distinct disorders have been identified, and their maternal–fetal interactions characterized, will we be able to improve our ability to diagnose and treat the pregnant patient and fetus with a gastrointestinal disorder. We will begin with a discussion of peptic ulcer as an example of the several approaches utilized by the geneticist to delineate the genetic basis of common disorders. It emphasizes how the genetic heterogeneity approach has added considerably to our knowledge of this group of disorders. We will then discuss what is known about peptic ulcer and pregnancy. A discussion of several other common GI diseases, where there is information about both the genetics and maternal–fetal interaction, will follow.

IV. PEPTIC ULCER

A peptic ulcer is a circumscribed loss of tissue occurring in those parts of the gastrointestinal tract exposed to acid and pepsin—primarily, the lower esophagus, stomach, and upper intestine (duodenum) (Grossman, 1979, 1980). An ulcer is generally thought to occur when there is an imbalance between the "aggressive" forces of acid and pepsin and the less defined "defensive" forces of mucosal resistance and regeneration.

Peptic ulcer, like many common chronic diseases, develops as a result of the interaction of a genetic component with environmental factors. This genetic predisposition varies from individual to individual; all persons are not equally susceptible. A major goal of the geneticist is to identify those individuals who are genetically predisposed and thus have a greater likelihood of developing peptic ulcer. We then attempt to identify environmental factors, such as diet, stress, drugs, viral infections, etc., that cause the predisposed individuals to progress to overt clinical disease. Thus, the more we know about the genetics of peptic ulcer, the better able we may be to treat and prevent the disease by modifying specific environmental factors.

Genetic studies of peptic ulcer have been hampered by a number of difficulties. Peptic ulcer is one of the most common chronic diseases, occurring in some 5–10% of the population during their lifespan (Grossman, 1979; Langman, 1979). When a relative of a patient with peptic ulcer develops the disease, might it be due to shared genetic predisposition, exposure to the same environmental cause, or the coincidental occurrence of a common disorder? Distinguishing among these possibilities is often difficult. There is marked variability in the age of onset of peptic ulcer; it is impossible to know at any given time whether a clinically unaffected person carries the genetic predisposition and will become affected. Genetic studies of peptic ulcer are hampered by the varying definitions of the term affected as used by different investigators. "Affected" has ranged from a person who has abdominal pain to one with an endoscopically demonstrated crater. The greatest obstacle limiting genetic studies has been ignorance of the basic defects in various kinds of peptic ulcer, and correspondingly, the lack of genetic markers that would disclose clinically unaffected individuals with the hereditary predisposition.

Despite these difficulties, it has been known for many years that genetic factors play a role in the etiology of peptic ulcer. The evidence consists of family studies, twin studies, and blood group studies.

A. Early Genetic Evidence

To determine familial aggregation or grouping of a disorder, the incidence of the disease in relatives of patients is compared with the incidence in the general population. If the incidence in relatives is increased, this often provides the first clue that there may be a genetic component to a disease. Because familial aggregation does not distinguish between common genetic and common environmental influences, this finding needs to be supported by additional evidence. Family studies have consistently shown that peptic ulcer disease is two to three times as frequent in the first degree relatives of peptic ulcer patients as it is in relatives of controls. Because these differences persist across generations and social classes, genetic factors were presumed to explain these findings (reviewed in McConnell, 1966, 1980a; Rotter, 1980a).

Twin studies help resolve if familial aggregation is due to common genes or a common environment. Traditional twin studies compare the frequency of concordance (both members of the twin pair affected) for monozygotic (identical) twins with that for dizygotic (nonidentical) twins. Since monozygotic twins share all genes, they theoretically should be concordant for disorders with pure genetic etiologies. Dizygotic twins are no more alike genetically than any pair of siblings. Twin studies have supported the evidence of family studies that genetic factors are involved in the pathogenesis of peptic ulcer disease. Although the reported percentages varied in different studies, the concordance in monozygotic twins was always higher than that in dizygotic twins (reviewed in Gotlieb-Jensen, 1972; Rotter, 1980a; McConnell, 1980a).

A third line of evidence for genetic factors in peptic ulcer are the blood group associations. The basis for studying blood groups and other gene markers is that a positive association with traits known to be inherited in a Mendelian pattern, such as ABO blood groups, means that these genetic traits may predispose to the disease. Peptic ulcer has been reported to be associated with a number of such genetic traits (reviewed in McConnell, 1966, 1980a; Mourant *et al.,* 1978; Rotter, 1980a). The best established are the blood group O and nonsecretor status associations with duodenal ulcer; individuals with these traits have an increased risk for ulcer of some 30% and 50%, respectively, over the general population. However, the pathophysiologic relationship of these associations to peptic ulcer remains a mystery, probably because the magnitude of the association is so small. A number of other less well-studied genetic traits have also been reported to be associated with peptic ulcer; these include alpha-1-antitrypsin deficiency, certain HLA antigens, urinary pepsinogen phenotype, taster-nontaster, and G6PD deficiency (Mourant *et al.,* 1978; Rotter, 1980a).

B. Genetic Interpretations

The inheritance of peptic ulcer cannot be explained by a simple autosomal or sex-linked, dominant or recessive defect. Until recently, peptic ulcer was considered a classical polygenic disorder (McConnell, 1966; Cowan, 1973). Polygenic (multifactorial) disorders are thought to be due to the interaction of several additive genes with environmental factors. The genetic component in these illnesses thus reflects a combined contribution of many genes, resulting is a continuum of genetic predisposition to illness. The gene markers, blood group O and nonsecretor status provide some support for the polygenic hypothesis because each of these genes seems to contribute a small but measurable tendency toward peptic ulcer, and when present together the risk is correspondingly greater. Although there may be a polygenic contribution to peptic ulcer, accumulating evidence indicates that the major genetic factors and the differing physiologic observations can be best accounted for by an alternative explanation, genetic heterogeneity (Rotter and Rimoin, 1977; Rotter, 1980a; Rotter *et al.,* 1980). Many common disorders once thought polygenic, such as peptic ulcer, are now recognized as a group of different disorders that may share a common clinical manifestation—in this case, an ulcer crater in the lining of the gastrointestinal tract. The term for this situation is *heterogeneity:* what may appear to be one disease is found to have a variety of causes, both genetic and nongenetic. For example, a century ago all anemia was considered one disease. Today we know that anemia is a symptom or laboratory finding reflecting a varity of underlying syndromes. Some causes are genetic, such as sickle cell anemia or red cell enzyme deficiencies. Other causes of anemia are clearly nongenetic, such as iron deficiency or vitamin B_{12} deficiency. Only within the last few years has this approach been established in common disorders such as atherosclerosis, diabetes, and, most recently, peptic ulcer.

Several lines of evidence support the hypothesis of heterogeneity in peptic ulcer. They include (1) genetic separation of gastric and duodenal ulcer; (2) rare genetic syndrome with peptic ulceration; (3) variable incidence and types of ulcer in different ethnic groups; (4) physiologic differences between ulcer patients, and most powerfully, (5) family studies that demonstrate the genetic basis of certain physiologic abnormalities (Rotter and Rimoin, 1977; Rotter, 1980a; Rotter *et al.*, 1980).

C. Evidence for Genetic Heterogeneity

The separation of gastric and duodenal ulcer was the earliest evidence for genetic heterogeneity of peptic ulcer. In addition to showing familial aggregation, the early family studies demonstrated that gastric and duodenal ulcers segregated independently, that is, relatives of patients with gastric ulcer tended to have gastric ulcer and relatives of duodenal ulcer patients tended to have duodenal ulcers (Doll and Kellock, 1951; Monson, 1970). This is strong evidence that gastric and duodenal ulcer are independent disorders. Twin studies and blood group studies supported this separation (Gotlieb-Jensen, 1972; McConnell, 1966, 1980a; Mourant *et al.*, 1978). As genetic studies separated these disorders, physiologic studies, such as the measurement of acid secretion or serum pepsinogen I levels, supported this division (Rotter, 1980a, 1981). Further, there is evidence that combined gastric and duodenal ulcer might be a separate genetic entity (Doll and Kellock, 1951; Bonnevie, 1975; Lam and Lai, 1978).

Rare, distinct genetic syndromes associated with peptic ulcer also demonstrate genetic heterogeneity. These rare syndromes suggest that common peptic ulcer may in fact comprise several disorders (Table XI) (Rotter, 1980a,b). Likewise whereas many of the common GI malformations and disorders occur as isolated defects, they may also be a component of various syndromes. Therefore, it is important to look for associated abnormalities that may define a more specific disorder. (For example, see Pinsky (1978), regarding the anorectal malformations.)

The best known example of a Mendelian disorder with peptic ulcer as a feature is multiple endocrine adenoma (neoplasia) syndrome, type I (MEA I, Wermer-syndrome), characterized by pituitary, parathyroid, and pancreatic adenomas

TABLE XI.

Genetic Syndromes Featuring Peptic Ulcer[a]

Syndromes	Location of ulcer	Pathogenetic mechanism	Associated clinical findings	Pattern of inheritance
Relationship established				
Multiple endocrine adenoma, type I	DU	Excess gastrin	Pituitary–acromegaly (GH) infertility (prolactin) Parathyroid–renal stones, hypercalcemia (PTH) Pancreas–hypoglycemia (insulin), diabetes (glucagon)	Autosomal dominant
Sporadic gastrinoma	DU	Excess gastrin		Somatic mutation
Systemic mastocytosis	DU	Excess histamine	Skin lesions, urticaria	? (Both dominant and recessive pedigrees reported)
Ulcer-tremor-nystagmus	DU	Unknown	Essential tremor, nystagmus, narcolepsy	Autosomal dominant
Amyloidosis, type IV	DU, ?GU	Unknown	Sensorimotor neuropathy, nephropathy, hearing loss, cataracts	Autosomal dominant
Relationship suggested, but not established				
Hyperparathyroidism[b]	DU	?Calcium increasing gastric secretion	Hypercalcemia, renal stones	Familial form—autosomal dominant
Cystic fibrosis	DU	?Deficient pancreatic secretion	Pulmonary disease, malabsorption	Autosomal recessive
Alpha-1-antitrypsin deficiency	DU, GU	?Deficient mucosal protection	Pulmonary disease, cirrhosis	Autosomal recessive
Carcinoid syndrome[b]	?GU	?Excess-5-hydroxy-tryptophan	Flushing, cyanosis, diarrhea, bronchoconstriction	?
Stiff skin syndrome	DU	Unknown	Stiff skin, joint enlargement, renal stones, diabetes	Autosomal dominant
Pachydermoperiostosis	DU	Unknown	Clubbing, thickened skin, increased periosteum of distal extremities	Autosomal dominant

[a] For references, see Rotter (1980a), van Allen *et al.* (1969), and Lam *et al.* (1981).

[b] May occur only as a component of MEA I. Increased ulcer may be due to unrecognized gastrinomas.

(Schimke, 1976). The pancreatic tumors may secrete gastrin resulting in a severe ulcer diathesis, the Zollinger-Ellison syndrome (Isenberg *et al.,* 1973). Until the association of ulcer disease with the other endocrine tumors was recognized, the dominant pattern of inheritance elucidated, and the biochemical marker of increased plasma gastrin levels characterized, this specific entity was lost among the mass of peptic ulcer patients. Gastrinoma of the pancreas and associated ulcer disease may also occur as a sporadic somatic mutation without familial aggregation and without endocrine tumors in other organs. Sporadic gastrinomas and those associated with familial MEA I are about equally common (Lamers *et al.,* 1978). Systemic mastocytosis is another multisystem syndrome complicated by peptic ulcer—in this instance because of histamine excess. Both dominant and recessive pedigrees have been described (Rotter, 1980a). Another rare dominant syndrome, which consists of essential tremor, congenital nystagmus, a narcolepsy-like sleep disturbance, and a severe duodenal ulceration, was recently described (Neuhauser *et al.,* 1976). Other genetic syndromes in which an increased incidence of ulcer has been suggested include a form of amyloidosis (van Allen *et al.,* 1969), hyperparathyroidism, alpha-1-antitrypsin deficiency, cystic fibrosis, carcinoid syndrome, a recently described ''stiff-skin syndrome,'' and pachydermoperiostosis (Rotter, 1980a; Lam *et al.,* 1981). The recognition of these disorders is important not just for the implications for genetic heterogeneity, but also as illustrations that different pathogenetic mechanisms (e.g., excess gastrin or histamine) can produce peptic ulcer. Patients with these syndromes should receive specific genetic counseling and therapy (Table XI). In MEA I, for example, the alert physician, after making an accurate diagnosis, should screen asymptomatic family members for subclinical disease. The endocrine tumors can thus be detected early, and many of their manifestations prevented or ameliorated.

The marked ethnic variability in the prevalence and clinical features of a disorder also suggests heterogeneity. For the gastrointestinal disorders, such variability has been well documented in peptic ulcer and gallstone disease (Grossman, 1979; Moshal, 1980; Rotter, 1980a; Bennion and Knowler, 1980). Clinical differences within a disorder can also suggest heterogeneity; it has been suggested that there are subsets of duodenal ulcer patients based on clinical criteria such as age of onset (Lam and Ong, 1976). In addition, physiologic differences between ulcer patients, such as acid secretion, serum pepsinogen I, gastrin response, and gastric emptying, provide evidence for heterogeneity (Rotter and Rimoin, 1977; Rotter 1980a; Lam, 1980).

D. Genetic Studies Using Physiologic Abnormalities as Subclinical Markers

A powerful method for demonstrating genetic heterogeneity directly is to utilize family studies to test whether these physiological abnormalities have a genetic basis. Such studies have been done for several of the physiological traits: (1) serum pepsinogen I, (2) gastric emptying, and (3) gastrin response to a protein meal (Rotter, 1980a,b).

The most extensive genetic studies of a physiological abnormality have utilized the radioimmunoassay of serum pepsinogen I. Human serum pepsinogens, the precursors of pepsin, have been separated into two immunochemically distinct groups (Samloff, 1979, 1980). Pepsinogen I is a protein of fast electrophoretic mobility which is confined to the acid secreting part of the stomach. Pepsinogen II has a slower mobility and is found throughout the stomach and in the first part of the duodenum.

Sibling and twin studies have shown that about half of duodenal ulcer patients have hyperpepsinogenemia I and the other half have normopepsinogenemia I (Rotter *et al,* 1977a, 1979a). Both forms are genetic, since both have an increased risk to sibs for ulcer. These studies and extended family studies have also demonstrated that the familial aggregation of elevated pepsinogen I is consistent with autosomal dominant inheritance; about half the siblings of a patient with hyperpepsinogenemia I also have the trait; multigenerational families have also been reported (Rotter *et al.*, 1979b). These studies have delineated a major genetic factor, elevated serum pepsinogen I, that can identify individuals at risk and that may account for the major genetic predisposition of a substantial portion of duodenal ulcer patients.

This does not mean there are only two types of duodenal ulcer. Based on studies utilizing other physiologic characteristics, further subdivisions are emerging in the hyperpepsinogenemic I and normopepsinogenemic I duodenal ulcer groups. There appears to be a subgroup of normopepsinogenemic I patients in whom rapid gastric emptying may be the inherited physiological abnormality predisposing to ulcer (Rotter *et al.*, 1979c). Familial hyperpepsinogenemic I duodenal ulcer seems to be separable into different groups. A group of patients with a markedly exaggerated antral gastrin response to a protein meal, known as antral G cell hyperfunction, have also been shown to have hyperpepsinogenemia I. These abnormalities in both gastrin and pepsinogen I were shown to have a familial basis (Calam *et al.*, 1979).

The combined genetic–physiologic studies do not exhaust the heterogeneity of the ulcer syndrome. Childhood duodenal ulcer may be a separate genetic entity from adult duodenal ulcer, analogous to the separation of juvenile type from maturity type diabetes (Cowan, 1973). The possible association between HLA antigens and duodenal ulcer, and the report of antibodies to secretory IgA in certain duodenal ulcer patients, suggest that there may be an immunologic form of peptic ulcer (Rotter *et al.*, 1977b; Kwitko and Shearman, 1978). Another means of detecting heterogeneity is by examining the association of ulcer with other diseases within families. The association of ulcer with certain chronic diseases may be due to the inheritance of common defect that predisposes to both diseases (Rotter *et al.*, 1979d; Rotter, 1980a,b). This may account for observations of the familial aggregation of ulcer disease and renal stones (without hyperparathyroidism), ulcer and coronary artery disease, and ulcer and chronic obstructive pulmonary disease. In the latter case, it has often been assumed that the ulcer develops secondary to the pulmonary disease, but it has recently been observed that in most cases the ulcer disease precedes the pulmonary disease (Rotter *et al.*, 1981).

E. Genetic Classification and Implications for Counseling

Thus, the accumulating evidence suggests that the genetic predisposition to peptic ulcer is due not to the cumulative effect of multiple, additive predisposing genes each with a small effect (the polygenic hypothesis), but rather that there are multiple forms of peptic ulcer, each with a different genetic basis. A still evolving classification of peptic ulcer genetic heterogeneity is given in Table XII (Rotter *et al.*, 1979d; Rotter, 1980a,b). It should be noted that polygenic inheritance and genetic heterogeneity are not necessarily mutually exclusive. There may well be a polygenic background upon which major predisposing genes act. Alternatively, one or more of the various heterogeneous forms may be found to be polygenic.

There appear to be different forms of peptic ulcer for which delineation and classification is ongoing; thus, it is difficult to accurately counsel patients and

TABLE XII

Proposed Classification of Peptic Ulcer Genetic Heterogeneity

I.	Peptic ulcer associated with rare genetic syndromes (see also Table XI)
	Multiple endocrine adenomatosis, type I (gastrinoma)
	Systemic mastocytosis
	Tremor-nystagmus-ulcer syndrome
	Amyloidosis, type IV
II.	Gastric ulcer
III.	Combined gastric and duodenal ulcer
IV.	Hyperpepsinogenemic I duodenal ulcer
	Without postprandial hypergastrinemia
	With postprandial hypergastrinemia
V.	Normopepsinogenemic I duodenal ulcer
	Without rapid gastric emptying
	With rapid gastric emptying
VI.	Childhood duodenal ulcer
VII.	Immunologic form of duodenal ulcer[a]
VIII.	Peptic ulcer associated with other chronic disease[a]
	Peptic ulcer and chronic lung disease
	Duodenal ulcer and renal stones (without hyperparathyroidism)
	Duodenal ulcer and coronary artery disease

[a] VII and VIII are tentative subdivisions. See Rotter *et al.* (1979d, 1980a) and Rotter (1980a, 1981) for further details.

their families. Overall, we know empirically that some 20% of first degree relatives (parents, siblings, and offspring) will also experience peptic ulcer disease in their lifetimes. But this risk undoubtedly varies depending on the specific disorder for which the relative is at risk. Since it is may be difficult to assign individual patients to one category or another in Table XII, we often must fall back to the empiric recurrence risks mentioned above.

F. Peptic Ulcer and Pregnancy

With this proposed classification scheme in mind, the question now becomes which of these disorders is a concern when discussing maternal–fetal interaction? In the brief discussion of peptic ulcer and pregnancy to follow, it will become apparent that the evidence presented does not separate the particular ulcer disorders. In other words, to answer questions regarding ulcer and pregnancy the disorders must be lumped together. Thus, any firm statements will have to await studies that identify the individual type of peptic ulcer in the pregnant patient.

It has been reported that the majority of women with peptic ulcer disease undergo improvement or complete remission of their symptomatology during pregnancy (Clark, 1953). Of interest, in the same study, a recurrence of symptoms occurred within several months after parturition in most of the women. These observations must be treated with caution, since ulcer disease is notorious as a disorder characterized by remission and relapse, and the relation of symptomatology to the existence of active ulcer is not as clear as we once believed. This reported decrease in ulcer symptoms has been attributed to decreased acid secretion, which reaches its nadir in the midportion of pregnancy (Murray *et al.*, 1957; Hunt and Murray, 1958). This evidence is not unequivocal, as blood pepsin and pepsinogen I values have been reported not to decrease during pregnancy (Gryboski and Spiro, 1956; Waldum *et al.*, 1980). Of interest, Hunt and Murray (1958) showed a marked increase in acid secretion during lactation, and this has led Barnes (1974) to recommend against breast feeding in women with prior ulcer complications of bleeding or perforation. In any case, complications of peptic ulcer necesitating surgery, specifically massive bleeding and perforation, do occur during pregnancy and the indication for operative intervention are similar to that of the nonpregnant state (Barnes, 1974; Miller, 1977). Attempts at conservative therapy of perforations have had uniformly dismal outcomes.

V. PYLORIC STENOSIS

A. Definition and Pathophysiology

Idiopathic hypertrophic pyloric stenosis is a disorder characterized by projectile vomiting and the finding of a palpable pyloric tumor (reviewed in Dodge,

1980). It occurs predominantly in early infancy, usually within the first 3 months of life. The obstruction is due both to the hypertrophy of the circular smooth muscle of the pylorus (the pyloric "tumor"), which is often accompanied by variable degrees of edema and hypertrophy of the mucosa, and a concomitant disturbance of motility. Treatment usually consists of Ramstedt's operation, involving an incision in the pyloric tumor without interrupting the mucosa. The clinical spectrum of this disorder is still to be fully defined, as there are children who have similar symptomatology but without a palpable tumor (often designated by the term pylorospasm), and conversely identical twins of affected infants have been described who have no clinical symptoms or signs, but in whom X-ray studies demonstrate pyloric hypertrophy and incoordination (Dodge, 1973, 1980). The preponderance of affected males is well known. This disorder is clearly familial and is most interesting in our context, because it is one of the few gastrointestinal disorders where a maternal effect appears well documented.

The pathophysiologic basis of this disorder is still to be determined. In a real sense, it is usually not congenital, since it rarely has even been detected at birth, either clinically or at autopsy. In addition, there are well-documented cases of infants who were operated on for another indication and who subsequently developed pyloric stenosis (Dodge 1973, 1980). A variety of authors have reported a decreased number of pyloric ganglion cells, with some attributing these findings to a degenerative process, others to a failure in development, and others to prolonged parasympathetic stimulation (Dodge, 1980).

Hypertrophic pyloric stenosis can be reproduced in puppies by injecting their pregnant mothers with pentagastrin, the five amino acid peptide that contains most of the physiologic properties of the gastrointestinal hormone gastrin (Dodge, 1972a; Dodge and Karim, 1976). The results are even more dramatic if the injections are continued after birth into the offspring. Since gastrin is known to be trophic to the gastrointestinal tract, this provides one theoretical mechanism for the pyloric hypertrophy. A variety of studies have examined serum gastrin levels, both fasting and postprandial, in infants with pyloric stenosis, and the majority report elevated levels compared to levels in normal infants (Dodge, 1980). It becomes difficult to distinguish cause and effect, since as a consequence of obstruction, the resultant gastric distension is a strong stimulus to gastrin secretion. The elevated levels do tend to persist after surgery, and elevated levels were documented after intraduodenal feeding in one study. No studies have reported on gastrin levels in the mothers themselves, either before or after birth. Because of the evidence for a maternal effect (see below), and the animal studies mentioned above, this possibility should be pursued by studying fasting and postprandial gastrin levels in women who have given birth to affected offspring. Since the highest group at risk are the offspring of mothers who themselves had pyloric stenosis (as discussed below), it would seem to be particularly worthwhile to study such mothers prospectively (Rotter, 1978). In addition, if the gastrin hypothesis is indeed a pathogenetic mechanism in this disorder, women with gastrinomas, either sporadic or as a component of multiple endocrine adenomatosis type I (see Section IV,B), should be at particularly high risk, and the reports of the outcome of such pregnancies would be extremely interesting (Rotter, 1978).

B. Genetics

There is clearly an increased incidence of pyloric stenosis among the family members of affected individuals, yet the pattern of familial aggregation does not fit any simple Mendelian pattern of inheritance. This risk is very dependent on the sex of the index patients, and the sex of their relatives at risk. It ranges from 2 to 3% of first degree female relatives of male patients, to some 4 to 9% of male relatives of male patients, to 4 and 10%, respectively of sisters and brothers of female patients, to the highest risk in daughters (7%) and sons (19%) of female patients (Carter and Evans, 1969; Carter, 1972; Dodge, 1973). This high risk to the offspring of females who are themselves affected in infancy remains one of the most intriguing aspects of this disorder. Twin data confirm that genetic factors are important in this familial aggregation, as the concordance is certainly higher in monozygotic than like-sexed dizygotic twins (Carter, 1972; Dodge, 1973, 1980).

A strong excess of affected males has been reported in all series of pyloric stenosis cases, ranging from 3 to 6 affected males for each affected female (Spence and Gladstien, 1978). The incidence of pyloric stenosis varies extensively over the globe, being found in highest frequency, 2–4 per 1000 in Northern Europe, and decreasing to ½ per 1000 in Blacks and Hawaiian Japanese, to a complete absence in Hawaiian Chinese. There also appear to be temporal variations of the incidence of pyloric stenosis, both longitudinal and seasonal, though the latter is in greater dispute. Birth order effects have also been claimed, namely an excess of first borns, though this is still a matter of some controversy (Spence and Gladstien, 1978; Gladstien and Spence, 1978; Dodge, 1980).

Pyloric stenosis has been reported to be associated with a number of disorders and malformation syndromes (Dodge, 1972a, 1973; 1980). An increased incidence of pyloric stenosis has been claimed in infants with diaphragmatic and hiatus hernia, esophageal atresia, and high intestinal obstruction. An increased occurrence of pyloric stenosis has also been suggested in thalidomide and rubella embryopathies, Turner syndrome, Smith–Lemli–Opitz syndrome, FG syndrome, abnormalities of chromosomes 18 and 21, Cornelia de Lange syndrome, phenylketonuria, adult polycystic kidney disease, and offspring of mothers with myasthenia gravis (Dodge, 1972a, 1973, 1980; Loh *et al.*, 1977). These associations may hint at genetic and etiologic heterogeneity of the disorder. An association with peptic ulcer later in life has been claimed, but control data were lacking (Bendix and Necheles, 1947), and in a subsequent study the prevalence of ulcer did not appear increased (Berglund and Rabo, 1973). Blood group studies have given inconsistent results, and there is no evidence of an HLA association.

As the familial aggregation of pyloric stenosis did not fit any simple Mendelian pattern of inheritance, its inheritance was explained by the polygenic model, that is the combined contribution of many genes of small individual effect acting together to predispose to the disorder (Carter and Evans, 1969; Carter, 1972). With enough predisposing genes (and environmental factors) an individual would be beyond the threshold for the disorder and be affected. In the case of pyloric stenosis, the observation that females are affected less frequently than males would imply in terms of the polygenic model that their threshold for the disease is higher. Therefore, in order to be affected females would have to have more disease predisposing genes, on the average, than would male patients. The genetic consequence of this polygenic model is that the relatives of female patients would on the average receive more of these disease predisposing genes than the relatives of male patients, and therefore the frequency of the disease would be predicted to be higher in relatives of female than male patients. This increased risk to relatives of female patients is precisely what is observed. However, even early on, the observation of an excess of affected offspring to affected mothers, even beyond that predicted by the sex threshold effect of the polygenic model, indicated the existence of maternal effect (Carter, 1972).

Although our analytic capabilities to test the polygenic–multifactorial model have increased, analyses regarding pyloric stenosis have been conflicting and often inconclusive (Kidd and Spence, 1976; Spence and Gladstien, 1978; Gladstien *et al.*, 1978; Lalouel *et al.*, 1977). Kidd and Spence (1976) emphasized the increased risk to offspring of affected mothers and suggested an interaction of maternal influence, possibly gastrin, and fetal sensitivity. All such analyses must be tempered by the observations mentioned above of asymptomatic but presumably affected cotwins of affected twins. Utilizing a newly developed method to analyze twin data, Marshall and Knox (1980) concluded that there was a high risk not only in particular sibships, but in particular pregnancies. Thus only particular pregnancies would be at high risk analogous to neonatal hemolytic disease. If genetic factors are responsible for this risk, they felt these would operate on the mother. Supporting this concept is the observation that the aggregation of this disease seems greater in maternal than paternal relatives (Dodge, 1973).

Thus a variety of epidemiologic data and analyses support a maternal contribution to the occurrence of pyloric stenosis. Dodge (1972b) hypothesized that one mechanism by which this might occur would be through maternal stress leading to production of a humoral factor, possibly gastrin, that would subsequently produce pyloric hypertrophy in the infant. This would presumbably explain the excess of first born affected, since a first pregnancy is frequently a more stressful event. In his Belfast series, Dodge (1972b) found that 42% of mothers of affected infants had experienced events that are defined as unequivocally stressful to normal individuals, compared to 8% of the controls. Mindful of the problem of selective recall in a parent searching for a reason for an abnormality in an infant, Revill and Dodge (1978) in a subsequent study in Cardiff used not only normal controls, but also mothers of patients with spina bifida. Using the Life Events Inventory, which enables a quantitative assessment of stressful events, they observed a significantly higher number of such events among the pyloric stenosis mothers, but not among the spina bifida mothers, compared to controls. Anxiety levels were also higher. Thus, these were not neurotic overreacting individuals, but individuals reacting normally to a greater number of stressful events. These data are probably among the best in existence for maternal psychologic events having a defined effect on the infant. Since we are currently undergoing an explosive growth in our knowledge regarding the interrelationship of neuropeptides in the brain and gut, there are many candidate neurohumural substances that might mediate this interaction, without necessarily involving maternal hypergastrinemia. It is of interest to speculate that the environmental mechanism, stress, and the maternal genetic factors both might operate through this common final pathway, affecting those infants who are genetically sensitive to the proposed trophic hormone.

VI. COELIAC DISEASE

Coeliac disease, also known as gluten sensitive enteropathy (GSE), or nontropical sprue, is a disorder characterized by intestinal malabsorption. This results from flattened intestinal mucosal villi due to the toxic effects of gluten, a wheat protein (reviewed in Stokes *et al.*, 1973; Strober *et al.*, 1975; Strober, 1976; McConnell, 1981). The basic pathogenesis appears to be an immunopathic process induced by gluten, which damages the mature intestinal epithelial cells that have gluten on their surface (Strober *et al.*, 1975; Strober, 1976, 1980). The resulting loss of absorptive area leads to malabsorption, which may be overt or subclinical. Coeliac disease usually presents with one of the features of the generalized malabsorption syndrome. This may range from anemia secondary to folic acid or vitamin B12 deficiency, bleeding abnormalities due to vitamin K deficiency, edema consequent to hypoproteinemia and protein-losing enteropathy, neuropathy, hypocalcemia and osteomalacia, and even hypokalemia leading to nephropathy (Strober, 1976). This malnourished state can presumably interfere with pregnancy and fertility (see below).

Since the advent of peroral jejunal biopsy, the definition of coeliac disease consists of the demonstration of a flat jejunal mucosa and of sensitivity to gluten.

The latter is usually determined by clinical, biochemical, and morphologic improvement on a gluten-free diet (Strober, 1976; MacKintosh and Asquith, 1978; McConnell, 1981). Some would add the requirement of deterioration, at least morphologically, on the reintroduction of gluten. The development of the organ-culture technique as *in vitro* test of gluten toxicity may help in this regard (Strober *et al.,* 1975; Falchuk *et al.,* 1978). Coeliac disease is most prevalent in European populations, with population estimates of 1 in 1200 to 1 in 6000 (Stokes *et al.,* 1973; Shipman *et al.,* 1975); it is especially prevalent in Ireland (Mylotte *et al.,* 1972).

A. Genetics

The earliest evidence that genetics factors could be of significance in coeliac disease consisted of isolated reports of multiple cases occurring within a family. Formal family studies in the 1950s, using clinical criteria to delineate which family members were affected, clearly demonstrated that GSE occurred more frequently in the relatives of patients than in the general population (reviewed in Ellis, 1980). These studies undoubtedly underestimated the number of affected relatives, since it is now known that the disease can wax and wane both clinically and biochemically, though the intestinal biopsy will probably remain abnormal. Later studies in the 1960s and 1970s, utilizing small intestinal biopsy as the criteria for affected, demonstrated that some 10% of first degree relatives were affected, and a large proportion of these were asymptomatic (Ellis, 1980; McConnell, 1981). In addition, most reported monozygotic twin pairs were concordant for the disorder, emphasizing the importance of genetic factors (Strober, 1976; Ellis, 1980). Despite this evidence, as for many disorders, the mode of inheritance remained unclear. At one time or another, both multifactorial inheritance and a single major gene with reduced penetrance had been proposed.

A major advance in our knowledge of the genetics of this disorder was made in 1972, when it was discovered that GSE was associated with the histocompatibility antigen B8 (Falchuk *et al.,* 1972; Stokes *et al.,* 1972). HLA B8 was found in some 80% of coeliac patients, compared to a 20% frequency in controls. The relative risk, the odds of acquiring the disease, were some 8–9 times higher for HLA B8 individuals, compared to those who are not B8. This does not mean that B8 directly contributes to the disease. This may be inferred by the observation that GSE occurs in non-B8 individuals, and that 20% or more of the general population carries this antigen. Rather, as has been hypothesized for many of these HLA associations, B8 is thought to serve as a marker for a closely linked gene in the HLA complex, as yet unidentified, that more directly determines susceptibility to coeliac disease. With additional studies HLA Al (thought secondary to B8), B13, Dw3(Drw3), and Dw7(Drw7) were also reported associated with coeliac disease (the B13 and Drw7 association seem principally to occur in Southern Europe). The D (Dr) antigen associations are felt to be even stronger than the B locus ones (Strober, 1977, 1980; Mackintosh and Asquith, 1978). These associations were important supporting evidence for the immunopathic hypothesis of the pathogenesis of coeliac disease, since the HLA B8 and Dw3 antigens are associated with a host of other disorders in which autoimmunity has been demonstrated to play a role, such as Addisons disease, insulin-dependent diabetes, Graves disease, and others (McMichael and McDevitt, 1977).

The importance of the HLA complex was also demonstrated by family studies, which usually demonstrated that the disorder segregated within families with the B8-Dw3 haplotypes (Mackintosh and Asquith, 1978). Nevertheless, the mode of inheritance still remained unclear. The observation of families where a grandparent and grandchild were affected, yet shared no HLA haplotypes in common, led Strober to hypothesize that at least two loci were necessary for disease susceptibility (Strober, 1976; Falchuk *et al.*, 1978). During this period, Mann *et al.* (1976), by using the sera of mothers of GSE patients, identified B-cell alloantigens that were specific for GSE. In a landmark paper, Peña, Strober, and co-workers demonstrated that these B-cell alloantigens segregated independently of the HLA complex, and proposed that the genetic basis of coeliac disease is a function of two genes, Drw3 (or a closely linked gene), and a non-HLA linked GSE-associated B-cell alloantigen (Peña *et al.*, 1978). They proposed at that time that Drw3 was necessary only in a single dose (dominantly inherited) and the GSE-associated B-cell alloantigen in a double dose (recessively inherited). The dominant–recessive model was postulated because of the observation that, in the families studied, individuals with coeliac disease were often heterozygous for Drw3, as inferred from the typing data on other family members. Subsequently it has been suggested that a recessive–recessive model might be more compatible with population prevalence and observed family data (Greenberg and Rotter, 1981).

This demonstration that genes at two different loci were most likely responsible for coeliac disease was a major advance. It may very well be the first human disorder where two genes have been demonstrated to be necessary for disease expression. Even then, however, considerable complexities remain. If there truly are discordant monozygotic twins for GSE, then additional factors other than dietary gluten are necessary for disease expression (Strober, 1976; Ellis, 1980; McConnell, 1981). There is suggestive evidence for heterogeneity within GSE. For example, it has been proposed that there are more affected relatives of B8-positive than B8-negative probands (Ellis, 1980). B8-positive and -negative patients may differ in their reactivity in organ culture (Strober, 1980). There may be differences in childhood and adult coeliac disease. The frequency of B8 has been reported to be somewhat lower in childhood than adult cases (65% compared to 85%) (Strober, 1976). Other genetic polymorphisms have been reported to be associated with disease onset in early childhood as contrasted to older cases. Ellis (1980) observed that in children with celiac disease having an age of onset before 9 years, there was an increased frequency of the heterozygous state of alpha-1-antitrypsin deficiency and of the urinary pepsinogen isozyme phenotype B. In addition, there is a disease closely related to coeliac disease, dermatitis herpetiformis. This is a vesicular skin disease characterized by IgA deposition in the basement membrane of skin. Most patients have an enteropathy similar to that of coeliac disease including a response to gluten withdrawal. These individuals also have an increased frequency of HLA antigens B8 and Drw3, and some evidence of the non-HLA B-cell alloantigens (Strober, 1980). What determines the differences between dermatitis herpetiformis with enteropathy, and coeliac disease alone, is yet to be determined.

B. Pregnancy

Untreated GSE appears to have a deleterious effect on fertility and at the same time increases the incidence of spontaneous abortion. Women with untreated coeliac disease have presented with years of infertility, which resolved relatively soon after disease remission was instituted by a gluten-free diet (Morris *et al.*, 1970). In these cases, GSE was diagnosed during childhood but a gluten-free diet had been discontinued by the patient. There is also evidence that untreated pregnant women with coeliac disease have an increased incidence of spontaneous abortions. Ogborn (1975) reported that first trimester abortions occurred in 8 of 38 pregnancies in untreated coeliac patients, compared to 1 in 22 pregnancies when the same patients were placed on an appropriate gluten-free diet. This increased incidence of abortions had been attributed to folic acid deficiency (Joske and Martin, 1971; Martin *et al.*, 1965), but this seems hard to separate from the multiple nutritional deficiencies that occur in the malabsorbing coeliac patient. Certainly patients who present with macrocytic or megaloblastic anemia during pregnancy should be investigated for folate deficiency. If it is found, coeliac disease should be considered as a prime diagnostic possibility (Joske and Martin, 1971; Myellot, 1976). It should not be surprising that the additional nutritional demands of pregnancy would overburden an individual who might otherwise remain subclinically affected. This might explain the data suggesting that coeliac disease might be more frequent in women than men (MacDonald *et al.*, 1965). Certainly the pregnant coeliac patient, and the woman with coeliac disease contemplating pregnancy, should be on a gluten-free diet. In addition, there are two reasons for keeping any coeliac patient on a lifelong gluten-free diet. First, it prevents undetected, subclinical malabsorption. Second, there is evidence that untreated coeliac patients have an increased incidence of small intestinal malignancy, and there is the possibility that a gluten-free diet may possibly diminish this risk (Ellis, 1980; Harris *et al.*, 1967; Holmes *et al.*, 1976).

VII. INFLAMMATORY BOWEL DISEASE

Inflammatory bowel disease (IBD) is a group of disorders characterized by chronic, recurrent inflammation of the small and large intestines. This spectrum of disorders, the interrelationships of which remain ill defined, include ulcerative colitis, ulcerative proctitis, regional enteritis (classic Crohn's disease), jejunoileitis, ileocolitis, and Crohn's colitis (Janowitz, 1979; Kirsner, 1980). The course of these diseases is characterized by numerous complications, both gastrointestinal and extraintestinal. Their etiology remains unknown even after years of intensive investigation. Currently the most popular hypothesis is that they are immunologic disorders, which are possibly triggered by or reflect an inability to deal with as yet undefined infectious agents (Farmer, 1980; Kirsner, 1980; Sachar *et al.*, 1980). Presumbably, as will be discussed below, this susceptibility to an immunopathic process and/or infectious agents is genetically determined.

Ulcerative colitis is a disorder characterized by superficial ulcerations that start in the mucosa and submucosa of the large bowel. It becomes transmural only in fulminant colitis. The entire colon is most commonly affected. The diffuse inflammation and superficial ulcerations presumbably lead to the symptoms of diarrhea, rectal bleeding, crampy abdominal pain, fever, and loss of appetite and weight.

Crohn's disease, or regional enteritis, is classically thought of as a transmural inflammatory disease of the small intestine, with thickening of the entire bowel, narrowed lumen, deep ulcerations, adherent bowel loops, enlarged mesenteric nodes, and on histology, noncaseating granuloma are often seen. Crohn's disease is characterized by asymmetric involvement of the small intestine, and while the ileum is most classically affected, any part of the entire gastrointestinal tract, from mouth to rectum, may be involved.

The etiology of all these disorders remains unknown. Treatment is, therefore, empirical (reviewed in Donaldson, 1978; Cello and Meyer, 1978a,b; Farmer, 1980; Janowitz, 1979). There is evidence that Crohn's disease and ulcerative colitis are related, and possibly the two ends of a spectrum of disorders. This evidence includes individual patients who are unclassifiable or who have features of both disorders, the occurrence of both diseases in families (reviewed below), and the association of both disorders with ankylosing spondylitis (McConnell, 1980b; Kirsner, 1980). These disorders vary in frequency, with the highest incidence in Northern Europeans and in Jewish populations of the Western world.

In addition, while the incidence of ulcerative colitis appears to be stable over the last generation or even longer, the incidence of Crohn's disease appears to be increasing dramatically (Langman, 1979; Mendeloff, 1980). Any such rapid temporal trends cannot be due to changes in genetic factors alone, but implicate environmental influences, though they are likely important only in those who are genetically susceptible.

A. Genetics

Inflammatory bowel disease is a group of disorders in which genetic factors clearly are important, but whose mode of inheritance remains unclear. Inflammatory bowel disease clearly has a markedly increased frequency in family members of affected individuals compared to the general population (McConnell, 1972, 1980b; Kirsner, 1973, 1980; Lewkonia and McConnell, 1976). Studies of the frequency of a positive family history among patients with IBD range from 5 to 17% (Binder *et al.*, 1966; Singer *et al.*, 1971; Lewkonia and McConnell, 1976), to a high of 32% in patients with onset of disease before age 20 years (Farmer *et al.*, 1980). Both vertical (parent–child) and horizontal (sib–sib) aggregation has been observed, the latter somewhat more frequently. Extended pedigrees with multiple members affected have also been reported. While there is clearly increased familial aggregation, the observed family data does not satisfy a simple mendelian pattern of inheritance.

Familial aggregation can also be due to common environmental factors. Arguing against this in IBD is the extreme rarity of occurrence of IBD in spouses of affected individuals, the extended families with multiple members affected with IBD, and the many years between onset of disease in familial cases with different generations affected (McConnell, 1980b; Kirsner, 1980). Also arguing strongly for genetic determination of IBD is the association of IBD with ankylosing spondylitis, both in individuals and within families (reviewed in McConnell, 1972; Kirsner, 1973). Ankylosing spondylitis is a disorder in which genetic factors seem to be of primary importance. Those individuals with IBD and ankylosing spondylitis also have increased incidence of HLA B27, as do ankylosing spondylitis patients in general (Lewkonia and McConnell, 1976; McConnell, 1980b; Kirsner, 1980). Providing further evidence for genetic factors is the occurrence of inflammatory bowel disease in several genetic syndromes (reviewed below).

No gene markers, blood groups or enzyme polymorphisms, have been found to be consistently associated with either Crohn's disease or ulcerative colitis. Because of the association of certain HLA antigens with diseases of autoimmune origin, HLA association studies have been pursued vigorously, with essentially negative results (Strober, 1977; McConnell, 1980b; Kirsner, 1980; Peña, 1980). Such studies are usually performed by comparing the incidence of HLA antigens in unrelated patients and controls. Another approach is to not look for an association with any particular antigen, but investigate whether affected members within families share the entire HLA haplotype more commonly than would be expected. One such study of five families did reveal that affected siblings shared haplotypes more often than would be expected (Schwartz *et al.*, 1980), but this was not observed in two other studies of nine and five families, respectively (Peña *et al.*, 1980; Eade *et al.*, 1980). A variety of other immunogenetic approaches are still being pursued (Peña, 1980).

One of the most important results of the family studies of ulcerative colitis and Crohn's disease has been the observation that the two diseases occur together in the same family more often than expected (Kirsner and Spencer, 1963; Lewkonia and McConnell, 1976). In fact, the incidence of ulcerative colitis in relatives of Crohn's disease patients approaches that of Crohn's disease itself. These observations, along with the association of both disorders with ankylosing spondylitis, provides some of strongest evidence that these disorders are related. McConnell (1972, 1980b) has concluded from these data that inflammatory bowel disease is a multifactorial disorder, with Crohn's disease and ulcerative colitis sharing many of the same common genetic determinants in a polygenic system. The apparent increased risk to relatives in childhood IBD would support this hypothesis (Farmer *et al.*, 1980). McConnell interprets the family data to mean that Crohn's disease patients have more of these genetic determinants. He marshalls as supporting evidence the higher incidence of ulcerative colitis in Crohn's families than Crohn's in ulcerative colitis families, the higher incidence of a positive family history of IBD in Crohn's patients than in ulcerative colitis patients, and the reports that monozygotic twins with Crohn's show a higher concordance rate than those with ulcerative colitis. In order to explain the rapidly increasing incidence of Crohn's disease, McConnell (1980b) postulates an environmental factor common to both, and a recent additional factor predisposing to Crohn's disease.

While the polygenic–multifactorial hypothesis as delineated by McConnell may be the explanation for the genetic predisposition to much of inflammatory bowel disease, there is evidence for at least some genetic heterogeneity. This is due to the association of IBD with other diseases and most particularly with rare genetic syndromes. Besides the association with ankylosing spondylitis, there is an increased incidence of eczema, allergic rhinitis, hay fever, and asthma in IBD patients and their families (Hammer *et al.*, 1968). Inflammatory bowel disease has been described in a patient with deficiency of the second complement component (Slade *et al.*, 1978). A deficiency of the free secretory component of IgA has been noted in five patients with ulcerative colitis, including three members from one family (Engström *et al.*, 1978), though it is not clear if this was a primary or secondary abnormality. A distinct disorder may have been observed by Fried and Vure (1974), who reported a family with three of four children who died within 2 months of birth, over a period of 10 years, of a fulminant disorder with bloody diarrhea and swollen abdomen, and at autopsy, ulcerative colitis in two infants and pseudomembranous colitis in the third. They suggest this infantile colitis is a distinct autosomal recessive disorder. Schinella *et al.* (1980) described the occurrence of granulomatous colitis in four members of two families from Puerto Rico with Hermansky–Pudlak Syndrome, an autosomal recessive disorder consisting of tyrosinase positive albinism, a defect in the second phase of platelet aggregation, widespread accumulation of ceroidlike pigment in various tissues, and pulmonary fibrosis. Most interesting has been the recent recognition of an increased incidence of inflammatory bowel disease in patients with Turner syndrome (Weinrieb *et al.,* 1976; Price, 1979: Arulanantham *et al.,* 1980). Price (1979) found four individuals with IBD among 135 adult Turner patients. This association is of interest because a variety of autoimmune disorders have been found with increased frequency in Turner patients and their relatives (Simpson, 1976). These autoimmune associations may be even higher when the Turner syndrome is due to an abnormal X chromosome rather than a 45XO karyotype. Thus it is of particular interest that in 7 of 11 reported cases of Turner patients with IBD, the Turner syndrome was due to an abnormal X—in 6 cases an isochromosome Xq, and in one case a deleted Xp.

B. Pregnancy

There has been a long interest in the course of patients with inflammatory bowel during pregnancy, and the interested reader is referred to several reviews (Zetzel, 1975; Fielding, 1976; Miller, 1977). There have been eight major series reporting on the course of ulcerative colitis during pregnancy (Abramson *et al.,* 1951; Crohn *et al.,* 1956a; MacDougall, 1956; Banks *et al.,* 1957; De Dombal *et al.,* 1965, 1972; McEwan, 1972; Webb and Sedlack, 1974; Willoughby and Truelove, 1980) and six major series dealing with Crohn's disease (Crohn *et al.,* 1956b; Fielding and Cooke, 1970; Schofield *et al.,* 1970; Norton and Patterson, 1972; Homan and Thorbjarnarson, 1976). It should be noted at the onset that the majority of series upon which our knowledge is based, especially regarding ulcerative colitis, were reported some 20 or more years ago. As will be seen, the most recent series are the most encouraging.

Fertility is unimpaired in patients with ulcerative colitis. Initially based on impressions (MacDougall, 1956; Banks *et al.,* 1957; De Dombal *et al.,* 1972), this has now been well documented by Willoughby and Truelove (1980) who found involuntary infertility in only 6.8% of their patients. In addition the patients' mean family size was similar to that of the general British population. In contrast, there is evidence that fertility is impaired in Crohn's disease (Fielding and Cooke, 1970; De Dombal *et al.,* 1972; Homan and Thorbjarnarson, 1976). This is most often attributed to fallopian tube adhesions and appears to have a higher incidence in those patients with involvement of the colon as opposed to the small intestine. However, malabsorption and resultant B_{12} deficiency may be a contributing factor to the subfertility, as may simply the pain accompanying intercourse in the face of pelvic inflammation (Fielding, 1976; Miller, 1977).

The effect of ulcerative colitis or Crohn's on the outcome of pregnancy appears to be minimal, unless the disease activity is severe. In most series, the percentage of completed pregnancies, spontaneous abortions, stillbirths, and newborns with congenital anomalies appear to be comparable to the normal population (Fielding, 1976; Miller, 1977; Willoughby and Truelove, 1980). In ulcerative colitis, if the disease activity is severe, there is a less favorable outcome. In the older series this was evidenced by maternal and concurrent fetal mortality (Fielding, 1976; Miller, 1977). In the most recent series there was no maternal mortality and instead this was reflected in an increased incidence of spontaneous abortions (Willoughby and Truelove, 1980).

When discussing the effect of the occurrence of pregnancy on the course of the inflammatory bowel disease, it has been traditional since Abramson *et al.* (1951) to divide patients into four categories—those with established disease which is inactive at conception, those with active disease at conception, those who develop disease for the first time during pregnancy, and those who develop disease for the first time during the puerperium. It appears that some 30–40% of patients with inactive ulcerative colitis will relapse during pregnancy (Fielding, 1976; Miller, 1977; Willoughby and Truelove, 1980), but this is apparently no higher than the nonpregnant yearly relapse rate (De Dombal *et al.,* 1965). In regional enteritis patients, some 25% relapsed during pregnancy. Relapses usually occur in the first trimester and puerpium, though the latter is not so frequent in the most recent series (Willoughby and Truelove, 1980). This observation has been attributed to the protective effect of the rise of endogeneous steroid levels during the second and third trimester and the lower levels in the first trimester and puerpium (Zetzel, 1975). In patients with active disease at the time of conception, more than half will continue with active disease, with a large percentage of these worsening (Miller, 1977). Older series of ulcerative colitis patients had noted special concern if the colitis originated during pregnancy or the puerpium (Fielding, 1976; Miller, 1977). The bulk of maternal deaths occurred in these situations, some 10–15%. But it must be noted that the mortality of new onset ulcerative colitis in the 1950s was of the same order in the nonpregnant individual. In the most recent series of Willoughby and Truelove (1980), there were no maternal deaths and the outcome of pregnancy was quite good. They attribute this improved prognosis to the subsequent advances in medical therapy. There are only a few reports of the new onset of Crohn's disease during pregnancy or the puerpium. Martimbeau *et al.* (1975) review 10 cases and report a high fetal mortality and high incidence of surgery required for the mother.

The basic rule for managing these disorders in the pregnant patient is that they should be managed as in the nonpregnant patient (Zetzel, 1975; Fielding, 1976;

Miller, 1977; Willoughby and Truelove, 1980). The state of the art of the treatment of these disorders has been reviewed (Donaldson, 1978; Cello and Meyer, 1978a,b; Farmer, 1980). Since these are certainly complex disorders, it is probably best that they be managed by the obstetrician and experienced gastroenterologist in concert. Uncontrolled disease is undoubtedly more toxic to the mother and thus to the fetus than the appropriate medical therapy, including such drugs as sulfasalazine and steroids. Whereas there are theoretical concerns from animal studies regarding possible teratogenic effects of steroid use during pregnancy, these concerns have not been substantiated in human studies (reviewed in Zetzel, 1975; Miller, 1977). The same holds for sulfasalazine, with no real evidence of adverse effects (Miller, 1977; Willoughby and Truelove, 1980). In any case, if this therapy is needed for control of the bowel disease, this should considerably outweigh these other concerns. This probably even applies to immunosuppresive agents such as azathioprine since they are only used when the disease is potentially life threatening (Miller, 1977). This concept certainly applies to the indications for surgery, which should be the same as in the nonpregnant patient, and should be applied to the bowel, where the disease is, and not to the fetus (Fielding, 1976; Miller, 1977). As in the nonpregnant patient, the outlook for the patient requiring emergency surgery of this nature is not always good. Termination of pregnancy for disease control is rarely if ever indicated (Fielding, 1976; Willoughby and Truelove, 1980). The patient who has had prior surgery can usually manage quite well (reviewed in Hudson, 1972; Zetzel, 1975; Fielding, 1976; Miller, 1977). Most in fact can undergo vaginal delivery, though an episotomy is usually indicated in the patient who has had rectal surgery. The physician must be alert to the occasional complication of intestinal obstruction in the ileostomy patient, since in the face of the normal distended abdomen of pregnancy this diagnosis may be delayed.

In summary, there appears to be a general consensus regarding the management of inflammatory bowel disease in the pregnant patient (Zetzel, 1975; Fielding, 1976; Miller, 1977; Willoughby and Truelove, 1980). First, whenever possible, the woman with inflammatory bowel disease desiring a pregnancy should plan to become pregnant during an inactive phase of her disease. Second, the management of active disease in the pregnant patient should proceed in the same fashion and with the same vigor as in the nonpregnant patient.

VIII. GALLSTONES

A. Genetics and Pathophysiology

Gallstone disease is an excellent example of etiologic and probable genetic heterogeneity, with the existence of two entirely different major forms—bilirubin pigment stones, and cholesterol stones. Pigment stones are more common in the Orient, and cholesterol gallstones predominate in the West (van der Linden, 1973, 1980; Bennion and Knowler, 1980). The sequelae of gallstones is gallbladder disease. The stones apparently develop for many years prior to the onset of symptoms. This prolonged silent period has hampered studies of gallstone epidemiology and genetics. We know little of the pathophysiology and genetics of pigment stones, so this chapter will focus on cholesterol gallstones.

Cholesterol gallstones are believed to be the consequence of an excess of cholesterol in a supersaturated solution of bile. This can be due to either an excess of cholesterol or a reduction in bile acids, or both (Hofmann, 1980; Bennion and Knowler, 1980). Disturbed gallbladder motility may also play a role, possibly by allowing the growth and retention of early crystals. The delineation of the relationship between gallstones and supersaturated bile has provided considerable insight into studies of gallstone natural history, pathophysiology, epidemiology, and even genetics.

A number of factors have been found to be associated with an increased frequency of cholesterol gallstones. These include residence in the West, family history, obesity, bile acid malabsorption, clofibrate therapy, estrogen use, female sex, increasing age, and possibly pregnancy and diabetes mellitus (Bennion and Knowler, 1980). The majority of these factors, e.g., obesity, clofibrate, estrogen, act by increasing cholesterol secretion. Many of these influences may be genetic. Diabetes mellitus clearly has important genetic determinants, as likely does obesity. There are extreme ethnic differences in the incidence of gallstones and supersaturated bile, only some of which can be attributed to environment (Redinger and Small, 1972), and are thus likely to be due to genetic differences.

Several additional lines of evidence provide evidence that genetic factors are important in the etiology of cholelithiasis. Harvald and Hauge's (1956) study of the Danish twin registry did find greater concordance in monozygotic than dizygotic twin pairs. The most complete clinical studies have been performed by van der Linden and co-workers in Sweden. They demonstrated that the incidence of gallstones was significantly higher in the siblings of index patients compared to the patients' spouses, with the siblings age and sex matched to the spouse (van der Linden and Lindelof, 1965; van der Linden and Westlin, 1966). They also demonstrated that the familial incidence was particularly striking when the gallstones presented at a young age, with a marked increase in the percentage of parents affected (van der Linden and Simonson, 1973). Thus, these studies clearly demonstrated a genetic component to gallstone susceptibility. At least one pathophysiologic basis for this susceptibility was provided by the work of Danzinger *et al.* (1972), who examined the bile of the sisters of patients operated on for gallstones, and found it to be lithogenic more often than that in matched controls.

B. Pregnancy

While there is a generally acknowledged clinical association between pregnancy and gallstones, this has never been substantiated by appropriate epidemiologic studies, because none have been done (Bennion and Knowler, 1980). What has been documented is that oral contraceptives increase cholesterol saturation in bile (Bennion *et al.*, 1976), but this study has not been extended to pregnant women. Given the increase in endogeneous sex steriods in pregnancy, it is likely to also occur in the pregnant women, but this does require documentation. (However, there has been an observed increase of gallstone formation in patients with intrahepatic cholestasis of pregnancy, a disorder apparently due to an increased reactivity to sex steriods. See Section IX, A.) In addition, Braverman *et al.* (1980) have demonstrated by ultrasonography that after the first trimester gallbladder resting and residual volume were increased, and the rate and percentage of gallbladder emptying were diminished. This could promote gallstone formation by retaining incipient cholesterol crystals. Thus, two independent factors could increase the susceptibility to gallstone formation during pregnancy. With the availability of ultrasound and other techniques, the natural history and epidemiology of gallstones in pregnancy is now amenable to study.

As regards the symptomatic disease during pregnancy, the usual presentation is biliary colic as opposed to jaundice (Miller, 1977). A cholecystectomy is reportedly performed in the United States once in every 1000 to 3000 pregnancies. It was suggested by Greene *et al.* (1963) that cholecystectomy was followed by an increased incidence of spontaneous abortions. This is denied by Hill *et al.* (1975). They report that cholecystectomy is safe for both the mother and fetus, and recommend the second trimester as the optimum period.

IX. HEREDITARY HYPERBILIRUBINEMIAS

Disorders leading to jaundice during pregnancy are not especially common. Jaundice in pregnancy occurs in about 1 in 1500 gestations (Haemmerli, 1975). Of interest here are those familial disorders resulting in primary hyperbilirubinemia and hepatobiliary disease, affected by or affecting the maternal–fetal interaction of the pregnant state. Six specific disorders will be discussed: recurrent intrahepatic cholestasis of pregnancy, Dubin-Johnson syndrome, Gilbert syndrome, Crigler–Najjar syndromes (types I and II), and transient familial neonatal hyperbilirubinemia. In recurrent intrahepatic cholestasis of pregnancy the mother-to-be experiences jaundice and the effect on the fetus appears to be an increased chance of prematurity. Both Dubin-Johnson and Gilbert syndromes also occur in the nonpregnant state and their course in pregnancy is not well established. The Crigler–Najjar syndromes (types I and II) are severe disorders of nonhemolytic hyperbilirubinemia with survival into adulthood rarely reported in the type I condition. The last disorder, transient neonatal hyperbilirubinemia, is different from the others in that it is the infant, rather than the mother, who is affected. It is an example of how the biochemical maternal–infant interactions can continue into the postpartum period. For the interested reader, several detailed reviews discussing the topics of the familial hyperbilirubinemias and the subject of jaundice during pregnancy have been published (Thorling, 1955; Ikonen, 1964; Iber, 1965; Haemmerli, 1966, 1975; Arias, 1971; Barnes, 1974; Holzbach, 1976; Miller, 1977; Schwarz *et al.*, 1977; Schmid and McDonagh, 1978; Zimmerman, 1979; Javitt, 1980; Odell and Childs, 1980).

A. Intrahepatic Cholestasis of Pregnancy

Recurrent intrahepatic cholestasis of pregnancy is a relatively benign form of jaundice occurring only in pregnancy or upon administration of oral contraceptives. Most women experience the disorder as pruritus and mild icterus with pale stools and dark urine. The jaundice usually follows the onset of pruritus within 1 to 2 weeks, although it may be delayed longer. These symptoms do not regress before the pregnancy is completed. Intrahepatic cholestasis of pregnancy is estimated to occur in 1/2000 to 1/8000 gestations (Haemmerli, 1975) or about one-fifth of all gestations in which maternal jaundice occurs (Sherlock, 1975). The disorder has been primarily reported in Scandinavian and Chilean populations, but has also been documented to occur in North America, Europe, and Australia. Recurrent intrahepatic cholestasis of pregnancy is of interest here because its clinical symptomatology occurs only in pregnancy, there is evidence suggesting that the condition may not be fully benign to the neonate, and familial factors appear to play a role in the disorder's etiology.

During the period of clinical expression, the women have elevations in serum alkaline phosphatase and conjugated bilirubin. Further supporting these biochemical findings of cholestasis, serum 5′-nucleotidase and γ-glutamyltranspeptidase are also elevated. Serum transaminases reportedly are only moderately increased. In addition, cholesterol levels are elevated and the prothrombin time may also be prolonged (Haemmerli, 1975; Sherlock, 1975). The occurrence of pruritus appears to be associated with the rise of serum bile acid levels (Laatikainen and Ikonen, 1977). Laatikainen and Ikonen (1977) showed that serum transaminases and cholic acid (a primary bile acid) were the most sensitive routine laboratory tests to indicate disturbance of liver function in pregnant women with pruritus. In the last trimester affected women show an increased difficulty in the transfer of bromsulphalein (BSP) to the canaliculi over that of normal pregnant women (Sherlock, 1975). The liver is grossly normal in shape, size, consistency, and color. Both light and electron microscopy are nondistinctive, being the same as other forms of intrahepatic cholestasis. No permanent liver disease occurs, although there is an increased incidence of gallstone formation (Haemmerli, 1975).

Since many women with recurrent intrahepatic jaundice of pregnancy experience pruritus gravidarum alone in some of their pregnancies or when taking oral contraceptives, many investigators have suggested that pruritus gravidarum may be a forme fruste of intrahepatic cholestasis of pregnancy. Furthermore, female relatives of affected women have been known to experience only pruritus gravidarum in pregnancy or upon oral contraceptive usage. In pruritus gravidarum, the total serum bilirubin is normal although the direct-reacting fraction may be somewhat elevated. Other laboratory tests, which are abnormal in intrahepatic cholestasis of pregnancy, may be only slightly elevated (Haemmerli, 1975). Occasionally patients predisposed to intrahepatic cholestasis of pregnancy have similar, although milder, clinical signs of transient generalized pruritus intramenstrually (Holzbach, 1976).

The symptoms of intrahepatic cholestasis of pregnancy itself also tend to occur with oral contraceptive use both in women who have previously been affected in pregnancy as well as those who have no prior affected pregnancies (Boake *et al.*, 1965; Haemmerli, 1966, 1975; Thulin and Nermark, 1966; Drill, 1974). Orellana-Alcalde and Dominguez (1966) made the following conclusions regarding this interrelationship: (1) oral contraceptives are C_{17} alkyl substituted compounds, and other compounds of this group have been reported to induce cholestatic jaundice; (2) no increase in the incidence of intrahepatic cholestasis in the same age-group of women not receiving contraceptives, or in men, has been reported; (3) in all cases withdrawal of the drug was followed by recovery; (4) jaundice and abnormal laboratory findings reappeared when the taking of the drug was resumed. Clinical, biochemical, and microscopic finding resemble those found in intrahepatic cholestasis of pregnancy.

The etiology of recurrent intrahepatic cholestasis of pregnancy remains undetermined. It may be related to an unusual response to either a natural steroid or the production of an unknown cholestatic one (Sherlock, 1975). Drill (1974) suggested that there is a metabolic defect such that the liver cannot excrete bilirubin at a normal rate when exposed to the increased secretion of estrogens and progestogens during pregnancy or upon administration of oral contraceptives. Substantial evidence points toward estrogenic steroids as the pathogenic factors, where as progestogens have no such effect (Haemmerli, 1975). In addition to the evidence presented above, it has been noted as a rare occurrence that affected women develop the disorder in early menopause just prior to an abnormally delayed menstrual period, a point when endogenous estrogens may be very high (Haemmerli, 1966). Also, animal studies have shown that estrogens routinely reduce the maximum capacity to excrete both bile acids and conjugated bilirubin (Javitt, 1980). It is known that pregnant women with intrahepatic cholestasis do not have increased estrogen production. However, biliary excretion of estrogens is diminished, thus disrupting normal enterohepatic circulation and metabolism of estrogen (Haemmerli, 1975).

The evidence for a genetic basis to recurrent intrahepatic cholestasis of pregnancy is suggestive although not conclusive. There have been numerous case reports of sisters and daughter–mothers both with the disorder in its various forms, in either pregnancy or upon oral contraceptive administration (e.g., Fast and Roulston, 1964; Holzbach and Sanders, 1965; Somayaji *et al.*, 1968; Furhoff and Hellström, 1973; review by Haemmerli, 1975; Spitaels and Noshel, 1976). Various large series have reported both a high (Furhoff and Hellstrom, 1973; Dalén and Westerholm, 1974) and a low familial tendency (Svanborg and Ohlsson, 1959). Reyes *et al.* (1976) reported a large kindred in which 10 of 32 multiparous women have experienced recurrent idiopathic cholestasis of pregnancy. The predisposition appeared to be transmitted by either sex. The incidence of cholelithiasis in this family was also high. The various branches of this family have different standards of living suggesting that exogenous environmental factors other than steroid use may not be important in the disorder's occurrence. Given that both vertical (mother–daughter pairs) and horizontal (female siblings) aggregation have been observed, in this and other pedigrees, the mode of inheritance remains unclear. It may be that the disorder is heterogeneous, or that the genetic predisposition is due to disease alleles at more than one locus. Further support for a genetic basis is the fact that the disorder is most frequent in distinct populations, specifically Chile and Scandinavia (Haemmerli, 1975; Reyes *et al.*, 1978).

A familial tendency to intrahepatic cholestasis due to use of sex steroids has also been noted. Dalén and Westerholm (1974) investigated the occurrence of hepatic dysfunction in the sisters and mothers of women who were jaundiced by oral contraceptives. As a control group these investigators used women on the same oral contraceptives not experiencing jaundice. The sisters of the patients had more pruritus in pregnancy and with oral contraceptive use than the sisters of the controls. Additionally, the mothers of the patients were more likely to have had pruritus during pregnancy than the controls' mothers. They also found that gallstone symptoms and subsequent cholecystectomy were greater in the patients' sisters than in the controls' sisters. It has also been observed that jaundice from oral contraceptives appears to be more prevalent in those areas with a higher incidence of recurrent cholestasis of pregnancy (Orellana-Alcalde and Dominguez, 1966; Perez and Gorodisch, 1974).

De Pagter *et al.* (1976) have suggested that intrahepatic cholestasis of pregnancy is a particular manifestation of a more generalized condition, familial benign recurrent intrahepatic cholestasis, found in the nonpregnant state. Tygstrup and Jensen (1969) proposed the following diagnostic criteria for familial benign intrahepatic cholestasis: (1) episodes of pronounced jaundice with severe pruritus and biochemical signs of cholestasis; (2) liver biopsy showing bile plugs; (3) normal intrahepatic and extrahepatic bile ducts on direct cholangiography; (4) absence of a known jaundice producing factor such as drugs or pregnancy; (5) symptom-free intervals of several months a year. In this disorder jaundice episodes are characterized by a preicteric phase of 2 to 4 weeks of malaise, anorexia, and pruritus, initially occuring only at night, followed by an icteric phase of increasing jaundice without pain or fever.

The interrelation between intrahepatic cholestasis of pregnancy, jaundice from oral contraceptives, and familial benign intrahepatic cholestasis was the subject of a family study by De Pagter *et al.* (1976). This extended family had definitely three and perhaps four members in two generations with benign recurrent intrahepatic cholestasis. Nine other members had pruritus and/or jaundice during pregnancy. Two additional members experienced pruritus upon oral contraceptive usage. In reviewing the literature to that time, De Pagter *et al.* (1976) report that 37 of 57 patients with benign recurrent intrahepatic cholestasis were males, evidence against the hypothesis that estrogens are the only etiologic factor. Other such mixed pedigrees are not commonly reported. The genetics of familial benign recurrent intrahepatic cholestasis itself is still undetermined (McKusick, 1978). It has also been suggested that intrahepatic cholestasis of pregnancy may occur when a women is heterozygous for the allele that in the homozygous state produces fatal cholestasis (Byler's disease) (McKusick and Clayton, 1968). Mothers of patients with Byler disease have been known to have severe pruritus in pregnancy. However, there is only this suggestive evidence to support this relationship.

The effect of this condition upon fetal well-being is also unclear. Numerous investigators cite that the disorder is benign to both mother and child, but that prematurity is a common complication. Fetal damage or stillbirths do not seem to occur more frequently than in normal pregnancies (Svanborg and Ohlsson, 1959; Amelung and Deupmann, 1965; Moore, 1963; Haemmerli and Wyss, 1967; Fast and Roulston, 1964; Furhoff, 1974; Haemmerli, 1975; Reid *et al.*, 1976; Misra *et al.*, 1980). The tendency toward prematurity appears to be a feature of the individual woman and not the severity of the disorder (Haemmerli, 1975).

However, despite the bulk of evidence suggesting otherwise, there are occasional case reports and series suggesting that there may be some obstetrical risks to the fetus, perhaps related to the occurrence of prematurity. For example, McAllister and Waddell (1962) reported three pregnancies in a woman with recurring idiopathic jaundice in pregnancy, all three of which resulted in abnormal outcomes: one was premature; another resulted in a first trimester spontaneous abortion; the third being a premature delivery with the infant dying of acute peritonitis. The relationship between the poor obstetrical history and the recurrent jaundice in pregnancy is unclear. One of the most suggestive reports of fetal complications in intrahepatic cholestasis of pregnancy is that of Reid *et al.* (1976). Among 56 pregnancies, there were five intrauterine and one neonatal death, with six additional infants requiring urgent delivery for intrapartum asphyxia. The fetal risk in this series was not related to the measured severity of the maternal cholestasis. In a follow-up report of the infants in Thorling's (1955) study, Furhoff (1974) reported lower birth weights and increased likelihood of meconium-stained amniotic fluid in infants of mothers with recurrent jaundice of pregnancy having itching and/or jaundice lasting up to delivery. Laatikainen and Ikonen (1975) found signs of fetal distress—abnormal heart rate or heart rate pattern and/or meconium-stained amniotic fluid. In over one-third of pregnancies complicated by cholestasis they also observed an increased incidence of asphyxia or imminent asphyxia in these pregnancies as compared to controls. They suggest that the disturbed metabolism of bile, and resulting meconium-staining, may be related to this observed increased occurrence of fetal asphyxia. Other investigators have observed a rise in acute intense contractions in such pregnancies, which shorten labor duration with a possible increased threat to the fetus (Roszkowski and Pisarek-Miedzinska, 1968). In this series only 3 of the 12 multiparious women had a past history of pruritus in pregnancy.

If the increased incidence of prematurity in pregnancies complicated by recurrent intrahepatic cholestasis of pregnancy is real, what could be the causative factor(s)? It has been suggested that the same metabolic disturbance that causes prematurity may cause intrahepatic cholestasis (Haemmerli, 1975). Maternal cholestasis is known to result in an elevation of both nonsulfated and sulfated bile acid levels in the fetal circulation (Laatikainen *et al.*, 1978). There appears to be a downward gradient in these bile acid levels from the maternal to the fetal compartments (Laatikainen, 1975). Amniotic fluid of patients with intrahepatic cholestasis has higher levels of the primary bile acids than that of normal controls, diabetic women, or women with toxemia (Heikkinen *et al.*, 1980). Whether the bile acids enter the amniotic fluid by diffusion through umbilical cord or placental membranes or via the fetal circulation and hence to fetal urine and meconium remains to be established. Laatikainen (1975) found a more frequent occurrence of signs of fetal distress during pregnancy or labor in those cases where cord plasma bile acid levels were elevated. In their small series of pregnancies complicated with intrahepatic cholestasis, Heikkinen *et al.* (1980) found no correlation between the condition of the newborn infants and the amniotic fluid bile acid concentrations. The clinical significance of elevated bile acid levels, as an indication of maternal jaundice and potentially resulting fetal distress, still needs to be determined (see also Laatikinen and Ikonen, 1977; Laatikainen *et al.*, 1978). Elevated bile acid levels may stimulate fetal colonic motility resulting in the observed increased occurrence of meconium-stained amniotic fluid in those pregnancies with elevated maternal serum bile acid levels (Laatikainen and Ikonen, 1977).

The observed increase incidence of prematurity may be related to another factor. Maternal intrahepatic cholestasis has been observed to cause changes in fetoplacental steroid metabolism. Some of these changes can be regarded as a secondary to the changes in maternal steroid metabolism but other changes are suggested to be due to a "toxic" effect of maternal cholestasis on fetal steroid metabolism, specifically impairment of 16(α-) hydroxylation (Laatikainen *et al.*, 1974). A negative correlation between fetal bile acid levels and concentrations of 16-oxygenated steroid sulfates, which are important estriol precursors produced by the fetus, has been observed, suggesting that there may be a cause and effect relationship between elevated bile acid levels and altered steroid metabolism that may result in premature labor (Laatikainen, 1975).

B. Dubin–Johnson Syndrome

Dubin–Johnson syndrome is a disorder of chronic or intermittent benign familial idiopathic jaundice related to a defect in biliary excretion of bilirubin and other cholephilic compounds (Berthelot and Dhumeaux, 1978; Schmid and McDonagh, 1978). It was first described by Dubin and Johnson (1954) and Sprinz and Nelson (1954). The disease is slightly more common in males and may not be noted in females until the use of oral contraceptives or at pregnancy (Arias, 1969; Cohen *et al.*, 1972). Patients with Dubin–Johnson syndrome have elevated serum bilirubin, predominately conjugated, abnormal retention of sulfobromophthalein sodium dye (BSP), nonvisualization of the gallbladder, and abnormal melanin-like pigment in hepatic parenchymal cells (Butt *et al.*, 1966; Shani *et al.*, 1970a). Serum bile acid levels are normal. The nature of the hepatic pigment is still unclear (Swartz *et al.*, 1979; Arias and Blumberg, 1979). Individuals with Dubin–Johnson syndrome also have a distinctive pattern of urinary coproporphyrins (Wolkoff *et al.*, 1973, 1976; Kondo *et al.*, 1976). A related condition, Rotor syndrome, has identical features except there is no abnormal pigment in the liver. Because of the clinical similiarities, Rotor syndrome has long been considered a variant of the Dubin–Johnson syndrome (Arias, 1961). Recent studies of urinary coproporphyrin excretion (Wolkoff *et al.*, 1976) and BSP metabolism (Wolpert *et al.*, 1977) have shown that the two are separate pathophysiologic entities (see also Namihisa, 1976; Namihisa and Nambu, 1977).

The inheritance of the Dubin–Johnson syndrome appears to be firmly established as an autosomal recessive disorder. In an early family study Butt *et al.* (1966) suggested autosomal dominant inheritance with considerable variability or multifactorial inheritance as the mode of transmission. They identified several clinically unaffected relatives of affected individuals who had elevated unconjugated bilirubin levels and abnormal liver pigmentation. Shani *et al.* (1970a) presented an extensive collection of Israeli kindreds with Dubin–Johnson syndrome. The disorder was most common in Jews of Iranian descent with an estimated prevalence of 1 in 3000. These investigators were among the first who firmly proposed autosomal recessive inheritance as the mode of transmission. In their own families they noted a high degree of consanguinity, which resulted in affected parents having affected children. Approximately 30% of the sibs of the probands were affected, which would be expected of a recessive disorder with a high gene frequency, in which an affected parent will reproduce. In the obligate carriers serum bilirubin concentrations and BSP storage capacity and transport maximum were all normal (see also Shani *et al.*, 1970b; Cohen *et al.*, 1972). Wolkoff *et al.* (1973) examined the total urinary excretion of coproporphyrin isomers I and II in Dubin-Johnson syndrome patients, their parents, and normal controls. They found the percentage of isomer I excretion was greatest in the Dubin-Johnson syndrome patients, lowest in the normals, and intermediate in the parents, as would be expected if the disorder is autosomal recessive and obligate carriers expressed the disorder partially. Further support for the recessive model can be found in the genetic study of Edwards (1975). In this report liver function tests on all 20 parents and on 64 children of Dubin-Johnson syndrome patients were all normal. Seventeen of 44 siblings of the probands had conjugated hyperbilirubinemia and elevated serum BSP. Consanguinity in the parents was also more frequent than expected. These findings—single generation involvement, increased incidence among siblings of the affected individuals, and consanguinity—all are indicative of autosomal recessive inheritance. It is important to note that given the estimated frequency of Gilbert syndrome is about 5%, there is a chance occurrence of Gilbert syndrome gene(s) in kindreds identified as having the Dubin-Johnson or Rotor syndromes (Odell and Childs, 1980). This may confuse family studies that do not clearly discriminate between the different disorders (see discussion of Gilbert syndrome which follows).

When a woman with Dubin-Johnson syndrome becomes pregnant, her jaundice can be expected to become exacerbated, usually in the third trimester. However, the clinical course in pregnancy is variable, and there are reports of no increase in symptoms or even disease remission (Haverback and Wirtschafler, 1960; Haemmerli, 1966). There are suggestions that fetal outcome may be compromised, although the evidence is inconclusive. Soon after the Dubin-Johnson syndrome was first described, Nelson (1956) reported a full-term pregnancy resulting in a normal infant in a women diagnosed with the syndrome. Dubin (1958), in his second series of 53 cases with Dubin-Johnson syndrome, reported 17 pregnancies in seven women. Of these 17 pregnancies, eight resulted in abortions, one in a stillbirth, one in a neonatal death, one in a premature delivery, and six in normal deliveries. In his large review of jaundice in late pregnancy, Ikonen (1964) identified two women who may have had Dubin-Johnson syndrome. One had three pregnancies, one of which was spontaneously terminated, one which resulted in immature, nonsurviving twins, and one which was premature but resulted in a normal infant. Friedlaender and Osler (1967) observed one patient with Dubin-Johnson syndrome in a series of 135 patients with jaundice in pregnancy. In their patient, icterus became severe early in the pregnancy and stillbirth occurred early in the third trimester. In a large series of Dubin-Johnson syndrome families in Israel, Shani *et al.* (1970a) first diagnosed the syndrome in

11 women during their pregnancies but did not describe the pregnancy outcome in these women. Cohen *et al.* (1972) reported nine women with the Dubin–Johnson syndrome and jaundice in pregnancy, but again, no follow-up on fetal outcome was included. Di Zoglio and Cardillo (1973) present a single case report of pregnancy in a woman with Dubin–Johnson syndrome whose only pregnancy resulted in a normal, full-term infant. More data is needed before more definitive conclusions can be made regarding the pregnancy outcome in the affected female.

C. Gilbert Syndrome

Gilbert syndrome (hyperbilirubinemia I) is the occurrence of mild, chronic unexplained hyperbilirubinemia, primarily unconjugated. Individuals with Gilbert syndrome have normal liver function tests, normal liver histology, no evidence of hemolysis, but simply delayed clearance of bilirubin. The syndrome may very well simply be a variant of normal bilirubin metabolism. Clinically, most affected individuals have no symptoms, although subjectively they may complain of lethargy, upper abdominal pain, and indigestion (Powell *et al.*, 1967). Individuals with Gilbert syndrome have been shown to have unconjugated hyperbilirubinemia with reduced caloric intake (Felsher *et al.*, 1970; Owens and Sherlock, 1973). The onset of jaundice is usually between the age of 15 and 20 years, raising the possibility of a role of endogeneous sex steroids in the disease expression. The clinical course is usually benign and the diagnosis is typically one of exclusion (Berthelot and Dhumeaux, 1978). Numerous pathoetiological mechanisms may be involved, including reduced bilirubin UDP-glucuronyltransferase activity (Black and Billing, 1969), altered bilirubin kinetics, and slightly reduced red-blood cell survival (for review, see Bailey and Robinson, 1977; Schmid and McDonagh, 1978). The population frequency is approximately 5% (Berthelot and Dhumeaux, 1978; Owens and Evans, 1975).

There is suggestive evidence that Gilbert syndrome, or a subset of Gilbert syndrome, is inherited as an autosomal dominant disorder with reduced expressivity. Foulk *et al.* (1959) presented one of the first large family studies and reviewed earlier evidence suggesting a hereditary component. They found jaundice in both parents and siblings of the probands, and proposed that autosomal dominant inheritance may be likely. In another large family study, 27.5% of the sibs and 16.2% of the parents of the probands had elevated unconjugated serum bilirubin (Powell *et al.*, 1967). These family studies showed that the distribution of serum bilirubin levels in unaffected family members was bimodal with the higher mean near that of the clinically recognized affected family members. The male to female ratio in these family members was 4 to 1. (As regards family studies, it is worth noting that fluctuations in serum bilirubin levels do occur, and serial samples may be informative.) This deviation in the sex ratio may be explained by the population studies of Owens and Evans (1975). These investigators found that the mean bilirubin levels in females was lower than in males, although both distribution were bimodal. Hence, what may have been considered normal for a female in the study of Powell *et al.* (1967) may actually be elevated when corrected for the differences in the sexes. Bailey and Robinson (1977) in their study of individuals undergoing a health screening also noted that on the average men have higher serum bilirubin levels than females. Rather than observing a bimodality of bilirubin distributions, they reported the distributions to be skewed to the right. These latter investigators suggested that Gilbert syndrome may simply be the upper end of the normal range of bilirubin levels rather than a disease state and that the term constitutional hyperbilirubinemia is a more appropriate description. The interested reader is referred to Odell and Childs (1980) for an extensive discussion on the genetics of Gilbert syndrome.

Recent hepatic electron microscopy (EM) studies have shown that a significant subset of Gilbert syndrome individuals have gross hypertrophy of the smooth endoplasmic reticulum (Dawson *et al.,* 1979). Those individuals with these positive EM findings had a greater percentage bilirubin response to caloric restriction than those with normal EM or normal controls. This finding may potentially identify heterogeneity within the Gilbert phenotype, a subset of which may be more severely affected.

Although the clinical significance of Gilbert syndrome is still an open question, because it is such a common diagnosis, the pregnancy history of affected women is of interest. There have been only a few isolated reports of pregnancies in women diagnosed with Gilbert syndrome. In Meulengracht's (1947) early series, two women with the diagnosis of Gilbert syndrome were jaundice-free during gestation. Friedlaender and Osler (1967) reported three women, who had icterus when not pregnant, who were diagnosed as having Gilbert syndrome. They had a total of eight pregnancies—six of which resulted in normal infants and two which resulted in stillborns. Considering the high estimated prevalence of this condition, it is surprising that there are not more reports of its influence in pregnancy. It may very well be that Gilbert patients are incorrectly diagnosed and included in the large series of women with jaundice in pregnancy. Haemmerli (1975) in his extensive review of jaundice in pregnancy states that Gilbert syndrome is probably unaltered in pregnancy, although bilirubin levels in pregnant women with the disorder have not been studied. Whether Gilbert syndrome actually is a disease entity and how familial factors influence constitutional hyperbilirubinemia are still unanswered questions. The practical importance of Gilbert syndrome lies in an awareness of its existence, so that serious liver disease is not diagnosed in error (Editorial, 1978).

D. Crigler–Najjar Syndromes, Types I and II

Gilbert syndrome is not the only familial disorder featuring chronic unconjugated hyperbilirubinemia and abnormal hepatic glucuronyl transferase activity.

Crigler–Najjar syndrome Types I and II are clinically more severe than Gilbert syndrome, with greater degrees of jaundice and absence or near absence of hepatic bilirubin glucuronyl transferase activity (recently reviewed by Schmid and McDonagh, 1978; Odell and Childs, 1980). Indeed, numerous investigators have proposed that the separation of Gilbert syndrome and Crigler–Najjar syndrome Type II patients is arbitrary and that these three conditions may actually constitute a spectrum of unconjugated hyperbilirubinemias with Gilbert syndrome as the mildest form (Kreek and Sleisenger, 1968; Arias, 1971; Hunter *et al.*, 1973; Black *et al.*, 1974; Gollan *et al.*, 1975).

Crigler and Najjar (1952) first described what is now recognized as Crigler–Najjar syndrome Type I: severe unconjugated hyperbilirubinemia in the absence of hemolytic disease (serum bilirubin levels usually 25 to 35 mg/dl), infant death due to kernicterus within a few months of life, normal liver function tests, colorless bile, undetectable hepatic glucuronyl transferase activity (Childs *et al.*, 1959), and unresponsiveness to phenobarbital therapy (Blumenschein *et al.*, 1968; Crigler and Gold, 1969). Subsequently, Crigler–Najjar Type I patients have been observed who did not develop neurological findings until the time of puberty or early adulthood (Childs and Najjar, 1956; Blumenschein *et al.*, 1968; Blaschke *et al.*, 1974; Wolkoff *et al.*, 1979). There is considerable evidence that Type I is an autosomal recessive disorder with obligate heterozygotes having a partial deficiency of glucuronyl transferase but normal bilirubin levels. Such heterozygotes can be identified by reduced formation of menthol glucuronide (Childs *et al.*, 1959; Szabó *et al.*, 1962; Szabó and Ébrey, 1963).

Arias *et al.* (1969) have compiled the evidence supporting the existence of an additional disorder, similar to Crigler and Najjar's original cases, but with lower serum bilirubin levels (usually 15 mg/dl) and clinically far less severely affected, with kernicterus-like symptoms a rarity (see also Arias, 1962; Hunter *et al.*, 1973; Black *et al.*, 1974; Gollan *et al.*, 1975). This disorder has been termed Crigler–Najjar syndrome Type II (and is listed in the tables of Mendelian gastrointestinal disorders in this chapter as hyperbilirubinemia, Arias type). These patients have life-long jaundice, pigmented bile, undetectable or barely detectable glucuronyl transferase activity, and upon phenobaritol therapy experience a reduction in serum bilirubin levels (Yaffe *et al.*, 1966; Blumenschein *et al.*, 1968; Ertel and Newton, 1969; Blaschke *et al.*, 1974). There are reports of patients, apparently of the Type II variety, who develop neurological signs resembling bilirubin encephalopathy as adults (Jervis, 1959; Arias, 1962; Aziz and Siddiqui, 1967; Crigler and Gold, 1969; Gordon *et al.*, 1976). As regards the identification of potential maternal-fetal interactions, Crigler–Najjar Type I is of little interest since affected females are unlikely to reach reproductive age, let alone reproduce. On the other hand, Crigler–Najjar Type II is a less severe disorder and apparently reproductive fitness is unaffected.

Before discussing what is known about Crigler–Najjar syndrome Type II and pregnancy, the etiological relationship between it, Type I, and Gilbert syndrome needs to be explored. The occurrence in some families of individuals with both Type I and Type II phenotypes has not been documented (Arias, 1971). However, there are at least two reports of families that may blur the separation of the Type I variant from Type II and Gilbert syndrome. Sugar (1961) described an affected father and daughter who do not easily fit into the Type I or Type II variety. In this family, serum bilirubin levels were elevated into the Type I range, yet the only neurological finding was that the father had hearing difficulty. The daughter required several exchange transfusions as a neonate. The inheritance pattern would appear more consistent with autosomal dominant than autosomal recessive. Bloomer *et al.* (1971) reported the mother of a Type I female who had a bilirubin clearance similar to that in Gilbert syndrome. As will be outlined below, the bulk of evidence supports the contention that Type I is a separate disorder and Type II and Gilbert syndrome are interrelated.

Crigler–Najjar Type II had been proposed as an autosomal dominant condition with incomplete penetrance (Arias *et al.*, 1969) and as a recessive disorder requiring two abnormal alleles with the heterozygote state being Gilbert syndrome (Hunter *et al.*, 1963). The accumulated family data pertinent to this question is inconclusive, with a distinct possibility that several alleles at a single locus may be confounding the understanding of the various observed phenotypes. For example, Sleisenger *et al.* (1967) described an Irish kindred with abnormal menthol excretion tests inherited in a dominant fashion and partial deficiency of glucuronyl transferase in the two propositi. These observations are consistent with the heterozygote state for Type I, except that as mentioned above, such individuals have normal bilirubin levels. However, in this family, some of the members with abnormal excretion tests had lifelong jaundice as would be expected in the Type II variant. This pedigree can be interpreted as an example of dominant inheritance of the Type II variant with variable phenotypic expression. On the other hand, family studies of Type II patients have shown that first degree relatives have serum bilirubin levels or bilirubin clearance tests similar to those of Gilbert syndrome patients. Smith *et al.* (1967) studied several families ascertained through a patient with chronic, unconjugated hyperbilirubinemia. One of these families consisted of a child with raised unconjugated bilirubin levels, around 6.6 mg/dl, having parents and a sibling with elevated unconjugated bilirubin levels. They also found that those patients with levels of 2.5–4.2 mg/dl had one parent with elevated unconjugated serum bilirubin levels. Gollan *et al.* (1975) reported a family in which three brothers were affected with the Type II variety, a fourth had unconjugated bilirubin levels of 3.1 mg/dl with normal liver function tests consistent with Gilbert syndrome, and the parents were third cousins and never jaundiced (see also Arias, 1962; Kreek and Sleisenger, 1968; Hunter, 1973). Hunter *et al.* (1973) outlined several factors, which may result in the bilirubin levels of relatives of patients with Crigler–Najjar syndrome Type

II being reported as normal when they are actually elevated. These include method of analysis, food intake, time of sampling, sex differences, and definition of normality (see also Wetstone and Honeyman 1968; Owens and Sherlock, 1973).

In addition, the patterns of excreted bile pigments (Gollan *et al.*, 1975; Gordon *et al.*, 1976; Fevery *et al.*, 1977) and bilirubin clearance curves (Black *et al.*, 1974; Berk and Javitt, 1978) in Crigler–Najjar Type II patients are similar in pattern, but more severe than, those in Gilbert syndrome, suggesting that Crigler–Najjar Type II is a more severe form of the Gilbert phenotype. In contrast to Type I, in Type II families the oral menthol excretion test does not appear to be useful in discriminating between icteric patients and their clinically normal family members (Arias, 1962; Sleisenger *et al.*, 1967; Arias *et al.*, 1969). Type II patients respond to phenobarbitol therapy, although serum bilirubin levels cannot be lowered below that of Gilbert syndrome patients and they still retain undetectable glucuronyl transferase activity (Arias, *et al.*, 1969; Black *et al.*, 1974; Gollan *et al.*, 1975). In summary, although the evidence suggests an interrelationship between the etiology of Crigler–Najjar Type II and Gilbert syndrome, conclusive evidence is still lacking.

We have been able to identify only one female patient in the literature with apparent Crigler–Najjar syndrome Type II, for whom there is a pregnancy history (Kreek and Sleisenger, 1968). She was noted to be jaundiced at 8 days of life and had completely normal development and growth. She observed that her jaundice intensified during the one or two days preceeding through the first two days of her menses. Her serum bilirubin levels fell from a basal range of 5–8 mg/dl to 3.5 mg/dl upon phenobarbitol therapy. One pregnancy was reported in this woman. She was 21 years old at the time and did not experience any increase in her jaundice. The product of this pregnancy was a normal, full-term male who has never been jaundiced. Undoubtedly this is not the only instance of pregnancy in a woman with Crigler–Najjar Type II. Other such patients may be carrying a diagnosis of Gilbert syndrome or some other hyperbilirubinemia. Crigler–Najjar Type II has only recently been recognized and hence a paucity of reports is expected. The influence of this disorder on the maternal–fetal interaction is obviously unknown at this time. The delineation of whether Crigler–Najjar Type II is a severe form of Gilbert syndrome or a separate entity may help elucidate any deleterious effects on the fetus when the mother has chronic unconjugated hyperbilirubinemia.

E. Transient Familial Neonatal Hyperbilirubinemia

Transient neonatal unconjugated hyperbilirubinemia is not uncommon in normal full-term infants and can be exacerbated by numerous factors such as hemolysis, infection, drugs, and prematurity. There are cases of severe neonatal unconjugated hyperbilirubinemia in which none of the above factors can be implicated. The observation of repeated unexplained occurrences within select families led investigators in the late 1950s and early 1960s to further study this phenomenon (Lucey *et al.*, 1960). In premature infants, at least, this nondescript hyperbilirubinemia can be associated with the neurologic sequelae of bilirubin encephalopathy (kernicterus) (Brown and Zuelzer, 1957; Meyer, 1956).

Arias *et al.* (1965) report eight women who gave birth to a total of 25 infants, each of whom had severe transient neonatal hyperbilirubinemia. Four of the 24 infants died with kernicterus, with all of the remaining 20 in good health except for one who was diagnosed as having spastic quadriplegia. In this series, all the families were Caucasian and without known consanquinity. Transient familial neonatal hyperbilirubinemia may result from an increased amount of a substance in the serum of certain pregnant women that inhibits the hepatic conjugation of bilirubin. Beginning in the second trimester of pregnancy, mothers of infants with transient familial neonatal hyperbilirubinemia have significantly greater levels of this inhibitory factor activity than do normal women. The actual factor is unknown, but probably is a progestational steroid that inhibits glucuronyl transferase activity in the liver of neonates. At present there is no definitive evidence that this disorder is genetic. The most likely mode of inheritance, if this disorder is Mendelian, would be autosomal recessive (McKusick, 1978). If the trait were dominant we would expect the female offspring of these women to have infants with the same problem of transient nonhemolytic unconjugated hyperbilirubinemia. Therefore, as the currently affected infants reach reproductive age, their own obstetrical histories would help elucidate the mode of inheritance for this disorder. Transient nonhemolytic unconjugated hyperbilirubinemia appears to be an example of the mother's genotype influencing the phenotype of the infant. A similiar example of this maternal genotype infant phenotype interaction is in maternal phenylketonuria, where the offspring of phenylketonuric mothers are at high risk for retardation and other anomalies (Mabry *et al.*, 1963; see chapter by Schulman in this volume).

This transient familial neonatal hyperbilirubinemia is clinically different from transient nonhemolytic unconjugated hyperbilirubinemia associated with breast feeding that occurs in some infants (Newman and Gross, 1963). In this latter condition, only breastfeeding, and not bottle-feeding, results in unconjugated hyperbilirubinemia. The phenomenon of prolonged unconjugated hyperbilirubinemia in breast-fed but not bottle-fed infants may in itself be a genetic entity. Two-thirds of breast-fed siblings of breast fed index cases also had severe jaundice (Gartner and Arias, 1966). When only bottle-fed, siblings of jaundiced infants experienced no increase in unconjugated hyperbilirubinemia. Jaundiced breast-fed infants had rapid decrease of their hyperbilirubinemia when breast-feeding was halted. These investigators found that the breast milk of mothers of these infants contained pregnane-3-α, 20-β-diol that competitively inhibits hepatic glucuronyltransferase activity *in vitro* (Arias *et al.,* 1964; Gartner and Arias, 1966). This steroid has not been isolated from the breast milk of mothers with unaffected infants. Administration to unaffected infants results in unconjugated hyperbilirubinemia. However, other investigators have not found the pregnane-3-α, 20-β-diol to be the steroid causing the conjugation inhibition effect seen in breast milk of infants with breast-milk jaundice (Ramos *et al.,* 1966; Cole and Hargreaves, 1972). Kernicterus has not been observed in this disorder, probably because the jaundice is usually milder, and even if severe, does not develop until after the infant's blood–brain barrier has become impermeable to the unconjugated bilirubin. Winfield and MacFaul (1978) report a clinical epidemiological study in which 893 births were studied to determine the incidence of prolonged neonatal jaundice. Approximately one out of 50 breast-fed babies had prolonged jaundice while none of the bottle-fed babies were affected. Reportedly none of these jaundiced babies had siblings with a similar neonatal course. Cole and Hargreaves (1972) investigated the relationship between maternal inhibitory activity aganist bilirubin conjugation in breast milk and serum, and the degree of neonatal hyperbilirubinemia in normal pregnancies. They found no relationship between these two variables. They did not identify their inhibitory factor and it may be the same as that felt responsible for the breast milk jaundice.

X. SUMMARY

In this chapter, we have reviewed the genetics of the gastrointestinal disorders with attention to those conditions in which there is an interaction between the mother, the fetus, and the pregnant state. Whereas by definition most genetic disorders result from transmission of variant parental genes to the offspring (see Tables I–X), we have focused in this chapter on those GI genetic disorders in which there is an additional interaction with the pregnant state, with an effect either on the mother, the fetus, or the pregnancy itself. These categories of interactions include the following:

1. The pregnancy course or fetal outcome may be affected by the mother's pre-existing GI disease.
2. A clinical or subclinical GI disorder in the mother may be influenced by the pregnant state or the phenotype of the fetus.
3. The course of the pregnancy may be affected by GI abnormalities in the fetus.
4. The fetus' gastrointestinal integrity is influenced by a maternal factor. This interaction may continue into the postpartum period.

Such maternal–fetal interactions are undoubtedly more numerous than the extent of our present knowledge. In some disorders, more than one pathway apparently is active. We will conclude with a brief summary of the maternal–fetal interactions known for the disorders discussed in this chapter.

In the zinc-deficiency disorder acrodermatitis enteropathica (AE), affected infants are phenotypically normal until weaning. A factor in human breast milk apparently delays expression of the clinical symptoms. Hence, this disorder is an example of how a maternal factor transmitted in breast milk influences the phenotypic expression of the genetic disorder, in this case to the good. In addition, there are suggestions from animal and epidemiological evidence that maternal zinc deficiency may result in central nervous system defects. A greater knowledge of the pregnancy outcomes of women with AE would be an important source of evidence to help answer this question. Unfortunately, at this point, pregnancy reports in AE patients are too sparse for valid conclusions to be made regarding the incidence of malformations.

Fetuses with congenital chloride diarrhea uniformly are the products of pregnancies complicated by polyhydramnios. This apparently is the result of *in utero* fetal diarrhea. This disorder is an example of how the course of the pregnancy can be affected by a GI abnormality in the fetus.

Moving on to the more common disorders, peptic ulcer is rather unusual in that during pregnancy the maternal disease often improves, with a complete remission of symptoms frequently observed. As the various etiologies of the ulcer diathesis are identified, it will be valuable to our understanding of ulcer pathophysiology to determine if ulcers of varying etiologies respond differently when the woman is pregnant.

For pyloric stenosis, there is strong evidence of a maternal genetic effect since affected mothers are more likely to have an affected child than affected fathers. Additionally there appears to be a maternal environmental effect in that the emotional state of the woman during pregnancy (herself unaffected) may influence the occurrence of the disorder in her developing fetus.

Females with coeliac disease experience decreased fertility when their disease is active. Additionally, the increased nutritional demands of pregnancy can bring to light the disease in women who previously were only subclinically affected—another example of how the pregnant state alters the disease course.

The evidence for maternal-fetal interactions in inflammatory bowel disease (IBD) is different when the patients are separated into those with ulcerative colitis as compared to those with Crohn's disease. In ulcerative colitis, there is no reduced fertility and the relapse rate is about that of nonpregnant women, though with severe disease there appears to be an increase in spontaneous abortions. For Crohn's disease, the only well-established effect is decreased fertility attributed to fallopian tube adhesions. Relapses do occur, although at a rate not unlike that of nonpregnant women. Relapses in both disorders are most common in the first trimester and puerperium, presumably when there are less endogenous steroids providing a protective effect. In years gone by, maternal morbidity and mortality was not uncommon for patients with IBD. However, with developments in therapy, the maternal death rate should no longer be increased. The evidence for unfavorable fetal outcome—spontaneous abortions, stillbirths, and congenital anomalies—was likewise stronger in earlier reports than what is now observed in centers experienced with this disease.

There is a long-standing clinical association between pregnancy and gallstone formation. This finding is most likely related to increased levels of sex steroid and increased retention of incipient cholesterol crystals during pregnancy—another case in which pregnancy affects the disease course. Cholecystectomy, when necessary in pregnancy, appears to be safe to both mother and fetus.

The disorders of hereditary hyperbilirubinemia are excellent examples of maternal–fetal interactions. Recurrent intrahepatic cholestasis of pregnancy is, as its name implies, a disorder of benign jaundice occurring only in pregnancy or upon oral contraceptive therapy. There are indications that prematurity may be a consequence when the mother has this condition. Dubin–Johnson syndrome usually is exacerbated in pregnancy although the course is variable. Fetal outcome may be compromised—spontaneous abortions and stillbirths—although from the published reports the incidence of such outcomes may not be greater than in unaffected pregnancies. The disorders of chronic unconjugated hyperbilirubinemia, Gilbert syndrome, and Crigler–Najjar syndrome, Type II, are by all indications unaffected by pregnancy and do not measurably influence fetal prognosis. Transient familial neonatal hyperbilirubinemia is a condition in which the maternal genotype influences the fetus' phenotype. Mothers with the disorder appear to have an inhibitor to hepatic conjugation of bilirubin resulting in neonatal hyperbilirubinemia and possibly leading to kernicterus.

Thus, individual genetic GI diseases exhibit all the possible pathways of interaction between the mother, the fetus, and the pregnancy itself. Much more needs to be learned about these in all the GI disorders so that we can continually improve our care of the obstetric patient and her developing pregnancy.

ACKNOWLEDGMENTS

We wish to gratefully acknowledge the many helpful criticisms of Dr. Bruce Blumberg. We also thank Daune Thorington for her unflagging enthusiasm and secretarial assistance. This work was supported in part by the National Institute of Health Grants AM 17328 and 25330, Clinical Investigator Award AM 00523, and a Basil O'Connor Starter Research Award from the National Foundation March of Dimes.

REFERENCES

Abramson, D., Jankelson, I.R., and Milner, L.R. (1951). Pregnancy in idiopathic ulcerative colitis. *Am. J. Obstet. Gynecol.* **61,** 121–129.

Amelung, D., and Deupmann, F.J. (1965). Jaundice during pregnancy. *Ger. Med. Mth.* **10,** 264–269.

Arias, I.M. (1961). Studies of chronic familial non-hemolytic jaundice with conjugated bilirubin in the serum with and without an unidentified pigment in the liver cells. *Am. J. Med.* **31,** 510–518.

Arias, I.M. (1962). Chronic unconjugated hyperbilirubinemia without overt signs of hemolysis in adolescents and adults. *J. Clin. Invest.* **41**, 2233-2245.

Arias, I.M. (1969). Some effects of contraceptive steroids on hepatic function in normal women and in patients with acquired and inheritable defects in hepatic exectory function. *In* "Metabolic Effects on Gonadal Hormones and Contraceptive Steroids" (H.A. Salhanick, D.M. Kipnis, and R.L. Van der Wiele, eds.), p. 30. Plenum, New York.

Arias, I.M. (1971). Inheritable and congenital hyperbilirubinemia. Models for the study of drug metabolism. *N. Engl. J. Med.* **285**, 1416-1421.

Arias, I.M. and Blumberg, W. (1979). The pigment in Dubin-Johnson syndrome. *Gastroenterology* **77**, 820-821.

Arias, I.M., Gartner, L.M., Cohen, M., Ben Ezzer, J., Levi, A.J. (1969). Chronic nonhemolytic unconjugated hyperbilirubinemia with glucuronyl transferase deficiency. *Amer. J. Med.* **47**, 395-409.

Arias, I. M., Gartner, L.M., Seifter, S., and Furman, M. (1964). Prolonged neonatal unconjugated hyperbilirubinemia associated with breast feeding and a steroid, pregnane-3 (alpha), 20 (beta)-diol, in maternal milk which inhibits glucuronide formation in vitro. *J. Clin. Invest.* **43**, 2037-2047.

Arias, I.M., Wolfson, S., Lucey, J.F., and McKay, R.J. Jr. (1965). Transient familial neonatal hyperbilirubinemia. *J. Clin. Invest.* **44**, 1442-1450.

Arulanantham, K., Kramer, M.S., and Gryboski, J.D. (1980). The association of inflammatory bowel disease and X chromosomal abnormality. *Pediatrics* **66**, 63-67.

Aziz, M.A., Siddiqui, A.R. (1967). Congenital familial nonhemolytic hyperbilirubinemia in an adult with central nervous system derangement. *Gastroenterology* **52**, 254-258.

Bailey, A., and Robinson, D. (1977). Does Gilbert's disease exist? *Lancet* **1**, 931-933.

Banks, B.M., Korelitz, B.I., and Zetzel, L. (1957). The course of nonspecific ulcerative colitis: Review of twenty years' experience and late results. *Gastroenterology* **32**, 983-1012.

Barnes, C.G. (1974). "Medical Disorders in Obstetric Practice." Blackwell, Oxford.

Bendix, R.M., and Necheles, H. (1947). Hypertrophic pyloric stenosis: A follow up study. *J. Am. Med. Assoc.* **135**, 331-333.

Bennion, L.J., and Knowler, W.C. (1980). Epidemiology of gallstones. *In* "Genetics and Heterogeneity of Common Gastrointestinal Disorders" (J.I. Rotter, I.M. Samloff, and D.L. Rimoin, eds.), pp. 297-312. Academic Press, New York.

Bennion, L.J., Ginsberg, R.L., Garnick, M.B., and Bennett, P.H. (1976). Effects of oral contraceptives on the gallbladder bile of normal women. *N. Engl. J. Med.* **294**, 189-192.

Berglund, G., and Rabo, E. (1973). A long term follow up investigation of patients with hypertrophic pyloric stenosis with special reference to heredity and later morbidity. *Acta Paediatr. Scand.* **62**, 130-132.

Bergsma, D., McKusick, V.A., Bayless, T.M., Hussels, I.E., and Lichtenstein, J.R. (1972). G.I. tract including liver and pancreas. *Birth Defects, Orig. Artic. Ser.* **8**, No. 2.

Berk, P.D., Javitt, N.B. (1978). Hyperbilirubinemia and cholestasis. *Amer. J. Med.* **64**, 311-326.

Berthelot, P., and Dhumeaux, D. (1978). New insights into the classification and mechanisms of hereditary, chronic, non-haemolytic hyperbilirubinaemias. *Gut* **19**, 474-478.

Binder, V. Weeks, E., Olsen, J.H., Anthonisen, P., and Riis, P. (1966). A genetic study of ulcerative colitis. *Scand. J. Gastroenterol.* **1**, 49-56.

Black, M., and Billing, B.H. (1969). Hepatic bilirubin UDP-glucuronyl transferase activity in liver disease and Gilbert's syndrome. *N. Engl. J. Med.* **280**, 1266-1271.

Black, M., Fevery, J., Parker, D., Jacobson, J., Billing, B.H., Carson, E.R. (1974). Effect of phenobarbitone on plasma (^{14}C) bilirubin clearance in patients with unconjugated hyperbilirubinemia. *Clin. Sci. Mol. Med.* **46**, 1-17.

Blaschke, T.F., Berk, P.D., Scharschmidt, B.F., Guyther, J.R., Vergalla, J.M., Waggoner, J.G.

(1974). Crigler–Najjar syndrome: An unusual course with development of neurologic damage at age eighteen. *Pediat Res* **8,** 573–590.

Bloomer, J.R., Berk, P.D., Howe, R.B., Berlin, N.I. (1974). Bilirubin metabolism in congenital nonhemolytic jaundice. *Pediatr. Res.* **5,** 256–264.

Blumenschein, S.D., Kallen, R.J., Storey, B., Natzschka, J.C., Odell, G.B., Childs, B. (1968). Familial nonhemolytic jaundice with late onset of neurological damage. *Pediatrics* **42,** 786–792.

Boake, W.C., Schade, S.G., Morrissey, J.F., and Schaffner, F. (1965). Intrahepatic cholestatic jaundice of pregnancy followed by Enovid-induced cholestatic jaundice. Report of a case. *Ann. Intern. Med.* **63,** 302–308.

Bohane, T.D., Cutz, E., Hamilton, J.R., and Gall, D.G. (1977). Acrodermatitis enteropathica, zinc, and the Paneth cell. A case report with family studies. *Gastroenterology* **73,** 587–592.

Bonnevie, O. (1975). The incidence in Copenhagen Country of gastric and duodenal ulcers in the same patient. *Scand. J. Gastroenterol.* **10,** 529–536.

Braverman, D.Z., Johnson, M.L., and Kern, F. Jr., (1980). Effects of pregnancy and contraceptive steroids on gallbladder function. *N. Engl. J. Med.* **302,** 362–364.

Brown, A.K. and Zuelzer, W.W. (1957). Studies in hyperbilirubinemia-a. I. Hyperbilirubinemia of the newborn unrelated to isoimmunization. *Am. J. Dis. Child.* **93,** 263–273.

Butt, H.R., Anderson, V.E., Foulk, W.T., Baggenstoss, A.H., Schoenfield, L.J., and Dickson, E.R. (1966). Studies of chronic idiopathic jaundice (Dubin–Johnson syndrome). II. Evaluation of a large family with the trait. *Gastroenterology* **51,** 619–630.

Bynum, T.E. (1977). Hepatic and gastrointestinal disorders in pregnancy. *Med. Clin. North Am.* **61,** 129–138.

Calam, J., Taylor, I.L., Dockray, G.J., Simkin, E., Cooke, A., Rotter, J.I., and Samloff, I.M.(1979). Subgroup of duodenal ulcer patients with familial G-cell hyperfunction and hyperpepsinogenemia. I. *Gut* **20,** A934.

Carter, I.O. (1972). Genetics of infantile pyloric stenosis. *Birth Defects, Orig. Artic. Ser.* **8,** (2), 12–14.

Carter, C.O., and Evans, K.A. (1969). Inheritance of congenital pyloric stenosis. *J. Med. Genet.* **6,** 233–254.

Casey, C.E., Hambridge, K.M., Walravens, P.A., Silverman, A., and Neldner, K.H. (1978). Zinc binding in human duodenal secretions. *Lancet* **2,** 423.

Casey, C.E., Hambridge, K.M., and Walravens, P.A. (1979). Zinc binding in human duodenal secretions. *J. Pediatr. (St. Louis)* **95,** 1008–1010.

Cash, R., and Berger, C.K. (1969). Acrodermatitis enteropathica: Defective metabolism of unsaturated fatty acids. *J. Pediatr. (St. Louis)* **74,** 717–729.

Cello, J.P., and Meyer, J.H. (1978a). Ulcerative colitis. *In* "Gastrointestinal Disease" (M.H. Sleisenger and J.S. Fordtran, eds.), pp. 1597–1653. Saunders, Philadelphia, Pennsylvania.

Cello, J.P., and Meyer, J.H. (1978b). Crohn's disease of the colon. *In* "Gastrointestinal Disease" (M.H. Sleisenger and J.S. Fordtran, eds.), pp. 1658–1679. Saunders, Philadelphia.

Childs, B., Najjar, V.A. (1956). Familial nonhemolytic jaundice with kernicterus. *Pediatrics* **18,** 369–377.

Childs, B., Sidbury, J.B., Migeon, C.J. (1959). Glucuronic acid conjugation by patients with familial nonhemolytic jaundice and their relatives. *Pediatrics* **23,** 903–913.

Clark, D.H. (1953). Peptic ulcer in women. *Br. Med. J.* **1,** 1254–1257.

Cohen, L., Lewis, G., and Arias, I.M. (1972). Pregnancy, oral contraceptives and chronic familial jaundice with predominantly conjugated hyperbilirubinemia (Dubin-Johnson syndrome.) *Gastroenterology* **62,** 1182–1190.

Cole, A.P., and Hargreaves, T. (1972). Conjugation inhibitors and early neonatal hyperbilirubinaemia. *Arch. Dis. Child.* **47,** 415–418.

Cowan, W.K. (1973). Genetics of duodenal and gastric ulcer. *Clin. Gastroenterol.* **2,** 539–546.

Crigler, J.F., Jr., Najjar, V.A. (1952). Congenital familial nonhemolytic jaundice with kernicterus. *Pediatr.* **10,** 169–179.

Crigler, J.F., Jr., Gold, N.I. (1969). Effect of sodium phenobarbital on bilirubin metabolism in an infant with congenital, nonhemolytic, unconjugated hyperbilirubinemia, and kernicterus. *J. Clin. Invest.* **48,** 42–55.

Crohn, B.B., Yarnis, H., Crohn, E.B., Walter, R.I., and Gabrilove, L.J. (1956a). Ulcerative colitis and pregnancy. *Gastroenterology* **30,** 391–403.

Crohn, B.B., Yarnis, H., and Korelitz, B.I. (1956b). Regional ileitis complicating pregnancy. *Gastroenterology* **31,** 615–628.

Dalén, E., and Westerholm, B. (1974). Occurence of hepatic impairment in women jaundiced by oral contraceptives and in their mothers and sisters. *Acta. Med. Scand.* **195,** 459–463.

Danbolt, N., and Closs, K. (1942). Acrodermatitis enteropathica. *Acta Derm.-Venereol.* **23,** 127–169.

Danzinger, R.G., Gordon, A., Schoenfield, L.J., and Thistle, J.L. (1972). Lithogenic bile in siblings of young women with cholelithiasis. *Mayo Clin. Proc.* **47,** 762–766.

Dawson, J., Carr-Locke, D.L. Talbot, I.C., and Rosenthal, F.D. (1979). Gilbert's syndrome: Evidence of morphological heterogeneity. *Gut* **20,** 848–853.

De Dombal, F.T., Watts, J.M., Watkinson, G., and Goligher, J.C. (1965). Ulcerative colitis and pregnancy. *Lancet* **2,** 599–602.

De Dombal, F.T., Burton, I.L., and Goligher, J.C. (1972). Crohn's disease and pregnancy. *Br. Med. J.* **3,** 550–553.

De Pagter, A.G.F., van Berge Henegouwen, G.P., ten Bokkel Huinik, J.A., and Brandt, K-H. (1976). Familial benign recurrent intrahepatic cholestasis. Interrelation with intrahepatic cholestasis of pregnancy and from oral contraceptives? *Gastroenterology* **71,** 202–207.

Dillaha, C.J., Lorinez, A.L., and Aavik, O.R. (1953). Acrodermatitis enteropathica. Review of the literature and report of a case successfully treated with diodoquin. *J. Am. Med. Assoc.* **152,** 509–512.

Di Zoglio, J.D., and Cardillo, E. (1973). The Dubin-Johnson syndrome and pregnancy. *Obstet. Gynecol.* **42,** 560–563.

Dodge, J.A. (1972a). Infantile pyloric stenosis: A multifactorial condition. *Birth Defects, Orig. Artic Ser.* **8,** (2) 15–21.

Dodge, J.A. (1972b). Psychosomatic aspects of infantile pyloric stenosis. *J. Psychosom. Res.* **16,** 1–5.

Dodge, J.A. (1973). Genetics of hypertrophic pyloric stenosis. *Clin. Gastroenterol.* **2,** 523–538.

Dodge, J.A. (1980). Infantile hypertrophic pyloric stenosis, definition, physiology, and genetics. *In* "Genetics and Heterogeneity of Common Gastrointestinal Disorders" (J.I. Rotter, I.M. Samloff, and D.L. Rimion, eds.), pp. 419–440. Academic Press, New York.

Dodge, J.A., and Karim, A.A. (1976). Induction of pyloric hypertrophy by pentagastrin. *Gut* **17,** 280–284.

Doll, R., and Kellock, T.D. (1951). The separate inheritance of gastric and duodenal ulcers. *Ann. Eugen.* **16,** 231–240.

Donaldson, D.R. Jr. (1978). Crohn's disease of the small bowel. *In* "Gastrointestinal Disease" (M.H. Sleisenger and J.S. Fordtran, eds.), pp. 1052–1076. Saunders, Philadelphia.

Drill, V.A. (1974). Benign cholestatic jaundice of pregnancy and benign cholestatic jaundice from oral contraceptives. *Am. J. Obstet. Gynecol.* **119,** 165–174.

Dubin, I.N. (1958). Chronic idiopathic jaundice. *Am. J. Med.* **24,** 268–292.

Dubin, I.N., and Johnson, F.B. (1954). Chronic idiopathic jaundice with unidentified pigment in liver cells. *Medicine (Baltimore)* **33,** 155–197.

Eade, O.E., Moulton, C., MacPherson, B.R., St. Andre-Ukena, S., Albertini, R.J., and Beeken,

W.L. (1980). Discordant HLA haplotype segregation in familial Crohn's disease. *Gastroenterology* **79,** 271–275.

Eckhert, C.D., Sloan, M.V., Duncan, J.R., and Hurley, L.S. (1977). Zinc binding: A difference between human and bovine milk. *Science* **195,** 789–790.

Editorial (1978). Heredity jaundice. *Lancet* **2,** 926–927.

Edwards, R.H. (1975). Inheritance of the Dubin-Johnson-Sprinz syndrome. *Gastroenterology* **68,** 734–749.

Ellis, A. (1980). Coeliac disease: family and twin studies and disease association. *In* "Genetics and Heterogeneity of Common Gastrointestinal Disorders" (J.I. Rotter, I.M. Samloff, and D.L. Rimoin, eds.), pp. 235–242. Academic Press, New York.

Engstrom, J.F., Arvanitakis, C., Sagawa, A., and Abdou, N.I. (1978). Secretory immunoglobulin deficiency in a family with imflammatory bowel disease. *Gastroenterology* **74,** 747–751.

Epstein, S., and Vedder, J.S. (1960). Diagnosis: Acrodermatitis enteropathica (Danbolt-Closs) in five siblings. *Arch. Dermatol.* **82,** 189–190.

Ertel, I.J., Newton, W.A. (1969). Therapy in congenital hyperbilirubinemia: phenobarbital and diethylnicotinamide. *Pediatrics* **44,** 43–48.

Evans, G.W., and Johnson, P.E. (1977). Defective prostaglandin synthesis in acrodermatitis enteropathica. *Lancet* **1,** 52.

Falchuck, Z.M., Rogentine, G.N., and Strober, W. (1972). Predominance of histocompatibility antigen HL-A8 in patients with gluten-sensitive enteropathy. *J. Clin. Invest.* **51,** 1602–1605.

Falchuck, Z.M., Katz, A.J., Shwachman, H., Rogentine, G.N. and Strober, W. (1978). Gluten-sensitive enteropathy: Genetic analysis and organ culture study in 35 families. *Scand. J. Gastroenterol.* **13,** 839–843.

Farmer, R. G., ed. (1980). "Inflammatory Bowel Disease." *Clin. Gastroenterol.* **9** (2).

Farmer, R.G., Michener, W.M., and Mortimer, E.A. (1980). Studies of family history among patients with inflammatory bowel disease. *Clin. Gastroenterol.* **9,** 271–278.

Fast, B.B., and Roulston, T.M. (1964). Idiopathic jaundice of pregnancy. *Am. J. Obstet. Gynecol.* **88,** 314–321.

Felsher, B.F., Rickard, D., and Redeker, A.G. (1970). The reciprocal relation between caloric intake and the degree of hyperbilirubinemia in Gilbert's syndrome. *N. Engl. J. Med.* **283,** 170–172.

Fevery, J., Blanckaert, N., Heirwegh, K.P.M., Preáux, A-M., Berthelot, P. (1977). Unconjugated bilirubin and an increased proportion of bilirubin monoconjugates in the bile of patients with Gilbert's syndrome and Crigler–Najjar disease. *J. Clin. Invest.* **60,** 970–979.

Fielding, J.F. (1976). Inflammatory bowel disease and pregnancy. *Br. J. Hosp. Med.* **15,** 345–352.

Fielding, J.F., and Cooke, W.T. (1970). Pregnancy and Crohn's disease. *Br. Med. J.* **2,** 76–77.

Foulk, W.T., Butt, H.R., Owen, C.A. Jr., Whitcomb, F.F. Jr., and Mason, H.L. (1959). Constitutional hepatic dysfunction (Gilbert's disease): Its natural history and related syndromes. *Medicine (Baltimore)* **38,** 25–46.

Fried, K., and Vure, E. (1974). A lethal autosomal recessive entero-colitis of early infancy. *Clin. Genet.* **6,** 195–196.

Friedlaender, P., and Osler, M. (1967). Icterus and pregnancy. *Am. J. Obstet. Gynecol.* **97,** 894–900.

Furhoff, A-K., and Hellström, J. (1973). Jaundice in pregnancy. A follow-up study of the series of women originally reported by L. Thorling. *Acta Med. Scand.* **193,** 259–266.

Furhoff, A-K. (1974). Fate of children born to women with jaundice in pregnancy. *Arch. Gynaekol.* **217,** 165–172.

Garretts, M. and Molokhia, M. Acrodermatitis enteropathica without hypozincemia. *J. Pediatr. (St. Louis)* **91,** 492–494.

Gartner, L.M., and Arias, I.M. (1966). Studies of prolonged neonatal jaundice in the breast-fed infant. *J. Pediatr. (St. Louis)* **68,** 54–66.

Gladstien, K., and Spence, M. (1978). A statistical analysis of birth-order effects with application to data on pyloric stenosis. *Ann. Hum. Genet.* **42,** 213–217.

Gladstien, K., Lange, K., and Spence, M.A. (1978). A goodness-of-fit test for the polygenic threshold model: Application to pyloric stenosis. *Am. J. Med. Genet.* **2,** 7–13.

Gollan, J.L., Huang, S.N., Billing, B., Sherlock, S. (1975). Prolonged survival in three brothers with severe type 2 Crigler–Najjar syndrome. Ultrastructural and metabolic studies. *Gastroenterology* **68,** 1543–1555.

Gorden, P., and Levitin, H. (1973). Congenital alkalosis with diarrhea. A sequel to Darrow's original description. *Ann. Intern. Med.* **78,** 876–882.

Gordon, E.R., Shaffer, E.A., Sass-Kortsak, A. (1976). Bilirubin secretion and conjugation in the Crigler–Najjar syndrome type II. *Gastroenterology* **70,** 761–765.

Gotlieb-Jensen, K. (1972). "Peptic Ulcer: Genetic and Epidemiological Aspects Based on Twin Studies." Munksgaard, Copenhagen.

Greenberg, D.A., and Rotter, J.I. (1981). Two locus model for gluten sensitive enteropathy: Population genetic considerations. *Am. J. Med. Genet.* **8,** 205–214.

Greene, J. Rogers, A., and Rubin, L. (1963). Fetal loss after cholecystectomy during pregnancy. *Can. Med. Assoc. J.* **88,** 576–577.

Grossman, M.I. (1979). Peptic ulcer: Pathogenesis and pathophysiology. *In* "Textbook of Medicine" (P.B. Beeson, W. McDermott, and J.B. Wyngaarden, eds.), pp. 1502–1507. Saunders, Philadelphia.

Grossman, M.I. (1980). Peptic ulcer, definition and epidemiology. *In* "Genetics and Heterogeneity of Common Gastrointestinal Disorders" (J.I. Rotter, I.M. Samloff, and D.L. Rimoin, eds.), pp. 21–29. Academic Press, New York.

Gryboski, W.A., and Spiro, H.M. (1956). The effect of pregnancy on gastric secretion. *N. Engl. J. Med.* **255,** 1131–1134.

Guy, W.H. (1927). Epidermolysis bullosa. *Arch. Dermatol. Syphilol.* **15,** 30–42.

Haemmerli, U.P. (1966). Jaundice during pregnancy. *Acta Med Scand., Suppl.* **444,** 1–111.

Haemmerli, U.P. (1975). Jaundice during pregnancy. *In* "Diseases of the Liver" (L.Schiff, ed.) pp. 1336–1353. Lippincott, Philadelphia, Pennsylvania.

Haemmerli, U.P., and Wyss, H.I. (1967). Recurrent intrahepatic cholestasis of pregnancy. *Medicine (Baltimore)* **46,** 299–321.

Halsted, J.A., and Smith, J.C., Jr. (1970). Plasma-zinc in health and disease. *Lancet* **1,** 322–324.

Hambridge, K.M., and Droegemueller, W. (1974). Changes in plasma and hair concentrations of zinc, copper, chromium, and manganese during pregnancy. *Obstet. Gynecol.* **44,** 666–672.

Hambridge, K.M., Neldner, K.H., and Walravens, P.A. (1975). Zinc, acrodermatitis enteropathica and congenital malformations. *Lancet* **1,** 577–578.

Hambridge, K.M., Walravens, P.A., and Neldner, K.H. (1977). The role of zinc in the pathogenesis and treatment of acrodematitis enteropathica. *In* "Zinc Metabolism: Current Aspects in Health and Disease," pp. 329–340. Alan R. Liss, Inc., New York.

Hammer, B., Ashurst, P., and Naish, J. (1968). Diseases associated with ulcerative colitis and Crohn's disease. *Gut* **9,** 17–21.

Harris, O.D., Cooke, W.T., Thompson, H., and Waterhouse, J.A.H. (1967). Malignancy in adult coeliac disease and idiopathic steatorrhoea. *Am. J. Med.* **42,** 899–912.

Harvald, B., and Hauge, M. (1956). A catamnestic investigation of Danish twins. *Dan. Med. Bull.* **3,** 150–158.

Haverback, B.J., and Wirtschafter, S.K. (1960). Familial nonhemolytic jaundice with normal liver histology and conjugated bilirubin. *N. Engl. J. Med.* **262,** 113–117.

Heikkinen, J., Mäentausta, O., Tuimala, R., Ylöstalo, P., and Jänne, O. (1980). Amniotic fluid bile acids in normal and pathologic pregnancy. *Obstet. Gynecol.* **56,** 60-64.

Hill, L.M., Johnson, C.E., and Lee, R.A. (1975). Cholecystectomy in pregnancy. *Obstet. Gynecol.* **46,** 291-293.

Hirsh, F.S., Michel, B., and Strain, W.H. (1976). Gluconate zinc in acrodermatitis enteropathica. *Arch. Dermatol.* **112,** 475-478.

Hofmann, A.F. (1980). Gallstones, classification, diagnosis, and pathophysiology. *In* "Genetics and Heterogeneity of Common Gastrointestinal Disorders" (J.I. Rotter, I.M. Samloff, and D.L. Rimoin, eds.), pp. 293-295. Academic Press, New York.

Holmberg, C., Perheentupa, J., and Launiala, K. (1975). Colonic electrolyte transport in health and in congenital chloride diarrhea. *J. Clin. Invest.* **56,** 302-310.

Holmberg, C., Perheentupa, J., Launiala, K., and Hallman, N. (1977a). Congenital chloride diarrhea. *Arch. Dis. Child.* **52,** 255-267.

Holmberg, C., Perheentupa, J., and Pasternack, A. (1977b). The renal lesion in congenital chloride diarrhea. *J. Pediatr. (St. Louis)* **91,** 738-743.

Holmes, G.K.T. Stokes, P.L., Sorahan, T.M., Prior, P., Waterhouse, J.A.H. and Cooke, W.T. (1976). Coeliac disease, gluten-free diet, and malignancy. *Gut* **17,** 612-619.

Holzbach, R.T. (1976). Jaundice in pregnancy—1976. *Am. J. Med.* **61,** 367-376.

Holzbach, R.T., and Sanders, J.H. (1965). Recurrent intrahepatic cholestasis of pregnancy. *J. Am. Med. Assoc.* **193,** 542-544.

Homan, W.P., and Thorbjarnarson, B. (1976). Crohn disease and pregnancy. *Arch. Surg. (Chicago)* **111,** 545-547.

Hudson, C.N. (1972). Ileostomy in pregnancy. *Proc. R. Soc. Med.* **65,** 281-283.

Hunt, J.N., and Murray, F.A. (1958). Gastric function in pregnancy. *J. Obstet. Gynaecol. Br. Emp.* **65,** 78-83.

Hunter, J.O., Thompson, R.P.H., Dunn, P.M., Williams, R. (1973). Inheritance of type 2 Crigler-Najjar hyperbilirubinemia. *Gut* **14,** 46-49.

Hurley, L.S. (1977). Zinc deficiency in prenatal and neonatal development. *In* "Zinc Metabolism: Current Aspects in Health and Disease," pp.47-58. Alan R. Liss, Inc., New York.

Hurley, L.S. Eckhert, C.D., Duncan, J.R., and Sloan, M.V. (1977). Acrodermatitis enteropathica and human breast milk. *Lancet* **1,** 195.

Hurley, L.S., Lönnerdal, B., and Stanislowski, A.G. (1979). Zinc citrate, human milk and acrodermatitis enteropathica. *Lancet* **1,** 677-678.

Iber, F.L. (1965). Jaundice in pregnancy—a review. *Am. J. Obstet. Gynecol.* **91,** 721-753.

Ikonen, E. (1964). Jaundice in late pregnancy. *Acta Obstet. Gynecol. Scand.* **43,** Suppl. 5, 1-130.

Isenberg, J.I., Walsh, J.H., and Grossman, M.I. (1973). Zollinger-Ellison syndrome. *Gastroenterology* **65,** 140-165.

Janowitz, H.D. (1979). Chronic inflammatory diseases of the intestine. *In* "Textbook of Medicine" P.B. Beeson, W. McDermitt, and J.B. Wyngaarden, eds.) pp. 1560-1578. Philadelphia, Pennsylvania.

Javitt, N.B. (1980). Cholestatic liver disease: Mechanisms, diagnosis and therapy. *Adv. Intern. Med.* **25,** 147-168.

Jervis, G.A. (1959). Constitutional nonhemolytic hyperbilirubinemia with findings resembling kernicterus. *Arch. Neurol. Psychiatr.* **81,** 55-64.

Joske, R.A., and Martin, J.D. (1971). Coeliac disease presenting as recurrant abortion. *J. Obstet. Gynaecol. Br. Commonw.* **78,** 754-758.

Kayden, H.J., and Cox, R.P. (1973).. Evidence for normal metabolism and interconversions of unsaturated fatty acids in acrodermatitis enteropathica. *J. Pediatr. (St. Louis)* **83,** 993-998.

Kelly, R., Davidson, G.P., Townley, R.R.W., and Campbell, P.E. (1976). Reversible intestinal mucosal abnormality in acrodermatitis enteropathica. *Arch. Dis. Child.* **51,** 219-222.

Kidd, K.K., and Spence, M.A. (1976). Genetic analyses of pyloric stenosis suggesting a specific maternal effect. *J. Med. Genet.* **13,** 290–294.

Kirsner, J.B. (1973). Genetic aspects of inflammatory bowel disease. *Clin. Gastroenterol.* **2,** 557–575.

Kirsner, J.B. (1980). Inflammatory bowel disease—clinical, etiological, and genetic aspects. *In* "Genetics and Heterogeneity of Common Gastrointestinal Disorders" (J.I. Rotter, I.M. Samloff, and D.L. Rimoin, eds.), pp. 261–280. Academic Press, New York.

Kirsner, J.B., and Spencer, J.A. (1963). Family occurrences of ulcerative colitis, regional enteritis and ileocolitis. *Ann. Intern. Med.* **59,** 133–144.

Kondo, T., Kuchiba, K., and Shimizu, Y. (1976). Coproporphyrin isomers in Dubin-Johnson syndrome. *Gastroenterology* **70,** 1117–1120.

Kreek, M.J., Sleisenger, M.H. (1968). Reduction of serum-unconjugated-bilirubin with phenobarbitone in adult congenital non-haemolytic unconjugated hyperbilirubinemia. *Lancet* **2,** 73–78.

Krieger, I., and Evans, G.W. (1980). Acrodermatitis enteropathica without hypozincemia: Therapeutic effect of pancreatic enyzme preparation due to a zinc-binding ligand. *J. Pediatr. (St. Louis)* **96,** 32–35.

Kwitko, A., and Shearman, D.J.C. (1978). Antibodies to secretory IgA (SIgA) in duodenal ulcer disease. *Gut* **19,** A437.

Laatikainen, T. (1975). Fetal bile acid levels in pregnancies complicated by maternal intrahepatic cholestasis. *Am. J. Obstet. Gynecol.* **122,** 852–856.

Laatikainen, T., and Ikonen, E. (1975). Fetal prognosis in obstetric hepatosis. *Ann. Chir. Gynaecol. Fenn.* **64,** 155–164.

Laatikainen, T., and Ikonen, E. (1977). Serum bile acids in cholestasis of pregnancy. *Obstet. Gynecol.* **50,** 313–318.

Laatikainen, T.J., Peltonen, J.I., and Nylander, P.L. (1974). Effect of maternal intrahepatic cholestasis on fetal steroid metabolism. *J. Clin. Invest.* **53,** 1709–1715.

Laatikainen, T.J., Lehtonen, P.J., and Hesso, A.E. (1978). Fetal sulfated and nonsulfated bile acids in intrahepatic cholestasis of pregnancy. *J. Lab. Clin. Med.* **92,** 185–193.

Lalouel, J.M., Morton, N.E., MacLean, C.J., and Jackson, J. (1977). Recurrence risks in complex inheritance with special regard to pyloric stenosis. *J. Med. Genet.* **14,** 408–414.

Lam, S.K. (1980). Physiologic abnormalities and heterogeneity in peptic ulcer. *In* "Genetics and Heterogeneity of Common Gastrointestinal Disorders" (J.I. Rotter, I.M. Samloff, and D.L. Rimoin, eds.), pp. 67–80. Academic Press, New York.

Lam, S.K., and Ong, G.B. (1976). Duodenal ulcers, early and late onset. *Gut* **17,** 169–179.

Lam, S.K., and Lai, C.L. (1978). Gastric ulcers with and without associated duodenal ulcer have different pathophysiology. *Clin. Sci. Mol. Med.* **55,** 97–102.

Lam, S.K., Hui, K.K., Rotter, J.I., Samloff, I.M. (1981). Pachydermoperiostosis and peptic ulcer. Gastroenterology, in press.

Lamers, C.B., Stadil, F., and Van Tongeren, J.H. (1978). Prevalence of endocrine abnormalities in patients with the Zollinger-Ellison syndrome in their families. *Am. J. Med.* **64,** 607–612.

Langman, M.J.S. (1979). "The Epidemiology of Chronic Digestive Disease." Arnold, London.

Lewkonia, R.M., and McConnell, R.B. (1976). Familial inflammatory bowel disease—heredity or environment? *Gut* **17,** 235–243.

Loh, J.P., Haller, J.O., Kassner, E.G., Aloni, A., and Glassberg, K. (1977). Dominantly-inherited polycystic kidneys in infants: Association with hypertrophic pyloric stenosis. *Pediatr. Radiol.* **6,** 27–31.

Lombeck, I., von Bassewitz, D.B., Becker, K., Tinschmann, P., and Kästner, H. (1974). Ultrastructural findings in acrodermatitis enteropathica. *Pediatr. Res.* **8,** 82–88.

Lombeck, I., Schnippering, H.G., Ritzl, F., Feinendegen, L.E., and Bremer, H.J. (1975). Absorption of zinc in acrodermatitis enteropathica. *Lancet* **1,** 855.

Lucey, J.G., Arias, I.M., McKay, R.J., Jr. (1960). Transient familial neonatal hyperbilirubinemia. *Am. J. Dis. Child.* **100,** 787–789.

Mabry, C.C., Denniston, J.C., Nelson, T.L., and Son, C.D. (1963). Maternal phenylketonuria. A cause of mental retardation in children without metabolic defect. *N. Engl. J. Med.* **269,** 1404–1408.

McAllister, J.E., and Waddell, J.M. (1962). Recurring idiopathic jaundice in pregnancy. *Am. J. Obstet. Gynecol.* **84,** 62–64.

McConnell, R.B. (1966). "The Genetics of Gastrointestinal Disorders." Oxford Press, London and New York.

McConnell, R.B. (1972). Genetics of Crohn's disease. *Clin. Gastroenterol.* **1,** 321–334.

McConnell, R.B. editor (1973). "Genetics of Gastrointestinal Disorders," *Clin. Gastroenterol.*, Vol. 2(3).

McConnell, R.B. (1976). Genetics in gastroenterology. *In* "Gastroenterology," (H.L. Bockus, ed.), 3rd ed., Vol. 4, pp. 629–648. Saunders, Philadelphia.

McConnell, R.B. (1980a). Peptic ulcer, early genetic evidence—families, twins, and markers. *In* "Genetics and Heterogeneity of Common Gastrointestinal Disorders" (J.I. Rotter, I.M. Samloff, and D.L. Rimoin, eds.), pp. 31–41. Academic Press, New York.

McConnell, R.B. (1981). "The Genetics of Coeliac Disease" MTP Press Ltd., Lancaster, England (in press).

McConnell, R.B. (1980b). Inflammatory bowel disease: Newer views of genetic influence. *In* "Developments in Digestive Diseases" (J.E. Berk, ed.), Vol. 3, pp. 129–137. Lea & Febiger, Philadelphia, Pennsylvania.

McDonald, W.C., Dobbins, W.O., and Rubin, C.E. (1965). Studies of the familial nature of celiac sprue using biopsy of the small intestine. *N. Engl. J. Med.* **272,** 448–456.

MacDougall, I. (1956). Ulcerative colitis and pregnancy. *Lancet* **2,** 641–643.

McEwan, H.P. (1972). Ulcerative colitis in pregnancy. *Proc. R. Soc. Med.* **65,** 279–281.

Mackintosh, P., and Asquith, P. (1978). HLA and coeliac disease. *Br. Med. Bull.* **34,** 291–294.

McKusick, V.A. (1978). "Mendelian Inheritance in Man. Catalogs of Autosomal Dominant, Autosomal Recessive, and X-linked Phenotypes" John Hopkins Univ. Press, Baltimore, Maryland.

McKusick, V.A., and Clayton, R.J. (1968). Cholestasis of pregnancy. *N. Engl. J. Med.* **278,** 566–567.

McMichael, A., and McDevitt, H. (1977). The association between the HLA system and disease. *Prog. Med. Genet.***2,** 39–100.

Mann, D.L., Katz, S.I., Nelson, D.L., Abelson, L.D., and Strober, W. (1976). Specific B-cell antigens associated with gluten-sensitive enteropathy and dermatitis herpetiformis. *Lancet* **1,** 110–111.

Marshall, T., and Knox, E.G. (1980). Disease concordance and sex similarity in twins. *J. Epidemiol. Commun. Health* **34,** 1–8.

Martimbeau, P.W., Welch, J.S., and Weiland, L.H. (1975). Crohn's disease and pregnancy. *Am. J. Obstet. Gynecol.* **122,** 746–749.

Martin, R.H., Harper, T.A., and Kelso, W. (1965). Serum folic acid in recurrent abortions. *Lancet* **1,** 670–672.

Mendeloff, A.I. (1980). The epidemiology of inflammatory bowel disease. *Clin. Gastroenterol.* **9,** 259–270.

Meulengracht, E. (1947). A review of chronic intermittent juvenile jaundice. *Q. J. Med.* **16,** 83–98.

Meyer, T.C. (1956). A study of serum bilirubin levels in the newborn in relation to kernicterus and prematurity. *Arch. Dis. Child.* **31,** 75–80.

Miller, J.P. (1977). Disease of the liver and alimentary tract. *Clin. Obstet. Gynecol.* **4,** 297–317.

Misra, P.S., Evanov, F.A., Wessley, Z., Rosenblum, G.A., and Singh, P. (1980). Idiopathic intrahepatic cholestasis of pregnancy. *Am. J. Gastroenterol.* **73,** 54–59.

Monson, R.R. (1970). Familial factors in peptic ulcer, the occurrence of ulcer in relatives. *Am. J. Epidemiol.* **91,** 453–466.

Moore, H.C. (1963). Reccurent jaundice of pregnancy. A type of intrahepatic cholestasis. *Lancet* **2,** 57–64.

Morris, J.S., Adjukiewicz, A.B., and Read, A.E. (1970). Coeliac infertility: An indication for dietary gluten restriction? *Lancet* **1,** 213–214.

Moshal, M.G. (1980). Ethnic differences in dodenal ulceration. *In* "Genetics and Heterogeneity of Common Gastrointestinal Disorders" (J.I. Rotter, I.M. Samloff, and D.L. Rimoin, eds.), pp. 91–100. Academic Press, New York.

Mourant, A.E., Kopeć, A.C., and Domaniewska-Sobczak, A. (1978). "Blood Groups and Diseases: A Study of Associations of Diseases with Blood Groups and Other Polymorphisms." Oxford Univ. Press, London and New York.

Moynahan, E.J. (1974). Acrodermatitis enteropathica: A lethal inherited human zinc-deficiency disorder. *Lancet* **2,** 399–400.

Moynahan, E.J., and Barnes, P.M. (1973). Zinc deficiency and a synthetic diet for lactose intolerance. *Lancet* **1,** 676–677.

Murray, F.A., Erskine, J.P., and Fielding, J. (1957). Gastric secretion in pregnanry. *J. Obstet. Gynaecol. Br. Emp.* **64,** 373–381.

Myellot, M. (1976). Anaemia in pregnancy and the small intestine. *Br. J. Obstet. Gynaecol.* **83,** 411–412.

Mylotte, M.J., Egan-Mitchell, B., Fottrell, P.F., McNicholl, B., and McCarthy, C.F. (1972). Familial coeliac disease. *Q. J. Med.* **41,** 527–528.

Namihisa, T. (1976). Chronic benign hyperbilirubinemia. *Gastroenterology* **70,** 636.

Namihisa, T., and Nambu, M. (1977). Rotor's and Dubin-Johnson syndromes. *N. Engl. J. Med.* **297,** 560.

Neldner, K.H., and Hambridge, K.M. (1975). Zinc therapy of acrodermatitis enteropathica. *N. Engl. J. Med.* **292,** 879–882.

Neldner, K.H., Hagler, L., Wise, W.R., Stifel, F.B., Lufkin, E.G., and Herman, R.H. (1974). Acrodermatitis enteropathica. A clinical and biochemical survey. *Arch. Dermatol.* **110,** 711–721.

Nelson, R.S. (1956). Persistent nonhemolytic hyperbilirubinemia with unidentified pigment in liver cells. Report of case followed in pregnancy. *Gastroenterology* **30,** 301–304.

Neuhauser, G., Daly, R.F., Magnelli, N.C., Barreras, R.F., Donaldson, R.M., and Optiz, J.M. (1976). Essential tremor, nystagmus and duodenal ulceration. *Clin. Genet.* **9,** 81–91.

Newman, A.J., and Gross, S. (1963). Hyperbilirubinemia in breast-fed infants. *Pediatrics* **32,** 995–1001.

Norio, R., Perheentupa, J., Launiala, K., and Hallman, N. (1971). Congenital chloride diarrhea, an autosomal recessive disease. Genetic study of 14 Finnish and 12 other families. *Clin. Genet.* **2,** 182–192.

Norton, R.A., and Patterson, J.F. (1972). Pregnancy and regional enteritis. *Obstet. Gynecol. (N.Y.)* **40,** 711–712.

Odell, G.B., and Childs, B. (1980). Hereditary hyperbilirubinemias. *Prog. Med. Genet.* **4,** 103–134.

Ogborn, A.D.R. (1975). Pregnancy in patients with coeliac disease. *Br. J. Obstet. Gynaecol.* **82,** 293–296.

Orellana-Alcalde, J.M , and Dominguez, J.P. (1966). Jaundice and oral contraceptive drugs. *Lancet* **2,** 1278–1280.

Owens, D., and Evans, J. (1975). Population studies on Gilbert's syndrome. *J. Med. Genet.* **12,** 152–156.

Owens, D., and Sherlock, S. (1973). Diagnosis of Gilbert's syndrome: Role of the reduced caloric intake test. *Br. Med. J.* **3,** 559–563.

Passarge, E. (1979). Genetics of the gastrointestinal tract. *In* "Clinical Genetics: A Source Book for Physicians" (L.G. Jackson and R.N. Schimke, ed.), pp. 331–347. Wiley, New York.

Passarge, E. (1980). Genetics of common gastrointestinal malformations and the heterogeneity of Hirschsprung's disease. *In* "Genetics and Heterogeneity of Common Gastrointestinal Disorders" (J.I. Rotter, I.M. Samloff, and D.L. Rimoin, eds.), pp. 441–449. Academic Press, New York.

Peña, A.S. Immunogenetic aspects of inflammatory bowel disease. *In* "Genetics and Heterogeneity of Common Gastrointestinal Disorders" (J.I. Rotter, I.M. Samloff, and D.L. Rimoin, eds.), pp. 281–289. Academic Press, New York.

Peña, A.S., Mann, D.L., Hague, N.E., Heck, J.A., van Leeuwen, A., van Rood, J.J. and Strober, W. (1978). Genetic basis of gluten-sensitive enteropathy. *Gastroenterology* **75,** 230–235.

Peña, A.S., Biemond, I., Kuiper, G., Weterman, I.T., van Leeuwen, A., Schreuder, I., and van Rood, J.J. (1980). HLA antigen distribution and HLA haplotype segregation in Crohn's disease. *Tissue Antigens* **16,** 56–61.

Perez, V., and Gorodisch, S. (1974). Female sex hormones and the liver. *In* "The Liver and Its Diseases" (F. Schaffner, S. Sherlock, and C.M. Leevy, eds.), pp. 179–190. Intercontinental Medical Book Corp., New York.

Perheentupa, J., Eklund, J., and Kojo, N. (1965). Familial chloride diarrhoea ("Congenital alkalosis with diarrhoea"). *Acta Paediatr. Scand., Suppl.* **159,** 119–120.

Piletz, J.E., and Ganschow, R.E. (1978). Zinc deficiency in murine milk underlies expression of the lethal mile (1m) mutation. *Science* **199,** 181–183.

Pinsky, L. (1978). The syndromology of anorectal malformation (atresia, stenosis, ectopia). *Am. J. Med. Genet.* **1,** 461–474.

Piper, E.L. (1955). Acrodermatitis enteropathica in an adult. *Arch. Dermatol.* **76,** 221–224.

Portnoy, B., and Molokhia, M. (1975). Zinc and diodoquin in acrodermatitis enteropathica. *Lancet* **2,** 1156–1157.

Powell, L.W., Hemingway, E., Billing, B.H., and Sherlock, S. (1967). Idiopathic unconjugated hyperbilirubinemia (Gilbert's syndrome). A study of 42 families. *N. Engl. J. Med.* **277,** 1108–1112.

Price, W.H. (1979). A high incidence of chronic inflammatory bowel disease in patients with Turner's syndrome. *J. Med. Genet.* **16,** 263–266.

Ramos, A., Silverberg, M., and Stern, L. (1966). Pregnanediols and neonatal hyperbilirubinemia. *Am. J. Dis. Child.* **111,** 353–356.

Redinger, R.N., and Small, D.M. (1972) Bile composition, bile salt metabolism, and gallstones. *Arch. Intern. Med.* **130,** 618–630.

Reid, R., Ivey, K.J., Rencoret, R.H., and Storey, B. (1976). Fetal complications of obstetric cholestasis. *Br. Med. J.* **1,** 870–872.

Revill, S.I., and Dodge, J.A. (1978). Psychological determinants of infantile pyloric stenosis. *Arch. Dis. Child.* **53,** 66–68.

Reyes, H., Ribalta, J., and González-Cerón, M. (1976). Idiopathic cholestasis of pregnancy in a large kindred. *Gut* **17,** 709–713.

Reyes, H., Gonzáles-Cerón, M., Ribalta, J., Aburto, H., Matus, C., Schramm, G., Katz, R., and Medina, E. (1978). Prevalence of intrahepatic cholestasis of pregnancy in Chile. *Ann. Intern. Med.* **88,** 487–493.

Robertson, A.F., Schuerger, G.S., Brown, R.R., and Karp, W.B. (1973). Tryptophan metabolism in acrodermatitis enteropathica. *J. Pediatr. (St. Louis)* **83,** 1012–1016.

Roszkowski, I., and Pisarek-Miedzinska, D. (1968). Jaundice in pregnancy. II. Clinical course of pregnancy and delivery and condition of neonate. *Am. J. Obstet. Gynecol.* **101,** 500–503.

Rotter, J.I. (1978). Discussion on pyloric stenosis. *In* "Genetic Epidemiology" (N.E. Morton and C.S. Chung, ed.), pp. 351–352. Academic Press, New York.

Rotter, J.I. (1980a). Peptic ulcer disease—more than one gene, more than one disease. *Prog. Med. Genet.* **4,** 1–58.

Rotter, J.I. (1981). Gastric and duodenal ulcer are each many different diseases. *Dig. Dis. Sci.* **26,** 154–160.

Rotter, J.I. (1980b). Genetic approaches to ulcer heterogeneity. *In* "Genetics and Heterogeneity of Common Gastrointestinal Disorders" (J.I. Rotter, I.M. Samloff, and D.L. Rimoin, eds.), pp. 111–128. Academic Press, New York.

Rotter, J.I., and Rimoin, D.L. (1977). Peptic ulcer disease—a heterogeneous group of disorders? *Gastroenterology* **73,** 604–607.

Rotter, J.I., Rimoin, D.L., Samloff, I.M., McConnell, R.B., Gotlieb-Jensen, K., Gadeburg, O., and Hauge, M. (1977a). The genetics of peptic ulcer disease - elevated serum group I pepsinogen concentrations in siblings and twins of ulcer probands. *Gastroenterology* **72,** 1165.

Rotter, J.I., Rimoin, D.L., Gursky, J.M., Terasaki, P., and Sturdevant, R.A.L. (1977b). HLA-B5 associated with duodenal ulcer. *Gastroenterology* **73,** 435–437.

Rotter, J.I., Petersen, G.M., Samloff, I.M., McConnell, R.B., Ellis, A., Spence, M.A., and Rimoin, D.L. (1979a). Genetic heterogeneity of familial hyperpepsinogenemic I and normopepsinogenemic I duodenal ulcer disease. *Ann. Intern. Med.* **91,** 372–377.

Rotter, J.I., Sones, J.Q , Samloff, I.M., Gursky, J.M., Richardson, C.T., Walsh, J.H., and Rimoin, D.L. (1979b). Duodenal ulcer disease associated with elevated serum pepsinogen I: an inherited autosomal dominant disorder. *N. Engl. J. Med.* **300,** 63–66.

Rotter, J.I., Rubin, R , Meyer, J.H., Samloff, I.M., and Rimoin, D.L. (1979c). Rapid gastric emptying—an inherited pathophysiologic defect in duodenal ulcer? *Gastroenterology* **76,** 1229.

Rotter, J.I., Rimoin, D L., and Samloff, I.M. (1979d). Genetic heterogeneity in peptic ulcer. *Lancet* **1,** 1088–1089.

Rotter, J.I., Samloff, I.M., and Rimoin, D.L., eds. (1980). "Genetics and Heterogeneity of Common Gastrointestial Disorders." Academic Press, New York.

Rotter, J.I., Monson, R.R., and Grossman, M.I. (1981). Duodenal ulcer and pulmonary disease, which comes first? Submitted for publication.

Sachar, D.B., Auslander, M.O., and Walfish, J.S. (1980). Aetological theories of inflamatory bowel disease. *Clin. Gastroenterol.* **9,** 231–257.

Samloff, I.M. (1979). Serum pepsinogens I and II. *In* "Developments in Digestive Diseases" (J.E. Berk, ed), pp. 1–12. Lea & Febiger, Philadelphia, Pennsylvania.

Samloff, I.M. (1980). Pepsinogens and their relationship to peptic ulcer. *In* "Genetics and Heterogeneity of Common Gastrointestinal Disorders" (J.I. Rotter, I.M. Samloff, and D.L. Rimoin, eds.), pp. 43–49. Academic Press, New York.

Schimke, R.N. (1976). Multiple endocrine adenomatosis syndromes. *Adv. Intern. Med.* **21,** 249–265.

Schinella, R.A., Greco, M.A., Cobert, B.L., Denmark, L.W., and Cox, R.P. (1980). Hermansky-Pudlak syndrome with granulomatous colitis. *Ann. Intern. Med.* **92,** 20–23.

Schmid, R., and McDonagh, A.F. (1978). Hyperbilirubinemia. *In* "The Metabolic Basis of Inherited Disease" (J.B. Stanbury, J.B. Wyngaarden, and D.S. Fredrickson, eds.), 4th ed., pp. 1221–1259. McGraw-Hill, New York.

Schofield, P.F., Turnbull, R.B., and Hawk, W.A. (1970). Crohn's disease and pregnancy. *Br. Med. J.* **2,** 363–364.

Schwartz, S.E., Siegelbaum, S.P., Fazio, T.L., Hubbell, C., and Henry, J.B. (1980). Regional enteritis: Evidence for genetic transmission by HLA typing. *Ann. Intern. Med.* **93,** 424–427.
Schwarz, L.R., Schwenk, M. and Greim, H. (1977). Current concepts in intrahepatic cholestasis. *Acta Hepato-Gastroenterol.* **24,** 210–215.
Sever, L.E., and Emanuel, I. (1972). Is there a connection between maternal zinc deficiency and congenital malformations of the central nervous system in man? *Teratology* **7,** 117–118.
Shani, M., Seligsohn, U., Gilon, E., Sheba, C., and Adam, A. (1970a). Dubin-Johnson Syndrome in Israel. I. Clinical, laboratory and genetic aspect of 101 cases. *Q. J. Med.* **39,** 549-567.
Shani, M., Gilon, E., Ben-Ezzer, J., and Sheba, C. (1970b). Sulfobromaphthalein tolerance test in patients with Dubin-Johnson syndrome and their relatives. *Gastroenterology* **59,** 842-847.
Sherlock, S. (1975). ''Diseases of the Liver and Biliary System,'' pp. 570–578. Blackwell, Oxford.
Shipman R.T., Williams, A.L. Kay, R., and Townley, R.R.W. (1975). A family study of coeliac disease. *Aust. N. Z. J. Med.* **5,** 250–255.
Simpson, J.L. (1976). Gonadal dysgenesis. *In* ''Disorders of Sexual Differentiation: Etiology and Clinical Delineation,'' pp. 260–302. Academic Press, New York.
Singer, H.C., Anderson, J.G.D., Frischer, H., and Kirsner, J.B. (1971). Familial aspects of inflammatory bowel disease. *Gastroenterology* **61,** 423–430.
Slade, J.D., Luskin, A.T., Gewurz, H., Kraft, S.C., Kirsner, J.B., and Zeitz, H.J. (1978). Inherited deficiency of second component of complement and HLA haplotype A10, B18 associated with inflammatory bowel disease. *Ann. Intern. Med.* **88,** 796–798.
Sleisenger, M.H., Kahn, I., Barniville, H., Rubin, W., Ben Ezzer, J., Arias, I.M (1967). Non-hemolytic unconjugated hyperbilirubinemia with hepatic glucuronyl transferase deficiency: A genetic study in four generations. *Trans. Assoc. Amer. Physicians* **80,** 259–265.
Smith, D.W. (1976). ''Recognizable Patterns of Human Malformation.'' Saunders, Philadelphia, Pennsylvania.
Smith, P.M., Middleton, J.E., Williams, R. (1967). Studies on the familial incidence and clinical history of patients with chronic unconjugated hyperbilirubinemia. *Gut* **8,** 449–453.
Somayaji, B.N., Paton, A., Price, J.H., Harris, A.W., and Flewett, T.H. (1968). Norethisterone jaundice in two sisters. *Br. Med. J.* **2,** 281–283.
Spence, M.A., and Gladstien, K. (1978). Pyloric stenosis and simulation of Mendelism. *In* ''Genetic Epidemiology'' (N.E. Morton and C.S. Chung, eds.), pp. 331–351. Academic Press, New York.
Spitaels, J.M., and Moshal, M.G. (1976). Recurrent intrahepatic cholestasis. *S. Afr. Med. J.* **50,** 2090.
Sprinz, H., and Nelson, R.S. (1954). Persistent nonhemolytic hyperbilirubinemia associated with lipochrome-like pigment in liver cells: Report of four cases. *Ann. Intern. Med.* **41,** 952–962.
Steinberg, A.G., Bearn, A.G., Motulsky, A.G., and Childs, B., eds. (1980). ''Genetics of Gastrointestinal Disease.'' *Prog. Med. Genet.*, Vol. 4.
Stokes, P.L., Asquith, P., Holmes, G.K.T., Mackintosh, P., and Cooke, W.T. (1972). Histocompatibility antigens associated with adult coeliac disease. *Lancet* **2,** 162–164.
Stokes, P.L., Asquith, P., and Cooke, W.T. (1973). Genetics of coeliac disease. *Clin. Gastroenterol.* **2,** 547–556.
Strober, W. (1976). Gluten-sensitive enteropathy. *Clin. Gastroenterol.* **5,** 429–450.
Strober, W. (1977). Abnormalities of the HLA system and gastrointestinal disease. *In* ''HLA and Disease'' (J. Dausset and A. Svejgaard, eds.), pp. 168–185. Munksgaard, Copenhagen.
Strober, W. (1980). Genetic factors in gluten-sensitive enteropathy. *In* ''Genetics and Heterogeneity of Common Gastrointestinal Disorders'' (J.I. Rotter, I.M. Samloff, and D.L. Rimoin, eds.), pp. 243–259.Academic Press, New York.
Strober, W., Falchuk, Z.M., Rogentine, G.N., Nelson, D.L., and Klaeveman, H.L. (1975). The pathogenesis of gluten-sensitive enteropathy. *Ann. Intern. Med.* **83,** 242–256.

Sugar, P. (1961). Familial nonhemolytic jaundice. Congenital with kernicterus. *Arch. Intern. Med.* **108,** 121-127.

Svanborg, A., and Ohlsson, S. (1959). Recurrent jaundice of pregnancy: A clinical study of twenty-two cases. *Am. J. Med.* **27,** 40-49.

Swartz, H.M., Sarna T., and Varma, R.R. (1979). On the nature and excretion of the hepatic pigment in the Dubin-Johnson syndrome. *Gastroenterology* **76,** 958-964.

Szabó, L., Kovács, Z., Ébrey, P. (1962). Congenital nonhaemolytic jaundice. *Lancet* 1, 322.

Szabó, L., and Ébrey, P. (1963). Studies on the inheritance of Crigler-Najjars syndrome by the menthol test. *Acta. Paediatr. Hung.* **4,** 153-158.

Thorling, L. (1955). Jaundice in pregnancy: A clinical study. *Acta Med. Scand.* **151,** Suppl. 302, 1-123.

Thulin, K.E., and Nermark, J. (1966). Seven cases of jaundice in women taking an oral contraceptive, Anovlar. *Br. Med. J.* **1,** 584-586.

Tompkins, R.R., and Livingood, C.S. (1969). Acrodematitis enteropathica persisting into adulthood. *Arch. Dermatol.* **99,** 190-195.

Tygstrup, N., and Jensen, B. (1969). Intermittent intrahepatic cholestasis of unknown etiology in five young males from the Faroe Islands. *Acta Med. Scand.* **185,** 523-530.

van Allen, M.W., Frohlich, J.A., and Davis, J.R. (1969). Inherited predisposition to generalized amyloidosis, clinical and pathological study of a family with neuropathy, nephropathy, and peptic ulcer. *Neurology* **19,** 10-25.

van der Linden, W. (1973). Genetic factors in gallstone disease. *Clin. Gastroenterol.* **2,** 603-614.

van der Linden, W. (1980). Genetics of cholelithiasis. *In* "Genetics and Heterogeneity of Common Gastrointestinal Disorders" (J.I. Rotter, I.M. Samloff, and D.L. Rimoin, eds.), pp. 313-320. Academic Press, New York.

van der Linden, W., and Lindelof, G. (1965). The familial occurrence of gallstone disease. *Acta Genet. Stat. Med.* **15,** 159-164.

van der Linden, W., and Simonson, N. (1973). Familial occurrence of gallstone disease: Incidence in parents of young patients. *Hum. Hered.* **23,** 123-127.

van der Linden, W., and Westlin, N. (1966). The familial occurrence of gallstone disease II. Occurrence in husbands and wives. *Acta Genet. Stat. Med.* **16,** 377-382.

Vedder, J.S. (1956). Acrodermatitis enteropathica (Danbolt-Closs) in five siblings. *J. Pediatr. (St. Louis)* **48,** 212-219.

Verburg, D.J., Burd, L.I., Hoxtell, E.O., and Merrill, L.K. (1974). Acrodermatitis enteropathica and pregnancy. *Obstet. Gynecol.* **44,** 233-237.

Waldum, H.L., Straume, B.K., and Lundgren, R. (1980). Serum group I pepsinogens during pregnancy. *Scand. J. Gastroenterol.* **15,** 61-63.

Walravens, P.A., Hambidge, K.M., Weston, W., and Neldner, K.H. (1976). Plasma-zinc in acrodermatitis enteropathica. *Lancet* **1,** 488.

Walravens, P.A., Hambidge, K.M., Neldner, K.H., Silverman, A., van Doorninck, W.J., Mierau, G., and Favara, B. (1978). Zinc metabolism and acrodermatitis enteropathica. *J. Pediatr. (St. Louis)* **93,** 71-73.

Warkany, J., and Petering, H.G. (1972). Congenital malformations of the central nervous system in rats produced by maternal zinc deficiency. *Teratology* **5,** 319-334.

Webb, M.J., and Sedlack, R.E. (1974). Ulcerative colitis in pregnancy. *Med. Clin. North Am.* **58,** 823-827.

Weinrieb, I.J., Fineman, R.M., and Spiro, H.M. (1976). Turner syndrome and inflammatory bowel disease. *N. Engl. J. Med.* **294,** 1221-1222.

Wells, B.T., and Winkelmann, R.K. (1961). Acrodermatitis enteropathica. *Arch. Dermatol.* **84,** 40-52.

Wetstone, H.J. and Honeyman, M.S. (1968). Definition of normal blood bilirubin. *N. Engl. J. Med.* **278,** 564.

White, H.B., Jr., and Montalvo, J.M. (1973). Serum fatty acids before and after recovery from acrodermatitis enteropathica: Comparison of an infant with her family. *J. Pediatr. (St. Louis)* **83,** 999–1006.

Willoughby, C.P., and Truelove, S.C. (1980). Ulcerative colitis and pregnancy. *Gut* **21,** 469–474.

Winfield, C.R., and MacFaul, R. (1978). Clinical study of prolonged jaundice in breast- and bottle-fed babies. *Arch. Dis. Child.* **53,** 506–507.

Wolkoff, A.W., Chowdhury, J.R., Gartner, L.A., Rose, A.L., Biempica, L., Giblin, D.R., Fink, D., Arias, I.M. (1979). Crigler–Najjar syndrome (type I) in an adult male. *Gastroenterology* **75,** 840–848.

Wolkoff, A.W., Cohen, L.E., and Arias, I.M. (1973). Inheritance of the Dubin-Johnson syndrome. *N. Engl. J. Med.* **288,** 113–117.

Wolkoff, A.W., Wolpert, E., Pascasio, F.N., and Arias, I.M. (1976). Rotor's syndrome: A distinct inheritable pathophysiologic entity. *Am. J. Med.* **60,** 173–179.

Wolpert, E., Pascasio, F.M., Wolkoff, A.W., and Arias, I.M. (1977). Abnormal sulfobromophthalein metabolism in Rotor's syndrome and obligate heterozygotes. *N. Engl. J. Med.* **296,** 1099–1101.

Yaffe, S.J., Levy, G., Matsuzawa, T., Baliah, T. (1966). Enhancement of glucuronide-conjugating capacity in hyperbilirubinemic infant due to apparent enzyme induction by phenobarbital. *N. Engl. J. Med.* **275,** 1461–1466.

Zetzel, L. (1975). Fertility, pregnancy, and idiopathic inflammatory bowel disease. *In* "Inflammatory Bowel Disease" (J.B. Kirshner, R.G. Shorter, eds.), pp. 146–153. Lea & Febiger, Philadelphia, Pennsylvania.

Zimmerman, H.J. (1979). Intrahepatic cholestasis. *Arch. Intern. Med.* **139,** 1038–1045.

10

Diabetes Mellitus

MENACHEM NITZAN

Copyright © 1981 by Academic Press, Inc.
All rights of reproduction in any form reserved.
ISBN 0-12-630940-X

Diabetes mellitus is a metabolic disorder which is influenced by interacting genetic and environmental factors. Diabetes, among the commonest diseases afflicting mankind, is prevalent among most human races. Although it is difficult to ascertain the overall prevalence of this disease, the order of magnitude has been estimated to be at least 2% of the population in the United States (National Commission on Diabetes, 1976), with an additional equal percentage estimated to have undiagnosed disease. According to statistics for the years between 1965 and 1978, which show a rise of 50%, the incidence or recognition of diabetes seems to be increasing. The prevalence rate among children under 17 years of age is about 0.13%. Rates are consistently higher for females than for males. Gestational and chemical diabetes [class A according to White (1965)] probably occurs in 1–2% of pregnancies, and overt diabetes in 0.1–0.2% (Beard and Oakley, 1976).

The combination of diabetes mellitus and pregnancy represents a high risk situation for both mother and fetus. Before the introduction of insulin therapy, conception in diabetic women was uncommon. More than one-third of affected mothers died before or as a result of gestation, and infant survival was very low. At present, fertility in diabetic women is within the normal range (Farquhar, 1976). The immediate favorable influence of the insulin era on the course of diabetic pregnancies has been more significant in respect to maternal health and survival than to that of the fetus.

With the advent of insulin, maternal mortality among diabetic women has dropped from a rate of 25–60% in 1922, to less than 1% by 1970 (Daweke and Gütter, 1970). Nowadays, maternal survival is usual even in the presence of vascular complications, except for subjects with ischemic heart disease (Hare and White, 1977). Attention has since focused on perinatal mortality, which by 1970 still averaged 16.9% despite insulin treatment (Daweke and Gütter, 1970). Complications in the neonatal period occurred in approximately 50% of insulin-dependent, and in 10–20% of infants of gestational, diabetic women. Among these mothers intrauterine deaths occurred with higher frequencies than among normal mothers, the increase being noteworthy even among so-called prediabetics. However, since the advent of insulin fetal survival has steadily improved, and in this past decade many investigators have reported truly impressive results (see Gabbe, 1978). Many factors are involved in this improvement, including the availability of better insulin preparations, antimicrobial agents, and tests for the antenatal assessment of fetal condition.

I. DIAGNOSIS

The renal threshold for glucose is physiologically lower in pregnancy; thus, glycosuria is often detected in the presence of normal blood glucose concentrations. A glucose tolerance test should be carried out in subjects with antenatal glycosuria (see below), but in only one-quarter of these patients is an abnormal response to be expected (Brundenell and Beard, 1972). During late-gestation fasting glucose concentrations are lower than normal and gestational diabetics are usually normoglycemic in the fasting state. Therefore, unless glucose values are markedly elevated, fasting values cannot be utilized for diagnostic purposes.

Oral glucose tolerance test (OGTT) is the preferable diagnostic procedure in asymptomatic patients and should be performed whenever one of the following conditions exists: (a) first or second degree relatives with diabetes, especially if the subject is obese; (b) a previous oversized infant (over 4500 g); (c) a history of polyhydramnios, unexplained perinatal death, or certain congenital anomalies; and (d) glycosuria. Following preparation for a minimal period of three days by a diet containing at least 200 g of carbohydrate per day, a 3-hour oral glucose tolerance test is performed. In addition to a tendency to fasting hypoglycemia, the 2-and 3-hour values are characteristically higher during the third trimester of normal pregnancy. Therefore, different criteria for the diagnosis of diabetes in the gravid state should be used. O'Sullivan and associates (1974) suggested the following normal limits for OGTT (100 g glucose) during pregnancy:

	Blood glucose (mg/100 ml)
Fasting	90
1 hour	165
2 hours	145
3 hours	125

These values refer to venous whole blood glucose estimated by the Somoggyi-Nelson method. A diagnosis of diabetes is established if two or more values exceed those listed above. If plasma glucose is used, the values should be higher by 15%.

II. CLASSIFICATION

A classification system specifically designed for pregnant diabetics was originally proposed by White (1965). This system, which includes all degrees of severity, has proved useful in predicting the outcome of pregnancy and in individualizing medical and obstetric treatment. The different categories included in this division take into consideration factors present in the mother before the current gestation, particularly age of onset of diabetes, duration and severity of the diabetic state, and vascular complications.

White's (1965) Classification of Diabetes in Pregnancy

Class	Description
A	Gestational or chemical diabetic; asymptomatic
B	Overt diabetes, onset after 20 years of age, duration less than 10 years
C	Overt diabetes, onset before 20 years of age, duration 10–20 years
D	Overt diabetes, onset before 10 years of age or duration more than 20 years, benign retinopathy
E	Calcified pelvic vessels (not sought today)
F	Nephropathy (proteinuria, azotemia—not accounted for by pyelitis or nephritis)
R	Malignant retinopathy (retinitis proliferans)

In gestational diabetes glucose intolerance is present during the gravid state but reverts to normal postpartum. This clinical stage of diabetes is approximately six times more common than overt diabetes and may be detected in 1–2% of pregnant women (Sutherland *et al.*, 1975). Diagnosis of gestational diabetes is important because such women (1) have a greater risk of perinatal morbidity and mortality and (2) approximately 20–30% will develop overt diabetes 1–8 years following pregnancy (O'Sullivan *et al.*, 1974). Attempts to confirm the diagnosis during the postpartum period are complicated by the rapid decline in gestational hormones that are insulin antagonists. In chemical diabetes, asymptomatic but persistent abnormality in glucose tolerance or elevation in fasting blood sugar level may be present prior to or during gestation. In this condition carbohydrate metabolism does not revert to normal following delivery.

III. GENETICS

Familial aggregates in diabetes mellitus (DM) have been recognized for decades, but the mode (or modes) of inheritance has defied traditional genetic analysis (see Simpson, 1978; Rotter and Rimoin, 1980). Considered irrespective of age of onset or severity, 4–8% of parents of diabetic individuals also have DM; only about 2% of age-matched controls are similarly affected. Four to six percent of sibs of diabetic individuals are similarly affected, compared to fewer than 1% of controls. If glucose tolerance or cortisol-provocative tests are performed, 10–30% of first-degree relatives of diabetic probands (index cases) are similarly affected. About 80% of monozygotic twins are concordant for DM, whereas fewer than 20% of dizygotic twins are concordant. Familial aggregates thus exist; however, the likelihood that a first-degree relative will be affected is less than expected under genetic models for (1) a single autosomal dominant gene that is fully penetrant or (2) a single autosomal recessive gene.

Originally investigators postulated that a polygenic/multifactorial model satisfactorily explained the observed recurrence risks, but this conclusion was based mostly upon exclusion. Recent studies have greatly clarified the genetics of DM by definitively showing that DM is not one but rather several different genetic entities.

A. Evidence Suggesting Genetic Heterogeneity

Although previous investigators had suggested the possibility, Rimoin and co-workers (Rimoin, 1971; Rotter and Rimoin, 1978, 1980) deserve credit for clinically applying the concept of genetic heterogeneity to DM. Originally the suggestion was based upon observations of different forms of diabetes in different racial groups and in association with different syndromes. More recently direct evidence has become available.

Two lines of evidence provide direct evidence for genetic heterogeneity: (1) identification of a genetically well-defined group of patients whose DM is diagnosed relatively early (<25 years) yet behaves clinically in maturity-onset fashion [maturity-onset diabetes, youth (MODY)] and (2) association in linkage disequilibrium of certain histocompatibility antigens with juvenile onset diabetes (JOD) but not with maturity-onset diabetes (MOD).

1. Maturity-Onset Diabetes of Youth (MODY)

The MODY phenotype, initially described by Tattersall and Fajans (1975), occurs in individuals under age 25 years. In contrast to JOD, however, individuals with MODY manifest few overt symptoms, show little disease progression, and usually require no insulin. In one study Tattersall and Fajans (1975) identified 26 patients with MODY. In each patient the age of onset for DM was less than 25 years. None of the 26 required insulin until at least 2 years after initial diagnosis. Of the 26, 22 (85%) had a diabetic parent, usually with similar symptoms; 25 of 47 sibs (53%) were affected, all with similar phenotypes, and 46% of probands showed three generations of affected relatives. These data strongly suggest autosomal dominant inheritance. Delineation of MODY provides an explanation for some of the genetic confusion, inasmuch as MODY is inherited differently from JOD. Thus if a sample contained some probands with MODY and others with JOD, recurrence risks would naturally reflect the heterogeneity.

2. Distinction between JOD and MOD

The first direct evidence verifying this distinction was derived from twin studies. Tattersall and Pyke (1972) and Pyke and Nelson (1976) confirmed earlier reports that about 70% of monozygotic twins were concordant for DM; however, concordance was for the first time shown to be directly related to age of onset. Concordance among monozygotic twins was nearly 100% if onset occurred after 45 years. In addition, these twins usually developed DM within 3 years of each other. By contrast, concordance was only 50% if age of onset was less than 45 years. Some criticisms notwithstanding, differences in relative concordance between MOD and JOD diabetic twins suggests that MOD and JOD are distinct disorders. Environmental factors appear to be more important in JOD than in MOD, whereas genetic factors presumably are more important in MOD since almost 100% of MOD twins are concordant.

That JOD and MOD are genetically distinct was later strengthened by observations that different histocompatibility antigens were associated with the two disorders. Studies showed that the HLA antigens B8 and Bw15 were associated with JOD but not MOD (Rotter and Rimoin, 1978, 1980). Thus, JOD and MOD were distinct disorders. An individual with B8 or Bw15 was two to three times more likely to develop JOD than an individual lacking those antigens. Moreover, the likelihood than an individual will have JOD is greatest if both B8 and Bw15 are present, specifically a five- to sixfold increased risk over population incidence (Rotter and Rimoin, 1978). Other studies showed associations between Dw3 and JOD, and between Cw3 and JOD. Interestingly, the associations of Dw3 and Cw3 to JOD appear even stronger than those for B8 and Bw15. Moreover, Thomsen *et al.* (1975) noted in one family that not only were all diabetic B8 individuals Dw3 positive, but also diabetic members lacking B8 yet Dw3 positive. Similarly, in another family Schernthaner *et al.* (1975) detected an increased frequency of JOD among Cw3-positive individuals who lacked Bw15. These data supported the thesis that JOD predisposition was secondary to the presence of JOD-susceptible gene(s), which were in linkage disequilibrium not only to B8 and Bw15 but also to Dw3 and Cw3. Because HLA-C and HLA-D loci are located on opposite sides of HLA-B, it follows that gene(s) responsible for JOD susceptibility in B8 individuals are different from the gene(s) associated with Bw15 individuals.

HLA associations with MODY may or may not exist. Nelson and Pyke (1976) detected no associations, but Barbosa *et al.* (1976) reported in one family that 8 of 11 affected individuals had antigens A3 and Bw15. However, in the latter family both parents had MODY. If one considers the possibility that MODY could have been inherited from either parent, the association was not statistically significant.

3. *Heterogeneity within JOD*

Rotter and Rimoin (1978, 1980) gathered additional evidence for genetic heterogeneity among individuals with JOD by concluding that the form of JOD associated with HLA-B8 differs clinically from the form of JOD associated with HLA-Bw15. The HLA-B8-associated form appears to be characterized by (1) increased frequency of the Dw3 allele, (2) decreased frequency of B7, (3) increased frequency of microangiopathy, (4) poor antiinsulin antibody response to exogenous insulin, and (5) increased frequency of pancreatic islet cell antibodies. These observations suggested that the pathogenesis of the HLA-B8-associated form of JOD involved an autoimmune phenomenon. By contrast, the HLA-Bw15-associated form is characterized by (1) increased association with Cw3, (2) no decreased frequency of B7, (3) no microangiopathy, (4) normal antibody response to exogenous insulin, and (5) no increased frequency of pancreatic islet cell antibodies. Thus, the Bw15-associated form appears less likely to be autoimmune in pathogenesis than to be due to an inherited deficiency in the capacity for insulin synthesis or release.

4. Heterogeneity within MOD

Genetic heterogeneity could also exist in MOD, but relatively few studies addressing this question have been conducted. In one study, one subset of individuals with MOD showed HLA-B8 antigen, pancreatic islet cell antibodies, and insulin requirements despite age of onset for DM of more than 40 years (Rotter and Rimoin, 1978). This subset of MOD is thus phenotypically similar to, if not identical to, the HLA-B8-associated (autoimmune) form of JOD. Fajans *et al.* (1976) also reported heterogeneity among other individuals with MOD with respect to insulin response following administration of glucose; both under-responders and over-responders were observed. Heterogeneity may thus exist among MOD as well as among JOD, but fewer data are available for MOD.

B. Genetic Counseling

The genetic confusion concerning DM has been replaced by awareness that genetic heterogeneity not only exists but can explain many heretofore unexplained genetic observations. These observations provide the basis, albeit still not completely satisfactory, for genetic counseling. If a proband has MODY, probably an autosomal dominant trait, the likelihood that first degree relatives (sibs, parents, offspring) will be similarly affected is 50%. If a proband has JOD or MOD, disorders whose mode(s) of inheritance are unclear, counseling must be based upon empiric data. If a proband has JOD, the likelihood that any given sib will have the same disorder seems to be about 10%. The likelihood that offspring of a JOD proband will be affected seems to be only about 2%. If a proband has JOD the likelihood that a first degree relative will have MOD appears to be only slightly increased over the population at large, but precise figures are unavailable. Similarly, if a proband has MOD the likelihood that first-degree relatives will likewise have MOD is perhaps 5–10%. More specific risks for MOD can be offered if the age of the counselee is considered. The likelihood of having JOD is only slightly greater than the population at large.

IV. MATERNAL MANAGEMENT

A. Pathophysiology

During the first one-half of pregnancy the most important factor affecting maternal carbohydrate homeostasis is the continuous transfer of glucose to the fetus. In order to prevent the possibility of fasting hypoglycemia, insulin dosage should often be reduced. Diminished food intake due to nausea and vomiting may cause further reduction in insulin need. During the second one-half of gestation the diabetogenic action of the hormones [e.g., human placental lactogen (HPL)] secreted by the placenta becomes a dominant factor. In addition, during insulin deprivation in women whose pregnancy is complicated by insulin-dependent diabetes mellitus [White classes B–D according to White (1965)], glucagon is not suppressed with hyperglycemia following oral glucose load (Spellacy *et al.*, 1977). Thus, insulin requirements are usually increased, and as the peripheral resistance to insulin progresses, ketoacidosis may develop. The rapid postpartum decrease in the concentrations of estrogen, progesterone, as well as the continued suppression of growth hormone lead to a considerable reduction in insulin need.

Many factors operative during late gestation increase the body's demand for insulin. Therefore, pregnancy may be regarded as a diabetogenic event. The pregnant woman with a genetic predisposition may develop gestational diabetes late in pregnancy. Obesity constitutes an additional contributory factor, and the associated insulin resistance effects will increase the demand for β-cell secretion. Therefore, the obese pregnant woman with a diabetic family history is particularly prone to develop glucose intolerance (Felig, 1977).

B. Insulin Treatment

Reduction of blood glucose to values approaching the normal range is regarded as an important determinant in the prognosis of the fetus and newborn of the diabetic mother. Karlsson and Kjellmer (1972) have clearly demonstrated that rigid control of diabetes in pregnancy minimizes perinatal morbidity and mortality. These authors found that when mean concentrations of glucose were below 100 mg/100 ml, a perinatal mortality of only 3.8% was achieved. Normalization of circulating glucose concentrations could be achieved by prolonged hospitalization from weeks 30 to 32 of pregnancy, frequent monitoring of glucose levels, and administration of multiple doses of insulin. Episodes of maternal hypoglycemia should be avoided. Abell *et al.* (1976) have also shown that this complication may be associated with two- to threefold increase in both fetal mortality and central nervous system abnormalities. In order to prevent the potential hazard of hypoglycemia, Felig (1977), therefore, recommended that random blood glucose levels should be maintained below 150 mg/100 ml and fasting values below 110 mg/100 ml.

The retrospective study made by O'Sullivan *et al.* (1974) indicated that treatment with small amounts of insulin in gestational diabetes improved fetal salvage when maternal age exceeded 25 years. Khojandi and associates (1974), in contrast, found that results in patients receiving insulin therapy were no better than in those managed by diet alone, without any relation to age. Thus, at present, insulin administration is usually recommended only for gestational diabetics over the age of 25 years (Felig, 1977), provided glucose levels are not markedly increased.

C. Diet

Although dietary control is one of the most important components of management in diabetic pregnancies, caloric restriction and weight reduction should be discouraged. Dietary limitations may cause maternal hypoglycemia and hyperketonemia. Although increased fetal mortality, cerebral palsy and convulsive disorders have been related to the former (Abell *et al.*, 1976; Gillmer *et al.*, 1975), impaired neurophysiologic development has been correlated with the latter (Churchill *et al.*, 1969). Hytten and Thomson (1970) have shown that the overall incidence of perinatal complications such as preeclampsia, intrauterine death, and prematurity was lowest in association with an average total weight gain of 27.5 lb during gestation. Duhring (1977) encourages patients to gain at least 25 lb even if underweight or obese. He recommends a daily caloric intake ranging from 25 to 50 kcal/kg, depending on the activity and prepregnancy weight of the patient. The calories are divided into 1.5 g of protein/kg/day, a minimum of 3 g of carbohydrates/kg/day, and a sufficient amount of fat to make up the total. In order to achieve optimal balance, regularity of food intake should be maintained (Felig, 1977).

D. Ketoacidosis

The physiological metabolic responses to starvation are accelerated in late pregnancy; thus, the gravid woman is prone to develop starvation ketosis following an episode of vomiting or diminished food intake (Freinkel, 1972). It is of utmost importance to differentiate between starvation ketosis and diabetic ketoacidosis since the therapeutic approach is different (Felig, 1977). The former condition, which is not associated with hyperglycemia, responds promptly to glucose infusion. In the latter complication, intravenous infusions of fast-acting insulin, potassium supplements, and in the presence of severe metabolic acidosis (arterial pH below 7.1), bicarbonate should be administered. Unless the ketoacidosis is promptly corrected the fetus itself may become severely acidotic and often dies.

E. Hemoglobin A_1C as an Indicator of Fetal Status

Clinical and experimental observations indicate that glycohemoglobin (hemoglobin A_1C) is elevated in diabetic individuals, and its levels are directly related to the time and the average blood glucose over a period of weeks (Koenig *et al.*, 1976). This minor variant of Hb A is characterized by a single glucose moiety covalently bound to the N-terminal valine of the β chain (Bookchin and Gallop, 1968). Bunn *et al.* (1976) have shown that this hemoglobin is not genetically determined but synthesized presumably by nonenzymatic slow processes throughout the life span of the erythrocyte. Widness *et al.* (1978) employed this new method for evaluating the long-term glucose control of individuals with diabetes mellitus for the evaluation of pregnant diabetic women. They studied the relationship between Hb A_1C values during the third trimester of pregnancy and fetal size at birth. Glucose tolerance in subjects not insulin dependent was also examined relative to Hb A_1C and fetal size at birth. When corrected for gestational age, birth weights correlated in a significant linear regression with Hb A_1C. Third trimester maternal glucose tolerance (Kt) of women not insulin dependent correlated significantly with both Hb A_1C and birth weight for gestational age. Thus, Hb A_1C may be utilized to predict fetal size in pregnant diabetic women. Unlike other criteria of glucose intolerance, which may revert to normal immediately following delivery, the glycohemoglobin level may remain elevated for several weeks. Hence, postpartum mothers of stillborn, newborns with congenital malformations, or large-for-gestational-age infants, might be better evaluated for unsuspected gestational diabetes by this method (Widness *et al.*, 1978).

F. Delivery

Timing of delivery is one of the most important determinants of the outcome of the diabetic pregnancy. As term approaches there is an exponential rise in the number of stillborns. On the other hand, the neonatal death rate decreases exponentially the closer delivery is to term. If the infant is delivered preterm, he is exposed to the risks associated with prematurity, particularly hyaline membrane disease. Balancing risks of both prematurity and stillbirths traditionally resulted in the optimal time of delivery being 36–37 weeks gestation with allowances for different classes. However, present management is to assess each case on an individual basis, delaying delivery until maturity if fetal jeopardy remains small. In combination with a strict regulation of blood glucose throughout pregnancy, this approach has lowered preinatal mortality (Karlson and Kjellmer, 1972). Weiss and co-authors (1978) observed that it was possible to assess the fetal metabolic situation from the insulin content of the amniotic fluid. High and rising insulin levels at an early stage (26–28 weeks) may indicate fetal morbidity. Early adjustment may then be made in the fetus, thus promoting the development of a healthy child *in utero*.

These recent technological advance in assessment of fetal maturity and well being provide us with the means for optimal management of a pregnancy complicated by diabetes mellitus and ensure the best chances for a successful outcome.

The general obstetrical approach is to assess simultaneously fetal status, specifically placental function, and fetal maturity. If fetal status indicates no immediate risk of intrauterine death the pregnancy is allowed to continue until the fetus shows mature lungs. Fetal well being can most reliably be assessed by observing the fetal heart rate response to either spontaneous uterine contractions (nonstress test) or to oxytocin-induced contractions (stress test; oxytocin challenge test). In nonstress tests fetal well being is assured if the fetal heart rate accelerates in response to a contraction (reactive nonstress test). Absence of acceleration (nonreactive nonstress test) indicates that either the fetus is asleep or it is too ill to respond normally. Fetal status is likewise considered in danger if characteristic patterns of late deceleration occur in response to spontaneous or oxytocin-induced contractions (positive stress test or oxytocin challenge test). Negative nonstress or stress tests assures fetal well being, usually at least 1 week, unless maternal metabolic changes occur. Thus, often one need not interrupt pregnancy at 36–37 weeks, instead allowing continued fetal maturity until possibly 40 weeks. By contrast, a nonreactive nonstress test or positive oxytocin challenge requires intervention unless fetal lung maturity is not assured (immature lechithin/sphingomyelin ratio). In the latter instance, some investigators may delay delivery unless urinary excretion of estriols is also falling. Various management schemes have been proposed (Gabbe *et al.*, 1977; Gabbe, 1978).

V. MATERNAL AND FETAL COMPLICATIONS

A. Complications in Diabetic Pregnancy

In general, pregnancy complications in the presence of diabetes mellitus are similar to those in the nondiabetic population but occur with higher frequency and are most often consequences of maternal vascular complications. Obviously, these complications have an unequivocal effect on fetal and neonatal fate.

Preeclampsia, the diagnosis of which is based on elevated blood pressure, proteinuria, and edema or excessive weight gain, occurs in about 25% of pregnant diabetic women (Kyle, 1963; Pedersen, 1968). The incidence of eclampsia in diabetics may be estimated to be approximately 0.2% (Pedersen, 1968). Longstanding diabetes and renal lesions often lead to the development of preeclampsia and in particular to the occurrence of severe cases (Pedersen, 1968).

Polyhydramnios is also more common in diabetic pregnancies. The reported incidence ranges between 3% and 31% with an average of 19% (Kyle, 1963). Preeclampsia and polyhydramnios frequently coexist. The high frequency is probably related to the diabetic vascular changes.

Urinary tract infection and pyelonephritis are found more frequently in diabetic pregnant women (Pedersen, 1968). Screening for asymptomatic leucocyturia and bacteriuria in pregnant diabetics should be carried out carefully.

Obstetrical difficulties are more common in diabetics than in normal women. Frequently, babies are large and produce mechanical difficulties at delivery. The incidence of inductions as well as cesarean sections is high in diabetic pregnancies. After the discovery of insulin, maternal mortality decreased to low levels. However, the immediate favorable influence of the insulin era on the course of pregnancy in diabetes and on diabetes per se was more significant in respect to maternal experience than it was to fetal. Nowadays, maternal mortality occurring among diabetic mothers is usually due to vascular complications and is only slightly higher than among normal mothers.

As pregnancy extends beyond the 36th week, the incidence of intrauterine deaths increases markedly unless managed properly. The cause of stillbirth has not yet been elucidated; however, in some cases poor diabetic control and episodes of ketoacidosis may be involved. Placental insufficiency due to vascular changes as well as fetal hypoglycemia secondary to excessive insulin secretion have also been implicated (Burstein *et al.*, 1957; Oakley *et al.*, 1972). Women in the White diabetic classes E and F, as well as those with ketoacidotic episodes, toxemia, infections, and a history of a previous stillbirth have the least favorable prognosis for fetal survival.

B. Effects of Pregnancy on Diabetic Vascular Complications

In general, the gravid state does not add any great hazards to the long-term effects of diabetes, and aggravation of pre-existing vascular disease has been reported to occur in only about 3% of diabetic pregnancies (Pedersen, 1968). However, it is very difficult to ascribe the worsening in these vascular changes to the pregnancy as such.

Many diabetic women have retinopathy prior to gestation. Physicians are often asked for advice about planning or continuing pregnancy in the presence of this complication. Beetham (1950) suggested that patients presenting with severe proliferative retinopathy should not be permitted to undertake pregnancy. In contrast, Cassar *et al.* (1978) more recently expressed the opinion that although diabetic retinopathy may progress during pregnancy and needs careful supervision the prognosis is reasonable and has been improved by the advent of photocoagulation.

Glomerulosclerosis of varying degrees develops in subjects with longstanding diabetes. Thus, diabetic women with nephropathy of various degrees of severity pose a significant therapeutic problem (Pedersen, 1968). Preeclampsia is a frequent complication in such women. There is no proof, however, that pregnancy adversely affects the progression of the diabetic nephropathy (Pedersen, 1968).

C. Placental Changes in Diabetes

There is no doubt that the changes taking place in the placenta are of particular importance in the diabetic pregnancy. Considerable efforts have, therefore, been devoted to the elucidation of their role in fetal wastage.

The weight of the placenta of the diabetic woman is increased with oversized neonates and reduced with small infants; the umbilical cords associated with the overweight placentae are characteristically thick and succulent (Pedersen, 1968). Froelich and Fujikura (1966) reported the occurrence of a single umbilical artery in 6.4% of diabetic pregnancies, a figure five times greater than that of a control group. Under light microscopy the histology of the diabetic placenta is not remarkable (Cornblath, 1976). However, electron microscopy (Okudaira *et al.*, 1966) shows a significant increase of basement membrane material in terminal chorionic villi, involving both the epithelial and capillary basement membrane and the stroma. Although these phenomena were seen with greater frequency and severity than in normal placental tissue, as yet no specific morphological changes have been observed in the placenta of the diabetic mother.

Glycogen content has been shown to be increased at term in diabetics as compared to the normal placentas (Gabbe *et al.*, 1972). The ability of the placenta to accumulate glycogen during early pregnancy disappears in later stages (Villee, 1954). The persistence of glycogen in the placental tissue of the diabetic mother may reflect the continuation of an immature condition, or may result from the promotion of glycogen synthesis by the hyperinsulinemia of hyperglycemia (Pedersen, 1968; Gabbe *et al.*, 1972).

VI. CONGENITAL MALFORMATIONS

Congenital abnormalities are disproportionately common in children born to diabetic mothers (Pedersen *et al.*, 1964; Soler *et al.*, 1976; Day and Insley, 1976; Gabbe, 1977; Simpson, 1978). In the last few years several centers have reported a decline in overall perinatal mortality while the role of congenital malformation as the cause of fetal losses is becoming more significant. Obstetric and diabetic management has become more vigilant and longer periods of hospitalization of mothers at risk, especially during the third trimester of pregnancy, has become more common. Therefore, fetuses which may have been aborted in the past are sometimes carried to a viable stage and may now be contributing to the variety and number of fetal congenital malformations. This may explain the significant increase in the proportion of infants with congenital abnormalities born to diabetic mothers (Soler *et al.,* 1976). In several recent reports, anomalies have been responsible for up to 50% of perinatal mortality (Karlsson and Kjellner, 1972; Dundon *et al.,* 1974; Pedersen *et al.,* 1974; Soler *et al.,* 1976; Gabbe, 1977).

Koller (1953) has shown that malformation rate is associated with severity and duration of maternal diabetes and that the frequency of congenital anomalies decreases when good diabetic control is maintained throughout pregnancy, thus linking environmental changes with the teratogenic effects of diabetes. Disruption of the normal homeostatic mechanism and control by insulin administration exposes the fetus to the danger of hyperglycemia, ketoacidosis, hypoglycemia, hormonal disturbances, and other abnormal conditions (Reid, 1970). In addition, there are sufficient instances of alterations to the fine structure and biochemistry of the placenta (Powell *et al.,* 1967) to suggest that abnormal placental metabolism might itself deleteriously affect the fetus. Any of these manifestations of poorly controlled diabetes could have serious effects on the developing fetus. Several studies (Landauer, 1945; Lichenstein *et al.,* 1951; Duraiswami, 1950) have demonstrated increased malformation rate in the offspring of insulin-treated animals. Horii and associates (1966) found that alloxan-treated mice without added insulin had a higher incidence of malformed fetuses than mice with added insulin. In this connection, Reid (1970) has proposed that certain fetal cells may be especially sensitive to exogenous glucose levels; high blood glucose levels may switch sensitive cells to a glycolytic pattern of metabolism. Reversion to normal blood glucose levels could further widen the energy gap by reducing the rate of glycolysis before restoration of aerobic metabolism. Minute aberrations in carbohydrate metabolism during the first weeks of embryonic differentiation and organogenesis may be very critical (Taylor, 1954). Infants of diabetic men have no increased incidence of congenital anomalies (Rubin, 1958), nor does glucose tolerance indicate a statistically significant alteration in fathers of infants with congential malformations (Goldman, 1974). These observations lend support to the importance of environmental rather than the genetic determinants. However, heredity and environment could interact in varying degrees, rather than serving as mutually exclusive teratogenic factors.

The overall incidence of congenital malformations in infants of the diabetic mother is two- to fourfold greater than in infants born to normal women (Breidahl, 1966; Hubbell *et al.,* 1965; Soler *et al.,* 1976). The anomalies involve mainly the cardiovascular and skeletal systems but spinal, genitourinary, gastrointestinal central nervous system, and many more systems have been implicated.

The syndrome of phocomelic diabetic embryopathy or caudal dysplasia may be the congenital malformation most specific for diabetes mellitus (Gabbe, 1977). This has been established as a clinical entity occurring in the offspring of diabetic and, rarely, prediabetic mothers. This abnormality occurs in approximately 1% of infants of diabetic mothers; however, only 16% of infants with this syndrome are born to diabetic women (Passarge and Lenz, 1966).

Cardiovascular malformations were over-represented in several series (Soler *et al.*, 1976; Day and Insley, 1976). A review (Rowland *et al.*, 1973) of 470 infants of diabetic mothers followed at the Joslin Clinic, Boston, revealed a 4% incidence of proved congenital heart disease, five times the number expected to appear in the general population. Transposition of the great arteries, ventricular septal defect, and coarctation of the aorta together comprised over one-half of the 19 cases reported (Rowland *et al.*, 1973). No relationship between gestational factors and incidence of heart defects was observed. The newborn infant of a diabetic mother frequently mimicks the signs and symptoms of congenital heart disease (when, indeed, no heart disease exists). Although no single specific clue exists to identify congenital heart disease in infants of diabetic mothers, the abnormal electrocardiogram, presence of cyanosis unresponsive to administered oxygen without radiographic evidence of pulmonary disease, or significant murmur should alert the attending physician to the likelihood of heart disease (Rowland *et al.*, 1973). The "normal" cardiomegaly of the offspring of the diabetic woman (Neave, 1967) compounds the problem. Heart weight at post-mortem examination has been found to exceed two standard deviations above the normal in as many as one-half of infants of diabetic mothers without congenital heart disease (Miller, 1945). Earlier reports of increased glycogen deposition (Hurwitz and Irwing, 1937) have not been substantiated (Villee, 1954). Naeye (1965) described an increase in the mass of both myocardial nuclei and fiber sarcoplasm, suggesting an anabolic effect of hyperinsulinism.

Our understanding of the etiology of congenital malformations in infants of diabetic mothers is very limited. Exclusive of women with the mildest form of diabetes (White's class A), about 10% of pregnant diabetic women will have an infant with congenital malformation and in about 7% the abnormality will be major (Soler *et al.*, 1976). Malformations may precede the onset of overt diabetes mellitus and could result from subtle changes in maternal metabolism (Navarrete *et al.*, 1967). However, diabetic vascular complications would certainly increase the risk. Thus, in order to reduce congenital malformations, Molsted-Pedersen *et al.* (1964) discourage women in White's classification D-F from becoming pregnant. Others would terminate pregnancies in such women (Soler *et al.*, 1976). However, management suggestions should be individualized.

In summary, only by careful control of diabetes during pregnancy can there be hope of achieving improved fetal life and a neonate with brighter prospects for viability (Roversi *et al.*, 1973; Gabbe, 1977).

VII. NEONATAL COMPLICATIONS

The neonate of the diabetic mother is a product of abnormal intrauterine environment resulting from the altered maternal homeostasis.

A. General Features

Infants of mothers with mild to moderate diabetes are generally heavier and longer than infants of normal mothers and have plethoric and cushingoid face. This trend is observed from about the 32nd week of gestation. Increased total body fat and enlargement of the viscera are evident. The placenta as well as the umbilical cord are also significantly larger than in control neonates. Despite their size, which is usually "large for gestational age," infants of diabetic mothers are physiologically immature. The development of ossification centers does not correspond to weight and length but to gestational age or less (Osler and Pedersen, 1960). In the presence of severe vascular complications, infants of diabetic mothers may be "small for gestational age." These neonates are at a greater risk than the large obese infants. A combination of fetal hyperglycemia and hyperinsulinemia secondary to maternal hyperglycemia have long been implicated in the diabetic fetal macrosomia. These biochemical changes cause increased hepatic glucose uptake and glycogen synthesis, enhanced lipogenesis and accelerated protein synthesis, thus leading to excessive tissue formation and may be the underlying pathophysiological mechanism explaining neonatal gigantism. However, in response to separation from maternal circulation and energy reserves at birth, these infants mobilize fuels such as glucose and free fatty acids more slowly than normal infants (Adams, 1971). Careful antenatal management of diabetic mothers tends to help minimize excessive intrauterine weight gain as well as the general characteristic appearance of their offspring (Dundon *et al.*, 1974). However, correlations between maternal blood sugar and the infant birth weight have been found to be imperfect (Karlsson and Kjellmer, 1972). Therefore, to evaluate whether contributions to increased birth weight could arise via maternal fuels other than glucose, Metzger and associates (1976) studied plasma amino acids in gestational diabetes. After overnight fast during late pregnancy, circulating amino acids were found to be elevated in a fashion that seems to correlate with subsequent neonatal overweight. This antepartum surfeit of amino acid building blocks was suggested by these authors to be an important contributing factor in fetal macrosomia.

As far as their behaviour is concerned, during the first day of life infants of diabetic mothers may be lethargic, hypotonic, and have a poor sucking reflex.

Growth hormone has been implicated in the excessive growth of these infants. However, normal (Joassin *et al.*, 1967; Spellacy *et al.*, 1973) or diminished

(Laron *et al.*, 1967; Poonai *et al.*, 1975) values have been reported in neonates born to diabetic or gestational diabetic women. Hyperadrenocorticism has also been postulated as a contributory factor in gigantism and cushingoid appearance, yet no difference in cortisol production was observed between infants of diabetic mothers and controls (Aarskog, 1963).

B. Hypoglycemia

Hypoglycemia develops in infants of diabetic mothers in the first few hours of life more frequently than in infants of comparable gestation or weight born to normal mothers. This liability is seen mainly in neonates born to mothers requiring insulin (McCann *et al.*, 1966; Isles *et al.*, 1968; King *et al.*, 1969). Even in the presence of strict metabolic control during gestation, a greater decrease of blood glucose concentration in the immediate postnatal period is observed in these newborns (Pedersen *et al.*, 1974). This severe neonatal complication has been attributed by Pedersen and associates (1954) to maternal and fetal hyperglycemia during gestation. The excess glucose in the fetal circulation stimulates islet cell insulin release, leading to fecal and neonatal hyperinsulinism. Indeed, pathological findings in human fetuses of diabetic mothers include hypertrophy and hyperplasia of the pancreatic islets with disproportionate increase in the percentage of β cells (Van Assche and Gepts, 1971). Hypoglycemia during the first hours after birth probably results from a sudden interruption of glucose supply from the maternal circulation while neonatal hyperinsulinism still prevails. Because of the placental transfer of anti-insulin antibodies which interfere with the radioimmunoassay of insulin, it is difficult to document accurately neonatal hyperinsulinemia in infants of insulin-dependent diabetic mothers. The existence of hyperinsulinism could, however, be inferred from high immunoreactive insulin levels in IGDM or IDM who have had little or no insulin antibodies (Joassin *et al.*, 1967) and elevated free C-peptide levels (Sosenko *et al.*, 1978) as well as faster glucose disappearance rate (Kt) in neonates of insulin dependent diabetic mothers (Molsted-Pedersen, 1972). Lower free fatty acid levels and d-β- hydroxybutyrate determined in the first hour of life seems to further support the possibility of the presence of neonatal hyperinsulinemia (Persson *et al.*, 1973). The infants of diabetic mothers fail to stimulate pancreatic glucagon release at 2 hours of age or to increase urinary catecholamine excretion in the first 24 hours of life in response to spontaneous hypoglycemia (Bloom and Johnston, 1972; Stern *et al.*, 1968). Thus, the combined effect of increased insulin and diminished glucagon and catecholamine response might result in decreased hepatic glucose output both from glycogenolysis and gluconeogenesis (Kalhan *et al.*, 1977).

Blood glucose concentrations in infants of insulin-dependent diabetic mothers decline immediately following birth. In most infants the nadir is reached 1–2 hours after birth. These neonates seem to tolerate profound and prolonged hypoglycemia surprisingly well, and symptoms are observed only in 20–25%.

Most of these infants recover from their hypoglycemia during the first few hours of life, particularly if frequent feedings are given. No specific therapy is recommended for the asymptomatic infants. However, if blood glucose concentrations are excessively low (under 20 mg/100 ml for more than 1 hour), circulating glucose levels should be supported by glucose infusions (Cornblath and Schwartz, 1976). In symptomatic neonates correction of hypoglycemia can be achieved with a bolus injection of 25% glucose solution (0.5–1.0 g of glucose/kg of body weight over 2–4 minutes) and maintained with 10–15% solution (10 mg glucose/kg/minute). Glucose concentration in the infusate should be tapered gradually in order to reduce the prevailing hyperinsulinemia and to prevent reactive hypoglycemia after intravenous fluids are discontinued. If hypoglycemia recurs or is not corrected in 6 hours, hydrocortisone (5 mg/kg/day) should be started (Pildes, 1979).

C. Hyaline Membrane Disease

Respiratory distress syndrome due to hyaline membrane disease (HMD) is one of the most serious complications and a major cause of mortality among liveborn neonates of diabetic mothers. Robert and associates (1976) have shown that the incidence of HMD is five to six times greater in newborns of diabetic mothers even when gestational age is taken into consideration and regardless of mode of delivery. A delay in the ability of the fetal lung to synthesize lecithin, which is the major surface active material lining the alveoli, has been demonstrated in class A to class C diabetic pregnancies (Gluck and Kulovich, 1973). This delay in pulmonary maturation is the underlying mechanism for alveolar collapse and development of HMD. The synthesis of lecithin is enhanced by fetal glucocorticoids which are secreted during late gestation. Insulin has been shown to antagonize the production of this surfacant. Smith *et al.* (1975) examined lecithin synthesis in rabbit fetal lungs during late gestation and found that addition of insulin to the medium abolished the stimulatory effect of cortisol on this process. Delayed maturation of the lecithin/sphingomyelin (L/S) ratio in diabetic pregnancies has been reported by several groups (Polyshuck *et al.*, 1973; Yared *et al.*, 1973; Whitefield *et al.*, 1973). Thus, the fetal hyperinsulinism which characterizes this condition, may play an important role in the pathogenesis of HMD.

Lecithin appears earlier than normal in the amniotic fluid of severe diabetics with vascular complications (Gluck and Kulovich, 1973). A chest X-ray is necessary to confirm the diagnosis of HMD. A diffuse reticulogranular pattern and air bronchogram are characteristic findings of HMD.

D. Neonatal Free C-Peptide as Related to Complications

Fetal hyperinsulinism, which is secondary to maternal and fetal hyperglycemia, is postulated to be related to three major abnormalities in the infant of the diabetic mother (IDM): macrosomia, postnatal hypoglycemia, and hyaline membrane disease (HMD). To test these associations Sosenko *et al.* (1978) measured free C-peptide in cord serum of infants of gestational and insulin-requiring diabetics. This peptide is the connecting segment of the proinsulin molecule. When proinsulin is converted to insulin, the C-peptide and insulin are released in equimolar concentrations. The former is unaffected by maternal antibodies. Thus, free C-peptide levels measure endogenous insulin secretion independent of interfering antibodies. Hyperinsulinism in the IDM, as reflected by free C-peptide level, was shown to be directly related to macrosomia and postnatal hypoglycemia. The lower than expected incidence of HMD in this series precluded a definitive evaluation of the relationship or hyperinsulinism to HMD in the IDM.

E. Hypocalcemia

Hypocalcemia is considered to be present when to consecutive serum calcium levels are less than 8 mg/100 ml in full-term infants and less than 7 mg/100 ml in preterm infants (Tsang and Oh, 1970). Infants of diabetic mothers have been shown (Tsang *et al.*, 1972) to have a tendency to neonatal hypocalcemia even when gestational age and perinatal complicating factors are taken into consideration. Evidence that the maternal diabetic state plays a role in the pathogenesis of neonatal hypocalcemia includes the marked changes in serum calcium metabolism in infants of diabetic mothers, the correlation of serum calcium levels with severity of diabetes, and the significantly higher serum calcium levels in diabetic pregnant women as compared to nondiabetic controls (Tsang *et al.*, 1972). The lower serum calcium levels in the infants of diabetic mothers were associated with higher serum phosphate levels. The hyperphosphatemia usually precedes hypocalcemia by 24 hours, with a significant negative correlation between serum phosphorus and calcium levels. Tsang *et al.* (1972) speculated that relative maternal hyperparathyroidism may be a factor in the pathogenesis of neonatal hypocalcemia in infants of diabetic mothers. In a later study (Tsang *et al.*, 1975) serum parathyroid hormone, total and ionized calcium and phosphorus were determined serially from birth to 96 hours of age in infants and in their respective diabetic mothers at the time of delivery. The data were consistent with the thesis that functional hypoparathyroidism occurs in IDM. In spite of significant decreases in total and ionized Ca and rises in serum P, there was no significant increase in serum PTH levels from birth to 24–48 hours of age in infants of insulin-dependent diabetic mothers. Higher concentrations of ionized

Ca in maternal serum and cord blood were related to greater postnatal falls in ionized Ca and decreased parathyroid function in IDM. However, increased maternal serum PTH levels were not demonstrated in diabetic mothers. This may be explained by the relative insensitivity of the PTH radioimmunoassay. At any rate, it seems that the increased concentrations of calcium in maternal and fetal circulation play an important role in the suppression of the fetal parathyroid *in utero* which is the underlying mechanism of the neonatal hypocalcemia.

Symptomatic infants with hypocalcemia should be treated with intravenous calcium gluconate. A bolus (5–10 ml of 10% solution) is given rapidly under ECG tracing, and then a maintenance therapy 300–500 mg/kg/24 hours) is continued as long as the neonate cannot tolerate oral intake (Pildes, 1979).

Hypomagnesemia (serum < 1.5 mg/100 ml) may also be present in infants of diabetic mothers. Its presence should be considered in the presence of symptoms of neuromuscular excitability or whenever hypocalcemia does not respond to the regular treatment. The frequency and severity of the hypomagnesemia have been shown to correlate with the severity of maternal diabetes.

F. Effects of Maternal Ketoacidosis on IQ of Offspring

Churchill and Berendes (1969) and Churchill *et al.* (1969) have shown that diabetic ketoacidosis during pregnancy is associated with lower IQ of the offspring. The presence of acetonuria during gestation was a critical variable influencing intellectual status of the infant of a diabetic mother at 4 years of age. Intellectual impairment was more often present in offspring whose mothers had displayed acetonuria. The IQ of that sample was found to be unrelated to White's classification, gestational age, dosage of insulin received by the mother during pregnancy, or insulin reactions.

Stehbens *et al.* (1977) conducted a prospective study of infants born to women with diabetes mellitus. The children were examined at birth and followed at 1, 3, and 5 years of age. There was an adverse effect of ketoacidosis on intellectual status of the offspring, and the children born to acetone-positive diabetic mothers had a lower IQ at 5 years of age.

This finding could be explained by a direct toxic effect of ketoacids or by a disturbance in acid–base balance induced by the presence of ketoacids. On the other hand, it may be that the accumulation of ketobodies in the urine merely signifies an episode during which the nutrient supply of the mother has been reduced to levels insufficient for the needs of the fetus. In clinical practice, closer attention should be paid to ketonuria as an indicator of suboptimal fetal environment. A deficiency of nutrients late in pregnancy would be expected to have greater impact than early shortages, when fetal needs could be met by relatively small amounts of critical substances. A greater nutrient supply is, therefore, required for normal growth and development during the more advanced stages of pregnancy.

G. Other Neonatal Complications

In addition to the above mentioned neonatal abnormalities the following complications should be noted.

1. A high incidence of congestive heart failure even in the absence of cardiac anomalies. This condition seems to be related to poor control of diabetes (Pildes, 1973).
2. Hyperbilirubinemia, the underlying mechanism of which has not yet been elucidated, occurs more commonly among infants of diabetic mothers. Despite the frequent coexistence of predisposing factors such as hypoxia, acidosis, and hypoglycemia, kernicterus is a rare complication (Cornblath and Schwartz, 1976).
3. A tendency to neonatal polycythemia which is associated with enhanced hepatic erythropoiesis. Schwartz *et al.* (1976) have suggested that this might be a response to fetal hypoxia.
4. A higher incidence of neonatal renal vein thrombosis (Takeuchi and Benirschke, 1961). The pathogenesis is not clearly understood, although polycythemia and local stasis secondary to osmotic diuresis due to fetal hyperglycemia have been mentioned as contributory factors.
5. Immaturity of the central nervous system, as suggested by electroencephalographic studies (Schulte *et al.*, 1969).
6. A more frequent incidence of bacterial infections, usually staphylococcal, among infants of diabetic mothers. In recent years these have decreased, however, probably because of the availability of more effective control measures (Farquhar, 1976).

H. Long-Term Growth and Development

There is some controversy as to whether children of diabetic mothers differ from their normal peers in height and weight as they mature. White *et al.* (1953) found that they have a tendency to be taller than the average population. Weitz and Laron (1976) support this observation, while stating that these parameters are not related to obesity or a state of hyperinsulinism. In contrast, Hagbard *et al.* (1959) and Farquhar (1969) reported that children of diabetic mothers tended to be shorter and heavier than normal children. Breidahl (1966) also found that children of diabetic mothers were heavier than children of healthy mothers. There is no satisfactory explanation for the discrepancy in these studies.

Whereas the incidence of mental deficiency among children of diabetic mothers was reported by François *et al.* (1974) to be approximately 5%, less favorable findings were reported by Yssing (1975) as well as by Haworth and associates (1976). Definitive conclusions regarding the prognosis for psychomotor development must await detailed and well-controlled prospective studies.

Recent and on-going advances in improved maternal and neonatal management encourage the outlook for reduced perinatal morbidity and mortality and their long-term complications.

REFERENCES

Aarskog, D. (1963). Cortisol production rate in newborn infants of diabetic mothers. *J. Pediatr.* **62,** 807–814.

Abell, D. A., Beischer, N. A., Papas, A. J., and Willis, M. M. (1976). The association between abnormal glucose tolerance (hyperglycemia and hypoglycemia) and estriol excretion in pregnancy. *Am. J. Obstet. Gynecol.* **124,** 388–392.

Adam, P. A. J. (1971). Control of glucose metabolism in the human fetus and newborn infant. *Adv. Metab. Dis.* **5,** 183–275.

Barbosa, J., Noreen, H., Emme, L., *et al.* (1976). Histocompatibility (HLA) antigens and diabetic microangiopathy. *Tissue Antigens* **7,** 233–237.

Beard, R. W., and Oakley, N. W. (1976). The fetus of the diabetic. *In* "Fetal Physiology and Medicine. The Basis of Perinatology" (R. W. Beard and P. W. Nathanielsz, eds.), pp. 137–157. Saunders, Philadelphia, Pennsylvania.

Beetham, W. P. (1950). Diabetic retinopathy in pregnancy. *Trans. Am. Ophthalmol. Soc.* **48,** 205–216.

Biggs, J. S. G. (1975). Progress in fetal assessment. *Obstet. Gynecol.* **45,** 227–233.

Bloom, S. R., and Johnston, D. I. (1972). Failure of glucagon release in infants of diabetic mothers. *Br. Med. J.* **4,** 453–454.

Bookchin, R. M., and Gallop, P. M. (1968). Structure of hemoglobin A_{Ic}: Nature of the N-terminal β-chain blocking group. *Biochem. Biophys. Res. Commun.* **32,** 86–93.

Breidahl, H. D. (1966). The growth and development of children born to mothers with diabetes. *Med. J. Aust.* **1,** 268–270.

Brundenell, M., and Beard, R. (1972). Diabetes in pregnancy. *Clin. Endocrinol. Metab.* **1,** 673–695.

Bunn, H. F., Haney, D. N., Kamin, S., Gabbay, K. H., and Gallop, P. M. (1976). The biosynthesis of human hemoglobin A_{Ic}: Slow glycosylation of hemoglobin in vivo. *J. Clin. Invest.* **57,** 1652–1659.

Burstein, R., Soule, S. D., and Blumenthal, H. T. (1957). Histogenesis of pathological processes in placentas of metabolic disease in pregnancy. The diabetic state. *Am. J. Obstet. Gynecol.* **74,** 96–104.

Cassar, J., Kohner, E. M., Hamilton, A. M., Gordon, H., and Jopkin, G. F. (1978). Diabetic retinopathy and pregnancy. *Diabetologia* **15,** 105–111.

Churchill, J. A., and Berendes, H. W. (1969). Intelligence of children whose mothers had acentonuria during pregnancy. *Sci. Publ., Pan Am. Health Organ.* **185,** 30–35.

Churchill, J. A., Berendes, H. W., and Nemore, J. (1969). Neuropsychological deficits in children of diabetic mothers: A report from the collaborative study of cerebral palsy. *Am. J. Obstet. Gynecol.* **105**, 257–268.

Cornblath, M., and Schwartz, R., eds. (1976). "Disorders of Carbohydrate Metabolism in Infancy," pp. 115–154. Saunders, Philadelphia, Pennsylvania.

Creutzfeldt, W., Kobberling, J., and Neel, J. V., eds. (1976). "The Genetics of Diabetes Mellitus." Springer-Verlag, Berlin and New York.

Cudworth, A. G., and Woodrow, J. C. (1975a). HL-A system and diabetes mellitus. *Diabetes* **24**, 345–349.

Cudworth, A. G., and Woodrow, J. C. (1975b). Evidence for HL-A linked genes in "juvenile" diabetes mellitus. *Br. Med. J.* **3**, 133–135.

Daweke, H., and Gütter, K. A. (1970). Diabetes and Graviditat. *Verh. Dtsch. Ges. Inn. Med.* **76**, 341–358.

Day, R. E., and Insley, J. (1976). Maternal diabetes mellitus and congenital malformation. Survey of 205 cases. *Arch. Dis. Child.* **51**, 935–938.

Duhring, J. L. (1977). Diabetes in pregnancy: How to diagnose and treat it. *Contemp. Obstet. Gynecol.* **9**, 117–120.

Dundon, S., Murphy, A., Raftery, I., and Drury, M. I. (1974). Infants of diabetic mothers. *J. Ir. Med. Assoc.* **67**, 371–375.

Duraiswami, P. K. (1950). Insulin-induced skeletal abnormalities in developing chickens. *Br. Med. J.* **2**, 384–390.

Fajans, S. S., Floyd, J. C., Jr., Tattersall, R. B., *et al.* (1976). The various faces of diabetes in the young. *Arch. Intern. Med.* **136**, 194–202.

Farquhar, J. W. (1969). Prognosis for babies born to diabetic mothers in Edinburgh. *Arch. Dis. Child.* **44**, 36–47.

Farquhar, J. W. (1976). The infant of the diabetic mother. *Clin. Endocrinol. Metab.* **5**, 237–264.

Felig, P. (1977). Body fuel metabolism and diabetes mellitus in pregnancy. *Med. Clin. North Am.* **61**, 43–66.

François, R., Picaud, J.J., Ruitton-Ugliengo, A. *et al.* (1974). The newborn of diabetic mothers. *Biol. Neonate* **24**, 1–34.

Freinkel, N., Metzger, B. E., Nitzan, M., Hare, J. W., Shambaugh, G. E., Marshall, R. T., Surmaczynska, B. Z., and Nagel, T. C. (1972). Accelerated starvation and mechanisms for the conservation of maternal nitrogen during pregnancy. *Isr. J. Med. Sci.* **8**, 426–439.

Froelich, L. A., and Fujikura, T. (1966). Significance of a single umbilical artery. Report from the collaborative study of cerebral palsy. *Am. J. Obstet. Gynecol.* **94**, 274–279.

Fuhrmann, W. (1976). Problems in genetic counselling for diabetes mellitus. *In* "The Genetics of Diabetes Mellitus" (W. Creutzfeldt, J. Kobberling, and J. V. Neel, eds.), pp. 234–239. Springer-Verlag, Berlin and New York.

Gabbe, S. G. (1977). Congenital malformations in infants of diabetic mothers. *Obstet. & Gynecol. Surv.* **32**, 125–132.

Gabbe, S. G. (1978). Application of scientific rationale to the management of the pregnant diabetic. *Semin. Perinatol.* **2**, 361–371.

Gabbe, S. G., Demmer, L. M., Greep, R. O., and Villee, C. A. (1972). Placental glycogen metabolism in diabetes mellitus. *Diabetes* **21**, 1185–1191.

Gabbe, S. G., Mestman, J. H., and Freeman, R. K. (1977). Management and outcome of diabetes mellitus, classes B-R. *Am. J. Obstet. Gynecol.* **129**, 723–732.

Ganda, O. P., and Soeldner, S. S. (1977). Genetic, acquired and related factors in the etiology of diabetes mellitus. *Arch. Intern. Med.* **137**, 461–469.

Gillmer, M. D. G., Beard, R. W., Brooke, F. M., and Oakly, N. W. (1975). Carbohydrate metabolism in pregnancy. I. Diurnal plasma glucose profile in normal and diabetic women. *Br. Med. J.* **3**, 339–402.

Gluck, L., and Kulovich, M. V. (1973). Lecithin/sphingomyelin ratios in normal and abnormal pregnancy. *Am. J. Obstet. Gynecol.* **115,** 539–544.

Goldman, J. (1974). Carbohydrate metabolism in fathers of babies with congenital malformations. *Isr. J. Med. Sci.* **10,** 698–701.

Goldstein, S., Littlefield, J. W., and Soeldner, J. S. (1969). Diabetes mellitus and aging: Diminished planting efficiency of cultured human fibroblasts. *Proc. Natl. Acad. Sci. U.S.A.* **64,** 155–160.

Hagbard, L., Olow, I., and Reinand, T. (1959). A follow-up study of 514 children of diabetic mothers. *Acta Paediatr. Scand.* **48,** 184–197.

Hamlin, C. R., Kohn, R. R., and Luschin, J. H. (1975). Apparent accelerated aging of human collagen in diabetes mellitus. *Diabetes* **24,** 902–904.

Hare, J. W., and White, P. (1977). Pregnancy in diabetes complicated by vascular disease. *Diabetes* **26,** 953–955.

Haworth, J. C. (1974). Neonatal hypoglycemia: How much does it damage the brain? *Pediatrics* **54,** 3–4.

Haworth, J. C., McRae, K. N., and Dilling, L. A. (1976). Prognosis of infants of diabetic mothers in relation to neonatal hypoglycemia. *Dev. Med. Child Neurol.* **18,** 471–479.

Horii, K. I., Watanabe, G. I., and Ingalls, T. H. (1966). Experimental diabetes in pregnant mice: Prevention of congenital malformations in offspring by insulin. *Diabetes* **15,** 194–204.

Hubell, J. P., Muirhead, D. M., and Drorbaugh, J. E. (1965). The newborn infant of the diabetic mother. *Med. Clin. North Am.* **49,** 1035–1052.

Hurwitz, D., and Irving, F. C. (1937). Diabetes and pregnancy. *Am. J. Med. Sci.* **194,** 85–92.

Hytten, F. E., and Thomson, M. A. (1970). Maternal physiological adjustments. *In* "Committee on the Maternal Nutrition and the Course of Pregnancy." Natl. Acad. Sci., Washington, D.C.

Isles, T. E., Dickson, M., and Farquhar, J. W. (1968). Glucose tolerance and plasma insulin in newborn infants of normal and diabetic mothers. *Pediatr. Res.* **2,** 198–208.

Joassin, G., Parker, M. L., Pildes, R. S., and Cornblath, M. (1967). Infants of diabetic mothers. *Diabetes* **16,** 306–311.

Kalhan, S. C., Savin, S. M., and Adam, P. A. J. (1977). Attenuated glucose production rate in newborn infants of insulin-dependent diabetic mothers. *N. Engl. J. Med.* **296,** 375–376.

Karlsson, K., and Kjellmer, I. (1972). The outcome of diabetic pregnancies in relation to the mother's blood sugar level. *Am. J. Obstet. Gynecol.* **112,** 213–220.

Khojandi, M., Tsai, A. Y. M., and Tyson, J. E. (1974). Gestational diabetes: The dilemma of delivery. *Obstet. Gynecol.* **43,** 1–6.

King, K. C., Adam, P. A. J., Clemente, G. A., and Schwartz, R. (1969). Infants of diabetic mothers: Attenuated glucose uptake without hyperinsulinemia during continuous glucose infusion. *Pediatrics* **44,** 381–392.

Koenig, R. J., Peterson, C. M., Jones, R. L., Sandek, C., Lehrman, M., and Cerami, A. (1976). Correlation of glucose regulation and hemoglobin A_{1c} in diabetes mellitus. *N. Engl. J. Med.* **295,** 417–420.

Koller, O. (1953). Diabetes and pregnancy. *Acta Obstet. Gynecol. Scand.* **32,** 80–103.

Kyle, G. C. (1963). Diabetes and pregnancy. *Ann. Intern. Med.* **59,** Suppl. 3, 1–83.

Landauer, W. (1945). Rumplessness of chicken embryos produced by the injection of insulin and other chemicals. *J. Exp. Zool.* **98,** 65.

Langenbeck, U., and Jorgensen, G. (1976). The genetics of diabetes mellitus—a review of twin studies. *In* "The Genetics of Diabetes Mellitus" (W. Creutzfeldt, J. Kobberling, and J. V. Neel, eds.), pp. 21–25. Springer-Verlag, Berlin and New York.

Laron, Z., Mannheimer, S., Nitzan, M., and Goldman, J. (1967). Growth hormone, glucose and free fatty acid levels in mother and infant in normal, diabetic toxemic pregnancies. *Arch. Dis. Child.* **42,** 24–28.

Lichenstein, H., Guest, G. H., and Warkany, J. (1951). Abnormalities in offspring of white rats given protamine zinc insulin during pregnancy. *Proc. Soc. Exp. Biol. Med.* **78,** 398–402.

McCann, M. L., Chen, C. H., Katigbak, E. B., Kotchen, J. M., Likly, B. F., and Schwartz, R. (1966). Effects of fructose in hypoglucosemia in infants of diabetic mothers. *N. Engl. J. Med.* **275,** 1-7.

McDevitt, H. O., and Bodmer, W. F. (1974). HL-A immune-response genes and disease. *Lancet* **1,** 1269-1275.

Metzger, B. E., Phelps, R. L., and Freinkel, N. (1976). Correlation of plasma amino-acids with fetal macrosomia in gestational diabetes. *Clin. Res.* **24,** 502A.

Miller, H. C. (1945). Cardiac hypertrophy and extramedullary hematopoiesis in newborn infants of prediabetic mothers. *Am. J. Med. Sci.* **209,** 447-455.

Molsted-Pedersen, L. (1972). Aspects of carbohydrate metabolism in newborn infants of diabetic mothers. *Acta Endocrinol. (Copenhagen)* **69,** 174-189.

Naeye, R. (1965). Infants of diabetic mothers: A quantitative morphologic study. *Pediatrics* **35,** 980-988.

National Commission on Diabetes (1976). "The Long Range Plan to Combat Diabetes," Report to the Congress of the U.S., No. 76-1018. U.S. Department of Health, Education and Welfare, National Institutes of Health, Washington, D.C.

Navarrete, V. N., Torres, I. H., and Rivera, I. R. (1967). Maternal carbohydrate disorder and congenital malformations. *Diabetes* **16,** 127-130.

Neave, C. (1967). Congenital malformations in offspring of diabetics, Doctoral Thesis, Harvard School of Public Health, Cambridge, Massachusetts.

Neel, J. V. (1970). The genetics of diabetes mellitus. *Adv. Metab. Disord., Suppl.* **1,** 3-10.

Nelson, P. G., and Pyke, D. A. (1976). Genetic diabetes not linked to the HLA locus. *Br. Med. J.* **1,** 196-197.

Nerup, J., Platz, P., Anderson, O. O., Christy, M., Lyngsoe, J., Poulsen, J. E., Ryder, L. P., Nielsen, L. S., Thomsen, M., and Svejgaard, A. (1974). HL-A antigens and diabetes mellitus. *Lancet* **2,** 864-866.

Nora, J. J. (1971). Etiologic factors in congenital heart disease. *Pediatr. Clin. North Am.* **18,** 1059-1074.

Oakley, N. W., Beard, R. W., and Turner, R. C. (1972). Effects of sustained maternal hyperglycemia on the fetus in normal and diabetic pregnancies. *Br. Med. J.* **1,** 466-469.

Okudaira, Y., Hirota, K., Cohen, S., and Strauss, L. (1966). Ultrastructure of the human placenta in maternal diabetes mellitus. *Lab. Invest.* **15,** 910-926.

Osler, M., and Pedersen, J. (1960). The body composition of newborn infants of diabetic mothers. *Pediatrics* **26,** 985-992.

O'Sullivan, J. B., Mahan, C. M., Charles, D., and Dandrow, R. V. (1974). Medical treatment of gestational diabetic. *Obstet. Gynecol.* **43,** 817-821.

Passarge, E., and Lenz, W. (1966). Syndrome of caudal regression in infants of diabetic mothers: Observations of further cases. *Pediatrics* **37,** 672-675.

Pedersen, J. (1968). "The Pregnant Diabetic and Her Newborn: Problems and Management." Williams & Wilkins, Baltimore, Maryland.

Pedersen, J., Bojsen-Moller, B., and Poulsen, H. (1954). Blood sugar in newborn infants of diabetic mothers. *Acta Endocrinol. (Copenhagen)* **15,** 33-52.

Pedersen, J., Molsted-Pedersen, L., and Andersen, B. (1974). Assessors of fetal and perinatal mortality in diabetic pregnancy. Analysis of 1332 pregnancies in the Copenhagen series 1946-1972. *Diabetes* **23,** 302-305.

Pedersen, L. M., Tygstrup, I., and Pedersen, J. (1964). Congenital malformations in newborn infants of diabetic women. *Lancet* **1,** 1124-1126.

Persson, B., Gentz, J., and Kellum, M. (1973). Metabolic observations in infants of strictly con-

trolled diabetic mothers: Plasma levels of glucose, FFA, glycerol and D-β-hydroxybutyrate during the first two hours after birth. *Acta Paediatr. Scand.* **62,** 465–473.
Pildes, R. S. (1979). Infants of diabetic mothers. *N. Engl. J. Med.* **289,** 902–904.
Pildes, R. S., Cornblath, M., Warren, I., Page-El, E., di Menza, S., Merritt, D. M., and Peeva, A. (1974). A prospective controlled study of neonatal hypoglycemia. *Pediatrics* **54,** 5–14.
Polyshuck, W. Z., Anteby, S., Bar-On, H., and Stein, Y. (1973). Lecithin/sphingomyelin fluid in diabetic mothers: A warning of respiratory distaess in newborn? *Lancet* **1,** 36–37.
Poonai, A. P. V., Tang, K., and Poonai, P. V. (1975). Relation between glucose, insulin and growth hormone in the fetus during labor and at delivery. *Obstet. Gynecol.* **45,** 155–158.
Powell, E. D. U., Caulfield, J. B., and Field, R. A. (1967). Effects of exogenous insulin and alloxan diabetes on rat placenta. *Diabetes* **16,** 227–231.
Pyke, D. A., and Nelson, P. G. (1976). Diabetes in identical twins. *In* "The Genetics of Diabetes Mellitus" (W. Creutzfeldt, J. Kobberling, and J. V. Neel, eds.), pp. 194–202. Springer-Verlag, Berlin and New York.
Pyke, D. A., and Tattersall, M. B. (1973). Diabetic retinopathy in identical twins. *Diabetes* **22,** 613–618.
Ray, M., Freeman, R., Pine, S., and Hesselgesser, R. (1972). Clinical experience with the oxytocin challenge test. *Am. J. Obstet. Gynecol.* **114,** 1–9.
Reid, R. A. (1970). Diabetes and congenital abnormalities. *Lancet* **1,** 1030–1031.
Renold, A. E., Mintz, D. H., Muller, W. A., and Cahill, G. F. (1978). Diabetes mellitus. *In* "The Metabolic Basis of Inherited Disease" (J. B. Stanbury, J. B. Wyngaarden, and D. S. Fredrickson, eds.), pp. 80–109. McGraw-Hill, New York.
Rimoin, D. L. (1971). Inheritance of diabetes mellitus. *Med. Clin. North Am.* **55,** 807–819.
Robert, M. F., Neff, R. K., Hubbell, J. P., Taeusch, H. W., and Avery, M. E. (1976). Association between maternal diabetes and the respiratory distress syndrome in the newborn. *N. Engl. J. Med.* **294,** 357–360.
Rotter, J. I., and Rimoin, D. L. (1978). Heterogeneity in diabetes mellitus. Update. *Diabetes* **27,** 599–605.
Rotter, J. I., and Rimoin, D. L. (1980). Etiologies of diabetes genetics. *In* "Handbook of Diabetes Mellitus" (in press).
Roversi, G. D., Vittorio, C., Gargiulo, M., and Candiani, G. B. (1973). The intensive care of perinatal risk in pregnant diabetic (136 cases): A new therapeutic scheme for the best control of maternal disease. *J. Perinat. Med.* **1,** 114–124.
Rowland, T. W., Hubell, J. P., and Nadas, A. S. (1973). Congenital heart disease in infant of diabetic mothers. *J. Pediatr.* **83,** 815–820.
Rubin, A. (1958). Studies in human reproduction. II. The influence of diabetes mellitus in men upon reproduction. *Am. J. Obstet. Gynecol.* **76,** 25–29.
Schernthaner, G., Mayr, W. R., Pacher, M., *et al.* (1975). HLA-A8, W15, and T3 in juvenile-onset diabetes mellitus. *Horm. Metab. Res.* **7,** 521–522.
Schulte, F. J., Albert, G., and Michaelis, R. (1969). Brain and behavioral maturation in newborn infants of diabetic mothers. *Neuropaediatrie* **1,** 24–55.
Schwartz, H. C., King, K. C., Schwartz, A. L., Edmunds, D., and Schwartz, R. (1976). Effects of pregnancy on hemoglobin A_{1c} in normal, gestational diabetic and diabetic women. *Diabetes* **25,** 1118–1122.
Simpson, J. L. (1978). Genetics of diabetes mellitus (DM) and anomalies in offspring of diabetic mothers. *Semin. Perinatol.* **2,** 383–394.
Simpson, N. E. (1968). Diabetes in the families of diabetics. *Can. Med. Assoc. J.* **98,** 427–432.
Smith, B. T., Giroud, C. J. P., Robert, M., and Avery, M. E. (1975). Insulin antagonism of cortisol action on lecithin synthesis by cultured fetal lung cells. *J. Pediat.* **87,** 953–955.
Soler, N. G., Walsh, C. H., and Malins, J. M. (1976). Congenital malformations in infants of diabetic mothers. *Q. J. Med.* **45,** 303–313.

Sosenko, I., Kitzmiller, J., Loo, S., Gabbay, K., Blix, P., and Rubenstein, A. (1978). Free C-peptide levels in infants of diabetic mothers: Relationship to hypoglycemia, macrosomia, and hyaline membrane disease. *Pediatr. Res.* **12,** 513.

Spellacy, W. N., Buhi, W. C., Bradley, B., and Holsinger, K. K. (1973). Maternal fetal and amniotic fluid levels of glucose, insulin and growth hormone. *Obstet. Gynecol.* **41,** 323-331.

Spellacy, W. N., Buhi, W. C., and Birk, S. A. (1977). Studies of glucagon secretion in normal and insulin-dependent diabetic pregnancy women. *Gynecol. Invest.* **8,** 45.

Stehbens, J. A., Baker, G. L., and Kitchell, M. (1977). Outcome at ages 1, 3 and 5 years of children born to diabetic women. *Am. J. Obstet. Gynecol.* **127,** 408-413.

Stern, L., Ramos, A., and Leduc, J. (1968). Urinary catecholamine excretion in infants of diabetic mothers. *Pediatrics* **42,** 598-605.

Sutherland, H. W., Fisher, P. M., and Stower, J. M. (1975). Evaluation of maternal carbohydrate metabolism by the intravenous glucose tolerance test. *In* "Early Diabetes in Early Life" (R. A. Camerini-Davalos and H. S. Cole, eds.), pp. 365-377. Academic Press, New York.

Takeuchi, A., and Benirschke, K. (1961). Renal venous thrombosis of the newborn and its relation to maternal diabetes. Reports of 16 cases. *Biol. Neonate* **3,** 237-256.

Tattersall, R. B. (1974). Mild familial diabetes with dominant inheritance. *Q. J. Med.* **43,** 339-357.

Tattersall, R. B., and Fajans, S. S. (1975). Prevalence of diabetes and glucose intolerance in 199 offspring of 37 conjugal diabetic parents. *Diabetes* **24,** 452-462.

Tattersall, R. B., and Pyke, D. A. (1972). Diabetes in identical twins. *Lancet* **2,** 1120-1124.

Taylor, E. S. (1954). In: Panel discussion on diabetes in pregnancy. *Diabetes* **3,** 453-461.

Thomsen, M., Platz, P., Ortved Anderson, O., *et al.* (1975). MLC typing in juvenile diabetes mellitus and idiopathic Addison's disease. *Transplant. Rev.* **22,** 125-147.

Tsang, R. C., and Oh, W. (1970). Neonatal hypocalcemia in low birth weight infants. *Pediatrics* **45,** 773-781.

Tsang, R. C., Kleinman, L. I., Sutherland, J. M., and Light, I. J. (1972). Hypocalcemia in infants of diabetic mothers. *J. Pediatr.* **80,** 384-395.

Tsang, R. C., Chen, I. W., Friedman, M. A., *et al.* (1975). Parathyroid function in infants of diabetic mothers. *J. Pediatr.* **86,** 399-404.

Van Assche, F. A., and Gepts, W. (1971). The cytological composition of the foetal pancreas in normal and pathological conditions. *Diabetologia* **7,** 434-444.

Villee, C. A. (1954). The intermediary metabolism of human fetal tissues. *Cold Spring Harbor Symp. Quant. Biol.* **19,** 186.

Weiss, P. A. M., Lichtenegger, W., Winter, R., and Purstner, P. (1978). Insulin levels in amniotic fluid. Management of pregnancy in diabetes. *Obstet. Gynecol.* **51,** 393-397.

Weitz, R., and Laron, Z. (1976). Height and weight of children born to mothers with diabetes melltius. *Isr. J. Med. Sci.* **12,** 195-198.

White, P., Koshy, P., and Duckers, J. (1953). The management of pregnancy complicating diabetes and of children of diabetic mothers. *Med. Clin. North Am.* **37,** 1481-1496.

Whitefield, C. R., Sproule, W. B., and Brudenell, M. (1973). The amniotic fluid lecithin: Sphingomyelin area ratio (LSAR) in pregnancies complicated by diabetes. *J. Obstet. Gynaecol. Br. Commonw.* **80,** 918-922.

Widness, J. A., Schwartz, H. C., Thompson, D., King, K. C., Kahn, C. B., and Oh, W. (1978). Glycohemoglobin (HbA_{1c}): A predictor of birth weight in infants of diabetic mothers. *J. Pediatr.* **92,** 8-12.

Yared, L. E., McGeary, H. M., and Gaffney, T. J. (1973). The value of amniotic fluid lecithin: sphingomyelin ratio in the management of high risk pregnancy. *Aust. & N. Z. J. Obstet. & Gynecol.* **13,** 98-104.

Yssing, M. (1975). Long-term prognosis of children born to mothers diabetic when pregnant. *In* "Early Diabetes in Early Life" (R. A. Camerini-Davalos and H. S. Cole, eds.), pp. 575-586. Academic Press, New York.

11

Endocrine Aspects of Genetic Disorders in Pregnancy (Excluding Diabetes Mellitus)

JAMES R. SCHREIBER AND ROBERT W. REBAR

Copyright © 1981 by Academic Press, Inc.
All rights of reproduction in any form reserved.
ISBN 0-12-630940-X

I. INTRODUCTION

Disorders of the endocrine system influence fertility, the course of pregnancy, and the outcome of the fetus and newborn. In this chapter, we are concerned with inherited endocrine disorders other than diabetes mellitus (Chapter 10) that are compatible with pregnancy, with emphasis on the relationship between the endocrine abnormality and the pregnant state. Pregnancy alters the concentration of many circulating hormones, making the evaluation of endocrine gland function more difficult. These changes will be discussed. Finally, in certain disorders, therapy may involve a risk to the fetus and these dangers will be emphasized. We have chosen to include several disorders in which the etiology has not been definitively established but in which a genetic basis or tendency has been strongly suggested.

II. THE HYPOTHALAMIC–PITUITARY UNIT

Numerous genetic abnormalities have been reported which result in altered function of the hypothalamic–pituitary unit. Such disorders may affect a single trophic hormone or any combination of two or more pituitary hormones. They may result from disturbances in any part of the CNS–hypothalamic–pituitary–target organ complex. Thus, a syndrome of pituitary hormonal deficiency might result from a defect at any of several levels: within the higher centers of the central nervous system, the hypothalamus, or the pituitary gland.

A. Pituitary Function in Pregnancy

Before discussing those inherited hypothalamic–pituitary disorders in which pregnancy is possible, it is essential to understand the physiologic changes in pituitary function which occur during gestation. The interrelationships of the anterior pituitary gland with its target organs are summarized in Fig. 1. The anterior lobe of the pituitary gland secretes six distinct hormones: luteinizing (LH) and follicle-stimulating (FSH) hormones, important in reproductive function in follicular growth and development; prolactin, which plays a role in lacta-

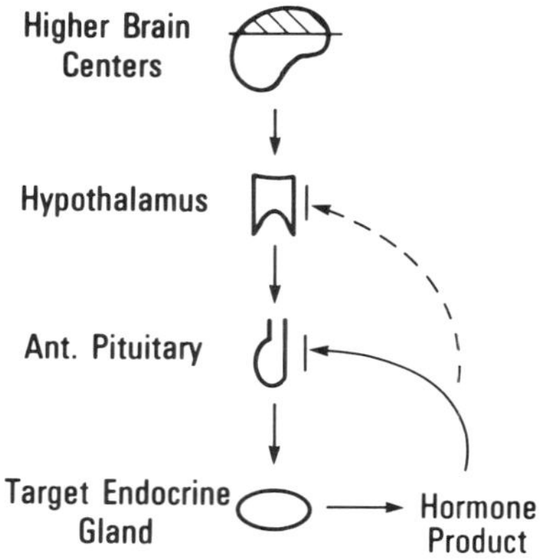

Fig. 1. General scheme of endocrine glands under pituitary control. The hypothalamus releases trophic hormones into the hypothalamic–pituitary portal system which stimulates the release of hormones from the anterior pituitary. These pituitary trophic hormones travel to the target endocrine organ via the general circulation and stimulate the synthesis and release of hormones. These hormone products have inhibitory feedback action on the anterior pituitary and perhaps the hypothalamus.

tion and may be important in ovarian function as well; adrenocorticotropic hormone (ACTH), which stimulates secretion of glucocorticoids by the adrenal cortex; growth hormone (GH), important in growth processes; and thyroid-stimulating hormone (TSH), which plays a prominent role in the regulation of thyroid function. The posterior lobe of the pituitary gland secretes oxytocin, important in milk let-down in lactating women, and vasopressin (also known as antidiuretic hormone), which plays a major role in water balance in the body.

During pregnancy, the anterior lobe of the pituitary gland increases dramatically to greater than twice the size of the nonpregnant pituitary, largely due to an increase in the number of lactotrophs (i.e., prolactin-secreting cells). Prolactin secretion increases dramatically during gestation. Numerous other hormonal changes in pituitary function occur in conjunction with these anatomic changes. Gonadotropin secretion is extremely low, presumably as a result of suppression by the increased levels of circulating estrogens and human chorionic gonadotropin (hCG). Because of cross reaction with human chorionic somatomammotropin (HCS), radioimmunoassay determination of growth hormone secretion during pregnancy is difficult. One group has reported diminished growth hormone responsiveness to both hypoglycemia and arginine during the third trimester (Tyson *et al.*, 1969). It is not yet clear if this blunted growth hormone responsiveness is a result of increased circulating HCS or of increased progesterone levels (Bhatia *et al.*, 1972). In contrast to the findings of Tyson *et al.* (1969), Yen *et al.* (1967) did not observe altered responsiveness of growth hormone to insulin-induced hypoglycemia during pregnancy.

TABLE I
Replacement Therapy for Hypopituitarism during Pregnancy

Deficiency	Replacement	Dosage
TSH	L-Thyroxine	0.15–0.2 mg/day
ACTH	Hydrocortisone[a]	10–15 $mg/m^2/day$[b] (2/3 in AM; 1/3 in PM)
GH	No therapy needed in adults	
Gonadotropin	No therapy necessary during pregnancy; ovulation induction needed for fertility	
Prolactin	No therapy necessary	

[a] Widely used glucocorticoid preparations and their potencies cited in Table III.
[b] Based on prepregnant body surface area.

TSH responsiveness to thyrotropin releasing hormone (TRH) is increased during pregnancy (Burrow, 1975). In contrast, ACTH responsiveness to metyrapone is diminished in pregnancy (Beck *et al.*, 1968). Whether the altered TSH and ACTH responsiveness is a direct result of increased circulating estrogen is not certain.

These alterations in pituitary function during pregnancy make evaluation of pituitary status difficult. LH cannot be easily distinguished from hCG in pregnant serum. Growth hormone secretion is difficult to evaluate. Low or even normal levels of circulating thyroid hormone (by nonpregnant standards) in association with low TSH levels may indicate hypothyroidism of central origin. The ACTH response to metyrapone may be difficult to interpret due to decreased pituitary responsiveness during gestation.

In almost all instances involving genetic abnormalities, the diagnosis of hypopituitarism will have been established before pregnancy. In fact, pregnancy will most likely have resulted following successful ovulation induction by the reproductive endocrinologist. In these instances, appropriate replacement therapy will have been instituted before pregnancy and should continue without interruption during gestation. As indicated in Table I, no replacement therapy for growth hormone, gonadotropin, or prolactin secretion is necessary in adults or during pregnancy. Evaluation of the adequacy of thyroid and glucocorticoid replacement must be considered in the light of normal alterations occuring during pregnancy and will be discussed in Sections IV and VI of this chapter.

B. Isolated Gonadotropin Deficiency (IGD)

A distinct genetic syndrome associating deficient sense of smell and hypogonadism was first described by Kallmann *et al.* in 1944 (Kallmann's syndrome). They described 11 cases of male hypogonadism with anosmia in three families with the trait being transmitted by the unaffected females and suggested that the hypogonadism was sex-linked. Some of their patients also had color blindness, deafness, and synkinesia, but these defects appeared to segregate independently of the hypogonadism.

Since the initial report, a number of similarly affected families have been described, including several with affected females (Ewer, 1968; Sparkes *et al.*, 1968; Tagatz *et al.*, 1970; Odell, 1966; Spitz *et al.*, 1974). Association of the hypogonadism and anosmia with midline facial defects, such as cleft lip and palate, has also now been documented (Rosen, 1965). The syndrome has been found to occur more frequently in males than in females; variable expression of the different traits has been documented, with hypogonadism the only characteristic manifested in some families. Autosomal recessive and male-limited autosomal dominant, as well as X-linked recessive inheritance, have all been suggested as the mode of inheritance in some families. Review of autopsy material from some affected patients with hypogonadism and anosmia has revealed agenesis or hypoplasia of the olfactory lobes, hypothalamus, and/or mamillary bodies (DeMorsier, 1954; Gauthier, 1960). This finding has led to use of the eponym "olfactogenital dysplasia." Together, these observations have led to the recognition that the isolated deficiency of gonadotropins is a heterogeneous syndrome, which may or may not be associated with other characteristics.

Regardless of etiology, gonadotropin deficiency results in hypogonadism with diminished ovarian steroid production. Circulating concentrations of LH and FSH are invariably low. The clinical features in women with IGD vary, however, with secondary sexual characteristics ranging from classic eunuchoidal features to moderate breast development (Odell, 1966; Spitz *et al.*, 1974). Primary amenorrhea is the rule. Goldenberg *et al.* (1976) have demonstrated that the ovaries in these patients rarely contain multiple follicles beyond the primordial stage of follicular maturation. However, these immature ovaries respond readily to exogenous gonadotropin (Gemzell, 1965; Thompson and Hansen, 1970). Both ovulation and pregnancy have resulted (Tagatz *et al.*, 1970; Spitz *et al.*, 1974). We have successfully induced pregnancy with exogenous gonadotropins in one woman with hypogonadism, hyposmia, and cleft lip (Fig. 2).

The availability of synthetic luteinizing hormone-releasing factor (LRF) has resulted in further clarification of the defect in this disorder and offers a potential therapeutic approach. Complete unresponsiveness to a single bolus of LRF is uncommon (Yen *et al.*, 1973). Even if the patient does not respond with release of LH and FSH, daily administration of LRF may result in gonadotropin release in some patients (Bell *et al.*, 1973). Thus, in most patients with IGD the primary site of the defect appears to be at the hypothalamic neuronal level for synthesis or release of LRF whereas pituitary gland dysfunction is less common. It is now apparent that it will soon be possible to use pulsatile LRF infusions to induce

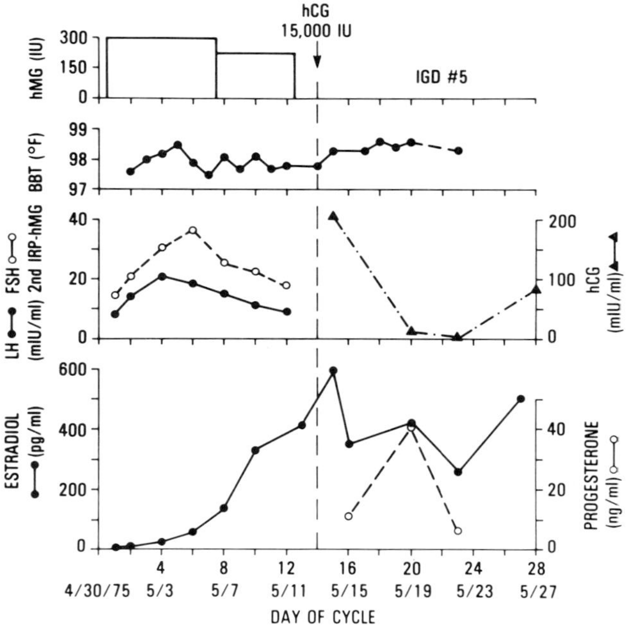

Fig. 2. Hormonal changes occuring during an ovulatory cycle induced with human menopausal gonadotropin–human chorionic gonadotropin (hMG-hCG) during which pregnancy resulted in a patient with isolated gonadotropin deficiency. The patient was subsequently delivered of a healthy full-term 3400 g male infant. The rise in serum levels of hCG on the 28th day of the cycle is diagnostic of pregnancy. This patient originally presented with hypogonadotropic hypogonadism, hyposmia, cleft lip, and symptoms of severe estrogen deficiency including emotional lability and dyspareunia at age 22 years (R. Rebar, J. Schreiber, and J. Vaitukaitis, unpublished).

both complete maturation of secondary sexual characteristics and ovulation in appropriate patients.

At present, it is possible to achieve complete sexual maturation by exogenous replacement of gonadal steroids. Ovulation can be induced with exogenous gonadotropins. Induction of maturation of secondary sexual characteristics is most effective when unopposed estrogen is administered in small doses. A progestogen, such as medroxyprogesterone acetate, may be given at 1- to 2-month intervals to induce withdrawal bleeding (Table II). Once sexual development is completed, combination oral contraceptive agents may be used for replacement therapy rather than sequential estrogen–progestin administration. Obviously, steroidal agents need to be discontinued before induction of ovulation. In our experience, most young individuals with hypogonadism complain of symptom of severe estrogen deficiency, including emotional lability, hot flushes, and dys-

TABLE II
Female Sex Steroid Replacement[a]

Estrogen:	10–20 μg ethinyl estradiol or 0.625 mg Premarin daily the first 21 days of each month.
Progesterone:	5–10 mg Provera (medroxyprogesterone acetate) every day during the last 5–7 days of estrogen.

[a]Oral contraceptive preparations are a satisfactory alternative in women with mature secondary sexual characteristics.

pareunia. This seems to be particularly true in women who have been given estrogen and then had it withdrawn. The woman whose pregnancy cycle is documented in Fig. 2 was under psychiatric care with marital difficulties related to her aversion of intercourse. Her life style changed dramatically after we made the diagnosis (low circulating estrogen and gonadotropins) and instituted estrogen replacement (Table II). No special endocrine management is required during pregnancy.

C. Familial Panhypopituitary Dwarfism (Asexual Ateliosis)

Although panhypopituitary dwarfism is usually sporadic, at least 20 kindreds with multiple affected family members have been reported (Rimoin and Schimke, 1971). The diagnosis of panhypopituitary dwarfism is based on the demonstration of growth hormone and other pituitary hormone deficiencies. Hereditary forms of the disorder must be distinguished from other forms of pituitary insufficiency, such as neoplasms, which must be treated appropriately before hormonal induction of pregnancy.

Most frequently associated with the growth hormone deficiency is a deficiency in gonadotropin secretion, resulting in hypogonadotropic hypogonadism. This is followed in order of frequency by ACTH and TSH deficiency (Brasel *et al.*, 1965; Rimoin *et al.*, 1968b). There is a variability of deficient hormones both within and among families. However, growth hormone and gonadotropin deficiency occur in all affected family members (Rimoin *et al.*, 1968b).

The presence of affected siblings of both sexes and a high frequency of consanguinity indicate that the disorder is generally inherited as an autosomal recessive trait. An X-linked recessive form has also been reported (Rimoin and Schimke, 1971).

A sufficient number of individuals has not yet been tested with hypothalamic releasing factors to define localization of the defect to the hypothalamus or pituitary gland. It seems reasonable to assume that the site of the defect varies among affected families. If this is true, gonadotropin deficiency should be amenable to treatment with pulsatile LRF infusions in the near future in individuals responding to LRF. Until then, maturity of secondary sexual characteristics can be achieved with gonadal steroids and ovulation can be induced with exogenous gonadotropins in affected females when pregnancy is desired. In childhood, normal linear growth requires exogenous purified human growth hormone. Treatment with gonadal steroids must be added when the patient is emotionally prepared and ultimate body height will not be adversely affected. Replacement of adrenal steroids and thyroid hormone is indicated in individuals unable to make these hormones.

Pregnancy and lactation can occur normally in GH-deficient individuals (Rimoin *et al.*, 1968a). As long as replacement therapy is continued during pregnancy, no unexpected difficulties should arise. Since growth hormone deficiency and short stature may be present despite appropriate therapy, cesarean section may be necessary if a contracted pelvis is present, as the infant will almost invariably be of normal size. A significant number of the infants will be carriers of this genetic trait.

D. Histiocytosis X

The common denominator of the illnesses grouped under the heading of histiocytosis X (Letterer–Siwe disease, Hand–Schüller–Christian disease, eosinophilic granuloma, reticuloendotheliosis) is a distinctive inflammatory foamy histiocytosis. Although the etiology is unknown, reports of affected siblings suggest autosomal recessive inheritance (Schoeck *et al.*, 1963). Diabetes insipidus is the most frequent endocrinologic complication of this disorder, but anterior pituitary deficiency, particularly of growth hormone and gonadotropins, is also common (Braunstein and Kohler, 1972; Vogel and Vogel, 1972). Autopsy reports have suggested that the pituitary insufficiency is due to histiocytic infiltration of the hypothalamus.

Endocrine therapy consists of replacing deficient hormones. Vasopressin and human growth hormone (in childhood) are indicated for treatment of diabetes insipidus and growth hormone deficiency. In females who are deficient in gonadotropins, estrogens will be necessary for maturation of secondary sexual characteristics. If this disorder results because of hypothalamic rather than pituitary deficiency, treatment with pulsatile LRF infusion should also prove feasible. Ovulation induction with gonadotropins (or LRF) is possible in affected women desiring pregnancy providing the disease is inactive and ovarian function is normal. Radiotherapy and chemotherapy are frequently utilized in treating the primary disease. These agents may result in ovarian ablation (Jacox, 1939; Koyama *et al.*, 1977). Some reports indicate that "ovarian failure" may be reversible (Siris *et al.*, 1976; Baker *et al.*, 1972; Koyama *et al.*, 1977). Though treated individuals developed menopausal symptoms and elevated FSH levels, which are usually associated with ovaries devoid of follicles, spontaneous menses returned in some patients and FSH levels fell to normal (Siris *et al.*, 1976).

In pregnancy complicated by hyperemesis, electrolyte balance must be closely observed. In women with growth hormone deficiency and a contracted pelvis, cesarean section may be necessary.

E. The Laurence–Moon–Bardet–Biedl Syndrome

The classic features of the syndrome of Laurence–Moon–Bardet–Biedl, inherited as an autosomal recessive trait, include retinitis pigmentosa, polydactyly, obesity, mental retardation, and hypogonadism. The hypogonadotropic hypogonadism has been reported to occur in 75–86% of the affected males and in 45–53% of the females (Klein and Ammann, 1969; Francke, 1950; Reinfrank and Nichols, 1964; Rimoin and Schimke, 1971). Affected females usually do not show signs of marked hypogonadism, and several females have been reported to have had children. No males to date have been fertile. In view of the hypogonadism, obesity and occasionally observed diabetes insipidus, it has been suggested that a hypothalamic lesion may be present in this syndrome. However, this hypothesis must be questioned in view of the fact that three male siblings with the syndrome all responded normally to synthetic LRF and clomiphene citrate (Toledo *et al.*, 1977). The normal response to a single bolus of LRF suggests that the hypothalamus is providing constant stimulation to the pituitary. In addition, basal levels of serum prolactin and GH were within the normal range and stimulatory and suppressive tests for these pituitary hormones were also normal in these patients. Basal serum LH levels were elevated in two of the three, while serum FSH levels were normal in all. Although plasma testosterone concentrations were found to be normal for each individual patient, repeated stimulation with hCG failed to further increase testosterone levels in the two elder brothers. These data indicate a peripheral gonadal disorder in these affected males rather than a central abnormality.

In view of the mental retardation, induction of ovulation is inappropriate in most of these hypogonadal patients. Contraceptive counseling will be necessary in fertile affected females.

F. Amenorrhea-Galactorrhea

The syndrome of amenorrhea-galactorrhea is generally sporadic, although Linquette *et al.* (1967) have described a family in which both mother and daughter developed amenorrhea and galactorrhea in association with pituitary adenomas. They suggested that this disorder might be inherited as an autosomal dominant trait in some instances. Wider *et al.* (1969) have reported galactorrhea after the use of oral contraceptives in three sisters with oligo-ovulation. They noted that the probability of the three sisters independently demonstrating galactorrhea was approximately 3.6×10^{-5}. No evidence of a pituitary adenoma was found in any of the three, although serum prolactin determinations were not available at the time of this study.

The amenorrhea–galactorrhea syndrome has also been described as a part of the syndrome of multiple endocrine adenomatosis (MEA type I, Wermer syndrome), associated with multifocal unremitting peptic ulcer disease (Zollinger-Ellison syndrome), islet cell adenomas of the pancreas, and hyperparathyroidism (LeBriggs and Powell, 1969; Ballard *et al.,* 1964). Currently, it is impossible to determine if the reported familial cases with only amenorrhea–galactorrhea are indeed a part of the MEA syndrome complex or whether they represent distinct clinical entities. The important point is that in any patient with a pituitary neoplasm every effort should be made to exclude involvement of other endocrine organs in both the patient and her close relatives.

Individuals with pituitary neoplasms and galactorrhea–amenorrhea frequently have hyperprolactinemia (Nader *et al.,* 1976). It has been postulated that the adenoma results because of a defect in dopamine regulation at the pituitary lactotroph; i.e., the dopamine fails to suppress prolactin secretion as it does in normal individuals (Lachelin *et al.,* 1977). Yet because of increased prolactin secretion, excess dopamine is released, inhibiting gonadotropin secretion and resulting in hypogonadotropinism and amenorrhea. Thus, in patients with this syndrome, determinations of serum prolactin and radiographic tomograms of the sella turcica are warranted. Treatment of pituitary tumors is now controversial but may involve the use of the dopamine receptor agonist 2α-bromoergocryptine or transphenoidal removal of the neoplasm. As many as 90% of patients undergoing such surgery stop lactating and resume spontaneous menses (Hardy, 1979). If hypogonadotropinism results after surgery, ovulation induction with exogenous gonadotropins may be necessary in individuals desiring pregnancy. Other pituitary hormonal deficits should be treated and replacement therapy continued during pregnancy.

G. Prader–Labhart–Willi Syndrome

Since the original description of this syndrome (Prader *et al.,* 1956), many cases have been described. Typical features include childhood onset obesity, decreased activity *in utero,* poor sucking reflex, hypotonia, acanthosis nigricans, and mental retardation. The males have a small penis, a hypoplastic flat scrotum, and cryptorchidism with delayed and diminished pubertal changes. The genitalia are normal in affected females, but menstrual irregularities may occur. Variable features include mild diabetes mellitus, hypertelorism, epicanthal folds, strabismus, low-set ears, high arched palate and micrognathia. At least one family with possible autosomal recessive inheritance has been described (DeFraites *et al.*, 1971). Pregnancy may occur in affected females, but should not be encouraged in view of the multiple abnormalities. Although the etiology of this syndrome is not known, we have elected to include it among those disorders with hypothalamic-pituitary dysfunction because of the insatiable appetite characteristically present in affected individuals, which is typical of a hypothalamic abnormality.

III. THE OVARY

A. Familial Polycystic Ovary (PCO) Syndrome

Polycystic ovary syndrome, originally described by Stein and Leventhal in 1935, is typified by excess androgen production and inappropriate gonadotropin release with elevated levels of circulating LH and normal to decreased levels of circulating FSH (Yen *et al.*, 1970). Hirsutism and irregular menses generally begin at puberty but virilization is rarely seen (Ross and Vande Wiele, 1974). On the basis of extensive studies, it is currently believed that the chronic anovulation with inappropriate gonadotropin secretion is due to inappropriate estrogen feedback to the hypothalamic-pituitary unit and not to an inherent defect in this unit (Rebar *et al.*, 1976; Yen *et al.*, 1976). Estrogens are produced in the periphery from conversion of excess androgens, which may be produced by the ovary (principally), adrenal cortex, or both. Morphologically, the ovaries in this syndrome are usually enlarged with a smooth and thickened ovarian capsule. The capsule is avascular in appearance and on cut section, the ovary exhibits numerous small subcapsular follicular cysts. Frequently, a corpus luteum or corpus albicans is present, indicating that ovulation may occur on an irregular basis. The cysts are usually lined with a few layers of granulosa cells and many are atretic, but the most striking feature is the hyperplasia of the theca interna (believed to be the site of excess androgen production) surrounding the many cystic follicles. Luteinization of the theca cells often is evident (Green and Goldzieher, 1960). It must be emphasized that such changes are not seen in the ovaries of all patients with PCO. Variations may be due to differing etiologies or to the varying clinical manifestations and laboratory findings in this syndrome.

Cooper *et al.* (1968) have reported the results of extensive studies of members of 18 families in which the Stein-Leventhal syndrome had appeared. They noted an increased frequency of PCO and/or oligomenorrhea as well as elevated androgen levels among mothers and sisters of the index patients. Culdoscopy of 8 of 12 sisters of cases showed ovarian changes consistent with PCO. Male relatives tended to be abnormally hirsute. Although the genetic pattern is not well understood, an autosomal dominant mode of transmission is suggested by their data. No chromosomal abnormalities were detected. The authors concluded that bilaterally enlarged cystic ovaries may not be an essential manifestation of the hereditary potential for the disease, but are probably the best single indicator.

Ovulation can be successfully induced in affected patients desiring pregnancy by a variety of methods. The administration of clomiphene citrate to induce ovulation is currently the method of choice with its proven efficacy (Kistner, 1975; Rust *et al.*, 1974). In patients failing to ovulate in response to clomiphene, wedge resection of the ovaries can be undertaken in an attempt to induce ovulation (Judd *et al.*, 1976). Since frequent tubo-ovarian adhesions occur following wedge resection, it is not the procedure of choice unless other methods have failed or there is concern that an ovarian tumor is the cause of the patient's hyperandrogenism (Buttram and Vaquero, 1975). Human menopausal gonadotropins can also be used to induce ovulation; however, the potential of serious side effects, including massive ovarian enlargements, ascites, and multiple pregnancies preclude their use except by experienced personnel in unusual circumstances (Gemzell, 1965). More recently, several investigators have demonstrated the efficacy of utilizing purified pituitary FSH to induce ovulation (Raj *et al.*, 1977).

Pregnancies in women with PCO and hirsutism, but not virilization, rarely result in the birth of androgenized infants. However, ovarian tumors associated with sclerocystic ovaries and virilization (Zourlas and Jones, 1969; Babaknia *et al.*, 1976) may result in virilized infants. It is important to exclude the diagnosis of ovarian neoplasm in patients with PCO.

B. Familial Hyperthecosis

Several cases of familial hyperthecosis have now been described (Givens *et al.*, 1971; Judd *et al.*, 1973; Wilroy *et al.*, 1975). Inheritance appears to be autosomal dominant or X-linked. In 1943 Fraenkel used the term *hyperthecosis* to describe isolated nests of luteinized cells scattered throughout the ovarian stroma. Pathologically, however, there appears to be a spectrum of disease from classic PCO to hyperthecosis. Most ovaries with hyperthecosis contain all the features of polycystic ovaries in addition to nests of luteinized stromal cells. Thus, it is possible that familial hyperthecosis is a part of the spectrum of familial PCO.

Patients with hyperthecosis are typically hirsute with irregular menses. Androgen production and circulating LH concentrations are generally elevated as well. Ovulation may be successfully induced by the same methods used to induce ovulation in PCO.

C. Gonadal Dysgenesis with Normal Chromosomal Complement

Numerous forms of gonadal dysgenesis exist, with an abnormal chromosomal complement being present in almost all forms. These are discussed in Chapter 13.

The syndrome of pure gonadal dysgenesis, which was clearly delineated by Sohval (1965), is present in individuals who classically have an XX or XY karyotype. They are phenotypic females of normal or tall stature with sexual infantilism, who lack the stigmata of Turner's syndrome. The internal and external genitalia are hypoplastic. Pigmented nevi, mild cubitus valgus and cafe-au-lait spots may be present in some. These individuals typically are first seen because of primary amenorrhea and are found to have high levels of circulating gonadotropins and low levels or circulating estrogen.

XX gonadal dysgenesis, inherited as an autosomal recessive trait, may be distinguished from testicular feminization (T fem) by a karyotype (46, XY for T fem) and from isolated gonadotropin deficiency by the high levels of serum and urinary gonadotropins (Simpson *et al.*, 1971). Treatment consists of appropriate estrogen–progesterone replacement. Gonadectomy is unnecessary.

XY gonadal dysgenesis, inherited as either an X-linked recessive or a male-limited autosomal dominant characteristic, may be differentiated from T fem by the presence of a uterus and fallopian tubes (Espiner *et al.*, 1970). Although the gonads generally show no evidence of testicular differentiation, the high risk of gonadal neoplasms, usually dysgerminoma or gonadoblastoma, makes it necessary to perform prophylactic gonadectomy (Teter and Boczkowski, 1967; Manuel *et al.*, 1976). Clitoromegaly is frequently present even when there are no other manifestations of androgen excess. A few cases have been reported with marked virilism and the source of androgen excess has now been shown to be the streak gonads (Rose *et al.*, 1974; Judd *et al.*, 1970). The potential for virilization mandates removal of the gonads as soon as the diagnosis is made. Estrogen–progesterone replacement therapy is indicated in these individuals.

In view of the finding that individuals with gonadal dysgenesis have the normal complement of germ cells at 20 weeks *in utero* but generally lose these before birth (Singh and Carr, 1966), it is feasible that pregnancy may result in rare instances in which all oogonia are not destroyed in 46,XX individuals. Any resulting pregnancy should proceed normally. In fact, several pregnancies in patients with 45,XO cell lines have been reported (Reyes *et al.*, 1976).

D. Noonan's Syndrome

In 1963, Noonan and Ehmke reported nine male and female patients with "a new syndrome" including valvular pulmonic stenosis, short stature, hypertelorism, antimongoloid slant of the palpebral fissures, micrognathia, ptosis, skeletal anomalies, and mental retardation. While these patients have some features of Turner's syndrome, they had a normal complement of chromosomes. Affected females and males have been described with several hundred affected individuals having been reported. Although the mode of transmission is still in question, recent studies have suggested autosomal dominant inheritance (Collins and Turner, 1973; Nora *et al.*, 1974; Qazi *et al.*, 1974).

Affected females usually have normal ovarian function with regular menses and fertility. Primary amenorrhea has also been reported. In males, cryptorchidism, delayed puberty, and hypogonadism may occur, although in some, reproductive function is entirely normal (Saez *et al.*, 1969).

The major problem during pregnancy in affected females may well be secondary to the existence of cardiac disease. Pulmonic stenosis with or without atrial septal defect, Ebstein's anomaly, coarctation of the aorta, and multiple valvular sclerosis have all been described (Noonan, 1968). Cardiac evaluation should be undertaken once the diagnosis is established so that appropriate medical and surgical therapy may be recommended before pregnancy.

IV. THE ADRENAL GLAND

A. Adrenal Function during Pregnancy

Cortisol is transported in the blood stream primarily in the bound state, with approximately 10% circulating in the free or unbound state. The major binding protein is transcortin or cortisol-binding globulin (CBG), which has a high affinity for cortisol but a relatively low capacity. Cortisol bound to CBG is biologically inactive because its high affinity for CBG makes it unavailable to target tissues. Thus, only the free cortisol and presumably that portion bound to albumin, which has a low affinity for cortisol, are biologically active.

During normal pregnancy there is a progressive rise in total plasma cortisol levels to a plateau in the second trimester which is elevated in comparison to normal nonpregnant concentrations (Gemzell, 1953; Assali *et al.*, 1955; Bayliss *et al.*, 1955). This increase is primarily due to an increase in the circulating concentration of CBG (Doe *et al.*, 1964). Studies, however, have revealed that levels of unbound (and albumin-bound) cortisol are definitely elevated in pregnancy and rise as pregnancy progresses (Rosenthal *et al.*, 1969). The normal diurnal variation seen in nonpregnant patients is retained during pregnancy, but the levels of plasma cortisol are higher at each time period (Cohen *et al.*, 1958). There is also an increase in urinary free cortisol excretion during pregnancy (Burke and Roulet, 1970), but the urinary excretion of 17-hydroxycorticosteriods decreases (Migeon *et al.*, 1968b) with the decrease apparently being due to altered metabolism. The cortisol production rate also decreases during gestation (Migeon *et al.*, 1968b). The rate of disappearance of cortisol from the plasma is decreased (Christy *et al.*, 1959) and there is an associated decrease in the metabolic clearance rate. The adrenal gland appears to be more responsive to ACTH during pregnancy with a greater rise in plasma cortisol levels occurring after ACTH administration (Jailer *et al.*, 1959). The pituitary gland, however, may be less responsive, in view of the diminished response to metyrapone (Beck *et al.*, 1968).

These findings raise the question as to whether adrenal function is increased during pregnancy. It has been demonstrated that during pregnancy progesterone competes with cortisol for binding sites on CBG (Rosenthal *et al.*, 1969). This observation may explain the high level of unbound plasma cortisol in pregnancy. Similarly, the elevated glucocorticoid levels in pregnancy might be a response to competition with progesterone for intracellular binding in the hypothalamus and pituitary (Baxter and Forsham, 1972). Thus, it is not known if adrenal function is truly increased at the cellular level.

Adrenal insufficiency in pregnancy presents with the same signs and symptoms seen in affected nonpregnant individuals. Weakness, fatigue, weight loss, anorexia, hypoglycemia, nervousness, and increased pigmentation (if the hypoadrenalism is not central in origin) may be present. In pregnancy, asthenia, marked pigmentation, intensified gastrointestinal symptoms, and weight loss are particularly common (Burrow, 1975), but unfortunately such symptoms are common even in normal pregnancies. Hypotension, unexplained abdominal pain and low serum sodium values with elevated levels of serum potassium may be useful diagnostic clues. In Brent's (1950) review of 39 cases of Addison's disease occurring during pregnancy, the untreated patients tended to tolerate the pregnancy without difficulty but developed problems in the immediate postpartum period. However, it must be remembered that acute adrenal insufficiency, also known as Addisonian crisis, can occur at any time in untreated individuals.

Biochemically, the diagnosis of adrenal insufficiency during pregnancy is complicated by the increased circulating CBG. Plasma cortisol levels may appear normal in comparison to the nonpregnant state. Cortisol, however, should rise in response to an infusion of ACTH if adrenal insufficiency is not present. A decreased excretion of 24-hour urinary free cortisol may be the most useful diagnostic test of hypoadrenalism during pregnancy.

Maintenance replacement therapy during pregnancy need be no different from therapy in nonpregnant women. In secondary hypoadrenalism with ACTH deficiency, inadequate glucocorticoid production results but mineralocorticoid secretion is not significantly impaired. Replacement with steroid preparations having mainly glucocorticoid activity is all that is necessary in individuals with hypothalamic–pituitary dysfunction. In Table III are listed the most commonly used preparations, their relative glucocorticoid and sodium-retaining potencies,

TABLE III
Widely Used Glucocorticoid Preparations

Preparation	Relative glucocorticoid potency	Relative sodium retaining potency	Approximate dosage (mg/m²/day)	Usual dosage in pregnancy (mg) (2/3 AM; 1/3 PM)
Dexamethasone	25	0	0.4–0.6	0.75; 0.25
6-α-Methyl-prednisolone	5	0.5	2–3	4; 2
Prednisone, prednisolone	4	0.8	2.5–4	5.0; 2.5
Hydrocortisone	1	1	10–15	20; 10
Cortisone acetate	0.8	0.8	12.5–20	25; 12.5

and the usual dosage for pregnant women. The preparations listed in Table III may be used interchangeably. Treatment with hydrocortisone may be most practical because it is least expensive. Since about 40% of the administered dose is excreted as urinary 17-hydroxycorticosteroids, the physician can determine if the patient is taking and absorbing the medication appropriately. In an attempt to mimic the normal circadian rhythm of circulating cortisol, it has become accepted to administer two-thirds of the daily steroid dose in the morning and the remaining one-third in the evening. However, it has not been shown that this practice is beneficial, and the medication can generally be given as a single daily dose without ill effect. Individuals with adrenal failure will also require mineralocorticoid replacement, most frequently with 0.05–0.1 mg 9α-fluorohydrocortisone daily. The presence of hypertension or edema may require reduction in this salt-retaining steriod. Patients should also carry identification, such as Medic-Alert, and an injectable corticosteroid, such as dexamethasone phosphate in 4 mg injectable syringes, for emergencies.

It is important to remind the patient to increase the daily intake of glucocorticoid during any stressful situation or infection. The nausea and vomiting of early pregnancy may necessitate use of intramuscular steroid preparations. The patient on glucocorticoid replacement can be allowed to have a spontaneous delivery. During labor the patient should be adequately hydrated and administered large doses of intramuscular steroid (such as 100 mg hydrocortisone at 12-hour intervals). Following delivery, high doses of parenteral steroids should be continued for a few days to prevent the occurrence of a postpartum adrenal crisis. Lastly, it must be remembered that maternal glucocorticoid therapy may cause a low urinary excretion of estriol in some cases so that these measurements are not reliable (Greene, 1973).

B. Disorders of Steroid Biosynthesis (Congenital Adrenal Hyperplasia)

Several enzyme deficiencies have now been documented in individuals with disordered steroid biosynthesis. Such deficiencies may be manifested in the adrenal gland as well as in the ovary (or testis). In this discussion we will concern ourselves only with those disorders in which pregnancy may occur in genetic females. A summary of the appropriate biosynthetic pathways with the enzyme deficiencies to be discussed is shown in Fig. 3 and the manifestations of the various deficiencies are summarized in Table IV.

Cortisol and aldosterone are the key steroid products secreted by the adrenal cortex. In genetic females, estrogens, synthesized from androgens, are the im-

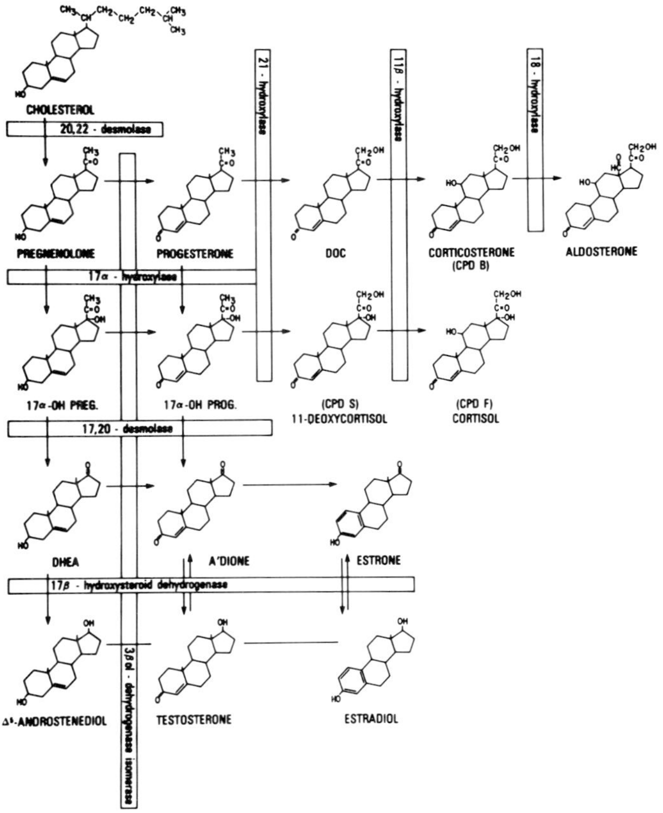

Fig. 3. Steroid biosynthetic pathways with enzymes noted for which deficiences have been delineated. Not all are associated with recognized defects in females.

TABLE IV
Biochemical and Clinical Manifestations of Enzyme Abnormalities in Steroid Biosynthesis

Enzyme deficiency	Increased	Decreased	Clinical findings
20,22-Desmolase	None	All steroids	Severe adrenal insufficiency Males: female external genitalia and a male duct system
3β-Hydroxysteroid dehydrogenase	DHEA and other Δ^5 steroids	Corticoids Aldosterone Estrogens Testosterone	Salt and water wasting Females: mild virilization Males: hypospadias
17α-Hydroxylase	Corticosterone DOC Progesterone	Cortisol Androgens Estrogens	Hypertension Hypokalemic alkalosis Females: immaturity Males: pseudohermaphroditism; gynecomastia at puberty
21-Hydroxylase	Androgens	Corticoids Aldosterone	1/3 salt losing Masculinization
11β-Hydroxylase	DOC 11-Deoxycortisol androgens	Corticoids Aldosterone	Hypertension Masculinization
18-Hydroxysteroid dehydrogenase	Corticosterone	Aldosterone	Salt losing
17,20-Desmolase	Progestins	Androgens Estrogens	Females: unreported Males: ambiguous external genitalia, hypospadias, male duct system; immaturity
17β-Hydroxysteroid dehydrogenase	Androstenedione Estrone	Testosterone Estradiol	Females: unreported Males: pseudohermaphroditism with pubertal virilism and gynecomastia

portant secretory product of the ovaries. Clinically, manifestations of abnormalities in steroid production result from deficient synthesis of adrenal steroids, inadequate ovarian estrogen production, and increased biologic activity of excessive steroid precursors resulting from an enzyme deficiency.

ACTH acts at the step between cholesterol and pregnenolone to stimulate adrenal steroidogenesis. The production of ACTH is regulated by the negative feedback effect of circulating cortisol. It is currently unclear if the effect of cortisol is exerted directly at the level of the pituitary gland, indirectly through control of release of corticotropin-releasing factor (CRF) by the hypothalamus, or both. In any event, when cortisol secretion is decreased, ACTH production is increased and hyperplasia of the adrenal glands results from increased stimulation. Increased production of steriods in the steps of the biosynthetic pathway immediately preceeding the enzyme block occurs as a result of excessive ACTH stimulation. Thus, steroids immediately preceding the enzyme deficiency are present in increased concentrations in the circulation. In some of the disorders, there is impaired mineralocorticoid production as well.

As a group, these disorders have been termed *congenital adrenal hyperplasia*. They may occur in males or females and, with few exceptions, are transmitted as autosomal recessive traits. The genetic basis for defective steroid synthesis is unknown. Although it may be assumed that a structural gene is at fault, it is possible that the graded reduction in activity of a single enzyme, such as is seen to occur in CAH, is due to abnormal control genes.

When the female is involved, masculinization of the external genitalia may occur with resulting female pseudohermaphroditism. By definition, patients with female pseudohermaphroditism have a female gonad (ovary), a normal female karyotype (46,XX), and varying degrees of external genital virilization. Thus, although abnormalities in sexual development may occur, the uterus, fallopian tubes, and ovaries are normal and reproductive function is often possible after appropriate replacement therapy has been instituted. Although female pseudohermaphroditism is most frequently caused by such enzyme deficiencies, it may also result from maternal ingestion of steroid preparations or androgen-producing tumors.

Five forms of congenital adrenal hyperplasia that may affect genital tract development in the genetic female have been delineated. These include 21-hydroxylase, 11-hydroxylase, 3β-hydroxysteroid dehydrogenase, 17α-hydroxylase, and 20,22-desmolase deficiencies. A rare sixth form, 18-hydroxysteroid dehydrogenase deficiency, leads to diminished aldosterone production but does not affect development of the genital tract.

1. 21-Hydroxylase Deficiency

The most common variant in the genetic female is due to deficiency of the 21-hydroxylase enzyme and accounts for approximately 90% of CAH patients (Grumbach and VanWyk, 1974). It is inherited as an autosomal recessive trait. The deficiency may be of varying severity, from partial to almost complete absence of the enzyme. Because of reduced cortisol production and resultant augmented ACTH secretion, increased quantities of 17-hydroxyprogesterone, just proximal to the enzymatic block, are produced. Increased levels of circulating 17-hydroxyprogesterone and of pregnanetriol in the urine are diagnostic. In addition, increased production of adrenal dehydroepiandrosterone and other androgens is found. When the block is severe, aldosterone production is also markedly diminished.

The androgen excess may first affect the female fetus *in utero,* resulting in variable masculinization of the external genitalia. The extent of the masculinization varies with the severity of the enzyme deficiency and with the stage of fetal development when excess androgen is first produced. Most commonly found at birth is enlargement of the clitoris with a common urogenital sinus into which both bladder and uterus empty. Any degree of external genital anomaly, from an essentially normal perineum to a phallic urethra with labioscrotal folds, may be seen (Fig. 4).

If untreated, heterosexual precocious puberty may be seen as a result of the androgen excess. In early childhood, the androgens may cause excessive long bone growth and the child may be taller than her peers. However, androgen excess ultimately results in premature closure of the epiphyses, so that the un-

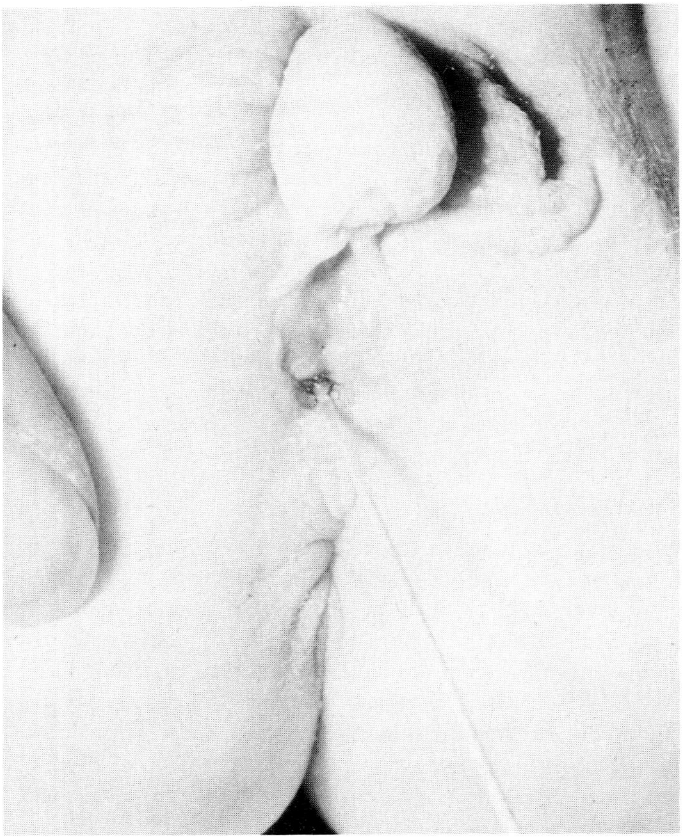

Fig. 4. Ambiguous genitalia in a female infant with 21-hydroxylase deficiency. The probe has been placed in the urogenital sinus. (Courtesy of Dr. Dorothy Hollingsworth, Reproductive Medicine, U.C.S.D.)

treated adult is short. In addition, androgens produce signs of masculinization. Furthermore, the androgens suppress gonadotropin secretion, leading to unstimulated ovaries and failure of the individual to develop female secondary sexual characteristics and menses.

In patients with severe defects, the lack of mineralocorticoid (aldosterone) results in the "salt-losing" form of the disorder with severe hyponatremia and marked dehydration. If not recognized and treated, the disorder can be fatal. If the defect is minimal, only hirsutism at the time of puberty and primary amenorrhea with failure to develop female sex characteristics may be seen and the patient's height may be normal.

Treatment consists of adequate cortisol replacement plus mineralocorticoid replacement in "salt-losing" forms. Plastic surgery is sometimes needed to repair the masculinized external genitalia.

2. 11β-Hydroxylase Deficiency

The 11β-hydroxylase enzyme appears to be unique to the adrenal gland in the human and is required as the final step in the biosynthesis of cortisol and corticosterone. Since aldosterone is synthesized from corticosterone, this steroid's production is sharply decreased as is cortisol's in individuals with this defect. Increased levels of plasma 11-deoxycortisol and 11-deoxycorticosterone (DOC) are produced since these steriods are immediately proximal to the enzymatic block. There is also increased urinary excretion of the tetrahydro derivatives of these two steroids and increased excretion of 11-deoxy-17-ketosteroids with resultant increased excretion of total urinary 17-ketosteroids.

Individuals with 11β-hydroxylase deficiency develop hypertension from overproduction of DOC and virilization from increased production of adrenal androgens. As with 21-hydroxylase deficiency, there may be rapid early somatic growth but eventual short stature because of early closure of the epiphyses. Clitoral enlargement and premature adrenarche are associated with failure to develop female secondary sexual characteristics and menses.

Treatment consists of replacement with glucocorticoids sufficient to suppress the excess secretion of ACTH. As with other forms of congenital adrenal hyperplasia, excessive glucocorticoid administration in childhood may inhibit linear growth. Mineralocorticoid substitution is not necessary because DOC production compensates for the diminished production of aldosterone.

3. 3β-Hydroxysteroid Dehydrogenase Deficiency

This rare disorder affects both adrenal and ovarian function and inhibits the conversion of the Δ^5 steroids to Δ^4 steroids. Most patients with this disorder have died in infancy. In the few surviving through puberty, excessive quantities of Δ^5-3-hydroxysteroid precursors of testosterone have been present as well as diminished quantities of both gluco- and mineralocorticoids.

Bongiovanni (1961, 1962) has reported that female infants may be moderately virilized at birth but the extent is often less than with other varieties. In addition, in most cases described there has been marked salt and water wasting. Failure to thrive in spite of apparently adequate replacement therapy is common and the outcome is often fatal in the first few months of life.

4. 17α-Hydroxylase Deficiency

In girls with the rare syndrome of 17α-hydroxylase deficiency who survive until the expected age of puberty, sexual infantilism and primary amenorrhea associated with increased serum and urinary gonadotropin concentrations have been noted (Biglieri *et al.*, 1966; Goldsmith *et al.*, 1967; Mallin, 1969). In such individuals, increased synthesis of DOC predisposes to hypertension with hypokalemic alkalosis. Furthermore, serum progesterone levels are elevated in the syndrome. In fact, the constellation of findings, including elevated serum progesterone concentrations, high serum gonadotropin levels, and hypertension with hypokalemic alkalosis responsive to glucocorticoid replacement therapy, is diagnostic. Biopsy of ovarian tissue in two girls with this syndrome showed numerous large cysts and primordial follicles with complete failure of orderly follicular maturation (Mallin, 1969). In view of the presence of ovarian follicles, it is possible that early recognition and appropriate treatment might allow pregnancy in affected patients. What would constitute appropriate therapy is not certain. However, we would suggest that intraovarian placement of estrogen implants might be necessary. Since follicular fluid estradiol levels are at least 1000-fold greater than serum concentrations (McNatty *et al.*, 1975a) and ovulation has been induced in the monkey following the placement of diethylstilbestrol in one ovary while the other ovary showed no stimulation after treatment with small doses of gonadotropin (G. Hodgen, personal communication), this approach appears to hold promise.

5. 20,22-Desmolase Deficiency

An extremely uncommon form of CAH, 20,22-desmolase deficiency, inherited as an autosomal recessive trait, has also been called lipoid hyperplasia because the enlarged adrenal gland contains excessive quantities of foam cells laden with cholesterol. Affected infants have severe adrenal insufficiency and the fatality rate is high despite apparently adequate steroid and sodium replacement. Long-term survival, however, has been reported (Camacho *et al.*, 1968). Affected females have normal external genitalia, while the males have female external genitalia and a male duct system. Both gluco- and mineralocorticoid replacement are necessary. Presuming that the ovaries have a similar defect, estrogen replacement is also likely to be necessary at the time of puberty in surviving females. With our current knowledge, induction of ovulation in affected females seems almost impossible. Since the ovaries should contain viable ova, techniques for inducing ovulation may be developed in the future.

6. *18-Hydroxysteroid Dehydrogenase Deficiency*

Although the number of reported cases with 18-hydroxylase deficiency is few, autosomal recessive inheritance seems likely (Rimoin and Schimke, 1971). In this disorder, formation of aldosterone from corticosterone is blocked and corticosterone and DOC are both present in the plasma in elevated concentrations (Visser and Cost, 1964). Although salt supplementation may be necessary early in life, no treatment may be required later if sodium intake is adequate. Pregnancy in affected females should proceed normally so long as electrolyte balance is maintained.

7. *Diagnosis, Treatment, and Pregnancy in Virilized Patients*

Ambiguous external genitalia at birth almost always suggest a diagnosis of CAH. As noted earlier, in genetic females there is usually no alteration of the internal duct system or the gonads. The urinary or plasma steroidal pattern will usually establish the diagnosis. It is imperative that the diagnosis be made as early as possible, so that suitable replacement therapy can be initiated. Treatment with cortisol or its analogs not only suppresses the abnormal steroids produced in excess but also avoids the possible consequences of cortisol deficiency. In the salt-losing forms it is also necessary to administer mineralocorticoids such as DOC or 9α-fluorohydrocortisone.

Several successful pregnancies in patients with treated virilizing CAH have now been reported (Yamashita and Kozakae, 1956; Wilson and Keating, 1958; Gans and Ser, 1959; Swyer and Bonham, 1961). Bongiovanni (1978) has reported 26 normal pregnancies in 21 women with 21-hydroxylase deficiency following the introduction of appropriate suppressive therapy. In 14 women therapy was not undertaken until after the age of 20 years; pregnancy occurred in all within 5 years. In all five women in whom therapy was not begun until age 30 years, pregnancy occurred within 18 months of treatment. Several of the pregnancies have followed plastic surgical procedures on the virilized external genitalia. The majority of patients have been delivered by cesarean section either because of the abnormal external genitalia or because of a small bony pelvis resulting from premature closure of the epiphyses. In the majority of cases, the patient was maintained on the same dosage of cortisol administered prior to the pregnancy. Since urinary excretion of pregnanetriol peaks during the third trimester because of placental production, increases in pregnanetriol excretion may not mean that the maintenance steroid dosage is inadequate. It is recommended that the glucocorticoid dosage be increased at delivery and during periods of stress or illness during the pregnancy (Bongiovanni and Root, 1963).

The majority of children delivered of women with CAH have been normal, without genital anomalies or clinical adrenocortical insufficiency. It must be remembered, however, that if the father is a heterozygote, one-half of all children born to an affected mother will also be affected; if the father is normal, all children will be heterozygotes. If the affected mother is not adequately treated during pregnancy, then virilization of a normal infant can occur. Exposure of a female fetus to endogenous or exogenous androgens can result in virilization ranging from mild clitoromegaly to complete labioscrotal fusion with a penile urethra (Mürset *et al.*, 1970; Haymond and Weldon, 1973; Ferrante, 1976; Bell, 1977). Additionally, since glucocorticoids can cross the placenta, the normal child must be observed closely following birth for signs of adenocortical insufficiency as a result of chronic suppression *in utero*. Most normal infants have received only routine care and done well; however, several infants have arbitrarily been administered small doses of steroids after birth (Bongiovanni and Root, 1963). Gans and Ser (1959) have recommended the prophylactic administration of 15 mg of cortisone acetate per day for 7 days and 6 IU of ACTH per day for 14 days to normal infants born of treated affected mothers. In view of the observation that the response of such infants to ACTH is poor at 11 days of age but normal at 10 weeks of age (Laron, 1959), such prophylactic therapy may well be justified.

8. Identification of Heterozygotes

In view of the possibility of asymptomatic heterozygotes transmitting the disorder to their offspring, attempts have been made to identify such carriers. Qazi *et al.* (1971) analyzed the urinary steroid pattern of 43 parents of children with CAH due to 21-hydroxylase deficiency. The fathers alone excreted significantly greater than normal amounts of 11-hydroxyandrosterone with the excretion of relatively large amounts of 11-oxygenated 17-ketosteroids being typical of the pattern found in 21-hydroxylase deficiency. Unfortunately, although statistically significant, the differences are not of sufficient magnitude to identify individual instances of heterozygosity. More recently, Lee and Gareis (1975) were able to identify six of seven parents of children with 21-hydroxylase deficiency as heterozygotes on the basis of their response to ACTH. These individuals demonstrated significantly elevated increases in 17-hydroxyprogesterone at 30 and 60 minutes following the rapid intravenous injection of 25 units of synthetic ACTH 1-24 in comparison to normal controls. Since one parent was not diagnosed as having a partial 21-hydroxylase deficiency on the basis of this test, the data suggest either that all heterozygotes cannot be identified on the basis of 17-hydroxyprogesterone levels following ACTH stimulation or that the occurrence of the enzyme deficiency in the affected offspring was the result of a new mutation.

Studies have also indicated a close genetic linkage between the locus for 21-hydroxylase deficiency and the locus for HLA-B (Levine *et al.*, 1978). Studies in 34 unrelated families have revealed that HLA genotyping of siblings of patients with CAH with 21-hydroxylase deficiency can be used to identify heterozygous carriers and normal siblings.

9. Prenatal Diagnosis

In fetuses at risk, prenatal diagnosis of CAH should be possible. Jeffcoate *et al.* (1965) proposed amniotic fluid analysis for the detection of CAH prenatally. This premise was based on the facts that fetal urine contributes to the amniotic fluid from the 20th week of gestation onward and that the fetal adrenal gland is functioning by the 12th to 14th week. They reported elevated concentrations of 17-ketosteroids and pregnanetriol in the amniotic fluid of a pregnancy that resulted in a normal appearing full-term female. Although these results were obtained in late pregnancy, they provided for prompt treatment of what proved to be an affected infant following delivery.

Subsequent determinations of pregnanetriol, 17-ketosteroids and estriol in amniotic fluid and maternal urine during pregnancies at risk for CAH have demonstrated that elevated levels in late pregnancy are diagnostic; however, values are not always abnormally increased in fetuses with CAH, and such determinations are often of no discriminatory value before the last few weeks of pregnancy (New, 1976).

Frasier *et al.* (1975) have attempted to make the prenatal diagnosis of CAH by measuring specific adrenal steroids rather than steroid metabolites in amniotic fluid. They have shown that 17-hydroxyprogesterone is significantly elevated in both amniotic fluid in late pregnancy and in cord blood of infants with CAH. More recently, it has been noted that 17-hydroxyprogesterone was even elevated in the maternal blood of an infant later born with CAH (Nagamani *et al.*, 1978). Nagamani *et al.* (1978) also noted increased amniotic fluid levels of 17-hydroxyprogesterone in four amniotic fluid specimens obtained between 14 and 34 weeks of a pregnancy, which resulted in the birth of a male infant with the salt-losing form of 21-hydroxylase deficiency. These last results await confirmation.

C. Adrenal Cortical Carcinoma

Familial adrenal cortical carcinoma independent of the multiple endocrine neoplasia syndromes is extremely uncommon. However, several cases have been reported. Mahloudji *et al.* (1971) reported an affected brother and sister who were the products of a second-cousin union. The affected female had marked hirsutism, an enlarged clitoris and fused labia. Nichols (1968) reported a brother and sister both masculinized since birth with adrenocortical tumors. The presence of normal parents in these females suggests the possibility of autosomal recessive inheritance. If the tumors are completely resected and any needed replacement therapy is instituted, then pregnancy should be possible in affected females, even if they have been virilized.

D. Syndromes with Adrenal Insufficiency

A number of familial syndromes involving adrenal insufficiency and possible other endocrine organs have now been described. Although apparently distinct, autoimmune phenomena appear to be common to all (Table V). Thus, these various syndromes may be a part of a common series of disorders (Spinner *et al.*, 1969).

1. Congenital Adrenal Hypoplasia

More than 30 cases of congenital adrenal hypoplasia have now been described (Rimoin and Schimke, 1971). Affected infants show low urinary steroid excretion unresponsive to ACTH. Males with congenital adrenal hypoplasia outnumber females by approximately 3:1, with most cases being sporadic (Uttley, 1968). Both X-linked and autosomal recessive inheritance have been suggested in different affected families, especially since affected female sibs have been described. Corticosteroid replacement is the only needed therapy. The presence of varying amounts of adrenal tissue occasionally may allow for prolonged survival even without treatment. Recent investigations have indicated that the diagnosis can be made prenatally (Hensleigh *et al.*, 1978). Fetal adrenal hypoplasia should be considered in pregnant patients with the disorder or with family histories of the condition who have drastically reduced maternal estriol excretion. Since

TABLE V
Apparent Autoimmune Endocrinopathies with Genetic Components

I.	Schmidt's syndrome (Addison's disease and chronic lymphocytic thyroiditis ± diabetes mellitus)
II.	Adrenocortical insufficiency, hypoparathyroidism, and mucocutaneous candidiasis
III.	Idiopathic hypoparathyroidism
IV.	Idiopathic ovarian failure
V.	Idiopathic Addison's disease (associated with hyperthyroidism and diabetes mellitus)
VI.	Autoimmune thyroid disorders (thyroiditis–thyrotoxicosis)
VII.	Pernicious anemia (thyrogastric antibodies ± diabetes mellitus)
VIII.	Diabetes mellitus and vitiligo (associated with any disorders in this table)

deteriorating adrenal function frequently follows an asymptomatic period during early neonatal life, antepartum diagnosis is important in the clinical management of these infants.

2. *Idiopathic Adrenal Insufficiency*

There is a marked male preponderance in individuals with simple Addison's disease with symptoms appearing prior to the age of 20 years (Spinner *et al.*, 1968). In those with onset after the age of 20 years, the sex incidence is more nearly equal. It is possible that some cases of infantile male Addison's disease result from congenital adrenal hypoplasia. However, even when those cases with symptoms arising in infancy are eliminated, a male preponderance still exists with numerous reports of affected brothers. On the basis of cases in affected families, it is possible that autosomal dominant, autosomal recessive, and X-linked recessive inheritance may be operative in certain cases (Rimoin and Schimke, 1971). Anti-adrenal antibodies have been demonstrated in several cases (Blizzard *et al.*, 1967; Irvine *et al.*, 1967). As with other forms of adrenal failure, treatment consists of glucocorticoid and possibly mineralocorticoid replacement with continuation through pregnancy in affected women.

3. *Isolated Glucocorticoid Deficiency without Hypoaldosteronism*

There are now several reports of familial glucocorticoid deficiency occurring without hypoaldosteronism, which are believed to be inherited as an autosomal recessive trait (Shepard *et al.*, 1959; Stempfel and Engel, 1960; Williams and Freeman, 1965; Migeon *et al.*, 1968a). This may well be part of the spectrum of congenital adrenal hypoplasia and/or familial Addison's disease since autopsies have revealed either no identifiable adrenal tissue or atrophic adrenal glands. Additionally, at least one child later developed signs of mineralocorticoid deficiency as well (Migeon *et al.*, 1968a). In childhood, the syndrome is characterized by feeding problems, hypoglycemic episodes, and hyperpigmentation of the skin early in life. Cortisol levels are low and do not increase following ACTH. Urinary excretion of 17-hydroxycorticosteroids is also diminished. On a low-sodium diet, serum electrolytes are normal and aldosterone increases normally when stimulated. Affected individuals should be observed closely for development of hypoaldosteronism. Women on replacement therapy should be capable of normal pregnancy.

4. *Adrenocortical Insufficiency, Hypoparathyroidism, and Moniliasis*

Addison's disease, hypoparathyroidism, and mucocutaneous candidiasis arising in childhood have now been observed in several families with autosomal recessive inheritance (Spinner *et al.*, 1969). Premature ovarian failure (with cessation of menses prior to age 40 years) has now been reported in several patients with this syndrome (Golonka and Goodman, 1968; Drury *et al.*, 1970; Kleerekoper *et al.*, 1974; Schachner, 1974). The existence of autoantibodies to endocrine tissues in at least some of these individuals has been demonstrated (Blizzard *et al.*, 1966; Irvine *et al.*, 1968; Block *et al.*, 1971; de Moraes-Ruehsen *et al.*, 1972; Vazquez and Kenny, 1973). More recently, the cytotoxic effects of serum from these patients on human granulosa cells in culture has been noted (McNatty *et al.*, 1975b). In view of these findings and the demonstration of antibodies to target organ receptors of other peptide hormones in myasthenia gravis and thyroid disease, it has been suggested that this disorder may be due to receptor antibodies or to a gonadotropin receptor defect (Lucky *et al.*, 1977). Lucky *et al.* (1977) have reported pubertal progression in some patients who had elevated levels of gonadotropin. These findings which usually suggest ovarian failure indicate that intact functioning follicles exist in some patients with this syndrome. In these cases pregnancy may be possible. Treatment consists of glucocorticoid and mineralocorticoid replacement for the adrenal insufficiency, calcium salts, and vitamin D for the hypoparathyroidism and estrogen replacement for the ovarian failure.

Several mothers with long-standing hypoparathyroidism have had infants with congenital hyperparathyroidism (Anast, 1976). The condition may result in neonatal death. It is believed that maternal hypocalcemia secondary to hypoparathyroidism results in decreased calcium transport from mother to fetus. With hypocalcemia *in utero,* there is hyperplasia of the fetal parathyroid glands with increased parathyroid hormone secretion. This hormone mobilizes calcium from the fetal skeleton. Increased amounts of calcium cross the placenta from the fetus to the mother and may alleviate maternal hypoparathyroidism in some instances. After birth, the infant's skeleton avidly takes up calcium, so that with time the bony lesions heal and the parathyroid hyperplasia subsides.

Treatment of hypoparathyroid mothers with calcium salts and vitamin D prevents these deleterious effects on the infant. It is reported that 50,000 to 200,000 units of vitamin D result in uncomplicated pregnancies with the birth of normal infants (Graham *et al.*, 1964). Unfortunately, treatment of patients with this disorder is not always easy. These women require intensive therapy from early childhood and often do not follow explicit instructions. In addition, the possibility that high levels of therapeutic vitamin D administered to the mother might result in vitamin D intoxication in the neonate must always be considered.

5. Pluriglandular Insufficiency (Schmidt Syndrome)

Adult onset Addison's disease has now been described in association with a host of other disorders. Of these, perhaps the Schmidt syndrome is the best known.

The association of nongranulomatous hypoadrenalism and chronic lymphocytic thyroiditis has been noted in more than 100 cases with the recognition of pleuriglandular endocrine failure of the parathyroid, pancreas, and gonads (Carpenter *et al.*, 1964; Strickland, 1969). Associated pernicious anemia and vitiligo have been described. Most familial cases of Schmidt syndrome have been reported in sibs, suggesting the disorder may be inherited as an autosomal recessive trait (Carpenter *et al.*, 1964). Variation among affected families exists, especially related to age of onset and degree of endocrine involvement. These observations raise the possibility of genetic heterogeneity (Spinner *et al.*, 1968). It has not been definitely established that all pluriglandular deficiency syndromes are inherited. For example, a virus with potential for endocrine gland destruction and autoantibody formation is an alternative etiologic explanation.

Although the Schmidt syndrome has many features in common with the complex of hypoparathyroidism, hypoadrenalism, and moniliasis, definite differences exist. Patients with the monilia-endocrinopathy complex usually show symptoms of either fungal disease or endocrine failure early in life while manifestations of the Schmidt syndrome generally appear in adulthood. The thyroid is involved in Schmidt syndrome by definition, but is typically unaffected in the monilia-endocrinopathy complex, although antithyroid antibodies may be present. Diabetes mellitus is seen frequently in the Schmidt syndrome but has not been reported in the childhood endocrine deficiency complex. It is possible, however, that affected children will develop diabetes later in life. Parathyroid hypofunction is uncommon in the Schmidt syndrome. Furthermore, although both syndromes usually occur in sibs, there seems to be no apparent overlap in any given family (Spinner *et al.*, 1968).

In the absence of gonadal failure, affected females are capable of pregnancy. With continued replacement therapy, resulting pregnancies should be relatively uneventful.

V. THE PARATHYROID GLANDS AND FAMILIAL HYPERPARATHYROIDISM

Hyperparathyroidism is predominantly a sporadic disease but familial occurrence has been recognized (Cutler *et al.*, 1964; Graber and Jacobs, 1968). The signs and symptoms are the same for both the familial and sporadic cases. These include nausea and vomiting, renal stones, abdominal pain, fatigue, bone pain, hallucinations, and bizarre behavior (Schneeberg, 1970). The familial syndrome has been reported in successive generations and includes both sexes, suggesting an autosomal dominant mode of inheritance (Wermer, 1954; Graber and Jacobs, 1968). The involvement of multiple parathyroid glands, rather than a single adenoma, occurs much more commonly in the familial cases (Greenbaum, 1960; Cutler *et al.*, 1964; Graber and Jacobs, 1968). Cutler *et al.* (1964) reported a 47% incidence of multiple gland involvement in familial hyperparathyroidism, which can be compared to a 15% incidence in sporadic cases. These authors emphasize that the finding of hyperparathyroidism in more than one member of a kindred should lead to the suspicion of multiple parathyroid gland involvement, necessitating a thorough neck exploration and careful postoperative follow-up with serum calcium measurements. While these parathyroid tumors are multifocal, they are benign with the histological appearance of generalized hyperplasia, often with nodular adenomatous areas of a single cell type (Wermer, 1954; Graber and Jacobs, 1968). Familial hyperparathyroidism can also be associated with adenomatosis of other endocrine glands, including the anterior pituitary and pancreatic islets (Wermer, 1954; Cutler *et al.*, 1964).

Hyperparathyroidism does not seem to result in a reduction of fertility (Wagner *et al.*, 1964; Johnstone *et al.*, 1972) and the effects of this endocrine disorder on both mother and child have been reported. Most often, tetany in the newborn has been associated with maternal parathyroid adenomas (Hartenstein and Gardner, 1966; Van Arsdel, 1955). The proper diagnosis is important since maternal hyperparathyroidism is detrimental to the fetus, causing increased fetal morbidity and mortality. Johnstone *et al.* (1972) reviewed 37 pregnancies in untreated hyperparathyroid patients and found 5 spontaneous abortions, 5 stillbirths, 4 neonatal deaths, and 18 infants with neonatal tetany. Ludwig (1962) reported a 50% complication rate including 10% stillbirths.

The diagnosis of hyperparathyroidism in pregnancy is often difficult because of the normal fall in total serum calcium concentration during pregnancy. Serum calcium begins to decrease during the second or third month and reaches a low point in the seventh or eighth month (Newman, 1947). The average decrease is about 0.6 mg/100 ml and is felt to be due to the hypoalbuminemia of pregnancy. Both the protein binding of calcium and the concentration of free calcium remain unchanged, resulting in a decrease in the concentration of total calcium (Kerr *et al.*, 1962; Burrow, 1975). In pregnant patients with hyperparathyroidism, serum calcium levels ranged from 9.5 to 20.6 mg/100 ml (Johnstone *et al.*, 1972) indicating that the nonpregnant upper limit of normal of about 10.6 mg/dl is not a useful guide. Other tests of parathyroid function may be helpful, such as serum phosphate, tubular resorption of phosphate, and urinary cyclic AMP levels, but the concentration of serum calcium is said to be the best single test (Burrow, 1975; Shaw *et al.*, 1977). Retrospectively, neonatal tetany may be the best clue for detection of maternal hyperparathyroidism, as it occurs in hypercalcemic women who are asymptomatic. All mothers whose babies develop neonatal tetany should be investigated for hyperparathyroidism (Hartenstein and Gardner, 1966).

Hyperparathyroidism in pregnancy is associated with maternal hypercalcemia, a high calcium concentration in amniotic fluid and fetal hypercalcemia greater than that of the mother (Wagner *et al.*, 1964). Calcium crosses the placenta by active transport and fetal serum calcium is 1–3 mg/100 ml higher than maternal calcium levels (Bodansky and Duff, 1939). Evidence suggests that maternal parathormone does not cross the placenta, but maternal hypercalcemia and hypophosphatemia lead to hypoplasia of the fetal parathyroid glands (Rasmussen, 1974). This hypoplasia leads to the neonatal tetany.

The neonate of a mother with hyperparathyroidism must be watched carefully. Prophylactic calcium administration may prevent neonatal tetany (Johnstone *et al.*, 1972). The infant should be given human milk since the increased phosphorus content of cow's milk is more likely to provoke tetany (Wagner *et al.*, 1964). Pregnant women should be treated surgically at the time of diagnosis since fetal and neonatal mortality and morbidity are high even in the asymptomatic woman (Johnstone *et al.*, 1972). A potentially fatal complication in the untreated mother is the development of a hyperparathyroid crisis during the immediate postpartum period (Schenker and Kallmer, 1965). Abortion is not medically indicated in a patient with hyperparathyroidism unless the disease has resulted in severe hypertension or renal failure.

VI. THE THYROID GLAND

A. Thyroid Function in Pregnancy

In pregnancy, the increased levels of circulating estrogen increase the serum concentration of thyroxine-binding globulin (TBG) resulting in an increase in total circulating thyroxine (Man *et al.*, 1969). The free thyroxine concentration, the biologically active component, remains the same as in the nonpregnant woman (Malkasian and Mayberry, 1970; Sterling and Brenner, 1966). TSH values are unchanged in pregnancy. The essentials of the control of thyroid function are summarized in Fig. 5.

The diagnosis of primary hypothyroidism in pregnancy is confirmed by elevated levels of serum TSH (> 10 μU/ml). Replacement therapy should be increased until TSH is suppressed into the normal range (< 10 μU/ml). For

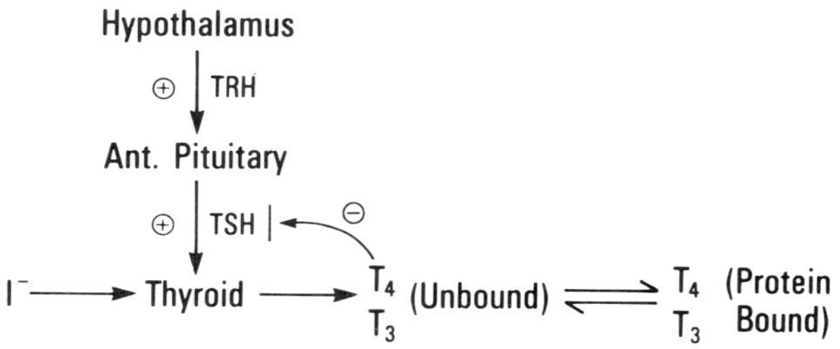

Fig. 5. The factors that regulate thyroid function. Unbound thyroid hormone, thyroxine (T_4), and triiodothyronine (T_3) inhibit secretion of thyroid stimulating hormone (TSH). TSH releasing hormone (TRH) sets the threshold in the pituitary at which this negative feedback occurs.

patients with secondary hypothyroidism (i.e., hypopituitarism), the measurement of TSH is not helpful in determining the adequacy of replacement. The clinical response and plasma T_4 and T_3 levels must be assessed.

The usual replacement dose for both primary and secondary hypothyroidism is approximately 0.15 mg L-thyroxine, or the equivalent strength of other preparations, per day. We prefer L-thyroxine (T_4; Synthroid) replacement. Relatively stable blood levels are achieved throughout the day with a single replacement dose. Treatment with thyroid extract may lead to over or under treatment, because different preparations contain variable amounts of T_4 and T_3. Replacement with triiodothyronine generally does not give a steady plasma level of hormone because of its short half-life. In addition, the laboratory assessment of adequate replacement with T_3 is difficult because of limited availability of accurate serum T_3 radioimmunoassays. Finally, since most of the circulating T_3 originates from deiodination of T_4, it is not necessary to give T_4-T_3 combinations.

B. Athyrotic Congenital Hypothyroidism

Most cases of congenital hypothyroidism (cretinism) are due to a defect in the embryonic development of the thyroid gland so that the gland is absent or aberrant in location. Most cases are sporadic but occasionally a familial occurrence has been reported which is compatible with an autosomal recessive mode of inheritance (Herrman, 1917; Ainger and Kelley, 1955). Ainger and Kelley (1955) reported three siblings with athyrotic cretinism who were products of consanguineous marriage. Herrman (1917) also reported three siblings in one family, all of whom had treatment instituted at an early age. Thyroid hormone replacement is the only effective therapy and if given early can result in relatively normal growth and development. Treated individuals can reproduce normally. When these individuals become pregnant, adequate thyroid replacement is important. Whether maternal thyroid hormone is absolutely necessary for fetal development is not clear, but Niswander and Gordon (1972) in a study of 244 untreated pregnant hypothyroid women reported a doubling of the stillbirth rate. Greenman *et al.* (1962) reported that women who had a low-serum butanol-extractable iodine (BEI) had a higher incidence of reproductive failure than women with normal thyroid function or those with adequate replacement therapy. Fertility is decreased in women with hypothroidism, but normal pregnancy with delivery of a healthy child has been reported in patients with hypothyroidism and even myxedema (Echt and Doss, 1963; Lachelin, 1970).

C. Familial Goiter Secondary to Defective Thyroxine Biosynthesis

Several defects in the biosynthesis of thyroxine have been reported (Stanbury, 1978). The main steps in thyroid hormone biosynthesis are depicted in Fig. 6.

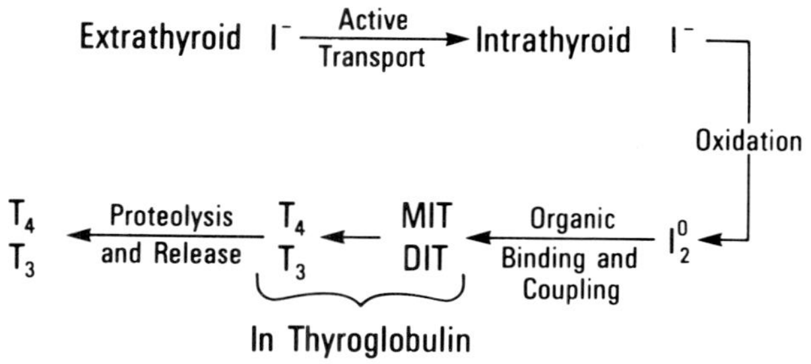

Fig. 6. The major steps in thyroid hormone biosynthesis. Iodide (I^-) is actively transported into thyroid cells. Iodide is oxidized to iodine and incorporated into inactive precursors, monoiodotyrosine (MIT) and diiodotyrosine (DIT). Iodotyrosines are coupled to form triiodothyronine (T_3) and thyroxine (T_4), the active thyroid hormones, and are linked to the colloidal thyroglobulin. When needed, T_4 and T_3 are cleaved from thyroglobulin by proteolysis and released into the general circulation.

Wolff *et al.* (1964) reported a child with congenital hypothyroidism whose thyroid, salivary glands, and gastric mucosa could not transport iodide. Consanguinity in the parents of one patient (Gilboa *et al.*, 1963) and a report of affected brothers in another (Stanbury and Chapman, 1960) support an autosomal recessive inheritance. The diagnosis is made by the demonstration of little, if any, uptake of radioiodide by thyroid tissue and a lack of concentration of iodide by salivary glands (Stanbury, 1978).

The inability of the thyroid to convert accumulated iodide to organically bound iodine has also been reported. This defect limits or prevents thyroid hormone synthesis (Stanbury *et al.*, 1955). The diagnosis is made by the rapid release of radioactive iodide after the administration of perchlorate. This defect is also often associated with low- or absent thyroid peroxidase activity (Niepomniszcze *et al.*, 1973). Again autosomal recessive inheritance is postulated.

Another biosynthetic defect involves the failure of coupling of iodotyrosines into thyroxine and triiodothyronine (Stanbury *et al.*, 1955). Inheritance is thought to be autosomal recessive. The inability to make T_3 and T_4 results in goiter and, if the block is complete, severe congenital hypothyroidism. In all these biosynthetic defects, thyroid hormone replacement will result in suppression of the goiter. In some cases, intelligence seems to improve after treatment. Treated women are fertile and pregnant women should have their therapy continued. Adequate replacement is indicated by normal levels of serum TSH. Untreated pregnant women will develop goiters, and may become hypothyroid. Hypothyroidism is associated with the increased risk of stillbirth mentioned earlier.

D. Thyroiditis

Hashimoto's thyroiditis is one of a group of closely related diseases resulting from self-destruction of the thyroid gland. The thyroid is usually enlarged, firm, and lobulated with the histologic features of parenchymal atrophy and lymphoid infiltration (Vickery and Hamlin, 1961). In many instances associated with hypothyroidism, the gland is not palpable. Hashimoto's thyroiditis is often associated with various antithyroid antibodies. However, there is often no correlation between the level of circulating antibody and the degree of thyroid involvement (DeGroot *et al.*, 1962), and it is unclear if the antibodies are a result of, or the cause of, the thyroiditis. Hashimoto's thyroiditis is almost certainly a primary immunologic disease and is probably due to a specific inherited defect in immunologic control. Both cell-mediated and humoral immunity seem necessary for full expression of this disorder (Volpe, 1975). Several authors have presented evidence for an autosomal dominant mode of inheritance (DeGroot *et al.*, 1962; Hall *et al.*, 1962), but Masi *et al.* (1965) have questioned this conclusion on epidemiologic grounds. Goiter is the outstanding clinical feature of this disease but may resolve later in the course of the illness. The thyroid occasionally enlarges rapidly and may be tender. Early in the illness there may be a mild hyperthyroidism, but later thyroid function tests may indicate euthyroidism or hypothyroidism. The diagnosis is confirmed by biopsy or the finding of high titers of thyroid autoantibodies. In some patients, especially young persons, the antibody titer is low and a thyroid biopsy is diagnostic. Treatment in patients who have become hypothyroid consists of thyroid hormone replacement. This therapy often decreases the size of the goiter and results in normal serum thyroid hormone levels. Thyroid replacement often will not reverse long standing goiter, possibly because of fibrosis of the gland (Ingbar and Woeber, 1974). Glucocorticoids have been given when there is pain and tenderness of the thyroid but are rarely necessary.

Hashimoto's thyroiditis must be distinguished from subacute thyroiditis. The etiology of this thyroid disorder seems to be viral; thyroid autoantibodies are found in a minority of patients and their presence is transitory (Volpe *et al.*, 1967).

Hashimoto's thyroiditis is common and affects women much more often than men. Euthyroid women are fertile and the effects of this disease on pregnancy have been reported. Blizzard *et al.* (1960) report that maternal antithyroid antibodies cross the placenta and occasionally result in the birth of a child with hypothyroidism. However, normal infants are also delivered in the presence of these antibodies, and Stevenson (1973) feels that the data are inconclusive implicating thyroid autoantibodies as a cause of neonatal hypothyroidism. Blizzard *et al.* (1960) stress, however, that the newborn infant of a woman with demonstrable circulating antithyroid antibodies should be watched closely for neonatal hypothyroidism. Klein *et al.* (1974) report that it is possible to detect neonatal hypothyroidism by the measurement of TSH in cord blood or in peripheral blood at 3 days of life. The mean cord serum concentration of TSH was 8.5 μU/ml with an upper limit of normal of about 50 μU/ml in the Klein *et al.* (1974) study. Similar results were reported by Delange *et al.* (1977) utilizing measurement of TSH at 5 days of age with an upper limit of normal for TSH of 12 μU/ml. Even with early detection and prompt replacement therapy, these children commonly show evidence of widespread minor impairment of brain function although intelligence is normal (MacFaul *et al.*, 1978).

E. Graves' Disease

Graves' disease is a disorder characterized by diffuse goiter, hypothyroidism, and infiltrative ophthalmopathy. In any affected individual, these features can be present singly or in combination. Several circulating antibodies to thyroid tissue are common in such patients. Most commonly present is an immunoglobulin G, the long-acting thyroid stimulator (LATS). However, LATS can be found in the serum of only about 50% of actively thyrotoxic patients (Adams, 1965). Another commonly seen immunoglobulin, called LATS-protector, can stimulate human thyroid tissue and has been found in almost 100% of thyrotoxic patients not demonstrating LATS (Adams and Kennedy, 1971; Shishiba *et al.*, 1973). Both LATS-protector and LATS are potent thyroid stimulators, with the former being species specific for the human and the latter stimulating many mammalian thyroids. Although the precise pathogenesis of Graves' disease is in question, knowledge of the etiology is important because neonatal thyrotoxicosis has been associated with the transplacental transfer of maternal antithyroid immunoglobulin (McKenzie, 1964).

The genetics of Graves' disease is complex but it has been reported to be transmitted, at least in some cases, as an autosomal recessive trait with incomplete penetrance (Bartels, 1941), and as a dominant trait (Hollingsworth and Mabry, 1976).

The treatment of hyperthyroidism in pregnancy is controversial. Everyone agrees that the mother should be euthyroid or slightly hyperthyroid, but the disagreement concerns the best method of achieving this goal. Some advocate subtotal thyroidectomy in order to prevent exposure of the fetus to antithyroid medications which have the potential risk of permanent fetal intellectual impairment (Hawe, 1965; Talbert *et al.*, 1970). However, the mother must be given antithyroid drugs just prior to surgery so that she is euthyroid during the actual procedure. Another suggested mode of therapy is the combined use of antithyroid drugs plus replacement thyroid hormone. This was suggested in the hope that the thyroid hormone would pass to the fetus and thus protect it from the antithyroid medication. However, the amount of thyroid hormone that crosses the placenta is limited and unpredictable in the individual case (Fisher *et al.*, 1964; Raiti *et al.*, 1967; Mestman *et al.*, 1974). Also, the amount of the antithyroid medication, which crosses the placenta readily, must be increased because of the supplemental thyroid hormone given to the mother. Thus, the fetus is exposed to higher levels of antithyroid drugs (Mestman *et al.*, 1974). We feel this mode of therapy is no longer indicated. A third treatment plan is to give antithyroid drug alone in the smallest doses possible to maintain the mother in a high euthyroid state (Mestman *et al.*, 1974). The fetal outcome has been equivalent to mothers treated with surgery. A treatment plan is to begin with 30–60 mg daily of methimazole (300–600 mg propylthiouracil). Improvement is expected in 4–8 weeks. The dose should be reduced according to the patient's thyroid function tests and resting pulse rate. Often the patient will undergo remission in the latter weeks of pregnancy, requiring no medication (Mestman *et al.*, 1974). The question of surgery versus antithyroid medication should be individualized. A key factor in the decision is the availability of an experienced thyroid surgeon. Those unwilling or unable to reliably continue medication should probably be treated surgically during the second trimester. If the woman is not euthyroid at term, the risk of thyroid storm during labor or in the immediate postpartum period is substantial.

The fetus and newborn infant of a mother with Graves' disease have several potential problems. Long-acting thyroid stimulator crosses the placenta and can result in thyrotoxicosis in the newborn, including exophthalmus, goiter, and hyperactive state (McKenzie, 1964; Martin and Matus, 1966; Stevenson, 1973). Long-acting thyroid stimulator persists in the mother after treatment by thyroidectomy or antithyroid drugs. The half-life of LATS has been estimated at 13 days. The newborn infant must be followed carefully and sometimes treated with digitalis, phenobarbital, and antithyroid drugs until the thyroid stimulator is no longer present. Newborn deaths from complications of the thyrotoxicosis have been reported in infants of mothers with Graves' disease who were euthyroid in pregnancy because of therapy (Skelton and Gans, 1955). If the mother is receiving no antithyroid medication in pregnancy, the baby's hyperthyroid symptoms will appear at birth. However, babies of mothers treated with antithyroid drugs can have a delayed onset of signs and symptoms since the baby clears the antithyroid drugs it received from the mother faster than the thyroid stimulator (Stevenson, 1973).

In summary, infants of mothers with Graves' disease can be potentially euthyroid (even in the presence of high levels of maternal LATS), hypothyroid (secondary to maternal therapy with propylthiouracil, methimazole, or radioiodine), or hyperthyroid (secondary to transplacental passage of thyroid stimulators). The infant's thyroid status should be assessed at birth by determining cord blood TSH, thyroxine, and LATS (and/or LATS-protector) and at 3 days of age by measuring TSH, thyroxine, triiodothyronine, and LATS. The following therapies for hyperthyroidism are contraindicated in the pregnant woman: iodine, radioiodine, and chronic propanolol.

VII. SUMMARY

It is not possible to generalize regarding the diverse endocrine disorders compatible with pregnancy considered in this chapter. It is fair to state, however, that as our ability to treat endocrine dysfunction effectively has increased, so has the ability of the affected women to become pregnant and bear viable infants. This increased reproductive capacity demands that we recognize those endocrine disorders that may be inherited and are cognizant of the problems that may arise during pregnancy in affected individuals. Potential difficulties for the developing fetus must be recognized as well.

REFERENCES

Adams, D. D. (1965). Pathogenesis of the hyperthyroidism of Graves' disease. *Br. Med. J.* **1,** 1015–1019.

Adams, D. D., and Kennedy, T. H. (1971). Evidence to suggest that LATS protector stimulates the human thyroid gland. *J. Clin. Endocrinol. Metab.* **33,** 47–51.

Ainger, L. E., and Kelley, V. C. (1955). Familial athyreotic cretinism: Report of 3 cases. *J. Clin. Endocrinol. Metab.* **15,** 469–475.

Anast, C. S. (1976). Parathyroid hormone during pregnancy and effect on offspring. *In* "Diabetes and Other Endocrine Disorders During Pregnancy and in the Newborn" (M. I. New and R. H. Fiser, Jr., eds.), pp. 235–248. Alan R. Liss, Inc., New York.

Assali, N. S., Garst, J. R., and Voskian, J. (1955). Blood levels of 17-hydroxycorticosteroids in normal and toxemic mothers. *J. Lab. Clin. Med.* **46,** 385–390.

Babaknia, A., Calfopoulos, P., and Jones, H. W., Jr. (1976). The Stein-Leventhal syndrome and coincidental ovarian tumors. *Obstet. Gynecol.* **47,** 223–224.

Baker, J. W., Morgan, R. L., Peckham, M. J., and Smithers, D. W. (1972). Preservation of ovarian function in patients requiring radiotherapy for paracervical and pelvic Hodgkin's disease. *Lancet* **1,** 1307–1308.

Ballard, H. S., Frame, B., and Hartsock, R. J. (1964). Familial multiple endocrine adenoma-peptic ulcer complex. *Medicine (Baltimore)* **43,** 481–516.

Bartels, E. D. (1941). "Heredity in Graves' Disease with Remarks on Heredity in Toxic Adenoma in the Thyroid, Non-toxic Goiter and Myxedema." Munksgaard, Copenhagen.

Baxter, J. D., and Forsham, P. H. (1972). Tissue effects of glucocorticoids. *Am. J. Med.* **53,** 573–589.

Bayliss, R. I. S., Browne, J. C. M., Round, B. P., and Steinbeck, A. W. (1955). Plasma-17-hydroxycorticoids in pregnancy. *Lancet* **1,** 62–64.

Beck, P., Eaton, C. J., Young, I. S., and Kupperman, H. S. (1968). Metyrapone response in pregnancy. *Am. J. Obstet. Gynecol.* **100,** 327–330.

Bell, J., Spitz, I., Slonim, A., Perlman, A., Segal, S., Palti, Z., and Rabinowitz, D. (1973). Heterogeneity of gonadotropin response to LH RH in hypogonadotropic hypogonadism. *J. Clin. Endocrinol. Metab.* **36,** 791–794.

Bell, R. J. M. (1977). Female virilization due to maternal Krukenberg tumour. *Lancet* **1,** 1162.

Bhatia, S. K., Moore, D., and Kalkhoff, R. K. (1972). Progesterone suppression of the plasma growth hormone response. *J. Clin. Endocrinol. Metab.* **35,** 364–369.

Biglieri, E. G., Herron, M. A., and Brust, N. (1966). 17-hydroxylation in man. *J. Clin. Invest.* **45,** 1946–1954.

Blizzard, R. M., Chandler, R. W., Landing, B. H., Pettit, M. D., and West, C. D. (1960). Maternal autoimmunization to thyroid as a probable cause of athyrotic cretinism. *N. Engl. J. Med.* **263,** 327–336.

Blizzard, R. M., Chee, D., and Davis, W. (1966). The incidence of parathyroid and other antibodies in the sera of patients with idiopathic hypoparathyroidism. *Clin. Exp. Immunol.* **1,** 119–128.

Blizzard, R. M., Chee, D., and Davis, W. (1967). The incidence of adrenal and other antibodies in the sera of patients with idiopathic adrenal insufficiency (Addison's disease). *Clin. Exp. Immunol.* **2,** 19–30.

Block, M. B., Pachman, L., Windhorst, D., and Goldfine, I. D. (1971). Immunological findings in familial juvenile endocrine deficiency syndrome associated with mucocutaneous candidiasis. *Am. J. Med. Sci.* **261,** 213–218.

Bodansky, M., and Duff, V. B. (1939). Regulation of the level of calcium in the serum during pregnancy. *J. Am. Med. Assoc.* **112,** 223–229.

Bongiovanni, A. M. (1961). Unusual steroid pattern in congenital adrenal hyperplasia deficiency of 3 β-hydroxysteroid dehydrogenase. *J. Clin. Endocrinol. Metab.* **21,** 860–862.

Bongiovanni, A. M. (1962). The adrenogenital syndrome with deficiency of 3 β-hydroxysteroid dehydrogenase, *J. Clin. Invest.* **41,** 2086–2092.

Bongiovanni, A. M. (1978). Congenital adrenal hyperplasia and related conditions. *In* "The Metabolic Basis of Inherited Disease" (J. B. Stanbury, J. B. Wyngaarden, and D. S. Fredrickson, eds.), 4th ed., pp. 868–893. McGraw-Hill, New York.

Bongiovanni, A. M., and Root, A. W. (1963). The adrenogenital syndrome. *N. Engl. J. Med.* **268,** 1391–1399.

Brasel, T. A., Wright, J. C., Wilkins, L., and Blizzard, R. M. (1965). An evaluation of seventy-five patients with hypopituitarism beginning in childhood. *Am. J. Med.* **38,** 484–498.

Braunstein, G. D., and Kohler, P. O. (1972). Pituitary function in Hand–Schuller–Christian disease. *N. Engl. J. Med.* **286,** 1225–1229.

Brent, F. (1950). Addison's disease and pregnancy. *Am. J. Surg.* **79,** 645–652.

Burke, C. W., and Roulet, F. (1970). Increased exposure of tissues to cortisol in late pregnancy. *Br. Med. J.* **1,** 657–659.

Burrow, G. N. (1975). Adrenal pituitary and thyroid disorders. *In* "Medical Complications During Pregnancy" (G. N. Burrow and T. F. Ferris, eds.), pp. 242–274. Saunders, Philadelphia, Pennsylvania.

Buttram, V. C., and Vaquero, C. (1975). Post-ovarian wedge resection adhesive disease. *Fertil. Steril.* **26,** 874–876.

Camacho, A. M., Kowarski, A., Migeon, C. J., and Brough, J. (1968). Congenital adrenal hyperplasia due to a deficiency in one of the enzymes involved in the biosynthesis of pregnenolone. *J. Clin. Endocrinol. Metab.* **28,** 153–161.

Carpenter, C. C. J., Solomon, N., Silverberg, S. G., Bledsoe, T., Northcutt, R. C., Klinenberg, J. R., Bennett, I. L., and Harvey, A. M. (1964). Schmidt's syndrome (thyroid and adrenal insufficiency): A review of the literature and a report of fifteen new cases including the instances of coexistent diabetes mellitus. *Medicine (Baltimore)* **43,** 153–180.

Christy, N. P., Wallace, E. Z., Gordon, W. E. L., and Jailer, J. W. (1959). On the rate of hydrocortisone clearance from plasma in pregnant women and in patients with Laennec's cirrhosis. *J. Clin. Invest.* **38,** 299–305.

Cohen, J., Stiefel, M., Reddy, W. J., and Laidlaw, J. C. (1958). The secretion and disposition of cortisol during pregnancy. *J. Clin. Endocrinol. Metab.* **18,** 1076–1092.

Collins, E., and Turner, G. (1973). The Noonan syndrome—a review of the clinical and genetic features of 27 cases. *J. Pediatr.* **83,** 941–950.

Cooper, H. E., Spellacy, W. N., Prem, K. A., and Cohen, W. D. (1968). Hereditary factors in the Stein-Leventhal syndrome. *Am. J. Obstet. Gynecol.* **100,** 371–387.

Cutler, R. E., Reiss, E., and Ackerman, L. V. (1964). Familial hyperparathyroidism. *N. Engl. J. Med.* **270,** 859–865.

DeFraites, E. B., Thurmon, T. F., and Farhadian, H. (1971). Familial Prader-Willi syndrome. *Birth Defects, Orig. Artic. Ser.* **7,** 123–126.

DeGroot, L. J., Hall, R., McDermott, W. V., and Davis, A. M. (1962). Hashimoto's thyroiditis. A genetically conditioned disease. *N. Engl. J. Med.* **267,** 267–273.

Delange, F., Camus, M., Winkler, M., Dodian, J., and Ermans, A. (1977). Serum thyrotrophin determination on day 5 of life as screening procedure of congenital hypothyroidism. *Arch. Dis. Child.* **52,** 89–96.

DeMorsier, G. (1954). Etudes sur les dysraphies cranio-encéphaliques. I. Agénésie des lobes olfactifs (télencéphaloschizis latéral) et/des commisseures calleuse et antérieure (télencéphaloschizis médian). La dysplasie olfactogenitale). *Schweiz. Arch. Neurol. Psychiatr.* **74,** 309–361.

de Moraes-Ruehsen, M., Blizzard, R. M., Garcia-Bunvel, R., and Jones, G. S. (1972). Autoimmunity and ovarian failure. *Am. J. Obstet. Gynecol.* **112,** 693–703.

Doe, R. P., Fernandez, R., and Seal, U. S. (1964). Measurement of corticosteroid-binding globulin in man. *J. Clin. Endocrinol. Metab.* **24,** 1029–1039.

Drury, M. I., Keelan, D. M., Timoney, F. J., and Irvine, W. J. (1970). Juvenile familial endocrinopathy. *Clin. Exp. Immunol.* **7,** 125–132.

Echt, L. R., and Doss, J. R. (1963). Myxedema in pregnancy. *Obstet. Gynecol.* **22,** 615–620.

Espiner, E. A., Veola, A. M. O., Sands, V. E., and Fitzgerald, P. H. (1970). Familial syndrome of streak gonads and normal male karyotype in five phenotypic females. *N. Engl. J. Med.* **283,** 6–11.

Ewer, R. W. (1968). Familial monotropic pituitary gonadotropin insufficiency. *J. Clin. Endocrinol. Metab.* **28,** 783–788.

Ferrante, L. (1976). Virilizzazione congenita in neonata di madre affetta da sindrome surrenogenitale per difetto di 21-idrossilasi. *Minerva Pediatr.* **28,** 1443–1444.

Fisher, D. A., Lehman, H., and Packey, C. (1964). Placental transport of thyroxine. *J. Clin. Endocrinol. Metab.* **24,** 393–400.

Fraenkel, L. (1943). Thecoma and hyperthecosis of the ovary. *J. Clin. Endocrinol. Metab.* **3,** 557–559.

Francke, C. (1950). The gonads in the Laurence-Moon-Biedl syndrome. *J. Clin. Endocrinol. Metab.* **10,** 108–112.

Frasier, S. D., Thorneycroft, I. H., Weiss, B. A., and Horton, R. (1975). Elevated amniotic fluid concentration of 17 α-hydroxyprogesterone in congenital adrenal hyperplasia. *J. Pediatr.* **86,** 310–312.

Gans, F., and Ser, J. (1959). Normal pregnancy and delivery in female pseudohermaphroditism. *Acta Endocrinol.* (*Copenhagen*) **30,** 424–434.

Gauthier, S. (1960). La dysplasie olfacto-genitale. *Acta Neuroveg.* **21,** 345–394.

Gemzell, C. A. (1953). Blood levels of 17-hydroxycorticosteroids in normal pregnancy. *J. Clin. Endocrinol. Metab.* **13,** 898–902.

Gemzell, C. A. (1965). Induction of ovulation with human gonadotropins. *Recent Prog. Horm. Res.* **21,** 179–204.

Gilboa, Y., Ber, A., Lewitus, Z., and Hasenfratz, J. (1963). Goitrous myxedema due to iodide trapping defect. *Arch. Intern. Med.* **112,** 212–215.

Givens, J. R., Wiser, W. L., Coleman, S. A., Wilroy, R. S., Anderson, R. N., and Fish, S. A. (1971). Familial ovarian hyperthecosis. A study of two families. *Am. J. Obstet. Gynecol.* **110,** 959–972.

Goldenberg, R. L., Powell, R. D., Rosen, S. W., Marshall, J. R., and Ross, G. T. (1976). Ovarian morphology in women with hypogonadotropic hypogonadism. *Am. J. Obstet. Gynecol.* **126,** 91–94.

Goldsmith, O., Solomon, D. H., and Horton, R. (1967). Hypogonadism and mineralocorticoid excess. *N. Engl. J. Med.* **277,** 673–677.

Golonka, J. E., and Goodman, A. D. (1968). Coexistence of primary ovarian insufficiency, primary adrenocortical insufficiency and idiopathic hypoparathyroidism. *J. Clin. Endocrinol. Metab.* **28,** 79–82.

Graber, A. L., and Jacobs, K. (1968). Familial hyperparathyroidism: medical and surgical consideration. *J. Am. Med. Assoc.* **204,** 542–544.

Graham, W. P., III, Gordon, G. S., Loken, H. F., Blum, A., and Halden, A. (1964). Effects of pregnancy and of the menstrual cycle on hypoparathyroidism. *J. Clin. Endocrinol. Metab.* **24,** 512–516.

Green, J. A., and Goldzieher, J. W. (1966). The polycystic ovary. IV. Light and electron microscope studies. *Am. J. Obstet. Gynecol.* **91,** 173–181.

Greene, J. W., Jr. (1973). Assessing maternal estriol excretion. *Contemp. Obstet. Gynecol.* **2,** 11–14.

Greenbaum, D. (1960). Multiple familial parathyroid adenomata. *Proc. R. Soc. Med.* **53,** 903.

Greenman, F. W., Gabrielson, M. O., Howard-Flanders, J., and Wessel, M. A. (1962). Thyroid dysfunction in pregnancy. *N. Engl. J. Med.* **267,** 426–431.

Grumbach, M. M., and VanWyk, J. J. (1974). Disorders of sex differentiation. *In* "Textbook of Endorinology" (R. H. Williams, ed.), 5th ed., pp. 476–501. Saunders, Philadelphia, Pennsylvania.

Hall, R., Saxena, K. M., and Owen, S. G. (1962). A study of the parents of patients with Hashimoto's disease. *Lancet* **2,** 1291–1292.

Hardy, J. (1979). The transsphenoidal surgical approach to the pituitary. *Hosp. Pract.* **14,** 81–89.

Hartenstein, H., and Gardner, L. (1966). Tetany of the newborn associated with maternal parathyroid adenoma. *N. Engl. J. Med.* **274,** 266–268.

Hawe, P. (1965). The management of thyrotoxiosis during pregnancy. *Br. J. Surg.* **52,** 731–734.

Haymond, M. W., and Weldon, V. V. (1973). Familial pseudohermaphroditism secondary to a maternal virilizing tumor. *J. Pediatr.* **82,** 682–686.

Hensleigh, P. A., Moore, W. V., Wilson, K., and Tulchinsky, D. (1978). Congenital X-linked adrenal hypoplasia. *Obstet. Gynecol.* **52,** 228–232.

Herrman, C. (1917). Three children with sporadic cretinism in one family. *Arch. Pediatr.* **34,** 831–836.

Hollingsworth, D. R., and Mabry, C. C. (1976). Congenital Graves disease. *Am. J. Dis. Child.* **130,** 148–155.

Ingbar, S. H., and Woeber, K. A. (1974). The thyroid gland. *In* "Textbook of Endocrinology" (R. H. Williams, ed.), 5th ed., pp. 45–232. Saunders, Philadelphia, Pennsylvania.

Irvine, W. J., Stewart, A. G., and Scarth, L. (1967). A clinical and immunological study of adrenocortical insufficiency (Addison's disease). *Clin. Exp. Immunol.* **2,** 31–70.

Irvine, W. J., Chan, M. M. W., Scarth, L., Kolb, F. O., Hartog, M., Bayliss, R. I. S., and Drury, M. I. (1968). Immunological aspects of premature ovarian failure associated with idiopathic Addison's disease. *Lancet* **2,** 883–887.

Jacox, H. W. (1939). Recovery following human ovarian irradiation. *Radiology* **32,** 538–545.

Jailer, J. W., Christy, N. P., Longson, D., Wallace, E. Z., and Gordon, W. E. L. (1959). Further observations on adrenal corticol function during pregnancy. *Am. J. Obstet. Gynecol.* **78,** 1–10.

Jeffcoate, T. N. A., Fliegner, J. R. H., Russell, S. H., Davis, J. C., and Wade, A. P. (1965). Diagnosis of the adrenogenital syndrome before birth. *Lancet* **2,** 553–555.

Johnstone, R. E., Kreindler, T., and Johnstone, R. E. (1972). Hyperparathyroidism during pregnancies. *Obstet. Gynecol.* **40,** 580–585.

Judd, H. L., Scully, R. E., Atkins, L., Neer, R. M., and Kliman, B. (1970). Pure gonadal dysgenesis with progressive virilization. *N. Engl. J. Med.* **282,** 881-885.

Judd, H. L., Scully, R. E., Herbst, A. L., Yen, S. S. C., Ingersol, F. M., and Kliman, B. (1973). Familial hyperthecosis: Comparison of endocrinologic and histologic findings with polycystic ovarian disease. *Am. J. Obstet. Gynecol.* **117,** 976-982.

Judd, H. L., Rigg, L. A., Anderson, D. C., and Yen, S. S. C. (1976). The effects of ovarian wedge resection on circulating gonadotropin and ovarian steroid levels in patients with polycystic ovary syndrome. *J. Clin. Endocrinol. Metab.* **43,** 347-355.

Kallmann, F. J., Schoenfeld, W. A., and Barrera, S. E. (1944). The genetic aspects of primary eunuchoidism. *Am. J. Ment. Defic.* **48,** 203-236.

Kerr, C., Loken, H. F., Glendening, M. D., Gordon, G. S., and Page, E. W. (1962). Calcium and phosphorous dynamics in pregnancy. *Am. J. Obstet. Gynecol.* **83,** 2-8.

Kistner, R. W. (1975). Induction of ovulation with clomiphene citrate. *In* "Progress in Infertility" (S. J. Behrman and R. W. Kistner, eds.), 2nd ed., pp. 509-536. Little, Brown, Boston, Massachusetts.

Kleerekoper, M., Basten, A., Penny, R., and Posen, S. (1974). Idiopathic hypoparathyroidism with primary ovarian failure. *Arch. Intern. Med.* **134,** 944-947.

Klein, A. H., Agustin, A. V., and Foley, T. P. (1974). Successful laboratory screening for congenital hypothyroidism. *Lancet* **2,** 77-79.

Klein, D., and Ammann, F. (1969). The syndrome of Laurence-Moon-Biedl and allied diseases in Switzerland. *J. Neurol. Sci.* **9,** 479-513.

Koyama, H., Wada, T., Nishizawa, Y., Iwanaga, T., Aoki, Y., Terasawa, T., Kosaki, G., Yamamoto, T., and Wada, A. (1977). Cyclophosphamide-induced ovarian failure and its therapeutic significance in patients with breast cancer. *Cancer* **39,** 1403-1409.

Lachelin, G. C. L. (1970). Myxoedema and pregnancy. *J. Obstet. Gynecol. Br. Commonw.* **77,** 77-79.

Lachelin, G. C. L., Abu-Fadil, S., and Yen, S. S. C. (1977). Functional delineation of hyperprolactinemic-amenorrhea. *J. Clin. Endocrinol. Metab.* **44,** 1163-1174.

Laron, Z. (1959). Observations on a baby born to a mother with congenital adrenal hyperplasia. *Am. J. Dis. Child.* **98,** 162-165.

LeBriggs, R., and Powell, J. R. (1969). Chiari-Frommel syndrome as a part of the Zollinger-Ellison major endocrine adenoma complex. *Calif. Med.* **111,** 92-96.

Lee, P. A., and Gareis, F. J. (1975). Evidence for partial 21-hydroxylase deficiency among heterozygote carriers of congenital adrenal hyperplasia. *J. Clin. Endocrinol. Metab.* **41,** 415-418.

Levine, L. S., Zachmann, M., New, M. I., Prader, A., Pollack, M. S., O'Neill, G. J., Yang, S. Y., Oberfield, S. E., and Dupont, B. (1978). Genetic mapping of the 21-hydroxylase-deficiency gene within the HLA linkage group. *N. Engl. J. Med.* **299,** 911-915.

Linquette, M., Herlant, M., Laine, E., Fossati, P., and Dupont-Lecompte, J. (1967). Adenome à prolactine chez une jeune fille dont a la mere était porteuse d'un adénome hypophysairc avec amenorrhee galactorrhee. *Ann. Endocrinol. (Paris)* **29,** 773-780.

Lucky, A. W., Rebar, R. W., Blizzard, R. M., and Goren, E. M. (1977). Prepubertal progression in the presence of elevated serum gonadotropins in girls with multiple endocrine deficiencies. *J. Clin. Endocrinol. Metab.* **45,** 673-678.

Ludwig, G. D. (1962). Hyperparathyroidism in relation to pregnancy. *N. Engl. J. Med.* **267,** 637-642.

MacFaul, R., Dorner, S., Brett, E. M., and Grant, D. B. (1978). Neurological abnormalities in patients treated for hypothyroidism from early life. *Arch. Dis. Child.* **53,** 611-619.

McKenzie, J. M. (1964). Neonatal Graves disease. *J. Clin. Endocrinol. Metab.* **24,** 660-668.

McNatty, K. P., Hunter, W. M., McNeilly, A. S., and Sawers, R. S. (1975a). Changes in the concentration of pituitary and steroid hormones in the follicular fluid of human graffian follicles throughout the menstrual cycle. *J. Endocrinol.* **64,** 555-571.

McNatty, K. P., Short, R. V., Barnes, E. W., and Irvine, W. J. (1975b). The cytotoxic effect of serum from patients with Addison's disease and autoimmune ovarian failure on human granulosa cells in culture. *Clin. Exp. Immunol.* **22,** 378-384.

Mahloudji, M., Ronaghy, H., and Dutz, W. (1971). Virilizing adrenal carcinoma in two sibs. *J. Med. Genet.* **8,** 160-163.

Malkasian, G. D., and Mayberry, W. E. (1970). Serum total and free thyroxine and thyrotropin in normal and pregnant women, neonates and women receiving progestogens. *Am. J. Obstet. Gynecol.* **108,** 1234-1238.

Mallin, S. R. (1969). Congenital adrenal hyperplasia secondary to 17-hydroxylase deficiency. *Ann. Intern. Med.* **70,** 69-75.

Man, E. B., Reid, W. A., Hellenger, A. E., and Jones, W. S. (1969). Thyroid function in human pregnancy. III. Serum thyroxine-binding prealbumin (TBDA) and thyroxine-binding globulin (TBG) of pregnant women ages 14 through 43 years. *Am. J. Obstet. Gynecol.* **103,** 338-347.

Manuel, J., Katayama, K. P., and Jones, H. W., Jr. (1976). The age of occurrence of gonadal tumors in intersex patients with a Y chromosome. *Am. J. Obstet. Gynecol.* **124,** 2923-300.

Martin, M., and Matus, R. M. (1966). Neonatal exophthalmus with maternal thyrotoxicosis. *Am. J. Dis. Child.* **111,** 545-547.

Masi, A., Hartmann, W. H., and Shulman, L. E. (1965). Hashimoto's disease. *J. Chronic Dis.* **18,** 1-22.

Mestman, J., Manning, P., and Hodgman, J. (1974). Hyperthyroidism and pregnancy. *Arch. Intern. Med.* **134,** 434-439.

Migeon, C. J., Kenny, F. M., Kowarski, A., Snipes, C. A., Spaulding, J. S., Finkelstein, J. W., and Blizzard, R. M. (1968a). The syndrome of congenital adrenocortical unresponsiveness to ACTH. Report of six cases. *Pediatr. Res.* **2,** 501-513.

Migeon, C. J., Kenny, F. M., and Taylor, F. H. (1968b). Cortisol production rate. VIII. Pregnancy. *J. Clin. Endocrinol. Metabl.* **28,** 661-666.

Mürset, G., Zachman, M., Prader, A., Fisher, J., and Labhart, A. (1970). Male external genitalia of a girl caused by a virilizing adrenal tumor in the mother. *Acta Endocrinol. (Copenhagen)* **65,** 627-638.

Nader, S., Mashiter, K., Doyle, F. H., and Joplin, G. F. (1976). Galactorrhoea, hyperprolactinaemia and pituitary tumours in the female. *Clin. Endocrinol.* **5,** 245-251.

Nagamani, M., McDonough, P. G., Ellegood, J. O., and Mahesh, V. B. (1978). Maternal and amniotic fluid 17 α-hydroxyprogesterone levels during pregnancy: Diagnosis of congenital adrenal hyperplasia in utero. *Am. J. Obstet. Gynecol.* **130,** 791-794.

New, M. I. (1976). Prenatal diagnosis of congenital adrenal hyperplasia. *In* "Diabetes and Other Endocrine Disorders During Pregnancy and in the Newborn" (M. I. New and R. H. Fiser, Jr., eds.), pp. 205-219. Alan R. Liss, Inc., New York.

Newman, R. L. (1947). Blood calcium: A normal curve for pregnancy, *Am. J. Obstet. Gynecol.* **53,** 817-822.

Nichols, J. (1968). Adrenal cortex. *In* "Endocrine Pathology" (J. Bloodworth, ed.), pp. 224-225. Williams & Wilkins, Baltimore, Maryland.

Niepomniszcze, H., Castells, S., DeGroot, L. J., Refetoffs, S., Kim, O. S., Rapoport, B., and Hati, R. (1973). Peroxidase defect in congenital goiter with complete organification block. *J. Clin. Endocrinol. Metab.* **36,** 347-357.

Niswander, K. R., and Gordon, M. (1972). "Women and their Pregnancies." Saunders, Philadelphia, Pennsylvania.

Noonan, J. A. (1968). Hypertelorism with Turner phenotype. *Am. J. Dis. Child.* **116,** 373-380.

Noonan, J. A., and Ehmke, D. A. (1963). Associated noncardiac malformations in children with congenital heart disease. *J. Pediatr.* **63,** 468–470.

Nora, J. J., Nora, A. H., Sinha, A. K., Spangler, R. D., and Lubs, H. A. (1974). The Ullrich-Noonan syndrome (Turner phenotype). *Am. J. Dis. Child.* **127,** 48–55.

Odell, W. D. (1966). Isolated deficiencies of anterior pituitary hormones. *J. Am. Med. Assoc.* **197,** 1006–1016.

Prader, A., Labhart, A., and Willi, H. (1956). Ein Syndrom von adipositas, kleinwuchs, kryptorchismus und oligophrenic nach myatonieartigem Zustand im Neugeboreuenalten. *Schweiz. Med. Wochenschr.* **86,** 1260–1261.

Qazi, Q. H., Arnon, R. G., Paydar, M. H., and Mapa, H. C. (1970). Familial occurrence of Noonan syndrome. *Am. J. Dis. Child.* **127,** 696–698.

Qazi, Q. H., Rabkin, M. D., and Frantz, A. G. (1971). Hypopituitarism: A study of growth hormone and other endocrine functions. *Ann. Intern. Med.* **64,** 1197–1207.

Raiti, S., Holzman, C. B., and Scott, R. L. (1967). Evidence for the placental transfer of tri-iodothyronine in human beings. *N. Engl. J. Med.* **277,** 456–459.

Raj, S. G., Berger, M. J., Grimes, E. M., and Taymor, M. L. (1977). The use of gonadotropins for the induction of ovulation in women with polycystic ovarian disease. *Fertil. Steril.* **28,** 1280–1284.

Rasmussen, H. (1974). Parathyroid hormone calcitonin and the calciferols. *In* "Textbook of Endorinology" (R. H. Williams, ed.), 5th ed., Chapter II, pp. 660–773. Saunders, Philadelphia, Pennsylvania.

Rebar, R., Judd, H. L., Yen, S. S. C., Rakoff, J., Vanden Berg, G., and Naftolin, F. (1976). Characterization of the inappropriate gonadotropin secretion in polycystic ovary syndrome. *J. Clin. Invest.* **57,** 1320–1329.

Reinfrank, R. F., and Nichols, F. L. (1964). Hypogonadotrophic hypogonadism in the Laurence-Moon syndrome. *J. Clin. Endocrinol. Metab.* **24,** 48–53.

Reyes, F. I., Koh, K. S., and Faiman, C. (1976). Fertility in women with gonadal dysgenesis. *Am. J. Obstet. Gynecol.* **126,** 668–670.

Rimoin, D. L., and Schimke, R. M. (1971). "Genetic Disorders of the Endocrine Glands." Mosby, St. Louis, Missouri.

Rimoin, D. L., Holzman, G. B., Merimee, T. J., Rabinowitz, D., Barnes, A. C., Tyson, J. E. A., and McKusick, V. A. (1968a). Lactation in the absence of human growth hormone. *J. Clin. Endocrinol. Metab.* **28,** 1183–1188.

Rimoin, D. L., Merimee, T. J., Rabinowitz, D., and McKusick, V. A. (1968b). Genetic aspects of clinical endocrinology. *Rec. Prog. Horm. Res.* **24,** 365–437.

Rose, L. I., Underwood, R. H., Williams, G. H., and Pinkus, G. S. (1974). Pure gonadal dysgenesis. *Am. J. Med.* **57,** 957–961.

Rosen, S. W. (1965). The syndrome of hypogonadism, anosmia, and midline cranial anomalies. *Proc. 47th Meet. Endocr. Soc.*

Rosenthal, H. E., Slaunwhite, W. R., Jr., and Sandberg, A. A. (1969). Transcortin: A corticosteroid-binding protein of plasma, cortisol and progesterone interplay and unbound levels of these steroids in pregnancy. *J. Clin. Endocrinol. Metab.* **29,** 352–367.

Ross, G. T., and WandeWiele, R. L. (1974). The ovaries. *In* "Textbook of Endocrinology" (R. H. Williams, ed.), 5th ed., pp. 368–422. Saunders, Philadelphia, Pennsylvania.

Rust, L. A., Israel, R., and Mishell, D. R. (1974). An individualized graduated therapeutic regimen for clomiphene citrate. *Am. J. Obstet. Gynecol.* **120,** 785–790.

Saez, J. M., Morera, A. M., and Bertrand, J. (1969). Testicular endocrine function in males with Noonan's syndrome. *Lancet* **2,** 1078–1079.

Schachner, S. H. (1974). Primary adrenal and ovarian failure. *Va. Med. Mon.* **101,** 638–639.

Schenker, J. G., and Kallmer, B. (1965). Fatal postpartum hyperparathyroid crisis due to primary chief cell hyperplasia of parathyroids. *Obstet. Gynecol.* **25,** 705–709.

Schneeberg, N. G. (1970). "Essentials of Clinical Endocrinology." Mosby, St. Louis, Missouri.

Schoeck, V. W., Peterson, R. D. A., and Good, R. A. (1963). Familial Letterer-Siwe disease. *Pediatrics* **32,** 1055–1063.

Shaw, J. W., Oldham, S. B., Rosoff, C., Bethune, J. E., and Fichman, M. D. (1977). Urinary cyclic AMP analyzed as a function of the serum calcium and parathyroid hormone in the differential diagnosis of hypercalcemia. *J. Clin. Invest.* **59,** 14–21.

Shepard, T. H., Landing, B. H., and Mason, D. G. (1959). Familial Addison's Disease. *Am. J. Dis. Child.* **97,** 154–162.

Shishiba, Y., Shimizu, T., Yoshimura, S., and Shizume, K. (1973). Direct evidence for human thyroidal stimulation by LATS-protector. *J. Clin. Endocrinol. Metab.* **36,** 517–521.

Simpson, J. L., Christakos, A. C., Horwith, M., and Silverman, F. S. (1971). Gonadal dysgenesis in individuals with apparently normal chromosome complements. *Birth Defects, Orig. Artic. Ser.* **7,** 215–228.

Singh, R. P., and Carr, D. H. (1966). The anatomy and histology of XO human embryos and fetuses. *Anat. Rec.* **155,** 369–381.

Siris, E. S., Leventhal, B. G., and Vaitukaitis, J. L. (1976). Effects of childhood leukemia and chemotherapy on puberty and reproductive function in girls. *N. Engl. J. Med.* **294,** 1143–1146.

Skelton, M. O., and Gans, B. (1955). Congenital thyrotoxicosis, hepatosplenomegaly and jaundice in two infants of exophthalmic mothers. *Arch. Dis. Child.* **30,** 460–464.

Sohval, A. R. (1965). The syndrome of pure gonadal dysgenesis. *Am. J. Med.* **38,** 615–625.

Sparkes, R. S., Simpson, R. W., and Paulsen, C. A. (1968). Familial hypogonadotropic hypogonadism with anosmia. *Arch. Intern. Med.* **121,** 534–538.

Spinner, M. W., Blizzard, R. M., and Childs, B. (1968). Clinical and genetic heterogeneity in idiopathic Addison's disease and hypoparathyroidism. *J. Clin. Endocrinol. Metab.* **28,** 795–804.

Spinner, M. W., Blizzard, R. M., Gibbs, J., Abbey, H., and Childs, B. (1969). Familial distributions of organ specific antibodies in the blood of patients with Addison's disease and hypoparathyroidism and their relatives. *Clin. Exp. Immunol.* **5,** 461–468.

Spitz, I. M., Diamant, Y., Rosen, E., Bell, J., Ben David, M., Polishuk, W., and Rabinowitz, D. (1974). Isolated gonadotropin deficiency. *N. Engl. J. Med.* **290,** 10–15.

Stanbury, J. B. (1978). Familial goiter. *In* "The Metabolic Basis of Inherited Disease" (J. B. Stanbury, J. B. Wyngaarden, and D. S. Fredrickson, eds.), 4th ed., pp. 206–239. McGraw-Hill, New York.

Stanbury, J. B., and Chapman, E. M. (1960). Congenital hypothyroidism with goitre. *Lancet* **1,** 1162–1165.

Stanbury, J. B., Ohela, K., and Pitt-Rivers, R. (1955). The metabolism of iodine in 2 goitrous cretins compared with that in 2 patients receiving methimazole. *J. Clin. Endocrinol. Metab.* **15,** 54–72.

Stein, I. F., and Leventhal, M. L. (1935). Amenorrhea associated with bilateral polycystic ovaries. *Am. J. Obstet. Gynecol.* **29,** 181–191.

Stempfel, R. S., and Engel, F. L. (1960). A congenital, familial syndrome of adrenocortical insufficiency without hypoaldosteronism. *J. Pediatr.* **57,** 443–451.

Sterling, K., and Brenner, M. A. (1966). Free thyroxine in human serum: Simplified measurement with the aid of magnesium precipitation. *J. Clin. Invest.* **45,** 153–163.

Stevenson, R. E. (1973). "The Fetus and Newly Born Infant." Mosby, St. Louis, Missouri.

Strickland, R. G. (1969). Pernicious anemia and polyendocrine deficiency. *Ann. Intern. Med.* **70,** 1001–1005.

Swyer, G. I. M., and Bonham, D. G. (1961). Successful pregnancy in a female pseudohermaphrodite. *Br. Med. J.* **1,** 1005–1006.

Tagatz, G., Fialkow, P. J., Smith, D., and Spadoni, L. (1970). Hypogonadotropic hypogonadism associated with anosmia in the female. *N. Engl. J. Med.* **283,** 1326–1329.

Talbert, L. M., Thomas, C. G., Holt, W. A., and Rankin, P. (1970). Hyperthyroidism during pregnancy. *Obstet. Gynecol.* **36,** 779–785.

Teter, J., and Boczkowski, K. (1967). Occurrence of tumors in dysgenetic gonads. *Cancer* **20,** 1301–1310.

Thompson, C. R., and Hansen, L. M. (1970). Peryonal (Menotropins): A summary of clinical experience in the induction of ovulation and pregnancy. *Fertil. Steril.* **21,** 844–853.

Toledo, S. P. A., Medeiros-Neto, G. A., Knobel, M., and Mattar, E. (1977). Evaluation of the hypothalamic-pituitary-gonadal function in the Bardet-Biedl syndrome. *Metab. Clin. Exp.* **26,** 1277–1291.

Tyson, J. E., Rabinowitz, D., Merimee, T. J., and Friesen, H. (1969). Response of plasma insulin and human growth hormone to arginine in pregnancy and postpartum females. *Am. J. Obstet. Gynecol.* **103,** 313–319.

Uttley, W. S. (1968). Familial congenital adrenal hypoplasia. *Arch. Dis. Child.* **43,** 724–730.

Van Arsdel, P. P. (1955). Maternal hyperparathyroidism as a cause of neonatal tetany. *J. Clin. Endocrinol. Metab.* **15,** 680–684.

Vasquez, A. M., and Kenny, F. M. (1973). Ovarian failure and antiovarian antibodies in association with hypoparathyroidism, moniliasis, and Addison's and Hashimoto disease. *Obstet. Gynecol.* **41,** 414–418.

Vickery, A. C., and Hamlin, E. (1961). Struma lymphomatosa (Hashimoto's thyroiditis). *N. Engl. J. Med.* **264,** 226–229.

Visser, H. K., and Cost, W. S. (1964). A new hereditary defect in the biosynthesis of aldosterone, suggesting an 18-oxidation defect. *Acta Endocrinol.* **47,** 589–612.

Vogel, J. M., and Vogel, P. (1972). Idiopathic histiocytosis. A discussion of eosinophilic granuloma, the Hand-Schuller-Christian syndrome, and the Letterer-Siwe syndrome. *Semin. Hematol.* **9,** 349–369.

Volpe, R. (1975). Thyroiditis. *Med. Clin. North Am.* **59,** 1163–1175.

Volpe, R., Row, V. V., and Ezrin, C. (1967). Circulating viral and thyroid antibodies in subacute thyroidits. *J. Clin. Endocrinol. Metab.* **27,** 1275–1284.

Wagner, G., Transbol, I., and Melchior, J. C. (1964). Hyperparathyroidism and pregnancy. *Acta Endocrinol. (Copenhagen)* **47,** 549–564.

Wermer, P. (1954). Genetic aspects of adenomatosis of endocrine glands. *Am. J. Med.* **16,** 363–371.

Wider, J. A., Marshall, J. R., and Ross, G. T. (1969). Familial galactorrhea in three sisters with oligo-ovulation. *J. Am. Med. Assoc.* **209,** 669–671.

Williams, H. E., and Freeman, M. (1965). Primary familial Addison's disease. *Aust. Pediatr. J.* **1,** 93–97.

Wilroy, R. S., Givens, J. R., Wiser, W. L., Coleman, S. A., Andersen, R. N., and Summitt, R. L. (1975). *In* "Genetic Forms of Hypogonadism" (D. Bergsma, ed.), pp. 81–85. Stratton Intercontinental Med. Book, New York.

Wilson, R. B., and Keating, F. R. (1958). Pregnancy following treatment of congenital adrenal hyperplasia with cortisone. *Am. J. Obstet. Gynecol.* **76,** 388–397.

Wolff, J., Thompson, R. H., and Robbins, J. (1964). Congenital goitrous cretinism due to the absence of iodine-concentrating ability. *J. Clin. Endocrinol. Metab.* **24,** 699–707.

Yamashita, T., and Kozakae, F. (1956). Pregnancy and delivery at term by long term cortisone treatment of a congenital adrenocortical hyperplasia. *Endocrinol. Jpn.* **3,** 176–180.

Yen, S. S. C., Samaan, N., and Pearson, O. H. (1967). Growth hormone levels in pregnancy. *J. Clin. Endocrinol. Metab.* **27,** 1341–1347.

Yen, S. S. C., Vela, P., and Rankin, J. (1970). Inappropriate secretion of follicle-stimulating hormone and luteinizing hormone in polycystic ovarian disease. *J. Clin. Endocrinol. Metab.* **30,** 435–442.

Yen, S. S. C., Rebar, R., VandenBerg, G., Ehara, Y., and Siler, T. (1973). Pituitary gonadotrophin responsiveness to synthetic LRF in subjects with normal and abnormal hypothalamic-pituitary-gonadal axis. *J. Reprod. Fertil.,* Suppl. **20,** 137–161.

Yen, S. S. C., Chaney, C., and Judd, H. L. (1976). Functional aberrations of the hypothalamic-pituitary system in polycystic ovary syndrome: Consideration of the pathogenesis. *In* "Endocrine Function of the Human Ovary" (V. H. T. James, M. Serio, and G. Giusti, eds.), p. 373. Academic Press, New York.

Zourlas, P. H., and Jones, H. W. (1969). Stein-Leventhal syndrome with masculinizing ovarian tumors. *Obstet. Gynecol.* **34,** 861–866.

12

Pregnancy, Genetic Counseling, and the Major Psychiatric Disorders

STEVEN D. TARGUM AND ELLIOT S. GERSHON

I. PREGNANCY, GENETIC COUNSELING, AND THE MAJOR PSYCHIATRIC DISORDERS

There is a significant interrelationship between pregnancy and the major psychiatric disorders (schizophrenia and affective illness). Individuals who have had a major psychiatric disorder are known to have an increased risk for the recurrence of such disorder should they become pregnant and give birth. The offspring of these individuals also have an increased risk for the development of a similar psychiatric disorder. The obstetrical course (pregnancy and child birth) may thus be more complicated and hazardous in psychiatrically ill mothers than in other women.

The perinatal period is known to be fraught with anxiety and the potential for emotional disturbance. For many it represents a maturational period in which conflicts about mothering, family, and generational relationships accompany

Copyright © 1981 by Academic Press, Inc.
All rights of reproduction in any form reserved.
ISBN 0-12-630940-X

changes in endocrine function and body imagery. Twenty to forty percent of women experience some postpartum "blues." However, the occurrence of postpartum psychoses appear to be an entirely distinct entity from the "blues" and most related to schizophrenia or affective psychoses. Prior to the development of antibiotics and modern obstetrical techniques, these psychoses could not be easily distinguished from toxic psychoses following delivery. Currently, it appears that the majority of postpartum psychoses can be classified as schizophrenia or affective psychoses, and reveal familial patterns of similar illness.

Individuals at an increased risk for the development of a major psychiatric disorder may consider in their decision to conceive offspring both the impact of the psychiatric disorder on their pregnancy as well as the future well being of their offspring. The assistance of genetic counselors in guiding and teaching these individuals has been given greater attention as the significance of genetic factors in the major psychiatric disorders has become better known and accepted.

II. THE MAJOR PSYCHIATRIC DISORDERS

The principal types of major psychiatric disorder, schizophrenia, and affective illness, have largely distinct clinical features which can generally lead to specific diagnoses of either. Some distinction can be found in the specificity of antidepressant drugs and lithium carbonate in the effective treatment of affective illness but not in schizophrenia, and the relative absence of schizophrenia in the relatives of affectively ill patients or the converse. The importance of diagnostic precision cannot be understated when genetic counseling is being considered. Research criteria have been developed (Spitzer *et al.*, 1978), and may be helpful in delineating the psychiatric syndromes which are under study. This helps to avoid the tendency to make diagnoses on the basis of a single symptom.

Most psychiatrists agree that schizophrenia is characterized by a prevalent thought disorder and marked by a tendency for the patient's associations to lose their continuity so that thinking may become confused, bizarre, or incorrect. There may be a limitation of emotional reactivity, considerable social isolation, and the appearance of delusions and hallucinatory experiences. Characteristically, there may be delusions of influence or control, experiences of external forces acting upon one's body, or experiences that one's thoughts are escaping and being read by others. Subtypes of this schizophrenic illness include symptoms like paranoia, catatonia, hebephrenia, inappropriate affect, and apathy. The subtypes of schizophrenic illness may not be related. Acute schizophrenic episodes may be more related to affective disorders than to other schizophrenic disorders. The chronic form of schizophrenia may be gradually deteriorating and lead to persistent impairment.

Primary affective disorders differ from schizophrenia in that the prominent feature is mood alteration (mania or depression) which may be episodic or chronic. To be diagnosed as a major affective illness, the disorder should be severe enough to cause impairment in normal social functions. The depressive syndromes are characterized by a dysphoric mood of despondency, hopelessness, worry, and four of the following eight symptoms: poor appetite or weight loss, sleep difficulty, fatigability or malaise, agitation or physical retardation, loss of interest in usual activities, feelings of self-reproach or guilt, complaints of slowed up thoughts or diminished ability to concentrate, and suicidal ideation.

Mania or hypomania is characterized by euphoria or a paranoid-angry state and three of the following six symptom categories: hyperactivity, pressured style of speech, racing thoughts, grandiosity, decreased need for sleep, and easy distractibility. The presence of hallucinations or delusions do not in themselves exclude the existence of a manic or depressive syndrome (Pope and Lipinski, 1978), although these must be congruent with mania or depression as described elsewhere (Spitzer *et al.*, 1978).

Major affective illness is defined as a primary affective disorder when it occurs in the absence of a previously existing diagnosable nonaffective psychiatric illness. Leonhard (1957) first proposed the subdivision of primary affective disorder into bipolar and unipolar subtypes. Bipolar illness characterizes patients who have both mania or hypomania and depression, whereas unipolar illness characterizes patients who have had only depression. This subdivision has improved genetic analyses and will be used in this chapter wherever possible.

III. GENETICS AND FAMILIAL RISK IN PSYCHIATRIC ILLNESS

The cumulative evidence suggests that genetic factors play a large role in the development of the major psychiatric disorders. Twin studies, family studies, and adoption studies have consistently shown a significantly higher risk rate of schizophrenia or affective illness in the first degree relatives of these patients (called probands) as compared to the general population.

Twin studies compare concordance rates between monozygotic (MZ) and dizygotic (DZ) twins. The actual rates determined have varied considerably due to the method of study and criteria of definition of the illness used. However, all studies reveal that the concordance rate in MZ twins is significantly greater than that of DZ twins (see Table I). Twin data for schizophrenia reported in recent studies found a 40–50% MZ concordance compared to a 9–15% DZ concordance rate (Kringlen, 1967; Allen *et al.*, 1972; Gottesman and Shields, 1972). Twin data for affective illness in six studies reviewed by Gershon *et al.* (1977) noted an overall MZ concordance rate of 69.2% compared to 13.3% DZ concordance. Price (1968) reviewed 12 MZ twin pairs who were reared apart and found a concordance rate for affective illness of 67%, implying that the MZ concordance is not due to a shared environment. The presence of discordant twin pairs

TABLE I
Some Twin Studies of Schizophrenia and Affective Illness

	Monozygotic twins		Dizygotic twins	
	Number of pairs	Percent concordance	Number of pairs	Percent concordance
Schizophrenia				
Luxenberger, 1934	17	77	48	2
Rosanoff *et al.*, 1934–1935	41	61	101	10
Essen-Moller, 1941	11	64	27	15
Kallmann, 1946	174	86	517	15
Inouye, 1962	55	60	17	12
Kringlen, 1967	69	(45)[a]	96	(15)
Fischer *et al.*, 1969	25	(56)	45	(26)
Tienari, 1971	20	(35)	23	(13)
Allen *et al.*, 1972[b]	121	(43)	131	(9)
Gottesman and Shields, 1972	26	(58)	34	(12)
Affective illness				
Luxenberger, 1930	4	75	13	0
Rosanoff *et al.*, 1934	23	70	67	16
Kallmann, 1952	27	93	55	24
Slater, 1953	8	50	30	24
Kringlen, 1967	6	33	20	0
Allen *et al.*, 1974	15	33	34	0
Harvald and Hauge, 1975	55	58	52	17
Bipolar illness				
Bertelsen *et al.*, 1977	27 (34)	74 (79)	36 (37)	17 (19)
Unipolar illness				
Bertelson *et al.*, 1977	28 (35)	43 (54)	16 (17)	19 (24)

[a] Parentheses indicate that the proband method was used to calculate concordance rates. This is a casewise approach based upon independently ascertained patients. The pairwise method is the traditional approach based upon the raw number of twin pairs observed, and tends to offer the lower limit of morbid risk estimates.

[b] Male pairs only.

suggests that factors other than genetics may influence the probability that the disorder will become manifest. However, assuming that the intrapair differences of environment are the same for MZ and DZ twins, the higher concordance rates for MZ twins does indicate that schizophrenia and affective illness are heritable.

Family studies of affective illness and schizophrenia have consistently reported a higher prevalence of these disorders in the first degree relatives of ill patients compared to the general population. In a review of seven major studies, Gershon *et al.* (1977) found that the incidence of major affective illness (both bipolar and unipolar illness) in the first degree relatives of bipolar patients was greater than 15%; and the incidence of affective illness in the first degree relatives of unipolar patients was greater than 6%. Bipolar and unipolar illness were noted in approximately equal distribution in the relatives of bipolar patients, whereas primarily unipolar illness was noted in the relatives of unipolar patients (see Table III). As reflected in Table II, numerous studies have ascertained the risk of psychiatric illness in the families of schizophrenic and affectively ill patients. The general population risk is approximately 1% for the development of schizophrenia (Kety *et al.*, 1978), and 2–3% for affective illness (Gershon *et al.*, 1977). The siblings of schizophrenic patients have a 5.4–12.2% risk for the development of schizophrenia, and higher if a parent is also ill. Parents of schizophrenics have a somewhat lower risk, possibly because schizophrenics represent a select group who are likely to marry and have children less frequently than the general population (Rosenthal, 1970).

The offspring of schizophrenic or affectively ill patients show considerable social and psychological maladjustment, and a range of psychiatric symptoms which may be related to the adult form of the illness (McKnew *et al.*, 1976; Rieder, 1973; Fowler *et al.*, 1977). The overt manifestation of a major psychiatric disorder may be only a part of the broad range of psychopathology from which the families of these patients may suffer.

Adoption studies have given the most convincing evidence for a genetic factor in the major psychiatric disorders. Recent studies of affective illness (Mendlewicz and Rainer, 1977; Cadoret, 1978) and the classic Danish adoption studies (Rosenthal *et al.*, 1971; Wender *et al.*, 1974; Kety *et al.*, 1978) have offered convincing data. Heston (1966) and Rosenthal *et al*. (1971) found no schizophrenia in the adopted-away offspring of normal control mothers, but frequencies of 9.6 and 7.7% in the adopted-away offspring of schizophrenic mothers. In all studies reported to date, the incidence of psychiatric illness similar to that of the index patient was noted in the biological rather than the adoptive side. The prevalence of schizophrenic psychopathology in relatives separated by adoption was not significantly different from the prevalence in the relatives not separated by adoption. Thus, the familial tendency in schizophrenia appears to be an expression of genetic and not of environmental factors. Further, the rearing family (adopting family) was found to have no higher prevalence of schizophrenic psychopathology than the respective control population or what would have been expected in the general population. Thus, schizophrenic psychopathology in the rearing family is not a requirement for the development of schizophrenia, although this does not rule out other environmental influences which may be necessary. Finally, the biological paternal half-siblings of schizophrenic adoptees showed a significantly higher prevalence of schizophrenic illness than did the controls. Since these individuals had the same father but different mothers than did their adopted-away schizophrenic half-siblings, they did not share the same uterine environment, early mothering experience, or postnatal environment. This suggests that only the genetic overlap shared with

TABLE II
Some Family and Adoption Studies of Schizophrenia and Affective Illness

	Percent morbid risk for schizophrenia in:			
	Offspring	Parents	Sibling	2° Relatives
Schizophrenia				
Schulz, 1932		2.6	6.6 (8.1)[a]	
Kallmann, 1938	13.9 (55.3)[b]	2.7	6.5 (11.7)[a]	2.6–2.7
Garrone, 1962	16.9	7.0	5.5 (33.7)[a]	
Book, 1953		12.0	9.0 (12.7)[a]	
Galatschjan, 1937		4.9	12.2 (23.3)[a]	3.6
Luxenberger, 1936		11.7	6.7 (12.2)[a]	
Slater, 1953		4.1	5.4	
Oppler, 1932	9.7			2.5
Heston, 1966	11 (17)[c]			
Rosenthal *et al.*, 1971	6			
Bleuler, 1974	7			
Schulsinger, 1976	9 (17)[c]			
Fowler *et al.*, 1977	4			
Reisby, 1967	6 (10.5)[c]			
Affective illness				
Stenstedt, 1952	(6.0)[c]	(5.3)[c]	(6.0)[c] (6.6)[a]	
Slater, 1953		13.6	8.8	
Weinberg and Lobstein, 1936	6.3	9.9	7.2	1.3–4.2
Kallmann, 1952		(23.4)[c]	(22.7)[c]	
Luxenberger, 1942	24.1		3.4 (16.1)[a]	
Rudin, 1923			7.4 (23.8)[a]	1.4
Elsasser, 1952	(34.5)[b]			

(continued)

the schizophrenic adoptee could account for the higher prevalence of schizophrenia among paternal half-siblings of adoptees (Kety *et al.*, 1978).

Precise genetic analyses in specific families has been impossible for the major psychiatric disorders, as reviewed elsewhere (Gershon *et al.*, 1977). Despite this limitation, the extensive data base of family and twin studies makes it possible to develop reliable risk estimates for the development of psychiatric illness in the first degree relatives of these patients. It is therefore reasonable to consider genetic counseling for these disorders.

IV. THE IMPACT OF THE MAJOR PSYCHIATRIC DISORDERS UPON PREGNANCY

Many investigators have suggested that prolonged emotional stress during pregnancy can directly result in perinatal and birth complications (PBCs) (see

TABLE II—*Continued*

Adoption Studies				
Schizophrenia	Percent morbid risk for schizophrenia in:			
	(a) ADOPTED-AWAY OFFSPRING			
	Risk in schizophrenic mothers		Risk in control mothers	
	Number	(%)	Number	(%)
Heston, 1966	47	9.6	50	0
Rosenthal et al., 1971	39	7.7	47	0
	(b) PARENTS OF ADOPTED-AWAY OFFSPRING			
	Schizophrenic adoptees		Control adoptees	
	percent risk in:		percent risk in:	
	Biological parents	*Adoptive parents*	*Biological parents*	*Adoptive parents*
Kety *et al.*, 1978	6.4	1.4	1.7	2.2
Affective illness	Percent morbid risk for affective illness in:			
	Parents of adopted-away offspring			
	Bipolar adoptees		*Control adoptees*	
	percent risk in:		percent risk in:	
	Biological parents	*Adoptive parents*	*Biological parents*	*Adoptive parents*
Mendlewicz and Rainer, 1977	31	12	2	10

[a] Schizophrenic sibling and one schizophrenic parent.
[b] Two schizophrenic parents.
[c] Age corrected morbid risk figures.

Mirdal *et al.*, 1977). The major psychiatric disorders have been implicated in reducing fertility as well as increasing perinatal and birth complications.

Several studies (Kallman, 1938; Norris, 1956; Essen-Moller, 1959; MacSorley, 1964) have established that schizophrenic patients admitted to mental hospitals have substantially reduced marital and reproductive rates compared to the prevailing rates of the general population at that time. Similar findings have been reported for affective illness (Landis and Page, 1938; Stenstedt, 1952; Norris, 1956). In a study of 390 psychotic patients diagnosed as schizophrenic, affectively ill, organic, or atypically psychotic, MacSorley (1964) found that the celibacy rates were more than 50% higher than that of both the general population and an independent control population drawn from the medical and surgical wards of a general hospital. The number of children produced by those patients who did marry was considerably less than that of the control group, and the percentage of childless marriages among patients was considerably higher. Siblings of the psychotic patients fell between the patient and control groups with

TABLE III
Morbidity Risks for the Development of Affective Illness in First-Degree Relatives Separated by Subtype (Bipolar–Unipolar)

	Number at risk (BZ)[a]	Percent morbid risk in relatives[b]	
		Bipolar illness	Unipolar illness
Bipolar illness			
Stenstedt, 1952	56[c]	5	8
Perris, 1966	509	11	0.6
Angst 1966	160	4	14
Goetzl *et al.*, 1974	141	4	21
Helzer and Winokur, 1974	131	5	12
Gershon *et al.*, 1975	341	4	7
Videbech, 1975	367	8	6
James and Chapman, 1975	265	6	13
Unipolar illness			
Stenstedt, 1952	432[c]	2	10
Perris, 1966	570	0.4	7
Angst, 1966	341	0.3	11
Gershon *et al.*, 1975	96	2	11
Videbech, 1975	596	2	9

[a] Indicates bezugziffers (BZ) calculated by Stromgren's method unless otherwise noted. BZ, Total number of persons corrected for age.
[b] First-degree relatives include parents, siblings, offspring.
[c] Bezugziffers calculated by Weinberg method.

respect to marriage, number of children per marriage, and percent of childless marriages.

Kallmann (1952) compared the marriage rates of affectively ill patients over the age of 45 years to those of their healthy siblings as well as controls. He found their respective rates in these groups to be 72.5%, 92.5%, and 87.5%, confirming the increased celibacy (consequently, reduced fertility) in patients with major affective disorders. Moreover, the affectively ill probands were more often childless and had fewer children. Among cases with affective psychosis, Odegaard (1961) found that the marriage rate was reduced 10% and fertility 20–25%. Rimmer and Jacobsen (1976) confirmed the earlier reports of a lower overall rate of fertility in male and female schizophrenics as compared to a matched control group. In their study, the mean number of children born to a group of 27 male and female schizophrenics was 0.259 as compared to 1.586 for 32 controls ($P < .0001$).

The marital and fertility rates of patients who have had major psychiatric disorders is clearly changing. Erlenmeyer-Kimling *et al.* (1969), in an examination of admissions done between 1934 and 1936 as compared to 1954 and 1956, observed that proportionally more patients of both sexes in the 1954–1956 sample were, or had been, married. Proportionally fewer had childless marriages than was true for the 1934–1936 group. The fertility rate for women increased from 58 to 87% of that of the general population of American women in the ages of childbearing. This change is attributed largely to the increased opportunity for schizophrenic patients to mate. As a consequence of medication and improved hospital management, schizophrenic patients are being released from the hospital and made more functional. Many children are born after the patient's first hospitalization. Thus, inquiries into the issue of reduced fertility among psychiatric patients must consider factors such as functional opportunity and availability for mating by these individuals.

Some retrospective studies of adult schizophrenics have found that patients had lower birth weights than their siblings (Lane and Albee, 1964; Pollin and Stabenau, 1968) and more frequent birth complications (Pollack and Woerner, 1966; Woerner *et al.*, 1973). Obstetrical complications or birth trauma are related to the development of neurological, physical, and behavioral disorders (see Dehorn and Strauss, 1977); consequently, in order to confirm a genetic correlation between lowered birth rates or increased perinatal and birth complications and psychiatric disorder, it would be of importance to demonstrate a significant difference in birth rates or PBCs in the offspring of psychiatric patients versus a comparison group. Some studies (Sobel, 1961; Paffenberger *et al.*, 1961; Sameroff and Zax, 1973; Rieder *et al.*, 1975) have found increased rates of perinatal and birth complications, or fetal and neonatal deaths, among the offspring of schizophrenics, but other studies have not (Mednick *et al.*, 1971; McNeil and Kaij, 1973; Mirdal *et al.*, 1974, 1977; Hanson *et al.*, 1976). Mirdal *et al.* (1977) studied the offspring of psychotic (both schizophrenic and affectively ill) fathers as well as mothers in order to distinguish between intrauterine and genotypic conditions. These authors found no difference in the rates of PBCs between the offspring of psychotic parents and normal parents, and no difference between offspring born to psychotic mothers compared to psychotic fathers. It would seem, from these data, that neither a possible fetal genotype for a major psychiatric disorder nor maternal effects of the disorder during pregnancy actually result in increased rates of PBCs. Further, Dehorn and Strauss (1977) have reported that the offspring of schizophrenics did not have lower birth weights than matched controls, and that those born a year or more before the onset of the parental disorder were actually heavier than control offspring.

Although PBCs or lower birth weight may not necessarily be a direct consequence of a schizophrenic diathesis, the children of schizophrenics may react differently to PBCs than the children of normal parents. Mednick *et al.* (1971) studied the offspring of 200 schizophrenic women and observed that PBCs could predict subsequent psychiatric illness. Using data from the Danish perinatal research project, they found that the PBCs of schizophrenic women were similar in number and kind to those of the controls, but that the offspring of these schizophrenic women had a higher incidence of retarded neurological development at 1 year, especially among those with low birth weight or PBCs. This suggests that the children who are genetically predisposed to schizophrenia may be more vulnerable to maternal perinatal events (e.g., toxemia, dystocia, fetal distress) than the children of normal parents, in the development of psychiatric illness later in life. Thus, the interaction of genetic predisposition, uterine factors, and early infantile environment could culminate in the overt manifestation of a major psychiatric disorder.

Illnesses like schizophrenia or affective illness might ultimately disappear if both fertility and fetal outcome were detrimentally affected. However, the incidence of these disorders has apparently remained the same. Some investigators have suggested that the relatives of these patients may, in fact, have a selective reproductive advantage, which enables them to sustain the population incidence of the illness (Kraus, 1973). Karlsson (1968), Hasenfus and Magaro (1976), Heston and Denney (1968) reported that the unaffected children, or siblings of schizophrenic patients are more artistic, creative, or intellectual than normal controls. However, these purported heterozygotes must possess greater fertility in order to demonstrate an actual selective advantage. Increased fertility in the siblings of schizophrenics compared to siblings of controls has not been demonstrated. Several studies (Kallman, 1938; Erlenmeyer-Kimling and Paradowski, 1966; Buck *et al.,* 1975; Rimmer and Jacobsen, 1976) report lower fertility or no difference in the fertility of the siblings of schizophrenics compared to siblings of controls. Some data (Rosenthal, 1970) actually suggest increased mortality and congenital malformations in the offspring of siblings of schizophrenics, and higher death rates in the relatives of schizophrenics (Kety *et al.*, 1978).

The influence of the major psychiatric disorders upon fertility and PBCs has not been fully elucidated. Nevertheless, psychiatric patients can and will proceed to bear children despite apparent reproductive disadvantages. Once conceived, these offspring are at a distinct environmental, as well as genetic, disadvantage relative to the offspring of normal mating pairs. The incidence of psychiatric disorder in the offspring of psychotic patients is high. As reviewed earlier, about 10–15% of the offspring of schizophrenic or affectively ill parents will develop illness similar to their parents. If both parents have been ill, risks approach 40–50% (Gottesman and Shields, 1972; Winokur and Clayton, 1967). The circumstances of their rearing may contribute to the high psychiatric morbidity. An increasing number of children born to schizophrenic patients will grow up in broken homes and contribute to the steadily increasing number of children who require protective custody because of neglect or maltreatment. Many battered children have been found to be reared in homes that housed a schizophrenic.

Dual marriages between schizophrenics exceed 2% of all marriages contracted by these patients (Erlenmeyer-Kimling and Paradowski, 1966). Nearly one-half of the dual marriage children born after the mother's first hospitalization (48.9%) were found to require the intervention of a child care agency (Rainer, 1966). Schizophrenic patients have minimal home building power due to frequent hospitalizations, limited integrative abilities, and the short-lived duration of such unions. Consequently, there is a high rate of disturbing effect on the offspring. In a consecutive series of 32 dual schizophrenic matings, Rainer (1966) found only two marriages in which the home milieu could be classified as unbroken or not chaotic. We would, therefore, discourage dual marriage or dual mating between two schizophrenics.

Follow-up studies of bipolar affectively ill patients reveals that over one-third of these patients remain functionally impaired with affective symptoms that interfere with careers, social relations, and family life (Winokur *et al.*, 1972). While they may have warm and empathic personal marital relations during well stages, these individuals tend to develop markedly dependent relationships. Consequently, their ability to rear children may be impeded.

The establishment of a healthy parent–child bond is disturbed by apparent psychiatric disorder. The disruption of the family creates more uncertainty for the offspring and may increase the risk for overt manifestation of the underlying predisposition for psychiatric disorder. The likelihood of an affected individual recognizing this limitation is limited by the characteristics of the illness itself. Affectively ill individuals use denial to a considerable extent in understanding and perceiving their illness. In a recent study (Targum *et al.*, in press), it was shown that bipolar affectively ill patients used denial to minimize the burden of their illness as compared to their well spouses. When asked if they would have married if they understood the circumstances of the illness previously, only 47% of spouses compared to 95% of patients said they would ($P < .05$). Similarly, only 42% of spouses compared to 95% of patients ($P < .05$) would have had children if they had known about the existence of mania or depression in their family prior to making these decisions.

In conclusion, it appears that the major psychiatric disorders have serious implications for both pregnancy and subsequent child rearing. Whereas obstetrical complications may not be induced by a schizophrenic or affective genome, these complications appear to contribute to the subsequent manifestation of the underlying predisposition. Further, schizophrenic or affectively ill individuals have reduced capability for rearing the subsequent offspring. Consequently, these offspring are at an increased genetic and environmental disadvantage relative to the offspring of normal individuals. The occurrence of similar illness in the offspring is, therefore, understandable. The estimate of morbid risk may be low in comparison to the amount of suffering and undefined psychopathology that ensues. To date, the data remain inconclusive that psychosis actually is correlated and produces a reduced fertility by virtue of physiological or genetic disadvantage. We must consider more than the number of living children when considering the effect of psychiatric illness on fertility and reproductive rate of psychiatric patients. In a particular genetic counseling situation, the counselor would be prudent to assume that the couple with pre-existing psychiatric illness (particularly psychosis) has an equal chance to conceive and bear a live infant as any other mating couple. The element, which continues to make these pregnancies high risk, is the increased liability for psychiatric sequelae in the mother (or father) and the genetic and environmental disadvantage conveyed to the offspring.

V. THE IMPACT OF PREGNANCY UPON THE MAJOR PSYCHIATRIC DISORDERS

The occurrence of psychotic reactions associated with childbirth are classified today as either major affective illness or schizophrenia. The incidence of toxic psychosis secondary to infection or other obstetrical factors is rare and more easily distinguished from functional psychotic disorders. Most authors concede that postpartum psychotic reactions are phenomenologically similar to the other nonpuerperal psychoses (Herzog and Detre, 1976). Depending on the criteria and definition of these reactions, the incidence of postpartum psychotic reactions varies from 1 in 400 to 1 in 1200 pregnancy deliveries. About one-half of these reactions occur within the first 2 weeks and two-thirds within the first 4 weeks postpartum (Protheroe, 1969). The first episode of a postpartum psychotic reaction can occur with any pregnancy; however, primiparas are at greatest risk (50–66%). The chance for a recurrence is equal for any subsequent pregnancy and varies from 13 to 20% after the initial episode (Martin, 1958; Protheroe, 1969). Reich and Winokur (1970) reported that for mothers with bipolar illness the recurrence rate for postpartum episodes can be as high as 50%. Most investigators report that from 10 to 15% of the mothers are single. In addition, Tetlow (1955) found a significant increase of pregnancies that were the result of an extramarital affair or of an incestuous union in the postpartum psychosis group.

The frequency of psychiatric illness in the first degree relatives of patients who have had postpartum psychosis is similar to that of the major psychiatric disorders. The morbidity risk for schizophrenia or affective illness among the first degree relatives of patients who have had postpartum psychotic reactions is greater than 10% (Herzog and Detre, 1976). No evidence suggests that postpartum psychosis is a specific or unique disorder with a distinct genetic predisposition.

Martin (1958) found a history of previous mental illness in 23% of his patients.

Fondeur *et al.* (1957), in a study using a control of matched nonpuerperal psychotic females, found an incidence of previous psychiatric illness in 30% of both groups. In a study of unipolar affective disorder, Baker *et al.* (1971) found that the clinical picture in progression of symptoms were similar in postpartum and nonpostpartum episodes. In patients whose depressive disease was diagnosed independently of the postpartum illness, the rate of morbidity risk for the development of a depressive episode in the postpartum state (7%) was not significantly different from the morbidity risk of any other time during the childbearing years (4%). Unipolar patients had much less frequent affective episodes within 6 months postpartum than did bipolar patients (Winokur and Ruangtrakool, 1966). Reich and Winokur (1970) found that 40% of patients with independently diagnosed bipolar affective disorder had postpartum affective episodes; 20% of the first degree family members of bipolar probands did, and 5% of the first degree family members of unipolar probands did. Whereas no increased risk for postpartum depression was observed in patients who had previous unipolar affective disorder, Reich and Winokur (1970) reported a 50% rate for postpartum illness in bipolar patients and 30% in their family members who had had previous postpartum illness. Several studies (Kendell *et al.*, 1976; Pugh *et al.*, 1963; Paffenberger and McCabe, 1966) have shown a rise in new episodes of functional psychosis in the first 3 months after delivery in patients with previous psychiatric illness. While it remains unclear if the crucial influence of pregnancy is psychological or physiological (endocrine), the evidence is clear that there is a raised incidence of psychiatric morbidity following childbirth. Interestingly, there is a reduced hospitalization rate during pregnancy (Pugh *et al.*, 1963; Kendell *et al.*, 1976).

The occurrence of postpartum psychosis may have psychological sequelae upon the offspring. Uddenberg and Englesson (1978) found that the offspring of mothers who had had postpartum psychotic reactions viewed their parents more negatively than did the offspring of normal mothers. Similarly, these mothers experienced much more subjective difficulty in child rearing than did the normal mothers. Sixteen mothers who had been severely mentally handicapped during their first postpartum period were studied. About one-half of these women reported repeated and/or prolonged periods of impaired mental health during their following years. Further, the women in this group reported a poor relationship to their marital partner more often than other women who had not had postpartum episodes. Adaptation to the parental role was most difficult for women who had had postpartum psychotic episodes. Their general attitudes to the child, as estimated by the interviewer in this study, were more often a rejecting one. It appears that women who are mentally disturbed postpartum, run a high risk of poor mental health in the future and that the mother–child relationship may often develop in an unfavorable way.

VI. PSYCHOTROPIC DRUGS AND PREGNANCY

Potential behavioral and teratogenic effects of the psychotropic drugs used during pregnancy creates an additional concern for the psychiatric patient. Lewis *et al.* (1977) cautioned that these agents alter the balance of neurotransmitter activity and may produce long-lasting functional disturbance when given to the developing brain. These authors suggested that neurotransmitters may act as neurohormones in the embryogenic regulation of cell proliferation of the nervous system, and that drugs, which interfere with this activity, may be causing subtle, but irreversible, deficits in cell number or in the formation of certain cell types. Reserpine, for example, has been shown to interfere with cell proliferation in the neonatal rat brain by decreasing the rate of DNA synthesis (Lewis *et al.*, 1977).

Phenothiazine derivatives are commonly used in the treatment of acute psychotic episodes and as prophylactic treatment in most schizophrenic patients. Although there have been a few isolated reports of congenital malformations occurring following the use of these drugs during pregnancy, phenothiazines do not necessarily harm the fetus (Ananth, 1975; Kalter, 1968). Several reports (Sobel, 1960; Dickel and Dixson, 1957; Kris, 1961) have reviewed cases of chlorpromazine use in human pregnancy and found no increase in fetal defects or wastage. Sobel (1960) studied 52 women treated with chlorpromazine during pregnancy and reported four cases of fetal damage, concluding that the incidence of fetal damage in the chlorpromazine-treated mothers (8% of his sample) was indistinguishable from that of the control group he studied (7%). Similar clinical reports in the literature find no apparent teratogenic effects in human infants whose mothers took trifluoperazine (Moriarty, 1963) or fluphenazine (Cleary, 1977). While these studies may rule out severe side effects in infants whose mothers took phenothiazines, a transient neonatal extrapyramidal syndrome marked by agitation, hypertonicity, vasomotor instability, uncoordinated suckling and swallowing movements has been noted as well as respiratory distress (Ananth, 1976). In addition, two studies (Nielson *et al.*, 1968; Abdullah and Miller, 1968) have reported chromosomal damage produced by phenothiazines. This latter finding, if substantiated by others, could have long-term clinical significance for the developing fetus. While animal studies suggest long-term effects on memory and learning, there have been no large follow-up studies of the infants born to mothers taking phenothiazines, which might permit significant conclusions (Ananth, 1976). In a follow-up of 14 mothers who used chlorpromazine during their pregnancies, no abnormalities were noted in the offspring at age 4 years (Kris, 1961).

Congenital anomalies have been associated with tricyclic antidepressants in one study. McBride (1972) reported three infants born with limb deformities whose mothers had taken imipramine or amitriptyline during the first trimester of pregnancy. In animal studies, imipramine has been found to be teratogenic in one study (Robson and Sullivan, 1963), but not in others (Larsen, 1963; Harper *et al.*, 1965; Stenger *et al.*, 1965). An extensive surveillance of hundreds of mothers who gave birth to children with congenital malformations found no support for an association between limb deformities and tricyclic antidepressant medications (Banister *et al.* 1972; Rachelefsky *et al.*, 1972; Idanpaan-Heikkila and Saxen, 1973). On the other hand, monoamine oxidase inhibitors (MAOIs) have been shown to cause relative infertility and reabsorption of animal fetuses (Poulson and Robson, 1964; Tuchmann-Duplessis and Mercier-Parot, 1963). Although defects in human fetuses have not been reported, this may be due to a markedly impaired fertility in women taking MAOIs, which make these drugs unsafe in pregnancy.

Lithium carbonate is used extensively in the treatment of affective illness. Because the lithium ion is in equilibrium between maternal and fetal circulation, transient lithium toxicity could occur in the fetus in a pregnant patient known to be on relatively stable therapeutic levels of the drug (Wilbanks *et al.*, 1970). Hypotonia and cyanosis have been reported in babies born to mothers on lithium (Tunnessen and Hertz, 1972; Stothers *et al.*, 1973; Schou *et al.*, 1973; Sjöquist *et al.*, 1972). Nars and Girard (1977) observed edema and a large goiter in a baby born to a mother whose goiter was exacerbated by lithium treatment. The most striking fetal effects have been cardiovascular malformations, occurring in 13 of 160 lithium-exposed babies; four had the rare Ebstein's anomaly (Weinstein and Goldfield, 1975; Schou, 1976). In addition, there have been more stillborns and neonatal deaths occurring in women who took lithium during pregnancy (Schou, 1976). Consequently, it is best to discontinue lithium carbonate in women who are attempting to conceive, and restarting treatment sometime after the first trimester of pregnancy. We advocate the re-introduction of lithium carbonate just prior to delivery or immediately postpartum as prophylaxis against the occurrence of postpartum psychosis. As described earlier, the incidence of postpartum affective illness in patients with a previously diagnosed bipolar illness can be as high as 50% (Reich and Winokur, 1970).

Perinatal exposure to diazepam has been associated with the development of cleft lip with or without cleft palate (Safra and Oakley, 1975; Saxen and Saxen, 1975). The use of this agent during pregnancy, particularly in the first trimester, is questionable. We discourage the use of benzodiazepine derivatives in light of the potential congenital malformations and the lack of demonstrated prophylactic benefit or necessity.

Psychotropic drugs used in pregnancy may have long-term behavioral side effects. Ananth (1976) has reviewed a number of animal studies, which reported diminished learning ability and heightened emotionality in the offspring of mothers who have received phenothiazines, benzodiazepines, barbiturates, or alcohol during gestation. The clinical significance of these data will remain unclear until controlled follow-up studies are available.

We may conclude that some psychotropic drugs used during pregnancy may have teratogenic effects and possible long-term behavioral side effects, particularly when used in the first trimester. It is wise to choose a conservative approach to the use of any drugs during pregnancy. Whereas the use of phenothiazines and lithium carbonate in the prophylactic management of the major psychiatric disorders is well established, the use of other psychotropic agents during pregnancy is less clearly indicated. It may be possible to withdraw phenothiazines or lithium carbonate in some patients who can be monitored carefully. It should be noted, however, that there may be delays of months before conception by these individuals, who have been reported to have reduced fertility. Consequently, the decision to withdraw prophylactic medication in these patients may involve many months. The withdrawal of these agents in pregnancy, a time of increased risk, may be more hazardous than the potential fetal effects which are still inconclusive and not invariably present in all cases. Whereas the clinical situation may dictate the use of phenothiazines during pregnancy, we would definitely discourage the use of lithium carbonate in the first trimester of pregnancy. In all cases, we advocate the re-institution of lithium carbonate just prior to delivery or immediately postpartum in order to avert instances of postpartum psychosis. The risk of postpartum psychosis is high in those patients who have had a history of previous psychotic illness, and the decision to discontinue medication must be weighed very carefully by these patients while considering conception.

VII. PSYCHIATRIC GENETIC COUNSELING

The major psychiatric disorders have multiple interactions with pregnancy, childbirth, and child rearing. The following may be significant:

1. Reduced fertility
2. Increased perinatal and birth complications and greater fetal wastage
3. Teratogenic and long-term behavioral side effects of psychotropic drugs used during pregnancy
4. Increased psychiatric illness in the postpartum period
5. Increased morbid risk to offspring for the development of a major psychiatric disorder
6. A poor milieu, which may include ineffective mothering, frequent hospitalization, family disruption, foster care
7. Economic disadvantage in families whose socioeconomic status is compromised by psychiatric illness

These factors complicate family planning decisions for individuals who have had a major psychiatric disorder. It is possible to consider the role and objectives of the genetic counselor in view of these multiple interactive factors.

As noted above, patients tend to minimize or deny the implications and burden of the disorder they have had. It is thus conceivable that those most in need of counseling would be least accessible to it. The authors have made available counseling to those who requested it and have not attempted to confront or inform other patients or family members. Whereas these families have a right to genetic information, they also have a right to ignore or deny it. Consequently, our initial step in psychiatric genetic counseling has been to assess the motivation of the consulting individual (consultand) for seeking such counsel.

Motivations for seeking psychiatric genetic counseling are variable. There exists the risk of emergent illness in the consultand or other family members as well as potential risk to offspring. In addition, the ability to comprehend genetic information may be limited both by the anxiety of the counseling situation and an impairment of integration secondary to the psychiatric diathesis (e.g., schizophrenia). The counselor should not presume to know the motive for consultations. Consultations might be triggered by the development of symptoms in the consultand, personal exposure to an ill family member, or re-evaluation of a family history of mental illness. A spouse may be concerned or fear illness within himself, or in the spouse, and may be interested in directing responsibility for mental illness to the other side of the family. Shame about family secrets or past history, or guilt about child rearing, may stimulate an individual to seek exoneration for responsibility by attributing the illness to genetics. The use of denial may have been shattered by the appearance of severe illness in a close relative, and the consultation may have been triggered by such an emotional reaction. Misconceptions may be rampant; the role of the counselor in understanding the dynamics of the motivation, and in reducing the anxiety and tension concomitant with the counseling situation, will do much to improve the outcome of the genetic counseling. The counselor must always be aware of the stigma attached to both genetic disease and mental illness. This double narcissistic blow can have a great damaging effect on the self-esteem of the individual who is seeking counseling. It is, therefore, necessary that counseling be in a confidential setting which offers a supportive continuous relationship in order to facilitate the resolution of these feelings. The establishment of a supportive therapeutic relationship should precede the dissemination of genetic information to the consultand.

An accurate diagnosis is critical in all genetic evaluations. There is growing recognition that symptoms alone do not adequately define schizophrenia or affective illness from a genetic point of view. The use of objective clinical criteria, as recently developed by Spitzer and Endicott (1975), assist in the establishment of an accurate psychiatric diagnosis. The diagnostic evaluation should consider the possibility of spectrum disorders for either schizophrenia or affective illness. The schizophrenia spectrum has included classical chronic schizophrenia, latent schizophrenia, acute schizophrenic reactions, schizoid personality, and inadequate personality (Rosenthal, 1975). The affective disorder spectrum has included bipolar and unipolar disorder, cyclothymic personality, minor depressive illness, and acute psychosis. It is of note that acute psychotic episodes may more often be related to affective illness than schizophrenia. In the study of Gershon *et al.* (1975), acute psychotic episodes were noted in the families of affectively ill probands. In the Danish adoption study, there were no instances of acute schizophrenic reactions among the relatives of chronic schizophrenic adoptees, and conversely, there was little schizophrenic illness among the biological relatives of adoptees who had been diagnosed as acutely schizophrenic (Kety *et al.*, 1978).

The use of empirical risk figures in the determination of risk for the development of illness is most appropriate when applied to children who have not yet been born. These figures indicate the risk over a normal lifetime; consequently, in an adult who has not developed schizophrenia there is a considerably lower chance of developing the disorder than the newborn. As Kety *et al.* (1978) have indicated, an adult of 30 years has approximately 50% of the risk he had at birth, and at 45 years the risk is indeed small. The determination of risk should consider the diagnoses in both parents. The existence of a spectrum disorder in the other parent may increase the risk to future offspring. Indeed, the incidence of assortative matings between two affected individuals has been shown to be high for both schizophrenia and affective illness (Dunner *et al.*, 1976; Gershon *et al.*, 1973). Finally, the examination of the consultands pedigree may improve the meaning of the risk estimate. A conspicuously "loaded" family history of homotypic mental illness indicates that the consultand has a greater risk and that a more guarded prognosis is advisable, whereas a negative family history in a large family suggests a more favorable prognosis and likely a lower risk. Family studies (Kety *et al.*, 1976) have revealed that some probands have families with considerable psychiatric illness whereas other may have none. One cannot conclude that there is no genetic basis for illness in the latter instances. The findings suggest that these disorders are heterogenous, as most investigators would agree (Tsuang, 1975). Currently, it would be spurious to convey certainty when quoting empirical risks for a psychiatric disorder, although the empirical risks are the best estimates we can offer.

An understanding of the nature and prognosis of the disease, and the perception of burden of that disease is as influential as the quoted risk in the genetic counseling situation. The consultand has a very personal view of the disease in question, and this view may not be in concordance with that of the genetic counselor. As such, specific genetically determined diseases can mean different things for different people. The estimate of burden of psychiatric disorders is complicated by the variable clinical manifestations, which can be present even within the same pedigree. A consultand's concept of disease is often formed by personal exposure to ill family members, and will be influenced by the extent of disruption caused by the illness. Some families may discount the burdensome aspects of psychiatric illness in light of positive aspects they attribute to the condition. In this respect, creativity has been associated with schizophrenia (Karlsson, 1968; Hasenfus and Magaro, 1976), and increased capacity in achievement noted in bipolar illness. Family members with milder forms of illness (e.g., cyclothymic personality) may be very successful people with no functional impairment noted. The possibility of personal failure, chronic illness, incapacity, or suicide will be weighed in accordance with the consultands experience and the family history. When psychiatric illness has been particularly disruptive, a higher estimate of burden may be assumed than in those families where the illness has had little impact. These variable estimates are useful since psychiatric entities are presumably due to multiple genotypes and are affected by environmental variance. Similar to the estimate of burden, the risk figure is also subject to personal interpretation. Some degree of mathematical sophistication is required to understand the meaning of risk (odds). The anxiety, which is part of counseling, may also impede the ability of consultands to comprehend the meaning of odds.

The communication of genetic information (the estimate of morbid risks and the nature and course of the illness in question) must be given in association with an understanding of the consultand's perception of burden. The counselor helps to alleviate anxiety and the emotional sequelae of the genetic diagnosis and the meaning of psychiatric illness to the individual. The counselor may then assist this individual to reconsider personal plans in view of the genetic information. There is no concrete advice that can be given to prospective parents regarding minimizing the risk for the development of psychiatric illness. Adoption studies have shown that parental environment has little impact upon the eventual development of psychiatric illness in those who are biologically predisposed (Kety *et al.*, 1978). From the viewpoint of secondary prevention, however, emergence of psychiatric illness in a family member may be diagnosed early and perhaps minimized by psychotherapy and pharmacotherapy.

Whereas the genetic counselor should not generally direct a consultand to make decisions about their future family planning, the high risks associated with mating between two schizophrenics are of great concern. In our opinion, marriage is ill-advised between two individuals with psychotic disorders. Similarly, patients with rapid cycling bipolar illness, chronic schizophrenics, or patients who have had multiple re-hospitalizations should not enter into marriage indiscriminately. The decision to conceive children is also questionable in these above situations, as well as for patients who have had a history of multiple suicide attempts or postpartum psychoses. We believe that voluntary therapeutic abortion is justifiable in these situations and may reduce the extent of social and personal burden for the family.

As gloomy as the counseling picture may be for some patients, they will still proceed to reproduce. On the other hand, there are some families who may be so frightened by their perceptions of burden that they may overestimate the meaning and misconstrue the odds for development of illness, and consequently be afraid to conceive children. Sorting out these factors and issues is a crucial role for the genetic counselor and may best be the function of the informed psychiatrist.

Another question concerns the risk of precipitating illness in those who have received genetic information. There are no data to show that schizophrenia or affective illness occurs as a result of self-fulfilling prophecies or by suggestion, however, the role of stress cannot be overestimated. Highly intelligent individuals may regress in the setting of genetic counseling, particularly when considering the double narcissistic blow of genetic illness and psychiatric illness. The support and understanding of the counselor and the willingness to give the patient time to resolve these conflicts may minimize the extent of counseling-related stress and problems.

The psychiatric genetic counselor must seek follow up in order to ascertain the outcome of counseling and to monitor for the emergence of psychiatric illness in previously well family members. The availability of a psychiatrist known to the family may reduce some of the hesitancy of family members who have been reluctant to seek psychiatric treatment previously.

VIII. CONCLUSION

The major psychiatric disorders (schizophrenia and affective illness) may adversely affect the fertility, the course and outcome of pregnancy, and the subsequent child rearing of the offspring. In addition, the major psychiatric disorders are transmitted to offspring, who have a considerable risk for the development of similar illness. Pregnancy can also precipitate or exacerbate an underlying psychiatric diathesis. Finally, the use of psychotropic drugs as prophylaxis against the manifestations of the major psychiatric disorders has potential teratogenic effects when used in pregnancy. These interactive factors make a pregnancy with concomitant major psychiatric disorder a high risk obstetrical situation. When individuals seek genetic counseling for the major psychiatric disorders, the counselor needs to understand the motivation for such referral and assist the individual in both comprehending and integrating the genetic information and in alleviating the tension and anxiety which accompanies the possibility of genetic and psychiatric illness in one's family. At present, the empirical risk figures drawn from large family studies can be used to guide the individual as to the morbid risk for development of psychiatric illness in his particular family. Childbearing should be weighed carefully, and perhaps discouraged, in matings between two patients with psychotic illness, rapidly cycling bipolar disorders, chronic schizophrenia, or among patients who have had multiple rehospitalizations, multiple suicide attempts, or a history of postpartum psychoses.

REFERENCES

Abdullah, S., and Miller, O. J. (1968). Effects of drugs on nucleic acid synthesis and cell division in vitro. *Dis. Nerv. Syst.* **29,** 829-833.

Allen, M. G., Cohen, S., and Pollin, W. (1972). Schizophrenia in veteran twins: A diagnostic review. *Am. J. Psychiatry* **128,** 939-945.

Allen, M. G., Cohen, S., Pollin, W., and Greenspan, S. I. (1974). Affective illness in veteran twins: A diagnostic review. *Am. J. Psychiatry* **131,** 1234-1239.

Ananth, J. (1975). Congenital malformations with psychopharmacologic agents. *Compr. Psychiatry* **16,** 437-445.

Ananth, J. (1976). Side effects on fetus and infant of psychotropic drug use during pregnancy. *Int. Pharmacopsychiatry* **11,** 246-260.

Angst, J. (1966). "Zur Atiologie und Nosologie endogener depressiver Psychosen." Springer-Verlag, Berlin and New York.

Baker, M., Dorzab, J., Winokur, G., and Cadoret, R. (1971). Depressive disease: The effect of the post-partum state. *Biol. Psychiatry* **3,** 357-365.

Banister, P., Dafoe, C., Smith, E. S. O., and Miller, J. (1972). Possible teratogenicity of tricyclic antidepressants. *Lancet* **1,** 838-839.

Bertelsen, A., Harvald, B., and Aauge, M. (1977). A Danish twin study of manic-depressive disorders. *Br. J. Psychiatry* **130,** 350-351.

Book, J. A. (1953). A genetic and neuropsychiatric investigation of a North Swedish population with special regard to schizophrenia and mental deficiency. *Acta Genet.* **4,** 1-139.

Buck, C., Hobbs, G. E., Simpson, H., and Wanklin, J. M. (1975). Fertility of the sibs of schizophrenic patients. *Br. J. Psychiatry* **130,** 350-351.

Cadoret, R. J. (1978). Evidence for genetic inheritance of primary affective disorders in adoptees. *Am. J. Psychiatry* **135,** 463-466.

Cleary, M. F. (1977). Fluphenazine decanoate during pregnancy. *Am. J. Psychiatry* **134,** 815-816.

Dehorn, A. B., and Strauss, M. E. (1977). Birthweights of male and female schizophrenics' offspring. *Acta Psychiatr. Scand.* **55,** 321-329.

Dickel, H. A., and Dixon, H. H. (1957). Inherent dangers in the use of tranquilizing drugs in anxiety states. *J. Am. Med. Assoc.* **163,** 422-426.

Dunner, D. L., Fleiss, J. L., Addanizio, G., and Fieve, R. R. (1976). Assortative mating in primary affective disorder. *Biol. Psychiatry* **11,** 43-51.

Elsasser, G. (1952). "Die Nachkommen geisteskranker Elternpaare." Thieme, Stuttgart.

Erlenmeyer-Kimling, L., and Paradowski, W. (1966). Selection and schizophrenia. *Am. Nat.* **100,** 651-654.

Erlenmeyer-Kimling, L., Nichol, S., Rainer, J. D., and Deming, W. E. (1969). Changes in fertility rates of schizophrenic patients in New York State. *Am. J. Psychiatry* **125,** 916-929.

Essen-Moller, E. (1941). "Psychiatrische Untersuchungen an einer Serie von Zwillingen." Munksgaard, Copenhagen.

Fischer, M., Harvald, B., and Hauge, M. (1969). A Danish twin study of schizophrenia. *Br. J. Psychiatry* **115,** 981-990.

Fondeur, M., Fixsen, C., Triebel, W. A., and White, M. A. (1957). Post-partum mental illness: Controlled study. *Arch. Neurol. Psychiatry* **77,** 503-512.

Fowler, R. C., Tsuang, M. T., and Cadoret, R. J. (1977). Psychiatric illness in the offspring of schizophrenics. *Compr. Psychiatry* **18,** 127-134.

Galatschjan, A. (1937). Die Vererbung der Schizophrenie. *Schweiz. Arch. Neurol. Psychiatr.* **39,** 291-319.

Garrone, G. (1962). Etude statistique et génétique de la schizophrenia à Génève de 1901 à 1950. *J. Genet. Hum.* **11,** 89-219.

Gershon, E. S., Dunner, D. L., Sturt, L., and Goodwin, F. K. (1973). Assortative mating in the affective disorders. *Biol. Psychiatry* **7,** 63–74.

Gershon, E. S., Mark, A., Cohen, N., Belizon, N., Baron, M., and Knobe, K. E. (1975). Transmitted factors in the morbid risk of affective disorders: A controlled study. *J. Psychiatr. Res.* **12,** 283–299.

Gershon, E. S., Targum, S. D., Kessler, L. R., Mazure, C. M., and Bunney, W. E., Jr. (1977). Genetic studies and biologic strategies in the affective disorders. *Prog. Med. Genet.* **11,** 101–164.

Goetzl, V., Green, R., Whybrow, P., and Jackson, R. (1974). X-linkage revisited. A further family study of manic-depressive illness. *Arch. Gen. Psychiatry* **31,** 665–672.

Gottesman, I. I., and Shields, J. (1972). "Schizophrenia and Genetics." Academic Press, New York.

Hanson, D. R., Gottesman, I. I., and Heston, L. L. (1976). Some possible childhood indicators of adult schizophrenia inferred from children of schizophrenics. *Br. J. Psychiatry* **129,** 142–154.

Harper, K. H., Palmer, A. K., and Davies, R. E. (1965). Effect of imipramine upon the pregnancy of laboratory animals. *Arznheim.-Forsch.* **15,** 1218–1221.

Hasenfus, N., and Magaro, P. (1976). Creativity and schizophrenia: An equality of empirical constructs. *Br. J. Psychiatry* **129,** 346–349.

Helzer, J. E., and Winokur, G. (1974). A family interview study of male manic-depressives. *Arch. Gen. Psychiatry* **31,** 73–77.

Herzog, A., and Detre, T. (1976). Psychotic reactions associated with childbirth. *Dis. Nerv. Syst.* **37,** 229–235.

Heston, L. L. (1966). Psychiatric disorders in foster home reared children of schizophrenic mothers. *Br. J. Psychiatry* **112,** 819–825.

Heston, L. L., and Denney, D. (1968). Interactions between early life experiences and biological factors in schizophrenia. *In* "The Transmission of Schizophrenia" (D. Rosenthal and S. S. Kety, eds.), pp. 363–376. Pergamon, Oxford.

Idanpaan-Heikkila, J., and Saxen, L. (1973). Possible teratogenicity of imipramine/chloropyramine. *Lancet* **2,** 281–283.

Inouye, E. (1962). Similarity and dissimilarity of schizophrenia in twins. *Proc., World Congr. Psychiatry, 3rd, 1961* Vol. 1, pp. 524–530.

James, N. M., and Chapman, C. J. (1975). A genetic study of bipolar affective disorder. *Br. J. Psychiatry* **126,** 449–456.

Kallman, F. J. (1938). "The Genetics of Schizophrenia." J. J. Augustin, New York.

Kallman, F. J. (1946). The genetic theory of schizophrenia. *Am. J. Psychiatry* **103,** 309–322.

Kallman, F. J. (1952). Genetic principles in manic-depressive psychoses: Depression. *Proc. 42nd Annu. Meet. Am. Psychopathol. Assoc.*

Kalter, H. (1968). "Teratology of the Central Nervous System." Univ. of Chicago Press, Chicago, Illinois.

Karlsson, J. L. (1968). Geneologic studies of schizophrenia. *In* "The Transmission of Schizophrenia" (D. Rosenthal and S. S. Kety, eds.), pp. 85–94. Pergamon, Oxford.

Kendell, R. E., Wainwright, S., Hailey, A., and Shannon, B. (1976). The influence of childbirth on psychiatric morbidity. *Psychol. Med.* **6,** 297–302.

Kety, S. S., Matthysse, S., and Kidd, K. K. (1978). Genetic counseling for schizophrenic patients and their families. *In* "Controversy in Psychiatry" (J. P. Brady and H. K. H. Brodie, eds.), pp. 776–784. Saunders, Philadelphia, Pennsylvania.

Kraus, R. F. (1973). Schizophrenia as a genetic polymorphism. *Psychiatr. Q.* **47,** 546–558.

Kringlen, E. (1967). "Heredity and Environment in the Functional Psychoses." Heinemann, London.

Kris, E. B. (1961). Children born to mothers maintained on pharmacotherapy during pregnancy and post-partum. *Recent Adv. Biol. Psychiatry.* **4,** 180–187.

Landis, C., and Page, J. D. (1938). "Modern Society and Mental Disease." Farrar & Rhinehart, New York.

Lane, E. A., and Albee, C. W. (1964). Early childhood intellectual differences between schizophrenics and their siblings. *J. Abnorm. Soc. Psychol.* **68,** 193–195.

Larsen, V. (1963). Teratogenic effects of thalidomide, imipramine Hcl and imipramine N-oxide Hcl on white danish rabbits. *Acta Pharmacol. Toxicol.* **20,** 186–200.

Leonhard, K. (1957). "Aufreilung der endogenon. Psychosen," 1st ed. Berlin.

Lewis, P. D., Patel, A. J., Bendek, G., and Balaz, R. (1977). Do drugs acting on the nervous system affect cell proliferation in the developing brain? *Lancet* **1,** 399–401.

Luxenberger, H. (1930). Psychiatrisch-neurologische Zwillingspathologie. *Z. Gesamte Neurol. Psychiatr.* **56,** 145–180.

Luxenberger, H. (1934). Die Manifestationswahrscheinlichkeit der Schizophrenie im Lichte der Zwillingsforschung. *Z. Psych. Hyg.* **7,** 174–184.

Luxenberger, H. (1936). Zur frage der Erbberatung in den Familien Schizophrener. *Med. Klin. (Munich)* **32,** 1136–1138.

Luxenberger, H. (1942). Das zirkulare Irresein. *In* "Handbuch der Erbrankheiten" (A. Gutt, ed.), Vol. 4. Thieme, Leipzig.

McBride, W. G. (1972). Limb deformities associated with iminodibenzyl hydrochloride. *Med. J. Aust.* **1,** 492

McKnew, D. H., Cytryn, L., Efron, A. M., Gershon, E. S., and Bunney, W. E., Jr. (1976). Offspring of patients with affective disorders. *Annu. Meet. Am. Psychiatr. Assoc.*

McNeill, T. F., and Kaij, L. (1973). Obstetric complications and physical size of offspring of schizophrenic, schizophrenic-like, and control mothers. *Br. J. Psychiatry* **123,** 341–348.

MacSorley, K. (1964). An investigation into the fertility rates of mentally ill patients. *Ann. Hum. Genet.* **27,** 247–256.

Martin, M. E. (1958). Puerperal mental illness, a follow up study of 75 cases. *Br. Med. J.* **2,** 737–777.

Mednick, S. A., Mura, E., Schulsinger, F., and Mednick, B. (1971). Perinatal conditions and infant development in children with schizophrenic parents. *Soc. Biol.* **18,** 103–113.

Mendlewicz, J., and Rainer, J. D. (1977). Adoption study supporting genetic transmission in manic-depressive illness. *Nature (London)* **268,** 327–329.

Mirdal, G. M., Mednick, S., Schulsinger, F., and Fuchs, F. (1974). Perinatal complications in children of schizophrenic mothers. *Acta Psychiatr. Scand.* **50,** 553–568.

Mirdal, G. M., Rosenthal, D., Wender, P. H., and Schulsinger, F. (1977). Perinatal complications in offspring of psychotic parents. *Br. J. Psychiatry* **130,** 495–505.

Moriarity, A. J. (1963). Trifluoperazine and congenital malformations. *Can. Med. Assoc. J.* **88,** 87.

Nars, P. W., and Girard, J. (1977). Lithium carbonate intake during pregnancy leading to large goiter in a premature infant. *Am. J. Dis. Child.* **131,** 924–925.

Nielson, J., Frederick, U., and Tsuboi, T. (1968). Chromosome abnormalities and psychotropic drugs. *Nature (London)* **218,** 488–489.

Norris, V. (1956). Mental illness in London: A statistical study of the influence of marriage on the hospital care of the mentally sick. *J. Ment. Sci.* **102,** 467.

Odegaard, O. (1961). The epidemiology of depressive psychosis: Depression. *Acta Psychiatr. Neurol. Scand., Suppl.* **162,** 33–38.

Oppler, W. (1932). Zum Problem der Erbprognosebestimmung, *Z. Gesamte Neurol. Psychiatr.* **141,** 549–616.

Paffenberger, R. S., and McCabe, L. J. (1966). The effect of obstetric and perinatal events on risk of mental illness in women of childbearing age. *Am. J. Public Health* **56,** 400–407.

Paffenberger, R. S., Steinmetz, C. H., Pooler, B. G., and Hyde, R. T. (1961). The picture puzzle of the post-partum psychoses. *Psychiatry* **13,** 161–173.

Perris, C. (1966). A study of bipolar (manic-depressive) and unipolar (recurrent depressive) psychoses. *Acta Psychiatr. Scand., Suppl.* **194,** 1–189.

Pollack, M., and Woerner, G. M. (1966). Pre and perinatal complications and childhood schizophrenia: A comparison of five controlled studies. *J. Child Psychol.* **7,** 235–247.

Pollin, W., and Stabenau, J. R. (1968). Biological, psychological, and historical differences in a series of monozygotic twins discordant for schizophrenia. *In* "The Transmission of Schizophrenia" (D. Rosenthal and S. S. Kety, eds.). Pergamon, Oxford.

Pope, H., and Lipinski, J. (1978). Differential diagnosis of schizophrenia and manic-depressive illness. *Arch. Gen. Psychiatry* **35,** 811–828.

Poulson, E., and Robson, J. M. (1964). Effect of phenelzine and some related compounds on pregnancy and on sexual development. *J. Endocrinol.* **30,** 205–215.

Price, J. (1968). The genetics of depressive behavior. *J. Psychiatr., Spec. Publ.* **2,** 37–54.

Protheroe, C. (1969). Puerperal psychoses: A long term study, 1927–1961. *Br. J. Psychiatry* **115,** 9–30.

Pugh, T. G., Jerath, B. K., Schmidt, W. M., and Reed, R. B. (1963). Rates of mental disease related to childbearing. *N. Engl. J. Med.* **268,** 1224–1228.

Rachelefsky, G. S., Flynt, J. W., Ebbin, A. J., and Wilson, M. G. (1972). Possible teratogenicity of tricyclic anti-depressants. *Lancet* **1,** 838.

Rainer, J. (1966). Genetic counseling and preventive psychiatry. *Ment. Hyg.* **50,** 593–595.

Reich, T., and Winokur, G. (1970). Post-partum psychoses in patients with manic depressive disease. *J. Nerv. Ment. Dis.* **151,** 60–69.

Reisby, N. (1967). Psychoses in children of schizophrenic mothers. *Acta Psychiatr. Scand.* **43,** 8–20.

Rieder, R. (1973). The offspring of schizophrenic parents: A review. *J. Nerv. Ment. Dis.* **157,** 179–190.

Rieder, R. O., Rosenthal, D., Wender, P., and Blumenthal, H. (1975). The offspring of schizophrenics: Fetal and neonatal deaths. *Arch. Gen. Psychiatry* **32,** 200–211.

Rimmer, J., and Jacobsen, B. (1976). Differential fertility of adopted schizophrenics and their half-siblings. *Psychiatr. Res.* **54,** 161–166.

Robson, J. M., and Sullivan, F. M. (1963). The production of fetal abnormalities in rabbits with imipramine. *Lancet* **1,** 638–639.

Rosanoff, A. J., Handy, L. M., and Plesset, I. R. (1934). The etiology of manic-depressive syndromes with special reference to their occurrence in twins. *Am. J. Psychiatry* **91,** 247–286.

Rosanoff, A. J., Handy, L. M., Plesset, I. R., and Brush, S. (1934–1935). The etiology of so-called schizophrenic psychoses. *Am. J. Psychiatry* **91,** 247–286.

Rosenthal, D. (1970). "Genetic Theory and Abnormal Behavior." McGraw-Hill, New York.

Rosenthal, D., Wender, P. H., Kety, S. S. (1971). The adopted-away offspring of schizophrenics. *Am. J. Psychiatry,* **128,** 307–311.

Rosenthal, D. (1975). The spectrum concept in schizophrenia and manic-depressive disorders. *In* "Biology of the Major Psychoses" (D. X. Freedman, ed.), pp. 19–25. Raven, New York.

Rosenthal, D., Wender, P. H., Kety, S. S., Welner, J., and Schulsinger, F. (1971). The adopted-away offspring of schizophrenics. *Am. J. Psychiatry* **128,** 307–311.

Rudin, E. (1923). Ober Vererbung geistiger Storungen. *Z. Gesamte Neurol. Psychiatr.* **81,** 459–496.

Safra, M. J., and Oakley, G. P. (1975). An association of cleft lip with or without cleft palate and prenatal exposure to valium. *Lancet* **2,** 478–479.

Sameroff, A., and Zax, M. (1973). Perinatal characteristics of the offspring of schizophrenic women. *J. Nerv. Ment. Dis.* **157,** 191–199.

Saxen, I., and Saxen, L. (1975). Association between maternal intake of a diazepam and oral clefts. *Lancet* **2,** 498.

Schou, M. (1976). Advances in lithium therapy. *Curr. Psychiatr. Ther.* **16,** 139–153.

Schou, M., Amidisen, A., and Streenstrup, O. R. (1973). Lithium and pregnancy. II. Hazards to women given lithium during pregnancy and delivery. *Br. Med. J.* **2,** 137–138.

Schulsinger, F. (1976). A ten year follow up of children of schizophrenic mothers; clinical assessment. *Acta Psychiatr. Scand.* **53,** 371–386.

Schulz, B. (1932). Zur Erbpathologie der Schizophrenie. *Z. Gesamte Neurol. Psychiatr.* **143,** 175–293.

Sjöquist, F., Bergfors, P. G., Borga, O., Lind, M., and Ygge, H. (1972). Plasma disappearance of nortriptyline in a newborn infant following placental transfer from an intoxicated mother. *J. Pediatr.* **80,** 496–500.

Slater, E. (1953). Genetic investigations in twins. *J. Ment. Sci.* **99,** 44–52.

Sobel, D. E. (1960). Fetal damage due to ECT, insulin coma, chlorpromazine, or reserpine. *Arch. Gen. Psychiatry* **2,** 606–611.

Sobel, D. E. (1961). Infant mortality and malformations in children of schizophrenic women. *Psychiatr. Q.* **35,** 60–64.

Spitzer, R. L., and Endicott, J. (1975). "Schedule for Affective Disorders and Schizophrenia-Lifetime Version." Biometrics Research, New York State Psychiatric Institute.

Spitzer, R. L., Endicott, J., and Robins, E. (1978). Research diagnostic criteria: Rationale and reliability. *Arch. Gen. Psychiatry* **35,** 773–782.

Stenger, E. G., Aeppli, L., and Fratta, I. (1965). Zur frage der Keimschadigenden wirkung von N-(gamma-dimethylaminopropyl)-iminodibenzyl-Hcl am tier. *Arzneim.-Forsch.* **15,** 1222–1224.

Stenstedt, A. (1952). A study in manic-depressive psychoses. Clinical, social, and genetic investigations. *Acta Psychiatr. Neurol. Scand., Suppl.* **79,** 1–111.

Stothers, J. K., Wilson, D. W., and Royston, N. (1973). Lithium toxicity in the newborn. *Br. Med. J.* **3,** 233–234.

Targum, S. D., Dibble, E. D., Davenport, Y. B., and Gershon, E. S. (1980). (1981). Family attitudes questionnaire: patients and spouses view bipolar illness. *Arch. Gen. Psychiatry,* in press. Etiology, familial risks, and chronic burden of bipolar illness: Contrasting perceptions of patient and spouse. Submitted for publication.

Tetlow, C. (1955). Psychoses of childbearing. *J. Ment. Sci.* **101,** 629–639.

Tienari, P. (1971). Schizophrenia and monozygotic twins. *Psychiatr. Fenn.* pp. 97–104.

Tsuang, M. T. (1975). Heterogeneity of schizophrenia. *Biol. Psychiatry* **10,** 465–474.

Tuchmann-Duplessis, H., and Mercier-Parot, L. (1963). Modifications du comportement sexual chez des descendants de rats traites par un inhibiteur des monoamine-oxidases. *C.R. Hebd. Seances Acad. Sci.* **256,** 2235–2237.

Tunnessen, W. W., and Hertz, C. C. (1972). Toxic effects of lithium in newborn infants: A commentary. *J. Pediatr.* **81,** 804–807.

Uddenberg, N., and Englesson, I. (1978). Prognosis of post-partum mental disturbance. *Acta Psychiatr. Scand.* **58,** 201–212.

Videbech, T. (1975). A study of genetic factors, childhood bereavement, and premorbid traits in patients with anancastic endogenous depression. *Acta Psychiatr. Scand.* **52,** 178.

Weinberg, I., and Lobstein, J. (1936). Beitrag zur Vererbung des manisch-depressiven Irreseins. *Psychiatr. Neurol. Bull.* (Amsterdam), **1**a(1):339–372.

Weinstein, M. R., and Goldfield, M. (1975). Cardiovascular malformations with lithium used during pregnancy. *Am. J. Psychiatry* **132,** 529–531.

Wender, P. H., Rosenthal, D., Kety, S. S., Schulsinger, F., and Weiner, J. (1974). Cross-fostering: A research strategy for clarifying the role of genetic and experiential factors in the etiology of schizophrenia. *Arch. Gen. Psychiatry* **30,** 121–128.

Wilbanks, G. D., Bressler, B., Peete, H. C., Cherny, W. B., and London, W. L. (1970). Toxic effects of lithium carbonate in a mother and newborn infant. *J. Am. Med. Assoc.* **213,** 856–867.

Winokur, G., and Clayton, P. (1967). Family history studies. I. Two types of affective disorders separated according to genetic and clinical factors. *Recent Adv. Biol. Psychiatry* **9,** 35–50.

Winokur, G., and Ruangtrakool. (1966). Post-partum impact on patients with independently diagnosed affective disorder. *J. Am. Med. Assoc.* **197,** 242–246.

Winokur, G., Morrison, J., Clancy, J., and Crowe, R. R. (1972). The Iowa 500. *Arch. Gen. Psychiatry* **27,** 462–464.

Woerner, M., Pollack, M., and Klein, D. (1973). Pregnancy and birth complications: A comparison of schizophrenic and personality disorder patients with their siblings. *Acta Psychiatr. Scand.* **49,** 712–721.

13

Pregnancies in Women with Chromosomal Abnormalities

JOE LEIGH SIMPSON

Copyright © 1981 by Academic Press, Inc.
All rights of reproduction in any form reserved.
ISBN 0-12-630940-X

Approximately 1.5% of liveborn infants show a change in chromosome number or structure. Some chromosome abnormalities (e.g., trisomies 13 or 18) so adversely affect an individual that limited life expectancy and severe anomalies preclude pregnancy by affected females. Other abnormalities affect the phenotype less severely and are compatible with pregnancy.

In contrast to most other genetic disorders discussed in this volume, chromosomal disorders are usually not characterized by consistent biochemical or physiological changes, even if mental retardation and somatic anomalies are present. However, offspring of women with chromosomal abnormalities may be chromosomally imbalanced as result of either normal or abnormal meiotic segregation. Inheritance of unbalanced gametes shall occupy most of our attention in this chapter.

I. ORIGIN OF CHROMOSOMAL ABNORMALITIES

A. Numerical Abnormalities

If a haploid gamete or diploid cell lacks its expected number of chromosomes (n or $2n$), aneuploidy exists. If one entire chromosome is missing ($2n - 1$), monosomy exists. The term polysomy is sometimes applied if one entire chromosome is a sex chromosome (e.g., 47,XXY).

In *primary trisomy,* the extra chromosome is one of the usual chromosomes of the complement ($2n + 1$ whole chromosome). In *secondary trisomy,* the extra chromosome is an isochromosome for one chromosome arm ($2n + 1$ isochromosome). In *tertiary trisomy,* which usually arises by 3:1 segregation in translocation heterozygotes (see below), the extra chromosome consists of two arms of nonhomologous chromosomes. We shall consider only primary trisomy in this section, for secondary and tertiary trisomy are best considered in context of structural abnormalities.

Primary trisomy may arise by several mechanisms, but in humans it usually

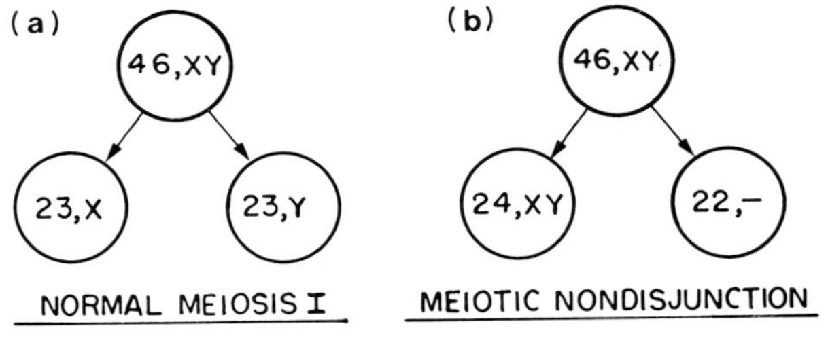

Fig. 1. The product of (a) normal meiosis and (b) meiosis in which nondisjunction produced 24,XY and 22,-gametes. (From Simpson, 1976.)

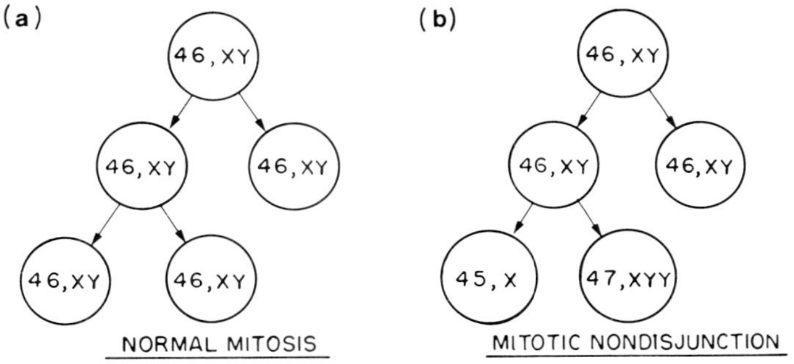

Fig. 2. The products of (a) normal mitosis and (b) mitosis characterized by nondisjunction of a Y chromosome. (From Simpson, 1976.)

arises *de novo* following meiotic or mitotic nondisjunction. Primary trisomy may originate following failure of homologous chromosomes to disjoin (meiosis I) or failure of sister chromatids to pass to complementary daughter cells (meiosis II or mitosis). Factors predisposing to nondisjunction in meiosis I include failure to synapsis, absence of chiasmata, or premature terminalization. Nondisjunction during meiosis produces aneuploid gametes, but the zygote will contain only one abnormal cell line. Nondisjunction during mitosis produces two or more cell lines (mosaicism) (Fig. 2). The causes for mitotic nondisjunction in humans are largely unknown.

A second mechanism for production of primary trisomy is normal meiotic segregation in a trisomic parent. Such a parent would be expected to produce n and $n + 1$ gametes in equal numbers. Fertilization involving an $n + 1$ gamete would yield a trisomic zygote (secondary nondisjunction). In plants triploids are also a good source or trisomies, but in humans triploidy is nearly always lethal; however, $2n/3n$ individuals have survived and could reproduce (Ferrier *et al.*, 1964; Jenkins *et al.*, 1971; Sellyei *et al.*, 1971). In plants, trisomies can also be recovered from monosomic individuals. Possibly this results from premature division and nondisjunction of the univalent chromosome at meiosis I, followed by passage to the same pole in meiosis II (Khush, 1973).

Monosomy arises by mechanisms similar to those described above, for the nondisjunctional events that lead to a trisomic gamete also produce a complementary monosomic gamete. In addition, sometimes one or more chromosomes fail to pass to daughter cells (anaphase lag). In particular, chromosomes lacking a centromere (acentric) are usually eliminated. Monosomy may also arise by normal segregation in a monosomic parent, who would be expected to produce both n and $n - 1$ gametes (secondary nondisjunction), or also by normal meiotic segregation in translocation heterozygotes (see Section, I,B). Haploid plants are a traditional source for monosomic plants, but in humans haploidy is lethal.

Similarly, triploid plants occasionally produce monosomic gametes, and intervarietal and interspecific crosses often yield monosomy as result of univalent formation at meiosis I.

Polyploidy is detected frequently among human abortuses but rarely among neonates. Triploidy ($3n = 69$) requires either two maternal or two paternal haploid complements. Possible mechanisms include fertilization of a single haploid ovum by two sperm (dispermy), fertilization of a single diploid ovum that failed to undergo meiosis I, reincorporation or suppression of one of the polar bodies (digyny), or a mitotic error at syngamy involving either male or female pronucleus. In humans dispermy is the most common mechanism (Kajii and Nikawa, 1977).

B. Structural Abnormalities

Minor structural variation (polymorphism) exists among human chromosomes, without apparent phenotypic consequence. However, clearly abnormal variations (traditionally termed *aberrations*) are usually associated with phenotypic abnormalities. If the variation is large enough to be recognized by light microscopy, one can assume that many genes are duplicated or deficient. A phenotypic abnormality might also result if the position of a gene with respect to its neighboring genes is altered (position effect).

A simple *deficiency* involves loss of one portion of one chromosome, either terminal or interstitial (Fig. 3). Deficiency in an autosome usually leads to

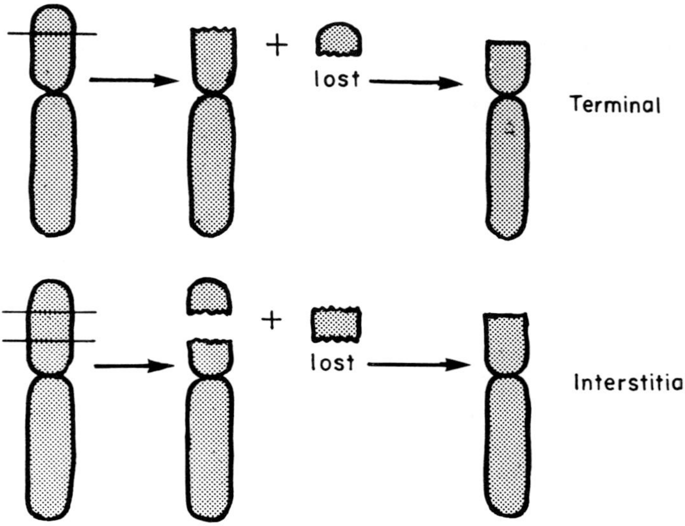

Fig. 3. The origin of deletions. (From Simpson, 1976.)

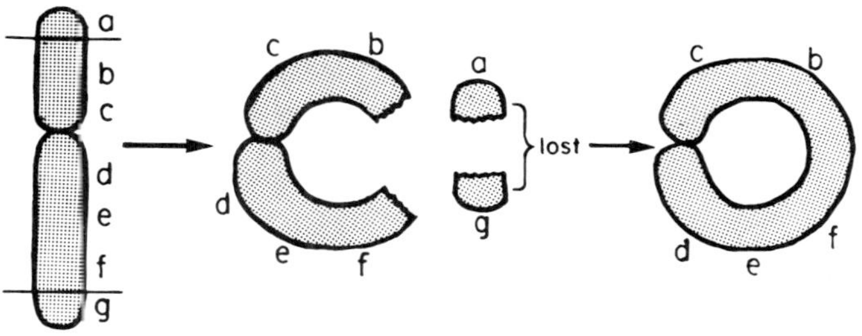

Fig. 4. The origin of a ring chromosome. (From Simpson, 1976.)

embryonic death or malformation, whereas deficiency in a sex chromosome is not necessarily so deleterious (Simpson, 1979a,b). Deficiencies may also arise following crossing over within an inversion loop (see below).

A *ring chromosome* arises following a break in both the long and the short arm. The centromeric ends fuse; the telomeric acentric fragments are lost (Fig. 4). In a ring chromosome portions of each are thus deficient. Ring chromosomes are unstable because their replication can lead to double-sized rings, dicentric rings, or loss of the entire chromosome.

Duplications may originate by unequal crossing over between sister chromatids. Duplications may also occur following meiotic segregation in a parent in which chromosomal translocation is present, or following crossing over within an inversion loop (see below). Duplications are usually less lethal than deficiencies; however, in humans duplications are usually associated with phenotypic abnormalities.

If following chromosome breakage material is exchanged between two or more chromosomes, a *translocation* occurs (Fig. 5). If no genetic material is lost, the individual is phenotypically normal and the translocation is said to be balanced. Such an individual is a translocation heterozygote. However, during

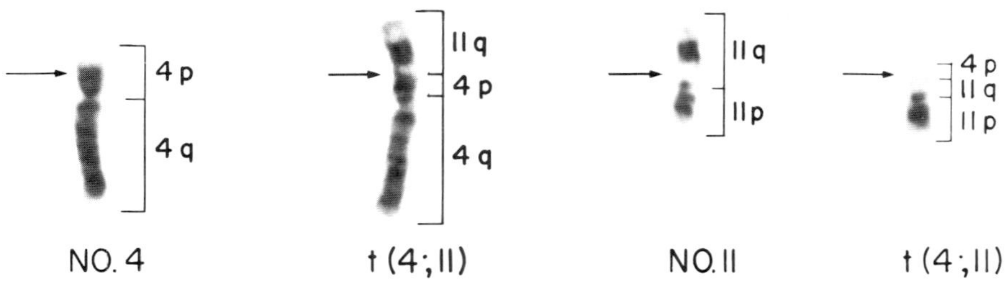

Fig. 5. Reciprocal translocations between chromosomes 4 and 11 detected in a woman who experienced eight spontaneous abortions. (From Simpson, 1980.)

meiosis I there are several ways in which the two translocation chromosomes and the two structurally normal chromosomes may segregate. Gametes would be chromosomally balanced if both translocation chromosomes pass to one gamete, whereas both normal chromosomes pass to the complementary gamete (*alternate segregation*) (Fig. 6). If either translocation chromosome and one of the normal chromosomes pass to the same pole, however, the resulting gametes are unbalanced as result of duplication or deficiency. If homologous centromeres pass to the same gamete, *adjacent segregation I* is said to occur. If nonhomologous centromers pass to the same gamete, *adjacent segregation II* occurs.

In an inversion (Fig. 7) the order of genes on a chromosome is altered, but

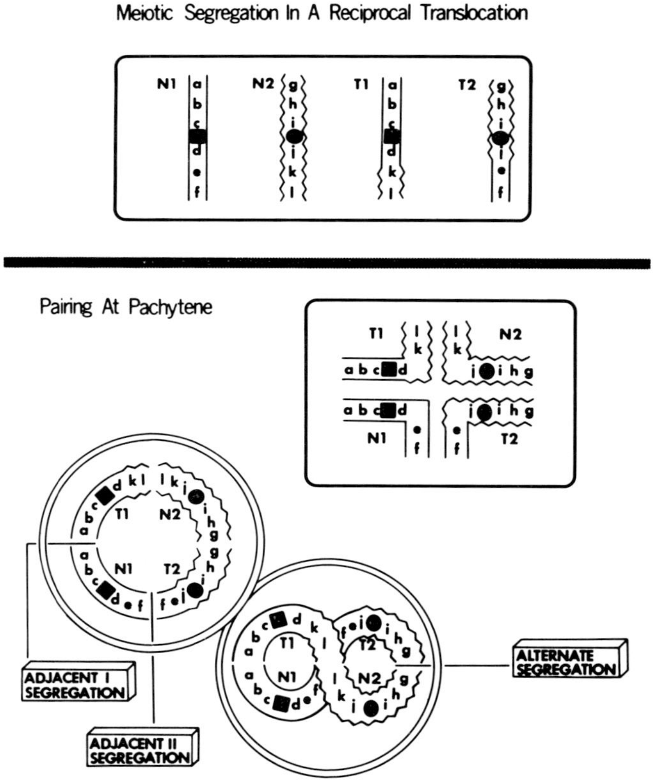

Fig. 6. Schematic diagram of alignment of two translocation chromosomes and two normal chromosomes at pachytene. If centromeres of chromosomes T1 and T2 pass to one pole, with N1 and N2 passing to the opposite, segregation is said to be alternate. If T1 and N1 (T2 and N2) pass to the same pole, adjacent I segregation occurs. If T1 and N2 (and T2 and N1) pass to the same pole, adjacent II segregation occurs.

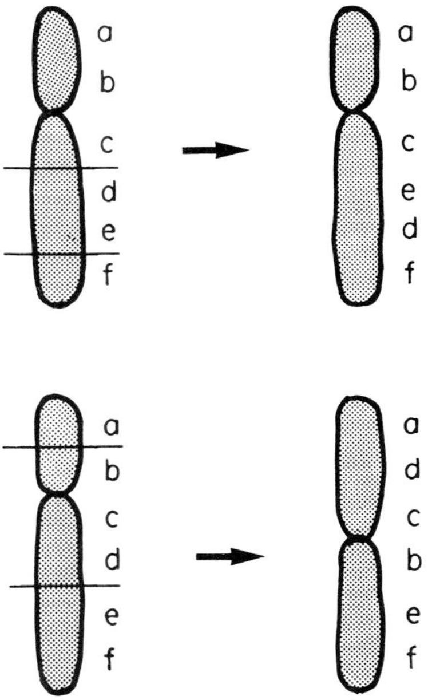

Fig. 7. Origin of paracentric (above) and pericentric (below) inversions. (From Simpson, 1976.)

neither duplication nor deficiency occurs. Two or more break points are required. If the segment between the breaks includes the centromere, the inversion is pericentric. If the segment does not include the centromere, the inversion is paracentric. Paracentric inversions are more difficult to detect than pericentric inversions, but the former can be recognized by chromosomal banding techniques, meiotic studies, or genetic studies. Although individuals with inversions are phenotypically normal, their progeny may be abnormal if a chromosome carrying an inversion undergoes a single crossing over event during meiosis I. The result of a crossover within a paracentric inversion would be one acentric, one dicentric and two normal chromatids (Fig. 8). If crossing over occurred at two sites within a paracentric loop there could be no deleterious effect. A single crossing over event within a pericentric inversion loop gives risk to two duplication-deficiency products and two normal products. Two crossovers within the loop would produce no abnormalities.

Isochromosomes are chromosomes with identical arms. They arise following horizontal, rather than longitudinal, division of the centromere (Fig. 9). One product is acentric, composed on one entire arm and thus lost at the next cell division; the remaining product is telocentric, capable of replicating itself at the next S period to form a metacentric chromosome. The two arms of an isochromosome are isologous, both structurally and genetically identical. An isochromo-

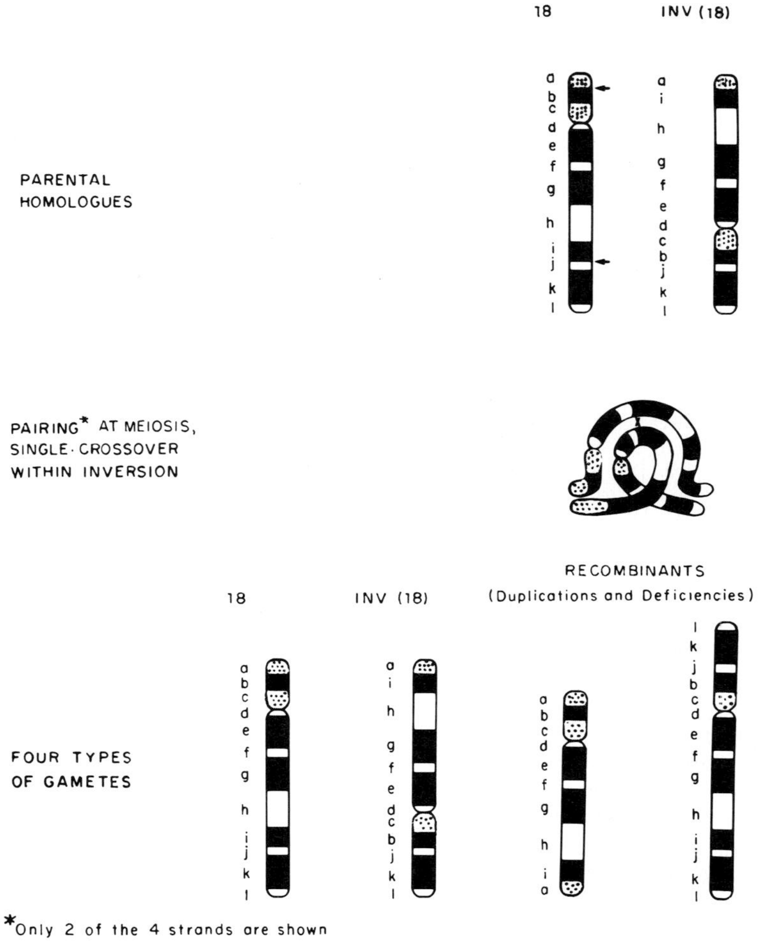

Fig. 8. The effects of a single crossing over within a pericentric inversion loop. Other mechanisms of crossing over need not involve formation of a loop.

some can be composed of identical long arms or identical short arms, depending upon which arm remains with the centromere. Thus, an isochromosome consists of complete duplication of one arm and complete deficiency of the other arm.

A *dicentric* chromosome has two centromeres. Its formation might begin with an isochromatid break (Fig. 10). The telomeric acentric fragment is lost, and sister chromatids join. Following longitudinal division of the centromere, the chromosome has two centromeres and both duplicated and deficient portions. A dicentric chromosome may also result from a single crossing over event within a

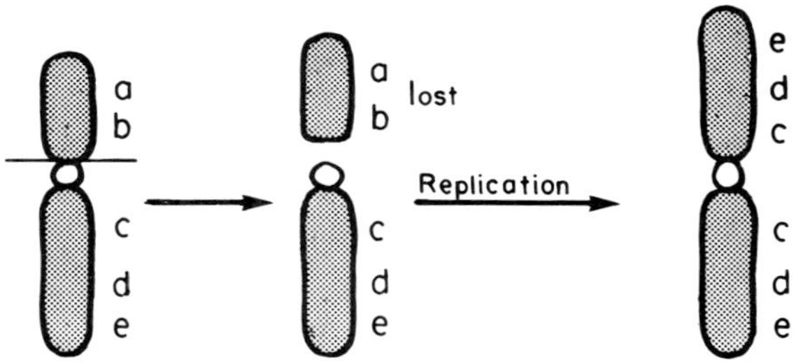

Fig. 9. The origin of an isochromosome. (From Simpson, 1976.)

paracentric loop. The presence of two centromeres (dicentric) leads to mitotic instability because the two centromeres can migrate to opposite poles during telophase. Stretching to the point of breaking might occur, followed by loss of chromosomal material or secondary rearrangements (breakage–fusion–bridge cycle).

C. Mosaicism and Chimerism

A mosaic individual has two or more genotypes (i.e., cell lines), each of which originate from a single zygote. Chromosomal mosaicism usually arises following mitotic nondisjunction or anaphase lag, with survival of at least two lines with different complements. A single nondisjunctional event can result in as many as three cell lines (Fig. 2), although failure of all daughter cells to survive often results in less complicated mosaicism. Conversely, some forms of mosaicism require errors in at least three cell divisions. Mosaicism may also follow production of an unstable (e.g., dicentric) chromosome, which secondarily leads to loss

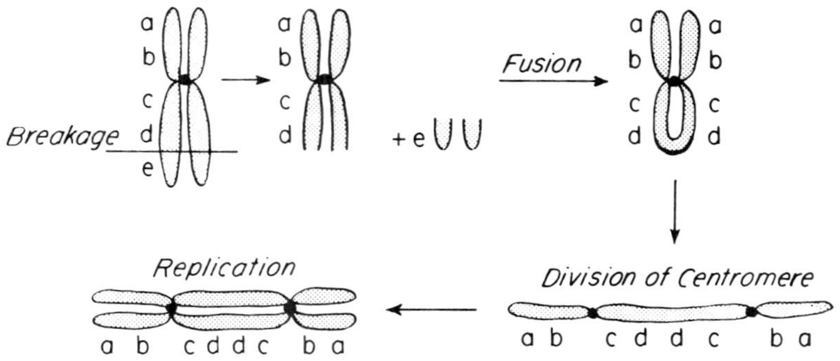

Fig. 10. Origin of a dicentric chromosome. (From Simpson, 1976.)

of a chromosome (breakage–fusion–bridge cycle) and development of a monosomic line.

In chimerism, a single individual has two or more genotypes. Each arises from different zygotes, which subsequently fuse.

Detection of mosaicism and chimerism is discussed elsewhere (Simpson, 1976).

II. INCIDENCE OF CHROMOSOME ABNORMALITIES

Prospective surveys of consecutively born neonates indicate that 0.6% show chromosomal abnormalities (Table I). The incidence apparently does not vary greatly among geographical location, although it is affected by such factors as maternal age. Trisomy 21 is most common, but polysomy X (47,XXX; 47,XXY) and polysomy Y (47,XYY) are also relatively common. Structural rearrangements in aggregate are also not rare, although any specific rearrangement is. Overall mutation rates for balanced and unbalanced structural rearrangements have been estimated at 1.78×10^{-4} and 1.1×10^{-4} per gamete per generation, respectively (Jacobs, 1977). (In this estimate unbalanced rearrangements included deletions and supernumerary chromosomes.) Frequencies of balanced D/D Robertsonian,* D/G Robertsonian,† and nonacrocentric reciprocal translocations are 0.72, 0.019, and 0.090 per 1000 births, whereas frequencies of unbalanced Robertsonian and reciprocal translocations are 0.007 and 0.012 per 1000 births (Hook and Hamerton, 1977). The frequencies of balanced and unbalanced inversions are 0.014 and 0.002 per 1000. The above rates might be higher if banding techniques had been utilized routinely in population surveys.

The prevalence of chromosomal abnormalities is much higher among stillborn infants and spontaneous abortuses than among liveborn infants, presumably reflecting selection against genetically disadvantaged embryos. About 50–60% of first trimester spontaneous abortions are chromosomally abnormal; 50% show autosomal trisomy, 25% polyploidy, 20% monosomy X, and 5% translocations (Boué *et al.*, 1975). Some 20% of second trimester abortuses are chromosomally abnormal (Ruzicska and Czeizel, 1970), usually because of autosomal trisomy or sex chromosomal mosaicism. About 5% of stillborn infants show chromosomal abnormalities, usually autosomal trisomy (Bauld *et al.*, 1974; Machin and Crolla, 1974; Kuleshov *et al.*, 1976). Chromosomal studies should thus be performed on all abortuses or stillborn infants, perhaps excluding those with neural tube defects. If one detects autosomal trisomy in an abortus, subsequent pregnancies should be monitored for fetal chromosomal abnormalities.

*D group, chromosomes 13, 14, and 15. †G group, chromosomes 21 and 22. Centric fusion translocations involving D and G chromosomes are called Robertsonian; other translocations are reciprocal.

TABLE I
Incidence of Chromosomal Abnormalities in Liveborn Infants[a]

Numerical aberrations	
Sex chromosomes	
47,XYY	1/1000 MB[b]
47,XXY	1/1000 MB
Other (males)	1/1350 MB
45,X	1/10,000 FB
47,XXX	1/1000 FB
Other (females)	1/2700 FB
Autosomes	
Trisomies	
13–15 (D Group)	1/20,000 LB
16–18 (E Group)	1/8000 LB
21–22 (G Group)	1/800 LB
Other	1/50,000 LB
Structural Aberrations	
Balanced	
Robertsonian	
t(Dq;Dq)	1/1500 LB
t(Dq;Gq)	1/5000 LB
Reciprocal translocations and insertional inversions	1/7000 LB
Unbalanced	
Robertsonian	1/14,000 LB
Reciprocal translocations and insertional inversions	1/8000 LB
Inversions	1/50,000 LB
Deletions	1/10,000 LB
Supernumeraries	1/5000 LB
Other	1/8000 LB
Total	1/160 LB

[a] Tabulated by Hook and Hamerton (1977).
[b] LB, livebirths; MB, male births; FB, female births.

III. PHYSIOLOGIC EFFECTS OF PREGNANCIES IN WOMEN WITH CHROMOSOMAL ABNORMALITIES

Hypogonadism is characteristic of 45,X females (Simpson, 1976), but few other specific physiologic or biochemical changes are consistently detected in individuals with chromosomal abnormalities. To be sure, autosomal trisomies, triploidy, and sometimes sex chromosome abnormalities cause mental retardation, multiple anomalies, intrauterine and postnatal growth retardation, and even death; however, the manner by which chromosomal imbalance exerts its effect is unknown. Thus, there are relatively few reasons to believe that pregnancies will be physiologically unusual in women with chromosomal abnormalities. Available data generally support this assumption.

On the other hand, deficient alleles (hemizygous) on triplicated alleles (trisomic) could theoretically exert a deleterious effect. Interestingly, about one-half of the chromosomally normal offspring of 47,XX,+21 women were said to show evidence of abnormal development (Scharrer *et al.,* 1975). The reported anomalies usually consist of craniofacial disproportion, skeletal anomalies, or hypospadias. They could represent merely biases of ascertainment or reporting, or they could indicate that pregnancy in women with trisomy 21 indeed leads to nonspecific deleterious fetal effects for unknown reasons. Additional data are necessary.

An occasional 45,X individual is fertile yet shows short stature and other somatic anomalies of the Turner stigmata (Simpson, 1979a,b). Somatic anomalies may exist despite normal gonadal functions, potentially interfering with pregnancy in predictable fashions. For example, short stature and a small pelvis may necessitate cesarean section. Renal anomalies may predispose to pyelonephritis; vertebral problems may cause ambulatory difficulties; cardiac anomalies may necessitate bed rest, careful antepartum and intrapartum fetal monitoring, and shortening of the second stage of labor. Other somatic disorders associated with 45,X are hypertension, autoimmune thyroiditis, and diabetes mellitus. Presence of such disorders in a pregnant 45,X individual could clearly affect pregnancy outcome, but clinical management is no different in 45,X than in 46,XX patients. A similar statement applies to pregnancies by women with a 47,XXX complement or autosomal imbalance.

IV. OFFSPRING OF WOMEN WITH AUTOSOMAL TRISOMY

Autosomal trisomy usually leads either to lethality or to such severe anomalies that pregnancy in trisomic individuals is unlikely. However, pregnancy in females with trisomy 21 has been reported, and pregnancy in trisomy 8 individuals would not be totally unexpected. Gonadal mosaicism could also occur in phenotypically normal individuals whose germ cells centained a trisomic line. Indeed, parental germinal mosaicism is an often assumed and sometimes proven (Kaffe *et al.,* 1974) explanation for recurrent aneuploid offspring (Hsu *et al.,* 1970).

A. Theoretical Considerations

The presence of three homologous chromosomes should lead to formation of a trivalent at pachytene of meiosis I. At any given point along the chromosome length, only two of the three chromosomes can pair; however, at another point, two different chromosomes can pair. Alternatively, one bivalent and one univalent can form the latter as a more likely occurrence if small chromosomes are involved. At anaphase I two chromosomes usually pass to one pole, whereas the other chromosome passes to the opposite pole, yielding $n + 1$ and n gametes. Irrespective of whether accompanied by a bivalent, a univalent is subject to several cytologic abnormalities, to be discussed in detail. The irregular behavior of univalents contributes to lower than expected (50%) proportions of $2n + 1$ progeny in plants, and similar phenomena probably occur in man.

Several principles derived from plant studies are worth summarizing because these principles are probably applicable to mammals. First, fewer than expected (50%) $2n + 1$ progeny are recovered, almost never the opposite. This is presumably due to irregular behavior of the univalent. Second, females transmit the additional chromosome more often than males. For example, the six trisomies of *Triticum durum* (Emmer wheat) are transmitted through males and females at average frequencies of 2% and 23%, respectively (Tsunewaki, 1964). Failure of transmission could result from elimination of the additional chromosome during meiosis, or from reduced viability of $n + 1$ gametes or $2n + 1$ embryos. Differences between transmission rates in males and females suggest that elimination during meiosis is the most important factor in plants. Small chromosomes are eliminated most frequently, presumably because their short length minimizes the opportunity for chismata formation. Third, trisomies for other chromosomes may occur, possibly due either to a generalized interference with disjunction or to induction of univalents in other chromosomes.

Because liveborn fertile trisomic animals are rare, the relevance of the above studies to humans cannot be proved. However, principles governing transmission of translocations apply equally to plants and small mammals, and there is little reason to suspect fundamental species differences.

B. Empiric Observations in Autosomal Trisomy

In only a few types of nonmosaic autosomal trisomy do individuals survive sufficiently long for pregnancy to be possible. Individuals with trisomy 13 and trisomy 18 rarely survive past infancy, and none have become pregnant. In fact, affected females often show ovarian dysgenesis (Russell and Altschuler, 1975; Kennedy *et al.*, 1977). Pregnancy could theoretically occur in individuals with trisomy 8, 9, or 22; however, the incapacitated nature of affected individuals makes pregnancy socially unlikely, even if ovarian function were normal.

At least 27 pregnant females with trisomy 21 have been reported (Van de Velde-Staquet, 1973). Nine of the 27 (33%) liveborn offspring had trisomy 21, with no other chromosomal abnormalities reported. Pregnancy and labor were not characterized by unusual features. The occurrence of trisomic offspring in fewer (33%) than expected (50%) proportion accord with observations in progeny of trisomic plants. As noted previously, chromosomally normal offspring of trisomy 21 mothers have been purported to show nonspecific malformations in greater than expected frequency (Scharrar *et al.,* 1975); however, additional data are necessary.

C. Empiric Observations in $2n/2n + 1$ Mosaicism

Parental autosomal mosaicism is a recognized cause for aneuploidy in offspring. The mosaicism is usually not detected until after birth of an aneuploid child (Kaffe *et al.,* 1974); thus, empiric data are difficult to derive. The mosaicism (e.g., 46,XX/47,XX,+21) is usually detectable in lymphocytes, but it could be restricted to gonads. Familial mosaicism is suspected (Hsu *et al.,* 1970) but unproved.

V. OFFSPRING OF 47,XXX AND 46,XX/47,XXX WOMEN

Meiosis in 47,XXX females should be governed by the same cytologic principles applicable for autosomal trisomy (see above). That is, either a single trivalent, or one bivalent and one univalent, might ordinarily segregate. A univalent is also subject to abnormal division (e.g., isochromosome). Thus, fewer chromosomally abnormal progeny than theoretically predicted might be observed. Meiotic studies have apparently not been performed in 47,XXX females. However, one explanation for the premature ovarian failure sometimes observed in 47,XXX individuals (Smith *et al.,* 1974) might be abnormal meiosis leading to oocyte atresia.

A. Empiric Observations

Because 47,XXX females are not rare (1:800 liveborn female births), their reproduction can be properly assessed only on the basis of cases ascertained in population surveys. Unfortunately, such data are not available. Thus, biases of ascertainment and reporting must be considered in interpreting published data.

The 47,XXX females should theoretically produce 23,X and 24,XX secondary oocytes, which at fertilization could theoretically yield 46,XX, 46,XY, 47,XXX, and 47,XXY zygotes in equal frequencies. However, most progeny are chromosomally normal. There is no explanation for these empiric observations, but analagous to the situation in trisomic parents one could postulate: (1) selection against abnormal embryos, or, more likely, (2) preferential segregation of 24,XX into a polar body; and (3) decreased fertilizability of 24,XX oocytes.

The above observations notwithstanding, 47,XXX and 46,XX/47,XXX women still show some increased risk for being delivered of chromosomally abnormal offspring. In 1969, Barr et al. tabulated that 1 of 20 reported 47,XXX women had a 47,XXY son; 3 of 8 46,XX/47,XXX women had 47,XXY or 46,XY/47XXY sons; and 1 of the 8 had a 46,XX/47,XXX daughter. Since then other observations have been reported. At least one 47,XXX female had a 47,XX,+21 child (Kadotoni *et al.*, 1970), whereas another had a 45,X child (Guzman-Toledano *et al.*, 1976). Three 46,XX/47,XXX women had offspring with Klinefelter syndrome (Baike, 1972; Geisler *et al.*, 1972; Tsuang *et al.*, 1975) and one 46,XX/47,XXX mother had a 48,XXYY son (Ziska *et al.*, 1975). In addition, 45,X/46,XX/47,XXX mothers have given birth to offspring with both sex chromosomal mosaicism (de Toni *et al.*, 1965; Giraud, 1973) and Down's syndrome (Armendares *et al.*, 1967; Hsu *et al.*, 1972) (see Section VI).

Although these reports suggest that women with a 47,XXX line show increased liability for chromosomally abnormal offspring,caution is warranted because biases of ascertainment and reporting clearly exist. Indeed, in many cases the maternal complement was detected only after birth of an abnormal proband (Dewhurst, 1978). It still seems prudent to offer antenatal chromosomal studies to 47,XXX and 46,XX/47,XXX women. Not only the possibility of 47,XXY and 47,XXX offspring, but also the possibility of other chromosomal abnormalities should be discussed.

VI. OFFSPRING OF 45,X; 45,X/46,XX; AND 45,X/46,XX/47,XXX WOMEN

Monosomy X in humans is usually but not always lethal, with surviving individuals displaying gonadal and somatic abnormalities (Simpson, 1976). The 45,X individuals were traditionally considered sterile, but it is now clear that occasional nonmosaic 45,X individuals not only menstruate but become pregnant. In addition, there are both theoretical and empirical reasons to suspect that offspring of 45,X women have an increased prevalence of chromosomal abnormalities.

A. Theoretical Considerations

In monosomic ($2n - 1$) plants, the monosomic chromosome forms a univalent at meiotic prophase. During meiosis I, this univalent remains separate on the equatorial plate until the bivalent (usually $n - 1$ in number) begin to disjoin. At this time the univalent moves to the equatorial plate, after which it may (1) lag and become lost, yielding two $n - 1$ first division products; (2) pass undivided to one pole to yield one n and one $n - 1$ second division product; (3) undergo abnormal transverse centromeric division to yield two or more telocentric chromosomes. If the univalent fails to divide during meiosis I (possibility 2), it may divide normally at meiosis II into two sister chromatids, each of which may pass to complementary n daughter cells. However, both chromatids may also pass undivided to a single cell, or transverse misdivision at the centromere may occur. If the univalent divides prematurely during meiosis I, the first division products may separate abnormally, especially if telocentrics or isochromosomes form. Trisomies and monosomies for chromosomes other than that involved in the univalent may also occur, possibly as result of indirect meiotic disturbances that induce univalent formation. For example, three univalents (rather than the expected one) were observed in 5, 10, 15, and 25% of four different monosomic wheat lines (Clausen and Cameron, 1944).

In view of the above, it is not surprising that far fewer than 50% of progeny of monosomic plants are monosomic. The relative frequencies of progeny with $2n$, $2n-1$, or structurally abnormal chromosomes vary according to species, but in almost all fewer than 50% of progeny are monosomic. A few examples will suffice. Lefever and Patterson (1964) reported that only 6.03% of gametes of a monosomic line of *Avena sativa* (oats) were monosomic. Singh and Wallace (1967) estimated that the frequency of monosomy in several monosomic *Avena byzantina* lines to be 9.3–21.4%. A study of pollen grains (haploid) of 20 wheat monosomics yielded average frequencies of 62% n; 30% n-1; and 8% consisting of telocentrics isochromosomes, n-2 or n-3 (Morrison, 1953). In other plants centromeric misdivision leading to structurally abnormal chromosomes is very common. For example, in *Triticum aestivum* (wheat) misdivision occurs in 39.7% of first division products (Sears, 1952). Monosomy for chromosome 4 of *Drosophila*, the only viable monosomy in animals, is transmitted to only 34% of progeny (Bridges, 1921), despite chromosome 4 being a small chromosome whose loss confers relatively little selective disadvantage.

These biological principles derived by plant cytogeneticists should not be altogether ignored, even though cell division in plants is fundamentally different from animals. The complexities resulting from division of plant univalents should thus caution one against simplistic approaches to the rare fertile 45,X human.

B. Empiric Observations

Pregnancies have been reported in at least nine 45,X patients (Bahner *et al.*, 1960; Nakashima and Robinson, 1971; Grace *et al.*, 1973; Lajborek-Czyz, 1976; Groll and Cooper, 1976; Reyes *et al.*, 1976; Philip and Sele, 1976; Shokeir, 1978; King *et al.*, 1978); eight 45,X/46,XX patients (Bomers-Marres, 1966; Kava and Klinger, 1968; Predescu *et al.*, 1969; Bishun *et al.*, 1964; Arakaki *et al.*, 1970; MacKay *et al.*, 1971; Siegelman, 1972; Lieber and Berger, 1973; Shokeir, 1978); five 45,X/47,XXX patients (Reyes *et al.*, 1976; Lewis *et al.*, 1963; McLean *et al.*, 1968); and eight 45,X/46,XX/47,XXX patients (Briggs *et al.*, 1963; de Toni *et al.*, 1965; Armendares *et al.*, 1967; Nielsen and Thomsen,

1968; Giraud, 1973; Hsu *et al.*, 1972). Pregnancy has also occurred in patients mosaic for both a 45,X line and a line with a structural abnormality of the X, and our group has recently studied a 45,X/46,X,i(Xq) patient delivered of a 46,XY son.

Table II shows outcome of pregnancy in 45,X patients and in patients with a 45,X cell line. The frequency of spontaneous abortion is about 30%. Among the liveborn offspring are five infants with sex chromosomal mosaicism, four with Down's syndrome, and six with abnormalities probably polygenic in etiology. The frequency of abnormalities and abortion would thus appear to be greatly increased. The ostensible increased frequencies in Down's syndrome and polygenic malformations are puzzling.

However, careful perusal of published reports suggests evidence for biases of reporting or ascertainment. Specifically, chromosomal studies in the mother were often initiated only after an abnormal proband was identified. Indeed, most pregnant mothers with a 45,X cell line had normal ovarian function and no somatic anomalies, and would ordinarily not have been subjected to cytogenetic studies. In fact, if one considers only the offspring of cases ascertained because of maternal infertility, the frequencies of anomalies and abortions do not appear increased. Further indicating biases are observations that offspring of 45,X/46,XX/47,XXX, and possibly also 45,X/46,XX women seem more likely to be abnormal than offspring of 45,X women. If abnormalities in offspring were indeed related to presence of 45,X line, precisely the opposite might be expected. Another important consideration is that mosaicism in many mothers was poorly documented and may not have even existed (see Simpson, 1972, for specific comments). Finally, families in which both mother and offspring showed sex chromosomal mosaicism suggest that progeny are abnormal not because of a maternal 45,X line but because of an inherited tendency toward mitotic nondisjunction.

Although the above caveats justify caution before concluding that a maternal 45,X line may lead to abnormal progeny, the theoretical considerations reviewed in the previous section provide ample reason to suspect that certain chromosomal abnormalities could occur, specifically 45,X and 45,Y. The latter would be lethal, apparently surviving only a few days (Sinow and Luthardt, 1978). The 45,X zygotes could either survive until birth, or, much more likely, often terminate in spontaneous abortion (Boué *et al.*, 1975). Offspring with autosomal trisomy should occur rarely, although one could postulate either a nonspecific deleterious effect upon meiosis or the existence of a gene predisposing to nondisjunction.

In conclusion, the frequency of chromosomal abnormalities in offspring of women with a 45,X line is probably not nearly so high as a literal interpretation of published data would indicate (Dewhurst, 1978). Nonetheless, it seems prudent to offer antenatal diagnosis for detection of fetal chromosomal abnormalities. Suitable empiric risk figures are not available.

TABLE II
Results of Pregnancies in Individuals with a 45,X Cell Line[a]

						Liveborn		
Complement	Number patients	Number pregnancies	Induced abortions	Spontaneous abortions	Stillborn without anomalies	Normal	Chromosomal abnormalities	Nonchromosomal anomaly
45,X	9	12	0	0	1	9	1 (47,+21)	1 (hydrocephaly)
45,X/46,XX	8	20	2	8	0	5	2 (47,+21; 45,X/46,XY)	3 (2 Neural tube defect, 1 multiple anomalies)
45,X/47,XXX	5	6	0	0	0	6	0	0
45,X/46,XX/47,XXX	8	23	0	9	1	5	6 (2 47,+21; 1 45,X; 3 X-Chromosome mosaics)	2 (Cardiac anomalies)

[a]Tabulated from published reports without regard to biases of ascertainment or reporting.

VII. OFFSPRING OF 47,XYY MEN

Reproduction by males is not emphasized in this volume. However, a few remarks concerning 47,XYY males seem appropriate.

A. Histologic and Cytologic Considerations

Although apparently most are fertile, 47,XYY males often show testicular abnormalities (Shakkabaek *et al.*, 1973; Baghdassarian *et al.*, 1975). Perhaps 80% of investigated subjects show abnormalities of seminiferous tubules, ranging from complete arrest of spermatogenesis to very minor changes. Tubules characterized by spermatogenic arrest are detected in 50% of 47,XYY males, and about 30% of tubules consist solely of Sertoli cells (Shakkebaek *et al.*, 1973). Sperm counts may be low even if testes are normal in size (Shakkebaek *et al.*, 1973). The XYY mice also show reduced testicular weight and impaired spermatogenesis (E. P. Evans *et al.*, 1978).

During male meiosis the short arms of the X and Y chromosome undergo end-to-end association, perhaps because of homologous statural loci (German *et al.*, 1973). In 47,XYY testes the following meiotic configurations have been observed: one X-Y bivalent; one X univalent and one Y univalent; one X univalent and one Y-Y bivalent; one X univalent (alone); and X, Y, and Y univalents (Hulten and Pearson, 1970; Luciani *et al.*, 1973). In one well-studied case Luciani *et al.* (1973) observed 92 cells with one X-Y bivalent, 10 cells with X and Y univalents, and one cell with (X, Y, Y) univalents. At metaphase II all 55 cells contained 23 chromosomes. Thus, there appeared to be far fewer XYY primary spermatocytes than expected. By contrast, in *mice* primary spermatocytes are unusually not XY but XYY (E. P. Evans *et al.*, 1978). In mice the most frequent association was an XY bivalent and a Y univalent; however, the following combinations were not infrequent: X,YY; XYY; X,Y,Y; XY alone. Fewer than expected secondary spermatocytes were also present in mice, suggesting loss between the first and second meiotic division. However, 20,X; 20,Y; 21,YY counts were all observed at second metaphase, in apparent equal frequency. Coupled with observations that testicular size is diminished and fertility decreased, the most likely conclusion is that the deleterious effect of the XYY constitution in mice is generalized and not limited to a particular stage of mitotic proliferation or meiosis.

In conclusion, both 47,XYY mice and 47,XYY humans show decreased spermatogenesis, but the reasons are not entirely certain. Selection against the supernumerary Y usually but not always (Hulten and Pearson, 1970) appears to occur in humans during mitotic proliferation, such that primary spermatocytes are 46,XY. However, this does not appear to occur in mice. Thus, the decreased spermatogenesis in XYY males may merely reflect generalized dysfunction.

B. Empiric Observations

Although 47,XYY men should theoretically produce 46,XX, 46,XY, 47, XYY, and 47, XXY offspring, few chromosomally abnormal offspring have been reported (Melnyk *et al.,* 1969). This is consistent with observations that primary spermatocytes are usually 46,XY (however, see above for contradictory animal studies). Offspring of 47,XYY males are thus usually normal because aneuploid gametes either never form, or form yet fail to be fertilized. Nonetheless, antenatal diagnosis would seem warranted in women impregnated by 47, XYY males.

VIII. OFFSPRING OF WOMEN WITH STRUCTURAL CHROMOSOMAL ABNORMALITIES

A. Chromosomal Variants

Human chromosomes may exhibit minor morphologic variation, e.g., prominent satellites on acrocentric chromosomes, apparently unrelated to phenotypic change. The concordance rates for variants is higher among monozygotic than dizygotic twins, confirming the stability and heritability of variants. With availability of C-, Q- and silver banding techniques, variants can be detected in almost every individual. For example, McKenzie and Lubs (1975) detected 166 C-band variants among 77 newborns, and Craig-Holmes *et al.* (1977) detected 55 C-band variants among 35 unrelated individuals. Among 400 7- and 8-year-old children, Lubs *et al.* (1977) detected an average of 3.3 Q- and 3.9 C-banded variants per child; however, some of the variants differed only slightly from normal. Frequency was directly proportional to the quality of cytogenetic preparations. The most common variants involve the centromere of chromosome 3, the satellites of acrocentric chromosomes, and the distal long arm of Y.

Variants segregate in autosomal dominant fashion. For example, a double satellited chromosome 21 investigated by our group was recovered in 20 of 31 offspring whose parent had the variant (Martin *et al.,* 1979). Other variants are likewise recovered in approximately 50% of offspring of affected individuals. Neither double-satellited chromosomes nor most other variants seem to interfere with meiotic segregation. Segregation distortion for the common 9qh variant has been claimed (Fitzgerald, 1973), but other investigators disagree (Magenis *et al.,* 1977). Not only do variants segregate in expected Mendelian fashion, but there is no solid evidence that normal meiosis or mitosis is affected adversely. The 9qh inversion variant has also been implicated in poor reproductive outcome, but again firm data are lacking.

Polymorphism for Yq deserves special mention because many investigators have claimed relationships between short Y or long Y chromosomes and fetal loss. However, other investigators erroneously concluded that long polymorphic Y chromosomes represented a structural abnormality. The distal 2/3 of Yq usually fluoresces brilliantly for Q-bands and stains positively (darkly) for C-bands; however, this region apparently contains no essential genetic information (see Simpson, 1976). The length of Y varies, with variation in length accounted for solely by fluorescent region of Yq (Tishler *et al.*, 1972). This variation is paternally inherited, and without phenotypic effect. Reports claiming a relationship between a long Y chromosome and fetal loss can usually be refuted by observing that the variant Y is present in other family members who experienced no fetal loss. However, an occasional report by experienced investigators (Patel and Lubs, 1977) leaves open the possibility that variant Y chromosomes could adversely affect fetal outcome.

B. Deletions and Duplications

Deletions or duplications affecting large chromosomal regions produce severe phenotypic abnormalities; thus, individuals with deletions or duplications rarely reproduce. If such individuals did reproduce, one would theoretically expect 50% of gametes to show the same abnormality as the parent. Analagous to other chromosomal abnormalities, however, the empiric likelihood of transmission would probably be less. Questions are most likely to arise in 46,X,del(Xp) patients, 25% of whom menstruate (Simpson, 1979a,b).

C. Dicentric and Ring Chromosomes

Dicentric and ring chromosomes are mitotically unstable. If the two centromeres of a dicentric pass to opposite poles at anaphase, the chromosome will inevitably break, leading to loss or secondary rearrangement. Similarly, ring chromosomes must break in order to divide, thus also being unstable. Offspring of parents with dicentric chromosomes may (1) receive the same dicentric as their parent; (2) inherit a rearranged chromosome; or (3) not receive either, thus being monosomic. No empiric data are available, but transmission of dicentric Y chromosomes has been observed (Starkman and Jaffe, 1967).

D. Translocations

Many different translocations occur, each probably with its own pattern of meiotic segregation. Thus, both theoretic and empiric data should ideally be available for each translocation. Unfortunately, paucity of data precludes analysis except by common types. We shall, therefore, consider the following types; D/G Robertsonian translocations, D/D Robertsonian translocations, and reciprocal translocations.

1. Theoretical Considerations

At meiosis I bivalents form between homologous chromosomes as result of chiasmata. If chiasmata form in each arm of chromosomes involved in translocation, the result is not a bivalent but a quadrivalent, specifically a ring of four chromosomes. If one arm has no chiasmata, a chain of three forms; the chromosome not involved constitutes a univalent. At metaphase a quadrivalent can undergo several possible alignments on the equatorial plate. The alignments determine segregation propensities, which in turn determine genetic consequences. There are three ways in which the four chromosomes involved in a translocation can segregate by groups of 2 (2:2 segregation). The three ways are termed *alternate segregation, adjacent segregation I, and adjacent segregation II*. As mentioned previously, in alternate segregation one first division product receives the normal pair of chromosomes, whereas the other receives both translocation chromosomes (Fig. 6); neither shows duplication or deficiency. In adjacent segregation I, homologous centromeres pass to the same pole; in adjacent segregation II, homologous centromeres pass to opposite poles. In both adjacent segregation I and adjacent segregation II, all gametes show either duplication or deficiency. If, however, crossing over occurs within an interstitial segment, two balanced and two unbalanced gametes will result. The proportion of unbalanced gametes thus not only reflects alignment but also frequency of crossing over within interstitial segments. Factors believed to favor alternate-segregation and, hence, balanced gametes, include equal-sized chromosomes, metacentric chromosomes, interchanges of equal length, and formation of chiasmata in terminal yet not interstitial segments.

In addition to 2:2 segregation, 3:1 discordant segregation may occur. A 3:1 segregation results in two abnormal first division products, one with an additional chromosome and one with a missing chromosome. Twenty theoretical combinations may result from either 3:1 segregation at meiosis I or from the indistinguishable phenomenon of nondisjunction at meiosis II, but not all combinations have actually been observed (Lindenbaum and Bobrow, 1975). If the additional chromosome consists of arms of nonhomologous chromosomes, tertiary trisomy is said to exist. The other trisomic possibility is the existence of not only two balanced translocation chromosomes but also one additional normal chromosome. Only a few of the 20 theoretical combinations have been observed in man, although the phenomenon is being recognized with increasing frequency (Sheres *et al.*, 1978). Factors favoring 3:1 segregation include disparities in length between translocated chromosomes, short interstitial segments, and involvement of acrocentric chromosomes. Each factor favors chain rather than ring configurations at meiosis I.

2. *D/G Translocations*

D/G Robertsonian translocations, usually between chromosomes 14 and 21, are clinically important because 2–3% of individuals with Down's syndrome are affected as result of a translocation (Mikkelsen, 1971). Neither males nor females with D/G translocations appear to show significant reduction in reproductive fitness, judged by (1) number of livebirths; (2) proportion of livebirths who survive until age 18 years; and (3) interval between livebirths (Jacobs, 1977; J. A. Evans *et al.*, 1978). Although in theory one-third of viable offspring of individuals with a t(14:21) should have Down's syndrome (Fig. 11), only 10% of viable offspring of female heterozygotes and only 2–3% of offspring of male heterozygotes actually do (Mikkelsen, 1971). Liveborn normal offspring have equal likelihood of being translocation heterozygotes, like one of their parents, or

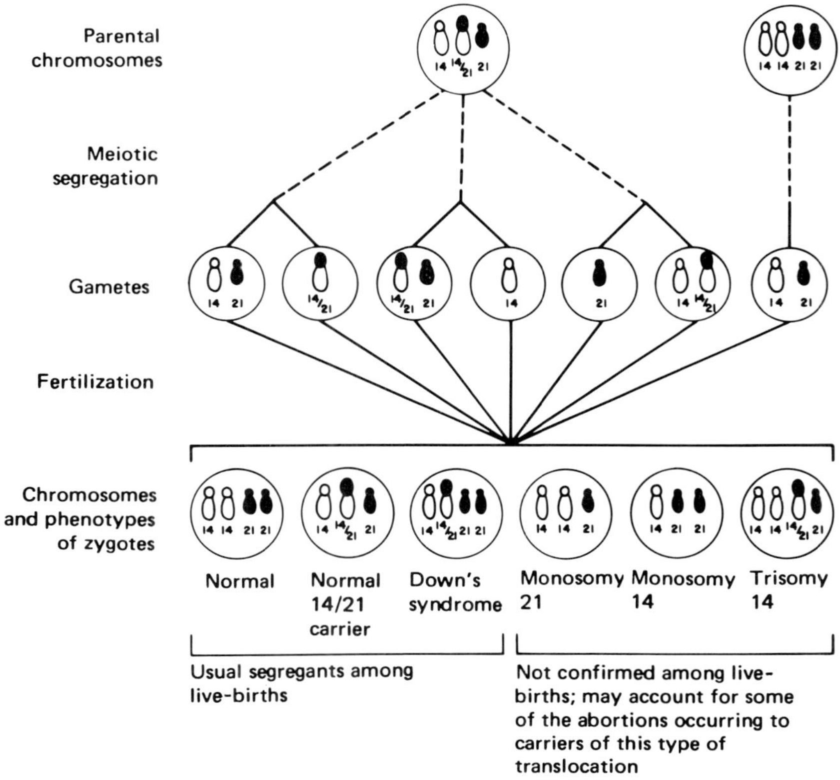

Fig. 11. Chromosome complement of gametes and zygotes resulting from normal meiotic segregation in an individual with a translocation between chromosomes 14 and 21. (From Gerbie and Simpson, 1976.)

being chromosomally normal. The frequency of clinically recognized spontaneous abortions is not increased among heterozygotes, although preimplantation and early implantation losses have not been assessed. The deficiency of expected versus observed could be accounted for by (1) preferential segregation of unbalanced products into polar bodies; (2) decreased fertilizability of unbalanced ova or sperm; (3) increased loss of unbalanced zygotes or early embryos; or (4) increased unrecognized loss of unbalanced late fetuses.

Unlike 14/21 translocations, no empiric risk figures are available for 13/21, 15/21, 13/22, 14/22, and 15/22 translocations. However, no infants with liveborn trisomies 13, 14, and 22 have been born to parents with such translocations, and trisomy 15 is considered lethal. Likewise, all autosomal monosomies are probably lethal. Nonetheless, individuals with the above translocations should be offered antenatal chromosomal studies, and counseled to expect risks potentially similar to those with 14/21 translocations.

3. D/D Translocations

The D/D translocations are the most common structural rearrangement in man. The frequency of balanced D/D translocations is 0.72 per 1000 livebirths, whereas the frequency of unbalanced translocations is 0.08 (Hook and Hamerton, 1977). D/D translocations often involve chromosomes 13 and 14, but other chromosomes may also be involved (13/15, 14/15, 13/13, 14/14, 15/15).

Both males and females heterozygous for D/D translocations show some evidence of diminished reproductive fitness, judged by significantly fewer liveborn children and more than expected fetal deaths (Jacobs, 1977). However, there is no significant increase in either the numbers of childhood deaths (Jacobs, 1977) or spontaneous abortions (J. A. Evans *et al.*, 1978). These findings suggest that meiosis in D/D translocations can lead to lethal unbalanced gametes (i.e., trisomy and monosomy for D group chromosomes). Indeed, most trisomy 13 conceptuses and almost all trisomy 14 and 15 conceptuses terminate in spontaneous abortion.

Individuals with D/D translocations who are ascertained through balanced probands show no apparent increased empiric risk of having liveborn anomalous offspring (Jacobs *et al.*, 1975; J. A. Evans *et al.*, 1978), although data are still not exhaustive. Unbalanced segregants presumably are either never fertilized or lost shortly thereafter. Liveborn progeny are equally likely to inherit the translocation chromosomes or the two normal chromosomes. Despite the generally favorably empiric prognosis, antenatal diagnosis should be offered couples in which a partner has a D/D translocation.

4. Non-Robertsonian Reciprocal Translocations

There are many different types of non-Robertsonian reciprocal translocations. Translocations involving all chromosomes have been described, and except possibly for chromosomes 11 and 13, there is no evidence that any particular chromosome is preferentially involved (Jacobs *et al.*, 1974, 1975; J. A. Evans *et al.*, 1978). Break points are more likely to occur in light G-banded regions than in dark regions. For each translocation empiric data should ideally be available, but at present analysis is possible only for the groups in aggregate. As if to emphasize the heterogeneity and, hence, unsoundness of this approach, the risk of anomalous liveborn infants depends upon whether the translocation was ascertained through a balanced or unbalanced proband.

In translocations ascertained through a balanced proband, reproductive fitness is slightly diminished (Jacobs, 1977; J. A. Evans *et al.*, 1978). Translocation heterozygotes show a significant decrease in livebirths and a significant increase in both fetal deaths and intervals between livebirths. However, the risk of having a liveborn child with chromosomal duplication or deficiency is extremely low, implying selection against unbalanced segregants. If, by contrast, a reciprocal translocation is ascertained through an abnormal (unbalanced) proband, not only is reproductive fitness decreased but the risk of abnormal liveborn offspring is increased (Jacobs, 1970, 1977; J. A. Evans *et al.*, 1978). Females with a reciprocal translocation ascertained in such fashion have approximately a 10% risk of being delivered of an abnormal liveborn infant; males have approximately a 2–5% risk. These risks are much less than the theoretical risks, which without crossing over would be 67% unbalanced gametes. (The two products of alternate segregation yield normal gametes, whereas both adjacent segregation I and adjacent segregation II without crossing over yield only abnormal gametes; thus, 4 of 6 are abnormal.) Liveborn normal offspring are equally likely to inherit either normal chromosomes (46,XY or 46,XX) or the same balanced translocation as the parent.

Counseling on the basis of pooled data naturally warrants caution because the general situation need not necessarily be representative of the specific chromosomes involved. Nonetheless, the empiric data listed above must suffice until more specific figures are available.

E. Inversions

In inversions the order of the genes is altered. Pericentric inversions occur 0.013 per 1000 livebirths (Hook and Hamerton, 1977). No accurate estimate is available for paracentric inversions because such rearrangements are difficult to detect without banding techniques, which were not used in population surveys.

Some small inversions are more properly considered merely to be variants. In chromosomes 1,3,9, and 10, small inversions near the centromere segregate in dominant fashion with no apparent deleterious effect upon offspring. These inversions probably involve only a short length of heterochromatin. Pericentric (metacentric) Y chromosomes also segregate with neither phenotypic effect nor ostensible relationship to fetal loss.

On the other hand, paracentric or pericentric inversions involving longer chromosomal segments predispose to abnormal progeny because crossing over within the inverted segment leads to unbalanced gametes. For example, a single crossover within a paracentric inversion leads to one dicentric, one acentric, and two normal products (chromatids). Acentric chromatids are lost, leading to monosomy for that chromosome. A single crossover within a pericentric inversion leads to one predominately duplicated, one predominately deficient, and two normal products. Dicentric and duplicated chromatids contain excess genetic information that can produce phenotypic abnormalities. If double crossing over occurs, the consequences depend upon the numbers of strands (2, 3, or 4) involved. Traditional cytogenetic dogmas dictate that crossing over within an inversion leads to formation of an inversion "loop" recognizable at meiosis. However, meiotic studies in humans with inversions show relatively few loops (Hauksdoltir *et al.*, 1972; Winsor *et al.*, 1978). Thus, alternative methods of pairing may occur in inversions.

The likelihood that a parent with an inversion will have anomalous liveborn offspring is a function of both the particular chromosome and the length of segment involved. In humans, inversion carriers show no significant increase in fetal deaths and no significant decrease in livebirths (Jacobs, 1977). Indeed, during evolution inversions serve to suppress crossing over, allowing favorable combinations of alleles to be retained on nonrecombinant products, which may be the only surviving products.

Because so many different potential products are recoverable from parents with inversions, counseling is less than satisfactory. Analogous to translocations, empiric data concerning reproductive outcome doubtless reflect the particular chromosome involved. Such data are usually not available, although inversions involving most chromosomes have now been reported (Winsor *et al.*, 1978). However, one general principle is that the longer the segment, the greater the likelihood of recombination and, hence, the greater the risk for unbalanced products. Tabulations of reported cases support this conclusion (Winsor *et al.*, 1978). Method of ascertainment is also important. An inversion ascertained through a balanced proband apparently carries a relatively low risk of producing offspring with a recombinant gamete (Jacobs, 1977). Presumably recombinant products either arise infrequently, or, more likely, are lethal. In families with low risks, the inverted segment usually constitutes less than 50% of total chromosomal length. The relatively short inverted segment apparently decreases the likelihood of recombination within the segment. On the other hand, an inversion ascertained through an anomalous proband carries a much higher risk for liveborn anomalous progeny. Such inversions encompass 50–90% of total chromosomal length. Usually only one type of recombinant product is recovered in liveborn offspring, but occasionally both deficient and duplicated products are recovered (Warter *et al.*, 1973).

In the future, empiric data will hopefully be available for each inversion. At present, however, estimates must suffice. If one pools data from all reported inversions and corrects for single ascertainment by elimination of probands, the likelihood of anomalous liveborn is about 10%, perhaps lower if a male carries the inversion (Sutherland *et al.*, 1976). That females have a higher risk of producing abnormal offspring is consistent with observations that recombination frequencies are usually higher in females than males (Weitkamp *et al.*, 1973). Antenatal diagnosis should be offered to inversion carriers, particularly if inversions are ascertained through an abnormal proband.

F. Nonspecific Effects of Translocation and Inversions

In addition to the abnormal gametes that predictably result from normal meiotic segregation, various investigators have claimed that translocations and inversions predispose to aneuploidy for chromosomes other than those involved in the translocation or inversion. For example, a 10/13 translocation might increase the frequency of trisomy 21. Indeed, quadrivalent or other unusual meiotic configurations could exert a generalized deleterious effect upon segregation of chromosomes not involved in a translocation or inversion.

At least 25 infants with trisomy 21 have had a parent with a translocation not involving chromosome 21. None of the 25 had an unbalanced translocation. In about one-half of these families, the child inherited the normal homologues. Although some investigators believe that the occurrence of these 25 infants with trisomy 21 indicates a casual relationship between translocations and aneuploidy for chromosomes not involved in the translocation, the concurrence of both trisomy 21 and a translocation involving a chromosome other than 21 would not be expected so infrequently that the two phenomena could not occasionally coexist by chance.

Inversions could also interfere with meiotic disjunction of other chromosomes. At least five cases of autosomal trisomy and three cases of 45,X have been associated with pericentric inversions for other chromosomes (Leonard *et al.*, 1975). Again, however, the theoretical frequency of observing two independent phenomena may or may not be so low that these eight cases would be unexpected.

In conclusion, translocations or inversions could exert a nonspecific adverse effect on meiotic segregation, but proof is lacking.

G. Supernumerary Chromosomes

So-called supernumerary chromosomes are usually of uncertain origin and significance, especially if small and metacentric. Precise origin should be sought through studies of parental karyotypes, for such chromosomes may have originated following 3:1 segregation in a parent with a balanced translocation. In the absence of precise origin, counseling concerning transmission and phenotypic significance is somewhat hazardous. Jacobs (1977) tabulated that males with a supernumerary chromosome have reduced reproductive fitness, although in her calculations females did not. The additional chromosome can be transmitted (Bernstein *et al.,* 1978), but no segregation ratios are available. If detected in amniotic fluid cells, counseling may prove difficult (Bernstein *et al.,* 1978). Additional data are necessary.

IX. PREGNANCIES BY TRUE HERMAPHRODITES

True hermaphrodites have both ovarian and testicular tissue. The etiology is usually unknown, and probably heterogeneous (Simpson, 1978). Most are 46,XX, but 46,XY, 46,XX/46,XY, 46XX/47,XXY, and other complements may be associated. Because most true hermaphrodites do not show chromosomal abnormalities, a discussion of reproduction by affected individuals may not be strictly appropriate in this chapter. Nonetheless, brief mention seems in order.

In aggregate, true hermaphrodites show external genitalia more male than female. Approximately two-thirds are raised as males, irrespective of complement. However, a uterus is usually present, and at puberty breast development usually occurs. Testicular tubules show degenerative changes whether present in an ovotestes or a separate testis; thus, fertility by true hermaphrodites raised as males seems precluded. On the other hand, ovarian follicles appear normal, even if present as part of an ovotestis. Cyclic menstruation may also occur. Thus, it is not unreasonable to expect occasional pregnancies in true hermaphrodites raised as females, especially if testicular tissue had been extirpated.

Narita *et al.* (1975) and Kim *et al.* (1979) described 46,XX true hermaphrodites who became pregnant after removal of a unilateral ovotestis, and Tegenkamp *et al.* (1979) reported a pregnancy without benefit of surgical reconstruction. The pregnancy by one individual terminated in a premature 1500 g morphologically normal infant (Kim *et al.,* 1979), an outcome that may or may not have been related to the true hermaphrodite status. A relationship seems unlikely, but subtle maternal uterine abnormalities leading to prematurity could exist undetected. The other true hermaphrodites were delivered of normal infants.

REFERENCES

Arakaki, D. T., and Waxman, S. H. (1970). Cytogenetics of spontaneous abortion. *Am. J. Obstet. Gynecol.* **107,** 1199–1204.

Armendares, S., Buentello, L., Sanchez, J., and Ortiz, M. (1967). XO/XX/XXX mosaicism without Turner stigmata. *Lancet* **2,** 840.

Baghdassarian, A., Bayard, F., Borgaonkar, D. S., Arnold, E. A., Solezk, K., and Migeon, C. J. (1975). Testicular function in XYY men. *Johns Hopkins Med. J.* **136,** 15–24.

Bahner, F., Schwarz, G., Harnden, D. G., Jacobs, P. A., Heinz, H. A., and Walter, K. (1960). A fertile female with XO sex chromosome constitution. *Lancet* **2,** 100–101.

Baike, A. G. (1972). XXY son of a possible XX-XXX mother. *Lancet* **1,** 697–698.

Barr, M. L., Sergovich, F. R., Carr, D. M., and Shaver, E. L. (1979). The triplo-X female: An appraisal based on a study of 12 cases and a review of the literature. *Can. Med. Assoc. J.* **101,** 247–258.

Bauld, R., Sutherland, G. R., and Bain, D. (1974). Chromosome studies in investigation of stillbirths and neonatal deaths. *Arch. Dis. Child.* **49,** 782–788.

Bernstein, R., Hakim, C., Hardwick, B., and Nurse, G. T. (1978). Significance of detection of extra metacentric microchromosome in amniotic cell culture. *J. Med. Genet.* **15,** 136–142.

Bishun, N. P., Rashad, M. N., Morton, W. R. M., Mannion, P. L., Neely, M. R., and Burke, G. (1964). Chromosomal mosaicism in a case of repeated abortion. *Lancet* **1,** 936.

Bomers-Marres, A. J. M. L. (1966). A fertile woman with an XO(45)/XX(46) mosaic. *Ned. Tijdschr. Geneeskd.* **110,** 1177–1181.

Boué, J., Boué, E., and Lazar, P. (1975). Retrospective and prospective epidemiological studies of 1,500 karyotyped spontaneous human abortions. *Teratology* **12,** 11–26.

Bridges, C. B. (1921). Genetical and cytological proof of nondisjunction of the fourth chromosome of Drosophila melanogaster. *Proc. Natl. Acad. Sci. U.S.A.* **7,** 186.

Briggs, D. K., Stimson, C. W., and Vinograd, J. (1963). Leukocyte anomaly, mental retardation, and dwarfism in a family with abnormal chromosomes. *J. Pediatr.* **63,** 21–28.

Clausen, R. E., and Cameron, D. R. (1944). Inheritance in Nicotiana tobacum. XVIII. Monosomic analysis. *Genetics* **29,** 447.

Craig-Holmes, A. P., Moore, F. B., and Shaw, M. W. (1977). Polymorphism of human C-band heterochromatin. II. Family studies with suggestive evidence for somatic crossing over. *Am. J. Hum. Genet.* **27,** 178–189.

de Toni, E., Massimo, L., Vianello, M. J., and Podesta, F. (1965). La casistica di un anno di attivita del Centro di Studi cromosomici della clinica Pediatrica di Genova nel campo delle malformazioni sessuali. *Minerva Pediatr.* **17,** 578–583.

Dewhurst, J. (1978). Fertility in 47,XXX and 45,X patients. *J. Med. Genet.* **15,** 132–135.

Evans, E. P., Beechey, C. V., and Burtenshaw, M. D. (1978). Meiosis and fertility in XYY mice. *Cytogenet. Cell Genet.* **20,** 249–263.

Evans, J. A., Canning, N., Hunter, A. G. W., Martsolf, J. T., Ray, M., Thompson, D. R., and Hamerton, J. L. (1978). A cytogenetic survey of 14,069 newborn infants. III. An analysis of the significance and cytologic behavior of the Robertsonian and reciprocal translocation. *Cytogenet. Cell Genet.* **20,** 96–123.

Ferrier, P., Ferrier, S., Stalder, G., Buhler, E., Bamatter, F., and Klein, D. (1964). Congenital asymmetry associated with diploid-triploid mosaicism and large satellites. *Lancet* **1,** 80–82.

Fitzgerald, P. H. (1973). The nature and inheritance of an elongated secondary constriction on chromosome 9 of man. *Cytogenet. Cell Genet.* **12,** 404–413.

Geisler, M., Svejcar, J., and Degenhart, K. (1972). XXY son of an XX/XXX mother. *Lancet* **1,** 955.

Gerbie, A. B., and Simpson, J. L. (1976). Antenatal detection of genetic disorders. *Postgrad. Med.* **59,** 159–166.

German, J., Simpson, J. L., and McLeMore, G. A. (1973). Abnormalities of human sex chromosomes. I. A ring Y without mosaicism. *Ann. Genet.* **16,** 225–231.

Giraud, F. (1973). La fertile des femmes: 45,X/46,XX et 45,X/46,XX/47,XXX. *Ann. Genet.* **15,** 256–258.

Grace, H. J., Quinlan, D. K., and Edge, W. E. B. (1973). 45,X lymphocyte karyotype in a fertile woman. *Am. J. Obstet. Gynecol.* **115,** 279–282.

Groll, M., and Cooper, M. (1976). Menstrual function in Turner's syndrome. *Obstet. Gynecol.* **47,** 225–226.

Guzman-Toledano, R., Ayala, A., Zarate, A., and Jimenez, M. (1976). Triple X female and Turner's syndrome offspring. *J. Med. Genet.* **13,** 516–517.

Hauksdoltir, H., Halldorsson, S., Jensson, O., Mikkelsen, M., and McDermott, A. (1972). Pericentric inversion of chromosome number 13 in a large family leading to duplication deficiency causing congenital malformations in three individuals. *J. Med. Genet.* **9,** 413–421.

Hook, E. B., and Hamerton, J. L. (1977). The frequency of chromosome abnormalities detected in consecutive newborn studies—differences between studies—Results by sex and severity of phenotypic involvement. *In* "Population Cytogenetics" (E. B. Hook and I. H. Porter, eds.), pp. 63–80. Academic Press, New York.

Hsu, L. Y. F., Hirschhorn, K., Goldstein, A., and Barcinski, M. A. (1970). Familial chromosomal mosaicism, genetic aspects. *Ann. Hum. Genet.* **33,** 343–349.

Hsu, L. Y. F., Palo-Garcia, F., Grossman, D., Kutinsky, E., and Hirschhorn, K. (1972). Fetal wastage and maternal mosaicism. *Obstet. Gynecol.* **40,** 98–103.

Hulten, M., and Pearson, P. L. (1970). Fluorescent evidence for spermatocytes with two Y chromosomes in an XYY male. *Ann. Hum. Genet.* **34,** 273–276.

Jacobs, P. A. (1970). The inheritance of randomly ascertained chromosome abnormalities. *In* "Human Population Cytogenetics" (P. A. Jacobs, W. H. Price, and P. Law, eds.), pp. 89–102. Edinburgh Univ. Press, Edinburgh.

Jacobs, P. A. (1977). Structural rearrangements of the chromosomes in man. *In* "Population Cytogenetics" (E. B. Mook and I. M. Porter, eds.), pp. 81–97. Academic Press, New York.

Jacobs, P. A., Buckton, K. E., Cunningham, C., and Newton, M. (1974). An analysis of the break points of structural rearrangements in man. *J. Med. Genet.* **11,** 50–68.

Jacobs, P. A., Frackiewicz, A., Law, P., Hilditch, C. J., and Morton, N. E. (1975). The effect of structural aberrations of the chromosomes on reproductive fitness in man. II. Results. *Clin. Genet.* **8,** 169–178.

Jenkins, M. E., Eisen, J., and Seguin, F. (1971). Congenital asymmetry and diploid-triploid mosacisim. *Am. J. Dis. Child.* **122,** 80–84.

Kadotoni, T., Ohama, K., and Makino, S. (1970). A case of 21-trisomic Down's syndrome from the triplo-X mother. *Proc. Jpn. Acad.* **46,** 709–712.

Kaffe, S., Hsu, L. Y. F., and Hirschhorn, K. (1974). Trisomy 21 mosaicism in a woman with two childrer with trisomy 21 Down's syndrome. *J. Med. Genet.* **11,** 378–379.

Kajii, T., and Nikawa, N. (1977). Origin of triploidy and tetraploidy in man: 11 cases with chromosome markers. *Cytogenet. Cell Genet.* **18,** 109–125.

Kava, H. W., and Klinger, H. P. (1968). Secondary infertility in a phenotypically normal 45,X/46,XX female. *Fertil. Steril.* **19,** 835–838.

Kennedy, J. F., Freeman, M. G., and Benirschke, K. (1977). Ovarian dysgenesis and chromosome abnormalities. *Obstet. Gynecol.* **50,** 13–20.

Khush, G. S. (1973). "Cytogenetics of Aneuploids." Academic Press, New York.

Kim, M. H., Gumpel, J. A., and Graff, P. (1979). Pregnancy in a true hermaphrodite. *Obstet. Gynecol.* **53,** Suppl., 40S–42S.

King, C. R., Magenis, E., and Bennett, S. (1978). Pregnancy and the Turner syndrome. *Obstet. Gynecol.* **52,** 617–624.

Kuleshov, N. P. (1976). Chromosomal anomalies in infants dying during the perinatal period and premature newborn. *Hum. Genet.* **31,** 151–160.

Lafever, H. N., and Patterson, F. L. (1964). Cytological and breeding behavior of an aneuploid line of *Avena sativa L. Crop Sci.* **4,** 535.

Lajborek-Czyz, I. (1976). A 45,X woman with a 47,XY,G+ son. *Clin. Genet.* **9,** 113–116.

Leonard, C. P., Hazael-Massieux, L., Bocquet, L., Larget-Piet, L., and Boué, J. (1975). Inversion pericentrique inv(2)(pllq13) dans des familles non apparantées. *Hum. Genet.* **28,** 121–128.

Lewis, F. J. W., Poulding, R. H., and Eastman, R. D. (1963). Acute leukaemia in an XO/XXX mosaic. *Lancet* **2,** 306.

Lieber, E., and Berger, J. (1973). Fertility in a 45,X/46,XX patient. *Lancet* **1,** 199.

Lindenbaum, R. H., and Bobrow, M. (1975). Reciprocal translocations in man. 3:1 meiotic disjunction resulting in 47- or 45-chromosome offspring. *J. Med. Genet.* **12,** 29–43.

Lubs, H. A., Patel, S. A., Kimberling, W. J., Brown, J., Cohen, M., Gerald, P., Hecht, F., Myrianthopoulos, N., and Summitt, R. L. (1977). Q and C banding polymorphisms in 7 and 8 year old children: Racial differences and clinical significance. *In* "Population Cytogenetics" (E. B. Hook and I. H. Porter, eds.), pp. 133–159. Academic Press, New York.

Luciani, J. M., Vagner-Capodano, A.-M., Devictor-Vuillet, M., Aubert, L. and Stahl, A. (1973). Presumptive fluorescent evidence for spermatocyte with X + Y + Y diakinetic univalent in an XYY male. *Clin. Genet.* **4,** 415–416.

Machin, G. A., and Crolla, J. A. (1974). Chromosome constitution of 500 infants dying during the perinatal period. *Hum. Genet.* **23,** 183–198.

Mackay, E. V., Adam, G. S., and Khoo, S. K. (1971). A successful pregnancy associated with sex chromosomal mosaicism of the XO/XX type. *Aust. & N. Z. J. Obstet. & Gynecol.* **11,** 259–261.

McKenzie, W. H., and Lubs, H. A. (1975). Human Q and C chromosomal variations: Distribution and incidence. *Cytogenet. Cell Genet.* **14,** 97–115.

Maclean, N., Court-Brown, W. M., Jacobs, P. A., Mantle, D. J., and Strong, J. A. (1968). A survey of sex chromatin abnormalities in mental hospitals. *J. Med. Genet.* **5,** 165–172.

Magenis, E., Palmer, C. G., Wang, L., Brown, M., Chamberlin, J., Parks, M., Merritt, A. D., Rivas, M., and Yu, P. L. (1977). Heritability of chromosome banding variants. *In* "Population Cytogenetics" (E. B. Hook and I. H. Porter, eds.), pp. 179–188. Academic Press, New York.

Martin, A. O., Miller, L., Simpson, J. L., Rzeszotarski, M. R., Elias, S., Sarto, G. E., and Patel, V. R. (1979). Localization of the nucleolar organizer by computer-aided analysis of a variant No. 21 in a human isolate. *Hum. Genet.* **48,** 211–219.

Melnyk, J., Thompson, H., Rucci, A. J., Vanasek, F., and Hayes, S. (1969). Failure of transmission of the extra chromosome in subjects with 47,XYY karyotype. *Lancet* **2,** 797–798.

Mikkelsen, M. (1971). Down's syndrome: Current stage of cytogenetic research. *Humangenetik* **12,** 1–28.

Morrison, J. W. (1953). Chromosome behavior in wheat monosomes. *Heredity* **7,** 203–217.

Nakashima, I., and Robinson, A. (1971). Fertility in a 45,X female. *Pediatrics* **47,** 771–773.

Narita, O., Manba, S., Nakanishi, T., and Ishizuka, N. (1975). Pregnancy and childbirth in a true hermaphrodite. *Obstet. Gynecol.* **45,** 593–595.

Nielsen, J., and Thomsen, N. (1968). A psychiatric cytogenetic study of a female patient with 45/46/47 chromosomes and sex chromosomes XO/XX/XXX. *Acta Psychiatr. Neurol. Scand.* **44,** 141–155.

Patel, S. R., and Lubs, H. A. (1977). A possible association of long Y chromosomes and fetal loss. *Hum. Genet.* **25,** 233–236.

Philip, J., and Sele, V. (1976). 45XO Turner's syndrome without evidence of mosaicism in a patient with two pregnancies. *Acta Obstet. Gynecol. Scand.* **55,** 283–286.

Predescu, V., Christodorescu, D., Tautu, C., Ciovirnache, N., and Constantinescu, E. (1969). Repeated abortions in a woman with XO/XXX mosaicism. *Lancet* **2,** 217.

Reyes, F. I., Koh, K. S., and Faiman, C. (1976). Fertility in women with gonadal dysgenesis. *Am. J. Obstet. Gynecol.* **126,** 668–670.

Russell, P., and Altschuler, G. (1975). The ovarian dysgenesis of trisomy 18. *Pathology* **7,** 149–155.

Ruzicska, P., and Czeizel, A. (1970). Cytogenetic studies on mid-trimester abortuses. *Humangenetik* **10,** 273–297.

Scharrar, S., Stengel-Rutkowski, S., Rodewald-Rudescu, A., Erdlen, E., and Zang, K. D. (1975). Reproduction in a female patient with Down's syndrome. *Hum. Genet.* **26,** 207–214.

Sears, E. R. (1952). Misdivision of univalents in common wheat. *Chromosoma* **4,** 535–550.

Sellyei, M., Derzsky, S., and Vass, L. (1971). 69XYY sporadic triploidy in an adult. *Humangenetik* **11,** 244–246.

Shakkabaek, N. E., Zeuthen, E., Nielsen, J., and Yde, H. (1973). Impaired spermatogenesis in XYY men: A report on 4 cases ascertained through a population study. *Fertil. Steril.* **24,** 390–395.

Sheres, J. M. J. C., Hustinx, T. W. J., TerHaar, B. G. A., Rutten, F. J., and Van der Mee-Wienen, H. H. A. M. (1978). 2:2 and 3:1 meiotic disjunctions in a carrier of a reciprocal 10/14 translocation. *Clin. Genet.* **13,** 481–485.

Shokeir, M. H. K. (1978). Pregnancy in five women with 45,X/46,XX and 45,X/47,XXX gonadal dysgenesis. *Birth Defects, Orig. Artic. Ser.* **14**(6C), 171–184.

Siegelman, M. (1972). XO/XX and XO/XY mosaicism: A study of a family. *Obstet. Gynecol.* **39,** 511–514.

Simpson, J. L. (1972). Familial mosaicism. *Obstet. Gynecol.* **40,** 768–769.

Simpson, J. L. (1976). "Disorders of Sexual Differentiation: Etiology and Clinical Delineation." Academic Press, New York.

Simpson, J. L. (1978). True hermaphroditism: Etiology and phenotypic considerations. *Birth Defects, Orig. Artic. Ser.* **14**(6C), 9–35.

Simpson, J. L. (1979a). Gonadal dysgenesis and sex chromosome abnormalities—Phenotypic/karyotypic correlations. *In* "Genetic Mechanisms of Sexual Development" (N. L. Vallet and I. Porter, eds.), pp. 365–405. Academic Press, New York.

Simpson, J. L. (1979b). Genetic aspects of reproductive failure: Deficiencies of oocyte number (gonadal dysgenesis) and abnormalities of oogenesis (spontaneous abortion). *In* "The Infertile Female" (J. R. Givens, ed.), pp. 293–325. Yearbook Publ., Chicago, Illinois.

Simpson, J. L. (1980). Genes, chromosomes, and reproductive failure. *Fertil. Steril.* **33,** 107–116.

Singh, R. M., and Wallace, A. T. (1967). Monosomics of Avena byzantina C. Koch. I. Karyotype and chromosome pairing studies. *Can. J. Genet. Cytol.* **9,** 87–96.

Sinow, R. M., and Luthardt, F. W. (1978). Chromosome analysis of 4-cell embryos from XO mice. *Am. J. Hum. Genet.* **30,** 94A.

Smith, H. C., Seale, J. P., and Posen, S. (1974). Premature ovarian failure in a triple X female. *J. Obstet. Gynaecol. Br. Commonw.* **81,** 405–409.

Starkman, M. N., and Jaffe, R. B. (1967). Chromosome aberrations in XO/XY mosaic individuals and their fathers. *Am. J. Obstet. Gynecol.* **99,** 1056–1066.

Sutherland, G. R., Gardiner, A. J., and Carter, R. F. (1976). Familial pericentric inversion of chromosome 19,inv(19)(p13q13) with a note on genetic counseling of pericentric inversion carriers. *Clin. Genet.* **10,** 54–59.

Tegenkamp, T. R., Brazzell, J. W., Tegenkamp, I., and Labidi, F. (1979). Pregnancy without benefit of reconstructive surgery in a bisexually active true hermaphrodite. *Am. J. Obstet. Gynecol.* **135,** 427–428.

Tishler, P. V., Lamborot-Manzur, M., and Atkins, L. (1972). Polymorphism of the human Y chromosome: Fluorescence microscopic studies on the sites of morphologic variation. *Clin. Genet.* **3,** 116–122.

Tsuang, M. T., Miller, J. R., and DeBault, L. E. (1975). Klinefelter's syndrome and maternal XX/XXX mosaicism. *J. Med. Genet.* **12,** 114.

Tsunewaki, K. (1964). Transmission of monosomes and trisomes in an Emmer wheat. *Wheat Inf. Serv.* **17-18,** 34.

Van de Velde-Staquet, M. R., Breynaert, R., Walbaum, R., Saint-Aubert, P., Farriaux, J. P., and Fontaine, G. (1973). La descendance des mères trisomiques 21. *J. Genet. Hum.* **21,** 187–206.

Warter, S., Ruch, V. J., and Lehman, M. (1973). Karyotype with chromosomal abnormality with various inherited defects in the offspring (recombination aneusomy). *Humangenetick* **20,** 355–359.

Weitkamp, L. R., Van Rood, J. J., Thorsby, E., Bias, W., Fotino, M., Lawler, S. D., Dausset, J., Mayr, W. R., Bodmer, J., Ward, F. E., Seignalet, J., Payne, R., Kissmeyer-Nielsen, F., Gatti, R. A., Sachs, J. A., and Lamm, L. U. (1973). The relation of parental sex and age to recombination in the HLA system. *Hum. Hered.* **23,** 197–205.

Winsor, E. J. T., Palmer, C. G., Ellis, P. M., Hunter, J. L. P., and Ferguson-Smith, M. A. (1978). Meiotic analysis of a pericentric inversion, inv(7)(p22q32), in the father of a child with a duplication-deletion of chromosome 7. *Cytogenet. Cell Genet.* **20,** 169–184.

Zizka, J., Balicek, P., and Nielsen, J. (1975). XXYY son of a triple-X mother. *Humangenetik* **26,** 159–160.

Index

B

C

H

I